THE SEPTAL NUCLEI

ADVANCES IN BEHAVIORAL BIOLOGY

Recent Volumes in this Series

Volume 10 • NEUROHUMORAL CODING OF BRAIN FUNCTION
Edited by R. D. Myers and René Raúl Drucker-Colín • 1974

Volume 11 • REPRODUCTIVE BEHAVIOR
Edited by William Montagna and William A. Sadler • 1974

Volume 12 • THE NEUROPSYCHOLOGY OF AGGRESSION
Edited by Richard E. Whalen • 1974

Volume 13 • ANEURAL ORGANISMS IN NEUROBIOLOGY
Edited by Edward M. Eisenstein • 1975

Volume 14 • NUTRITION AND MENTAL FUNCTIONS
Edited by George Serban • 1975

Volume 15 • SENSORY PHYSIOLOGY AND BEHAVIOR
Edited by Rachel Galun, Peter Hillman, Itzhak Parnas, and Robert Werman • 1975

Volume 16 • NEUROBIOLOGY OF AGING
Edited by J. M. Ordy and K. R. Brizzee • 1975

Volume 17 • ENVIRONMENTS AS THERAPY FOR BRAIN DYSFUNCTION
Edited by Roger N. Walsh and William T. Greenough • 1976

Volume 18 • NEURAL CONTROL OF LOCOMOTION
Edited by Richard M. Herman, Sten Grillner, Paul S. G. Stein, and Douglas G. Stuart • 1976

Volume 19 • THE BIOLOGY OF THE SCHIZOPHRENIC PROCESS
Edited by Stewart Wolf and Beatrice Bishop Berle • 1976

Volume 20 • THE SEPTAL NUCLEI
Edited by Jon F. DeFrance • 1976

A Continuation Order Plan is available for this series. A continuation order will bring delivery of each new volume immediately upon publication. Volumes are billed only upon actual shipment. For further information please contact the publisher.

THE SEPTAL NUCLEI

Edited by

Jon F. DeFrance

Wayne State University
Detroit, Michigan

SPRINGER SCIENCE+BUSINESS MEDIA, LLC

Library of Congress Cataloging in Publication Data

International Symposium on the Septal Nuclei, Wayne State University, 1974.
The septal nuclei.

(Advances in behavioral biology; v. 20)
Bibliography: p.
Includes index.
1. Septum (Brain)—Congresses. 2. Hippocampus (Brain)—Congresses. 3. Neuroendocrinology—Congresses. I. DeFrance, Jon F. II. Title.
QP383.2.S96 1974 599'.01'88 76-43348

DOI 10.1007/978-1-4684-3084-4

Proceedings of the International Symposium on The Septal Nuclei held at the Wayne State University, Detroit, Michigan, October 3–5, 1974

Acknowledgments

The success of the symposium was due to the collective effort of many people. Among these, Mr. Frank Seaver, Ms. Trudy Kiesewether, and Ms. Barbara Dubrinsky draw special praise for their pre-program planning and the actual conduct of the program. A special thanks must go to Dr. James Ranck for his many useful ideas and suggestions, and for functioning as my sounding board during the planning.

I want to express my appreciation to the Wayne State University School of Medicine and the Wayne State Fund for their generous financial support. I am also grateful for the emotional support and encouragement of Dr. Steven Kitai.

Finally, I want to thank all the participants. Their interest and enthusiasm made it all worth while.

JDF

Preface

The "septum pellucidum" has been described since antiquity. Even though people such as Vicq d' Azyr (12) and Burdach (3) pictured the septal region in their drawings, the nuclei associated with the pellucidum--the septal nuclei--were not described until Meynert (10).

Since Meynert's description, this portion of the limbic system has been troublesome in terms of nomenclature. Scientists about the turn of the twentieth century proposed quite diverse terminologies. For example, Elliot Smith first wrote of the precommissural area (4) and then later of the paraterminal body (5). These terms, however, were meant to extend beyond the septal nuclei of our understanding. Unger (11) and Herrick (8) proposed the familiar terms--nucleus lateralis septi and nucleus medialis septi; but again they were somewhat broader in definition than is accepted for current usage. These terms, however, were rejected by Johnston (9). In Johnston's great paper, he pointed out that the hippocampus seems to evolve out of large portions of the septal nuclei. It was appropriate then to borrow a term previously used by Elliot Smith (6), but in another context--primordium hippocampi. Johnston's primordium hippocampi corresponds to the lateral septal nucleus of current usage. He introduced the terms medial and lateral parolfactory area to refer to the remaining portions of the septal nuclei of Herrick (8). Hence, the lateral parolfactory area refers to the nucleus accumbens septi of Ariens Kappers (2). But, Johnston's terminology received little acceptance. The majority of investigators still favor the terms: medial septal nucleus and lateral septal nucleus. Fox (7) extended this notion to a medial and lateral septal region. The lateral septal region which included the nucleus accumbens septi.

The recent study of Andy and Stephan (1) has again changed the face of the terminology. They distinguish four basic septal groups--the medial, dorsal, lateral, and caudal. For the most part, the dorsal, lateral, and caudal septal groups of Andy and Stephan correspond to the lateral septal nucleus of Herrick (with the nucleus accumbens septi removed).

In spite of the problems of terminology, the septal nuclei have been the subject of intense investigation, spurred by their apparent role in emotional, motivational, and memory processes. A major aim of the current symposium is to bring recent research into focus so that a consensus can accrue with regard to the most appropriate terminology, and with that, a better understanding of the role of the septal nuclei in brain.

Detroit JDF

REFERENCES

1. Andy, O. J. and Stephan, H. The nuclear configuration of the septum of Galago demidovii. J. Comp. Neurol., 111 (1959) 503-545.
2. Ariens Kappers, C. U. The phylogenesis of the paleocortex compared with the evolution of the visual neocortex. Arch. Neurol. Psychiat. (Lond.) 4 (1909) 161-173.
3. Burdach, K. F. Vom Baue und Leben des Gehirns, Vol. 2, Leipzig, 1822.
4. Elliot Smith, G. The connections between olfactory bulb and hippocampus. Anat. Anz., 10 (1895) 470-474.
5. Elliot Smith, G. Notes upon the natural subdivision of the cerebral hemisphere. J. Anat. (Lond.) 35 (1901) 431-454.
6. Elliot Smith, G. On two morphology of the cerebral commissures in the Vertebrata. Trans. Linn. Soc. Lond. (Zool.) 8 (1903) 455-500.
7. Fox, C. A. Certain basal telencephalic centers in the cat. J. Comp. Neurol., 72 (1940) 1-62.
8. Herrick, C. J. The morphology of the forebrain in Amphibia and Reptilia. J. Comp. Neurol. 20 (1910) 413-547.
9. Johnston, J. B. The morphology of the septum, hippocampus, and pallial commissures in reptiles and mammals. J. Comp. Neurol., 23 (1913) 371-478.
10. Meynert, T. Der Bau der Grosshirnrinde und seine örthlichen Verschiedenheiten, nebst einem pathologisch - anatomischen Corollarium. Vierteljahresschr. Psychiat., 1 (1867) 77-93.
11. Unger, L. Untersuchungen über die Morphologie und Faserung des Reptiliengehirns. Anat. Hefte, 31 (1906) 271-348.
12. Vic d'Azyr, F. Traité de anatomie et de physiologie, Paris, 1786.

Contents

Part I

Anatomy

SEPTUM DEVELOPMENT IN PRIMATES

O. J. Andy[1] and H. Stephan[2]

University of Mississippi Medical Center, Jackson, Mississippi and the Max Planck Institute for Brain Research, Frankfurt, Germany

INTRODUCTION

It has been commonly believed that the septum underwent a reduction in size during evolution. Previous investigators (18,19, 20) thought that the septum became a functionless atrophic cortical structure in association with a general atrophic process of the olfactory system. More recent studies, based on utilizing insectivores as a reference since they represent the forerunners of the primate, reveal that the septum actually undergoes a progressive increase rather than a decrease in size in primate development (3,4,22). Among primates it attains its greatest degree of development in the human brain (Fig. 1).

METHODOLOGY

Brains of living animal forms are utilized. The evolutionary levels of the animals under consideration are established in accord with the degree of development and differentiation of the neocortex. The basal forms of insectivores consist of terrestrial and nonspecialized species. They possess the simplest cortical pattern and smallest cortical volume. The higher insectivores tend to be more specialized. They may be semi-aquatic, may burrow or may possess a specialized visual system. All animal brains are profused with Bouin's solution immediately after sacrifice.

---

[1]Supported in part by NIH grant 04510.

[2]Supported in part by Deutsche Forschungsgemeinschaft.

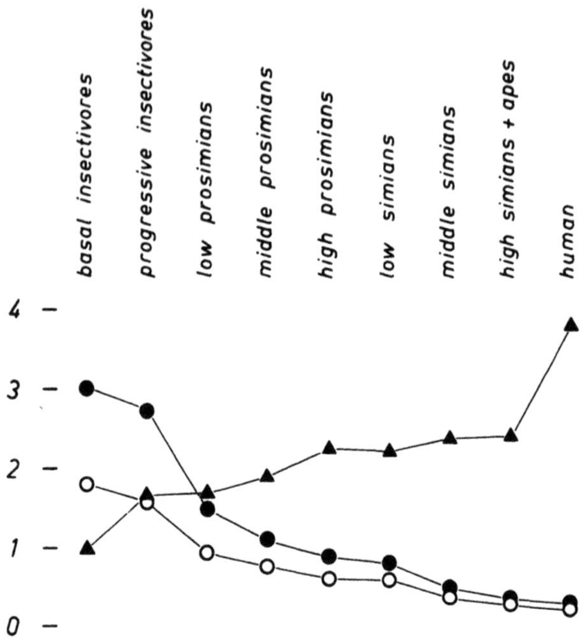

Figure 1. Comparison of the size of the septum at ascending levels of the primate scale by two different methods. The numbers on the left side indicate the percentage proportion of the septum related to the total brain (open circles) and telencephalon (solid circles). The same numbers represent progression indices for the septum in relation to body weight (triangles). These indices determine the enlargement multiple of the septum in the progressive group in comparison to that in basal insectivores (the latter have the index = 1). For more details refer to material, techniques and to discussion in Andy and Stephen (4).

The various insectivores and primate groups utilized are as follows:

Basal Insectivores

(Sorex, Crocidura, Suncus, Echinops, Hemicentetes, Setifer, Tenrec, Erinaceus)

Progressive Insectivores

(Solenodon, Nesogale, Limnogale, Potamogale, Neomys, Talpa, Galemys, Desmana, Chlorotalpa, Elephantulus, Rhynchocyon)

Tupaiidae

(Tupaia, Urogale)

Prosimians

(Microcebus, Cheirogaleus, Lepilemur, Hapalemur, Lemur, Avahi, Propithecus, Indri Daubentonia, Loris, Perodicticus, Galago, Tarsius)

Simians

Hapalidae

(Hapale, Leontocebus)

Aotes

Cebidae and Cercopithecidae

(Saimiri, Cebus, Ateles, Colobus, Macaca, Cercopithecus)

Pan

Homo

The brain of each animal is cut in either the frontal, horizontal or sagittal plane. The cytoarchitectonics are studied in serial paraffin sections in the frontal plane, 10 to 28 mm. thick. Every other section is stained with cresyl violet and Heidenhain Woelche stains in order to study nuclear and fiber components. All sections are stained in the smaller brains and every tenth section in the larger brains. Photographs magnified up to 35 X are made of 50 to 60 serial sections taken at equal intervals in each brain. For further details on technique refer to Stephan (23, 24).

Nuclear Cytoarchitectonics

In the insectivores the nuclear cytoarchitectonics of the septum is based upon detailed studies in 12 brains (2). Seven brains represent the sub-family Soricinae (red tooth shrew); five from the Genus Sorex Aranius (common shrew); one from the Genus Sorex Minutus (a dwarfed form of common shrew); and one from Neomys Fodiens (old world water shrew). Five brains are from the sub-family Crocidurinae (white tooth shrew); two from the Genus Crocidura

Russula (house shrew); three from the Genus Crocidura Occidentalis (an African musk shrew). The average body and brain weights in grams are as follows: Sorex Aranius, 10.3 and 0.2; Sorex Minutus, 5.3 and 0.11; Neomys Fodiens, 15.0 and 0.32; Crocidura Russula, 11.0 and 0.18; Crocidura Occidentalis, in males 32.0 and 0.45; in the females 24.0 and 0.43.

Detailed nuclear cytoarchitectonic studies representing the prosimian brain was performed on the Galago Demidovii (1). It belongs to a lower family of primates and possesses a body weight of 81.0 grams and a brain weight of 3.38 grams (Fig. 2). It lives in tropical Africa where it thrives on various forms of living matter and vegetation. Among higher primates the nuclear cytoarchitectonics was studied in the cercopithecus ascanius and colobus badius (25). Colobus badius body weight is 7,260 grams and brain weight 74.3 grams. Cerecopithecus ascanius body weight is 3,670 grams and brain weight 65.2 grams (Fig. 3).

Developmental size changes of the septum and its component nuclei are based on volumetric determinations, made by outlining various brain structures on the histologic serial photographs (Fig. 4). The various structures are then cut out and weighed. The volume of each structure is determined in accord with the weights of each structure translated to volume by taking into consideration the distances between the histologic photographs. Distances between

Figure 2. Galago demidovii from Durrell, G. M., 1955. Granckhsche Vertagshandlung, Stuttgart, 1955. Galago demidovii is a prosimian which represents a transition between the insectivores and the higher primates (overall body weight 10 grams, brain weight 3 grams). It has a relatively well differentiated brain and its septum demonstrates characteristics of both insectivores and higher primates. Taken from Andy and Stephan (1).

the histologic sections are calculated and thus by applying the formula:

$$V = \frac{WP \times WS \times D}{M2},$$

volumes for each division are derived. V = volume in cm^3; WP = average weight of the photograph paper in cm^2 per gram; WS = weight of the photographic paper of the cut-out structure in grams; D = distance of measured sections in cm; M^2 = square of linear magnification.

Anatomical Delineation of the Septum

Location. The septum arises from the commissure plate and extends into the medial wall of the hemispheres. Although it is contiguous with the cortex in the medial wall of the hemispheres, it does not possess a typical cortical organization.

Boundary. The boundaries of the septum are the anterior commisure which is posteriorly situated, the precommissure hippocampus anteriorly, the corpus callosum dorsally, the lateral ventricle laterally, and the nucleus accumbens and olfactory tubercle. The nucleus preopticus medianus is not included in the septum because the major part of the nucleus is in the preoptic area and only sends a very small projection into the anterior medial wall of the anterior commissure. The nucleus accumbens is not included in the septum because it is considered as a component of the striatum (5,6).

Figure 3. Left side: Cercopithecus ascanius; right side: Colobus badius (from J. A. Allen: Mammals from the American Mus. Congo Exp., Bull. Amer. Mus. Nat. Hist., 47, 1922-1925). Taken from Stephan and Andy (25).

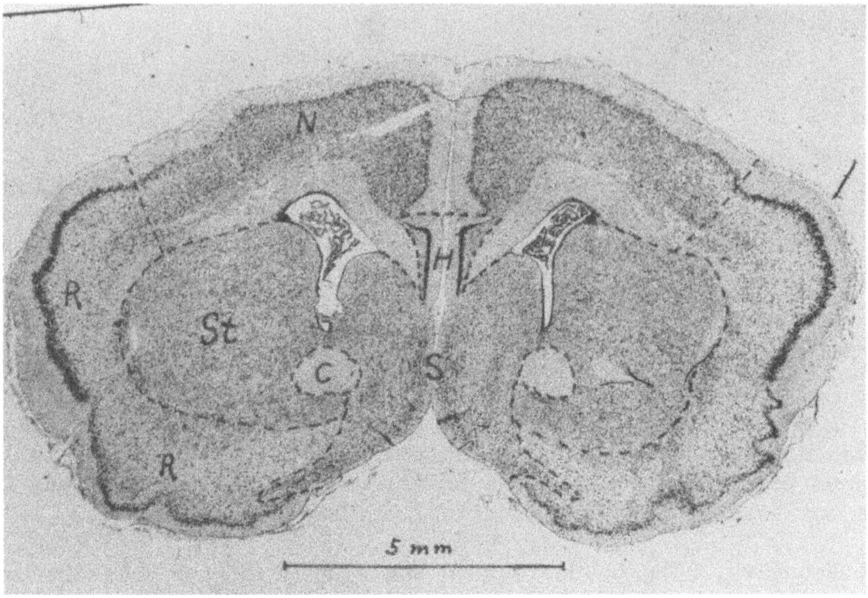

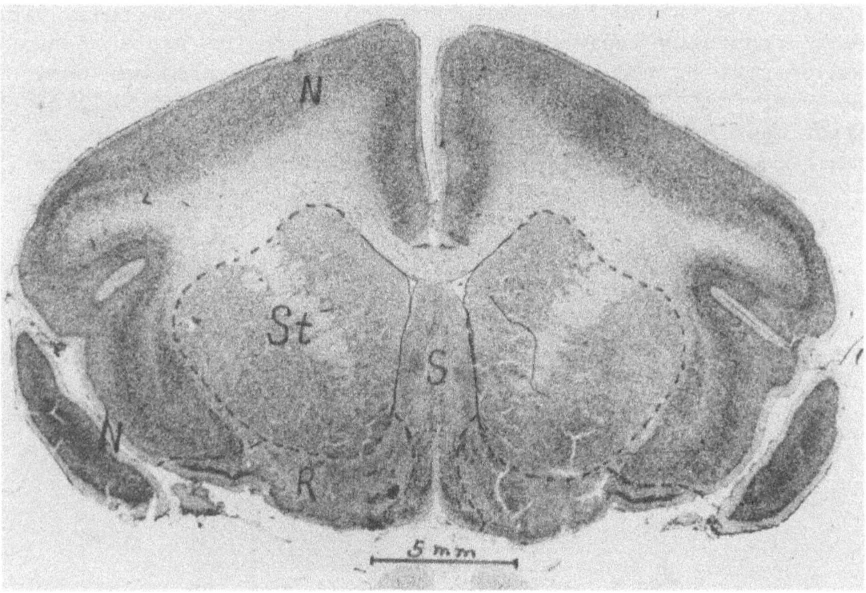

Figure 4. Upper figure: Tenrec 10005, slide 966; lower figure: Aoetes 387, slide 860. C-Commissura anterior; H-Hippocampus praecommissuralis; N-Neocortex; R-Rhinecephalon; S-Septum; St-Striatum. Note that upper figure was more enlarged to facilitate comparison of size relationships. Taken from Stephan and Andy (22).

Gross Configuration. In the lower insectivores the septum is relatively wide and flat, whereas in primates, especially in the higher ones, it is elongated in the vertical direction (Fig. 5). It is believed that the vertical narrow elongation is due to the tremendous development of the corpus callosum (Fig. 6) which draws the dorsal component of the septum anteriorly, dorsally and caudally (Fig. 7), thus resulting in the formation of the septum pellucidum.

In the human, the septum is divided into septum pellicidum and septum verum, the dorsal and ventral component respectively. The septum pellicidum which is especially well-developed and differentiated in the human brain, is composed of predominantly glia cells and fiber bundles. In contrast, the septum verum is primarily made up of neurons and fiber tracts. The relative anatomical position of the septal nuclei remains constant in progressing from insectivores to primates, irrespective of their relative levels in phylogenetic development. The major difference is in the relative shapes of the nuclei. In the lower forms, the nuclei tend to be plump whereas in the higher forms, especially in the higher primates, they are elongated vertically (Fig. 5).

Nuclear Configuration of the Septum

Illustrations of the following septal nuclei will be predominantly from the Galago demidovii because of a forerunner of higher primates it contains well differentiated nuclei. For details on the sagittal and horizontal relationships refer to Andy and Stephan (1). The nuclear organization of the septum consists of four major groups of nuclei; dorsal, ventral, medial and caudal. Each group is subdivided into nuclear components as follows:

Dorsal Group:

- Nucleus septalis dorsalis
  - Pars anterior
  - Pars externa
  - Pars intermedius
  - Pars interna

Ventral Group:

- Nucleus septalis lateralis
  - Pars externa
  - Pars interna

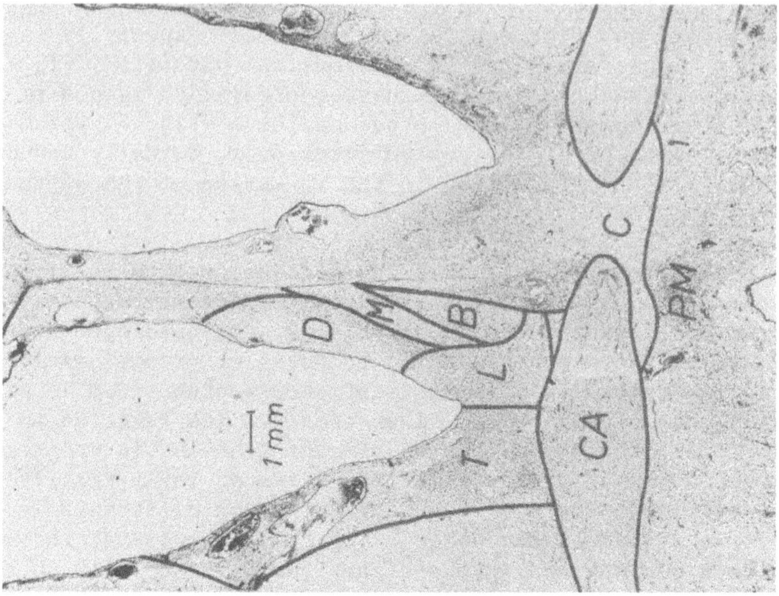
D
M
B
L
T
CA
C
PM
1mm

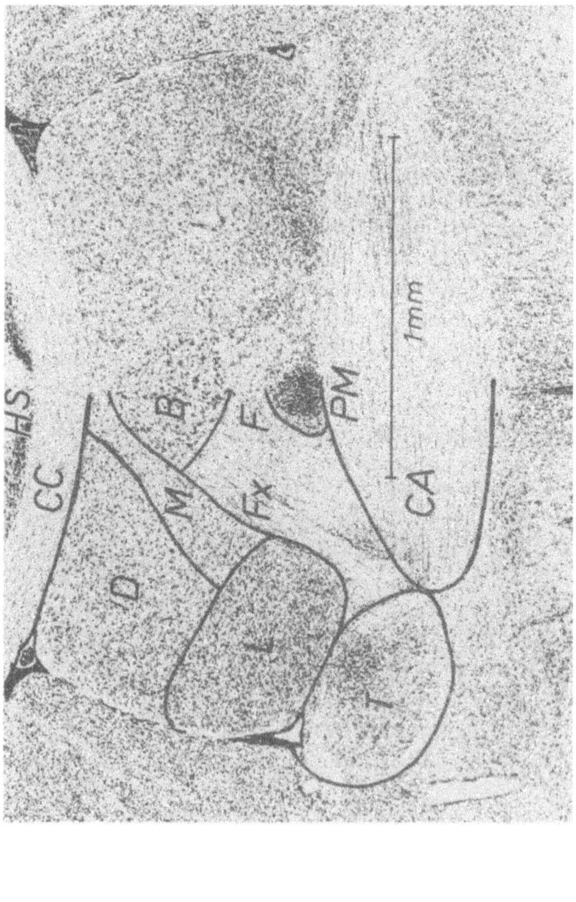
CC
HS
D
M
B
F
Fx
L
T
PM
CA
1mm

Figure 5. Frontal view of the septum in (a) Sorex araneus and (b) human. Note the relatively wide septum in Sorex and the very narrow septum with a septum pellicidum (immediately above nucleus septalis dorsalis) in the human; cresyl violet. Sorex araneus, brain 461, slide 152; enlargement 28.3x. Homo sapiens "Allocortex," slide 1935; enlargement 3.3x. Take from Andy and Stephan (3).

B	Diagonal band of Broca	Fx	Fornix
C	Bed nucleus of the anterior commissure	HS	Hippocampus, Pars supracommissuralis
DA	Commissura anterior	L	Nucleus septalis lateralis
CC	Corpus callosum	M	Nucleus septalis medialis
D	Nucleus septalis dorsalis	PM	Nucleus praeopticus medianus
F	Nucleus septalis fimbrialis	T	Bed nucleus of the stria terminalis

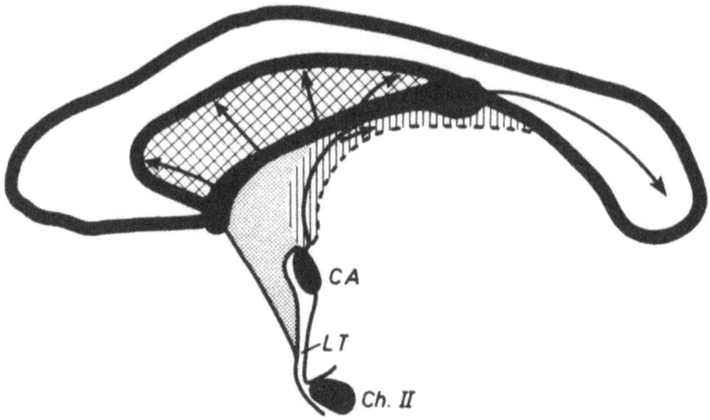

Figure 6. Schematic illustration in the sagittal plane of the formation of the septum pellucidum in the human. Dotted area represents the septum in Galago demidovii superimposed on the adjusted size of the septum verum in the human by coinciding the anterior commissure and the anterior borders of the fornix. Squares represent septum pellucidum and arrows indicate the directions of stretching. Vertical lines represent the fornix in the human. The portion in front of the solid line is the size of the fornix in the Galago. Taken from Andy and Stephan (4).

Medial Group:

- Nucleus septalis medialis
  - Pars anterior
  - Pars posterior
- Nucleus of the Diagonal Band of Broca
  - Pars dorsalis (septal component)
  - Pars ventralis (tubercular component)

Caudal Group:

- Nucleus septalis fimbrialis
- Nucleus septalis triangularis
- Bed nucleus of the anterior commissure
- Bed nucleus of stria terminalis
  - Pars anterior
  - Pars externa
  - Pars interna

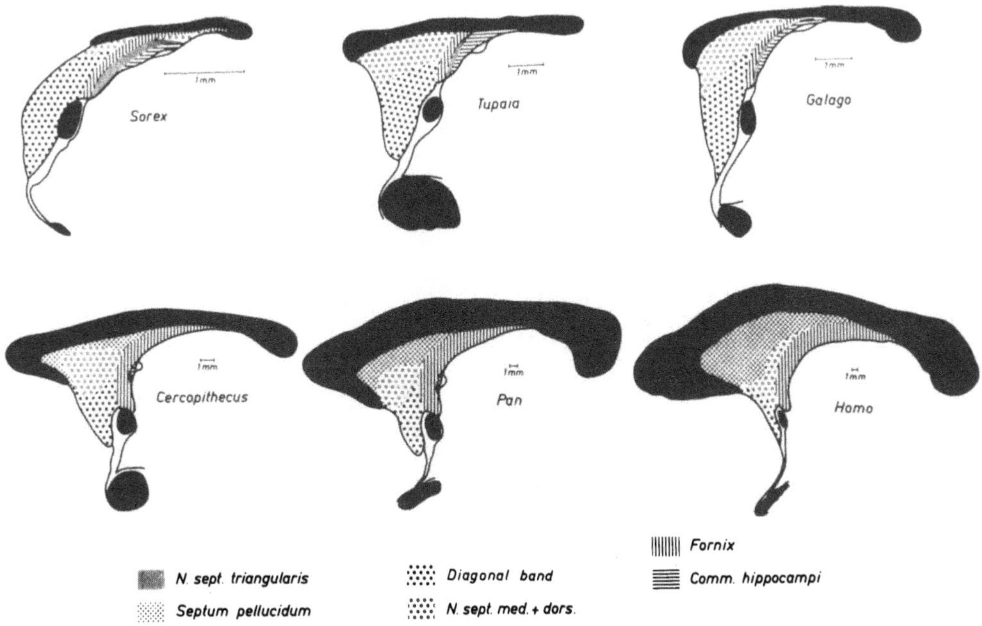

Figure 7. Diagrammatic sagittal views of the septum, including the corpus callosum and optic chiasm, representing various stages in primate evolution. The septum is at the rostral end of the brain stem just in front of the anterior commissure. This figure demonstrates the relatively small size of the corpus callosum in contrast to the total size of the septum in the lower forms such as the sorex, insectivore. Note the progressively enlarging corpus callosum with its relatively marked forward expansion and accompanying formation of the septum pellucidum in primates. In phylogeny the corpus callosum becomes progressively larger and the septum becomes relatively smaller, especially in the human brain. This type of comparison leaves the erroneous impression that the septum decreases in size during primate development.

Dorsal Group

The nucleus septalis dorsal in all forms is characterized by heterogeneity of cell shape, stain and distribution. The cells of this group tend to be large and spiny.

Pars anterior. This nucleus is not as well defined in insectivores and human as it is in the prosimian (Fig. 8). It is essentially made up of large round and faintly stained plump cells which are diffusely distributed.

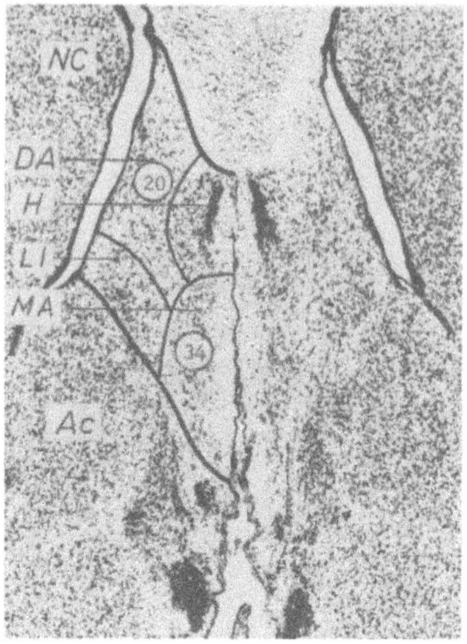

Figure 8. A frontal section through the rostral end of the septum going through the genu of the corpus callosum reveals the hippocampal continuation (H) just beneath the genue of the corpus callosum. Laterally, there is a nucleus septalis dorsalis anterior (DA) made up of large brown diffusely distributed lightly stained cells. Below that is a portion of the nucleus lateralis interna (LI). In the medial portion is the component of the nucleus septalis medialis anterior (MA). This slide also reveals the lower or ventral border of the septum along the nucleus accumbens (AC). The nucleus accumbens has not been included in the septum. Subsequent figures through Figure 13 consist of sequential presentation of more posterior frontal views of the septum with Galago demidovii. The reader is referred to the original article for details (1). Abbreviations for this and subsequent figures through Figure 14 follow:

A Nucleus septalis triangularis
Ac Nucleus accumbens
An Nucleus anterior dorsalis thalami
BD Nucleus of the diagonal band of Broca pars dorsalis
BV Nucleus of the diagonal band of Broca pars ventralis
C Bed nucleus of the anterior commissure
CA Commissura anterior
DA Nucleus septalis dorsalis pars anterior
DE Nucleus septalis dorsalis pars externa
DI Nucleus septalis dorsalis pars interna
IC Islands of Calleja
LE Nucleus septalis lateralis pars externa
LI Nucleus septalis lateralis pars interna
MA Nucleus septalis medialis pars anterior
MP Nucleus septalis medialis pars posterior
NC Nucleus caudatus
OT Olfactory tubercle
PM Nucleus preopticus medialis
PV Psalterium ventrale
S Subfornical organ (intercolumnar tubercle)

List of Abbreviations (continued)

DM Nucleus septalis dorsalis pars intermedius
F Nucleus septalis fimbrialis
Fx Fornix
H Anterior continuation of the hippocampus
SM Stria medullaris
TA Bed nucleus of the stria terminalis pars anterior
TE Bed nucleus of the stria terminalis pars externa
TI Bed nucleus of the stria terminalis pars interna

Pars externa. This component of the dorsal group is made up of the largest cells. They are faintly stained and diffusely scattered with relatively large amounts of intercellular space. Their shape tends to be round and polygonal. In the human it tends to occupy the entire width of the septum. Its transition into the pars intermedius is poorly defined. In the lower forms such as the insectivores and prosimians it is very well defined and developed with its lateral surface bordering the ventricle (Fig. 9).

Pars intermedius. The cells in this group are more heterogeneous in size and shape and tend to cluster haphazardly in little groups. This nucleus is located between pars externa and pars interna of the dorsal group. It is better developed in the lower than higher forms.

Pars interna. This group of cells forms a very conspicuous nucleus along the anterior dorsal medial border of the septum and runs diagonally from an antero-dorso-medial to a ventro-caucolateral direction (Figs. 9 and 10). The cells are densely packed into a narrow band close to the midline and tend to orient in a vertical direction at the midline. Anteriorly they are separated at the midline by a thickened layer of glia. Posteriorly this nucleus progresses later-ventrally and reaches the lateral border of the septum at about its midpoint. Medially it is bordered by the nucleus mediali septum and ventrally by the nucleus septalis lateralis. This nucleus is well developed throughout phylogeny.

Ventral Group

Nucleus septalis lateralis. It is located ventro-laterally in the septum and extends caudally to the level of the anterior commissure. Anteriorly it extends to the anterior third of the septum (Figs. 8-12).

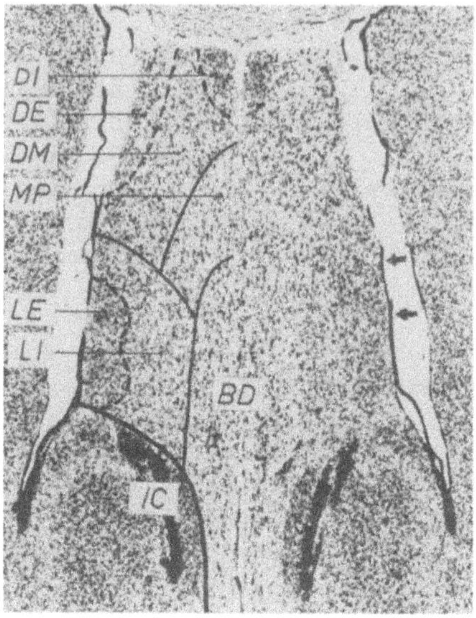

Figure 9. More posteriorly, additional components of the dorsal group become evident. Its most medial component is made up of darkly stained elongated and sharp cells called the dorsalis interna (DI) and the intermediate component (DM) is made up of a mixture of loosely distributed cells. The dorsalis externa (DE) is a group of large cells along the lateral border of the dorsal group. The nucleus dorsalis is, therefore, made up of four components--dorsalis anterior, dorsalis interna, dorsalis externa, and dorsalis intermedius. At this level, nucleus medialis posterior (MP), a component of the nucleus medialis septum is also evident. The anterior component (MA), was seen in Fig. 8. The nucleus of the diagonal band of Broca is divided into septal and tubercular components. These are also identified as dorsal (BD) and ventral components respectively. The dorsal or septal component contains nuclei oriented in the transverse direction. Note the islands of Calleja (IC) just beneath the lateral nucleus of the ventral group. This section shows a very nice differentiation between the loosely arranged cells of the lateralis interna and the compactly arranged cells of the nucleus lateralis externa.

Pars interna and externa. The cells in this nucleus are lightly stained, round, oval and small. They have faintly defined processes. The pars externa consist of a compact group of cellular elements. In the insectivores the transition from the compact pars externa to the loosely arranged pars interna is not as well defined as in the primates. It should also be noted that the dorsal transition into dorsal group may be poorly defined. In fact there occasionally apears to be a well defined nuclear group of cells between the dorsal and ventral groups which could be assigned to either one.

Medial Group

Nucleus septalis medialis. Pars anterior and posterior tend to be similar in the various forms (Figs. 8-12).

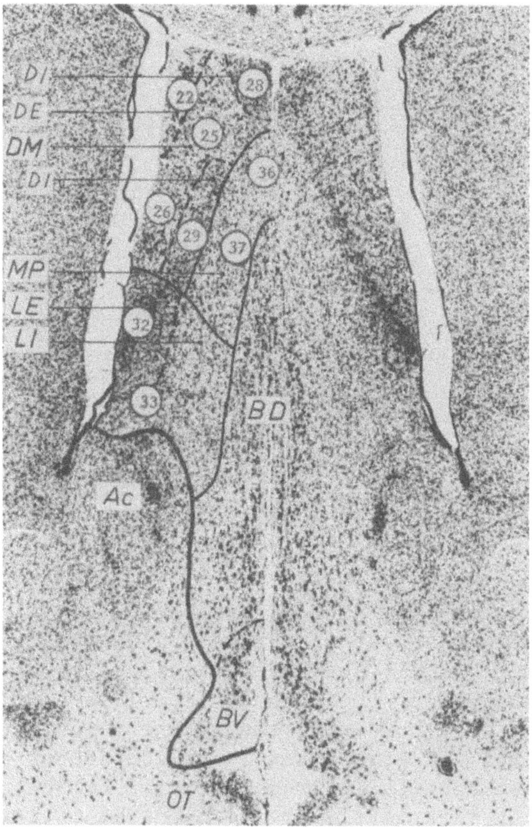

Figure 10. This frontal section reveals the ventral component of the nucleus of the diagonal band of Broca, also identified as tubercular component. Borders between nucleus septalis dorsalis and medialis are well differentiated whereas between nucleus dorsalis and lateralis, borders are not consistently well differentiated.

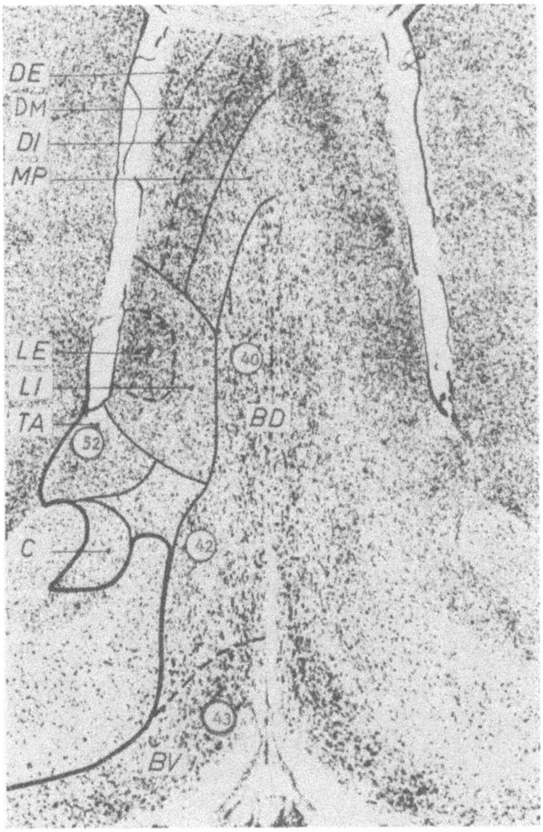

Figure 11. This frontal section reveals a cluster of cells between the nucleus dorsalis and the nucleus lateralis. Although it may be considered a separate group it has been incorporated with the dorsal group because of cell characteristics. This section also contains the bed nucleus of the anterior commissure (C) consisting of a thin layer of cells bordering the anterior surface of the anterior commissure. In some forms it is continuous across the midline and in other species it is interrupted by the nucleus preopticus medianum. At this level, the bed nucleus of the stria terminalis is also present. The bed nucleus of the stria terminalis is divided into three parts: One of these is pars anterior which is evident in this slide (TA). This is located just below the ventricular border and runs in the rostro-caudal direction as a tubular nucleus. Note the differentiation between the dorsal (BD) and ventricle (BV) components of the nucleus of the diagonal band of Broca.

Pars anterior. Cells of this nucleus are the most rostral in the medial part of the septum and represent the smallest of the two parts (Figs. 8 and 14). Ventrally it extends as a narrow band of cells between the nucleus accumbens and the cortex, and dorsally they meet the inferior border of the genu of the corpus callosum. Posteriorly it is continuous with the pars posterior of the same nucleus. The cells are large, oval, fusiform and polygonal in shape. The cell shapes and stain density vary. There are small and large cells which are darkly or lightly stained.

Pars posterior. Cells in the pars posterior are arranged in a vertical direction with many cell processes appearing in a sagittal plane but not in a frontal plane (Figs. 9-12). In the frontal plane the processes are oriented in a vertical direction.

Nucleus of the diagonal Band of Broca. This nucleus appears to form a large arc, oriented in front and around the anterior commissure (Figs. 9-12). It is divided into the dorsal and ventral components. The dorsal component extends above the level of the anterior commissure, and the ventral component extends below the level of the anterior commissure in a ventro-lateral direction into the diencephalon and tuberculum olfactorium. In the insectivores the septal and tubercular components are not as easily differentiated as in the prosimians and simians.

Pars dorsalis (septal component). This component of the diagonal band contains very large cells which are darkly stained and have multiple processes projecting in various directions. These cells are primarily oriented in a sagittal plane. In this plane there are multiple processes which are quite evident and the cell is quite plump. In the frontal plane, the cells appear narrow and elongated in a vertical direction with one or two processes.

Pars ventralis (tubercular component). The cells in this component tend to be large and darkly stained, but they are oriented predominantly in the frontal plane rather than in the sagittal plane. They have multiple spines, but not as many as in the cells of the dorsal component. In the human the cells of this component tend to be larger than those of the septal component. In the monkey this component contains cells oriented in both the sagittal and frontal plane whereas they are predominantly in the sagittal plane in the pars dorsalis.

Caudal Group

This group of nuclei appear to be more closely associated with fiber tracts entering and leaving the septum. The nuclei in the caudal group tend to remain similar in the insectivores and lower primates, but tend to undergo changes in primate evolution.

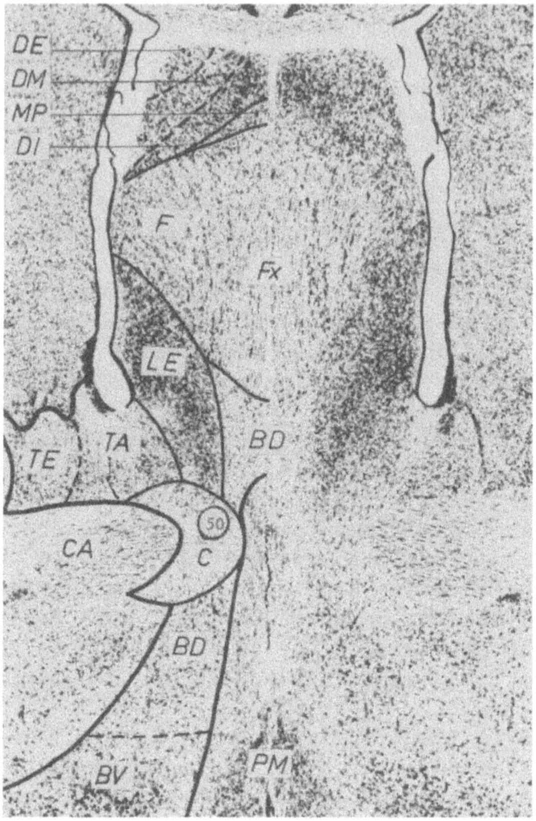

Figure 12. More posteriorly the nucleus fimbria fornix (F) is evident among the posterior group of septal nuclei. The cellular elements are loosely distributed through the fiber system of the fimbria. In addition, a portion of the bed nucleus of the stria terminalis is seen. One is the caudal limit of the pars anterior and the other is the beginning of the pars externa of the bed nucleus. Note that the nucleus preopticus medianum is present in the preoptic region and it is coursing upward toward the anterior commissure.

Nucleus septalis fimbrialis. The cells of this nucleus tend to be sandwiched between two layers of fibers, that of the fornix above and the hippocampal commissure below (Fig. 13). The cells are oval or spindle in shape with bipolar prolongations.

Nucleus septalis triangularis. The cells of this nucleus are concentrated along the midline between the fornix and ventral psalterium (Fig. 13). They are small, round and darkly stained. In the frontal view, the nucleus presents a triangular shape. This nucleus is relatively large and well defined in the insectivores, but becomes very small and progressively more difficult to detect in the higher primates.

Bed nucleus of the anterior commissure. The cells of this nucleus form a narrow band of vertically oriented cells in the frontal half of the anterior commissure (Fig. 12). Their continuation across the midline may be interrupted by the nucleus preopticus medianum (Figs. 12-14).

Bed nucleus of stria terminalis. This nucleus is associated with the termination of the stria terminalis fibers at the level of the anterior commissure and preoptic area (Figs. 12 and 13). It becomes progressively larger in primate evolution, although it is well defined in some of the lower forms. It can be separated into the three components.

Pars anterior. This nucleus is in the form of a rod which extends in the antero-posterior direction running along the ventral border of the lateral ventricle just above the anterior commissure (Fig. 12). The cells are round, densely packed and dark with faint unipolar extension.

Pars externa. The pars externa tend to be somewhat larger than other cells of the bed nucleus of stria terminalis. The cells have a pallisading arrangement with the stria terminalis running through the center of it. This portion of the nucleus extends around and ventrally to the anterior commissure into the preoptic area of the hypothalamus (Fig. 13).

Pars interna. This nuclear component is located medially to the pars externa and has smaller cells with no pallisading arrangement. It remains above the level of the anterior commissure and does not extend into the hypothalamus (Fig. 13).

Quantitative Development of the Septum in Phylogeny

The objective in this portion of the presentation is to demonstrate that the septum undergoes increased size in primate phylogeny and does not regress as does the olfactory system. The other objective is to determine which of the nuclear subdivisions becomes larger or smaller in phylogeny.

Septum development in relation to total brain volume (relative change). In phylogeny there is an evident marked reduction of the

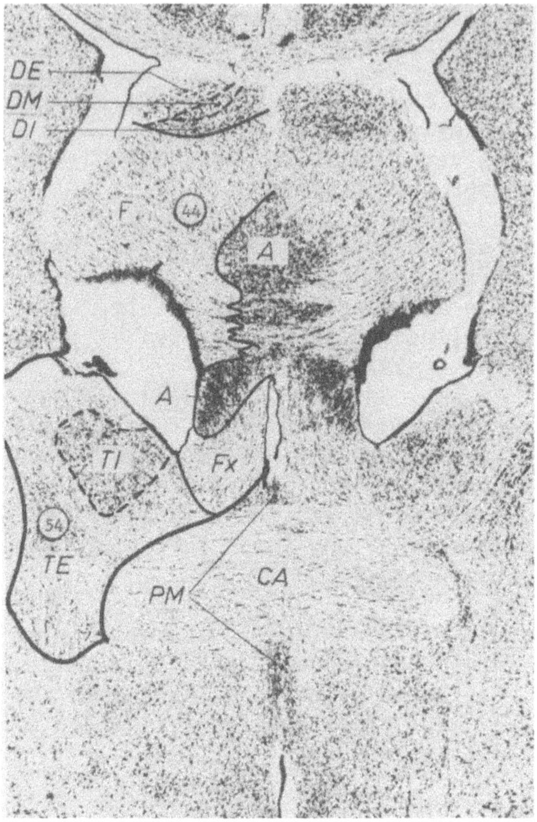

Figure 13. Further posteriorly the most posterior nucleus of the caudal group makes its appearance, the nucleus septalis triangularis (A). It is located in the midline between the fornix and the ventral hippocampal commissure. This section also reveals a portion of the bed nucleus of the stria terminalis, pars externa (TE) extending into the preoptic area and hypothalamus below the level of the anterior commissure (CA). In contrast, the bed nucleus of stria terminalis pars interna (TI) tends to remain above the level of the commissure. At this most posterior level, all four components of the caudal group are represented. These are: nucleus septalis fimbrialis, nucleus septalis triangularis, bed nucleus of the stria terminalis and the bed nucleus of the anterior commissure. The nucleus preopticus medialis (PM) is not included in the septal nuclei because the predominant mass of this nucleus is actually located in the hypothalamus and preoptic area, and only a small component extends into the midline of the septum, just anterior to the anterior commissure.

percentage of the total septum in relation to the total brain volume as one progresses from basal insectivores to higher primates and finally to man. The percentage volume of the septum in relation to the total brain volume from basal insectivores to Homo is as follows: Basal insectivores 1.79% progressive insectivores 1.57%, Tupaiidae 1.07%, prosimians 1.76%, simians 0.41%, Pan 0.22%, and Homo 0.20%. The apparent reduction of the septum is a result of the progressive and extreme enlargement of the neocortex in relation to other brain structures (Fig. 15). This obviously provides an erroneous conclusion in relation to whether or not the septum develops or regresses in phylogeny of the primate brain.

<u>Septum development in relation to body size (absolute change)</u>. In order to obtain a true picture of the phylogenetic changes in various brain structures in phylogeny, it is necessary to compare the size of the structure to an independent point of reference (23). The allometric method of relating brain volume to body weight was developed by Snell (21) and Dubois (8). The formula established by those investigators revealed that brain volume is proportional to body weight and for a group of animals in a closely related species a straight line relationship exists on a double logarithmic scale. Since the insectivores are the forerunners of the placental mammals, they are utilized as a reference for evaluating the development of brain structures in the primate. The non-specialized primitive terrestrial insectivores are used as basal forms of reference from which a regression line is constructed (Fig. 16). The increased size in higher forms is determined by measuring the distance from the regression line.

Growth Rates of the Septum Nuclei

<u>Growth rate of the total septum</u>. In prosimians the total septum is found to be 2.9 times larger than that in the basal insectivores. In the higher primate, it ranges from 1.6 to 2.8 times the basal forms. In the human brain it undergoes a four times increase in size. Consequently, the septum as a whole definitely shows a progressive increased growth rate in phylogeny. Note that in the human, the septum is furthest away from the regression line (Fig. 17).

It should also be noted that in relating the development of the septum to the Bulbus Olfactorius, the volume of the septum definitely increases in size in relation to the Bulbus Olfactorius (Fig. 18). In relating the hippocampus to the septum, the progressive insectivores, prosimians, and simians all lie along the regression line which is established by the basal insectivores (Fig. 19). This means that both the hippocampus and septum develop at a similar rate in phylogeny.

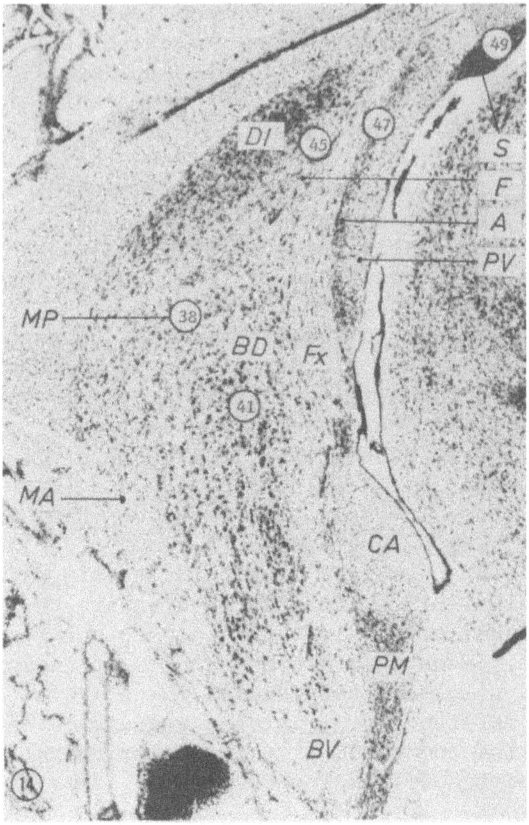

Figure 14. A sagittal section taken through a medial plane of the septum of the Galago demidovii. Note the postero-dorsal extent of the nucleus septalis dorsalis pars interna (DI) along the ventral border of the corpus callosum. The darkly stained cells are compactly organized. In contrast, nucleus septalis triangularis (A) consist of a thin layer of cells between the fornix and the ventral psalterium (PV) of the hippocampus. This section clearly demonstrates the nucleus preopticus medialis (PM) extending into the septum from the hypothalamus and preoptic area. Its stream of cells courses anterior to the anterior commissure.

The regression line obtained for the total septum volume in relation to body weight in a larger group of insectivores gave a slope of 0.62 (22). For illustrations 20 through 29, the calculations were based on a base line reference possessing a slope of

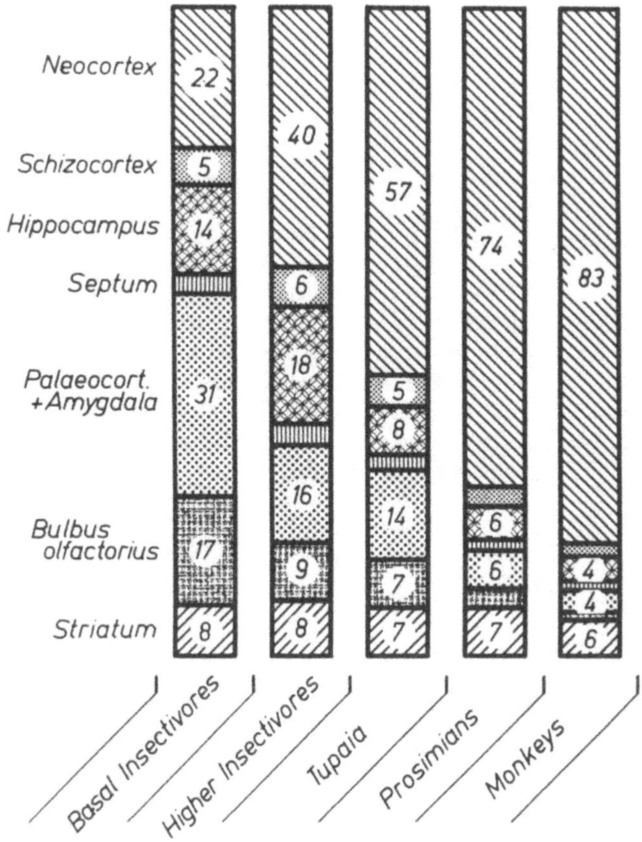

Figure 15. Bar graph illustrating the percentage values of the volume of various brain structure in relation to the total telencephalon in phylogenetic development. The septum goes from 2.0% in basal insectivores to 0.2% in monkeys. The hippocampus, a structure closely associated with the septum, makes up to 14% of the telencephalon in the basal insectivores and 4% in the monkeys. In contrast, in relation to the total telencephalon, the neocortex makes up 22% in insectivores and 83% in monkeys. This method of evaluating the volumetric changes of the septum in phylogeny leads to erroneous conclusions. The reasons for this erroneous conclusion is that the neocortex increases at such a tremendous rate, that other structures increasing at a relatively slower rate appear to become progressively smaller. In order to evaluate the actual or absolute development of a brain structure in phylogeny, it is necessary to refer that particular brain structure to an indifferent reference.

0.60, obtained from three basal insectivores. The respective growth rates of the total septum for Galago, cercopithecus and colobus are 1.9, 1.8 and 1.4 times the basal insectivores (Fig. 20). The differential growth rates of the following septal nuclei are based on the regression line with a slope of 0.60, derived from three basal insectivores. As mentioned above, this slope is closely similar to the slope of 0.62 derived from a larger group of basal insectivores.

Nucleus septalis lateralis. Based on the 0.60 regression line for the septum as a whole (Fig. 21), the growth rate for this nucleus is 3.1 times for Galago, 1.7 times for colobus and 2.1 times for cerecopithecus. The functional significance of this absolute increase in structure and size in the Galago in contrast to the higher primates is unknown. This difference may relate to some form of specialization in that animal. It should be noted that in this prosimian, the nucleus is much better differentiated than in the higher forms.

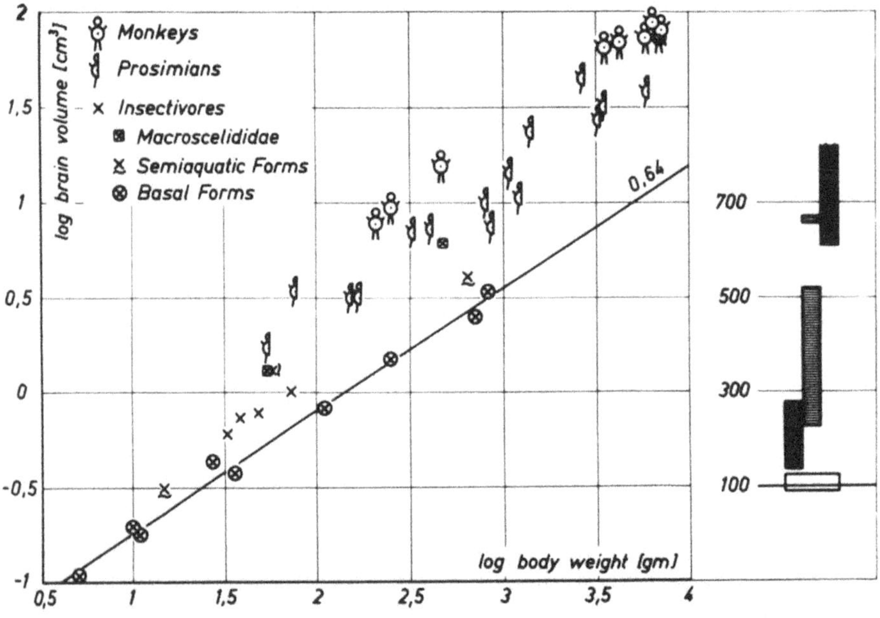

Figure 16. Total brain volumes plotted against body weight. Regression line was constructed from insectivores identified by encircled crosses (basal insectivores). Vertical graph (right side of figure), absolute growth of the total brain in relation to body weight, based on the regression line established from the basal insectivores. Monkeys (black bar), Prosimians (horizontal lined bar), Progressive insectivores (cross hatched bar), and basal insectivores (horizontal white bar).

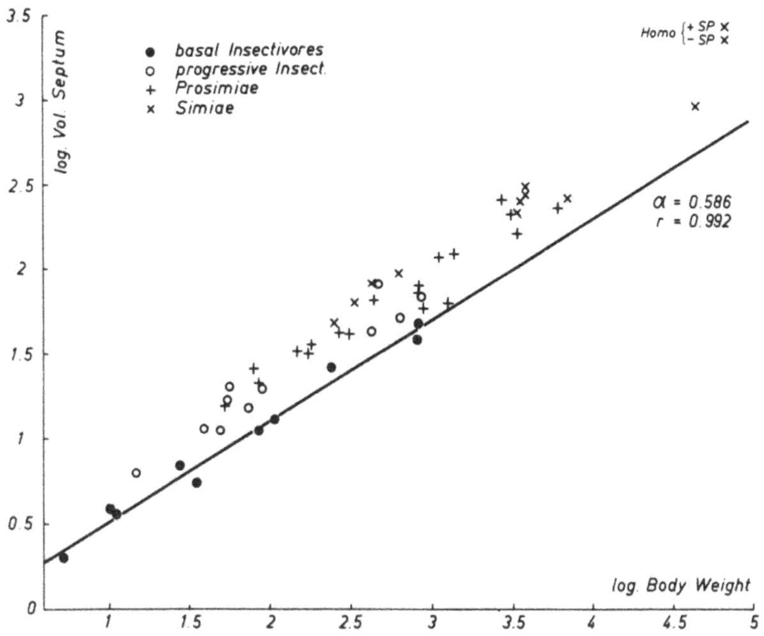

Figure 17. Total septum plotted against body weight. The regression line in this instance was established from the septum volume and body weight relationship in the Sorex Minutus, Sorex Araneus, Crocidura ressala, Crocidura occidentalis and Erinaceus europaeus (solid circles). The greater the distance above the regression line the larger the size of the septum. Note that in the progressive insectivores (open circles) the septum did become larger. The prosimians and simians (crossed and x's respectively) are further away from the baseline, thus indicating a further enlargement. The human septum (Homo) being furthest from the line represents the greatest development in phylogeny.

Nucleus of the diagonal Band of Broca. The growth rates are for Galago 2.1 times, cerecopithecus 1.8 times and colobus 1.3 times the basal forms (Fig. 22). The growth rate in this nucleus is somewhat similar to the total septum growth rate since the diagonal Band of Broca makes up approximately 1/3 of the whole septum.

Nucleus septalis triangularis. Based on the 0.60 regression line, the growth rates are 1.3 times for Galago, 0.7 times for cerecopithecus, and 0.5 times for colobus (Fig. 22). These figures represent a very definite decrease in size for the two higher

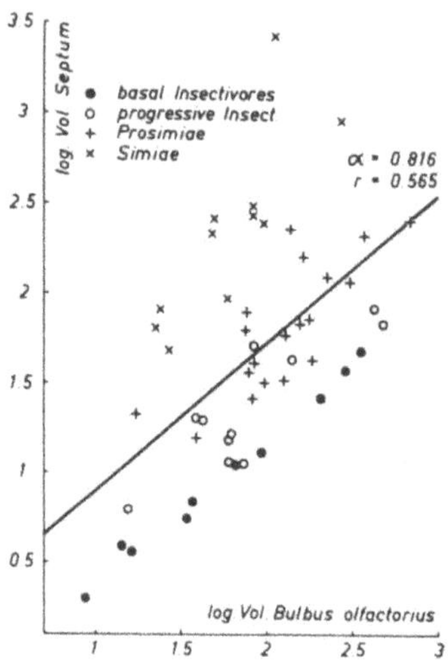

Figure 18. Septum plotted against olfactory bulb in a double logarithmic scale. Note the lack of size correlation (r = 0.565) between these two structures.

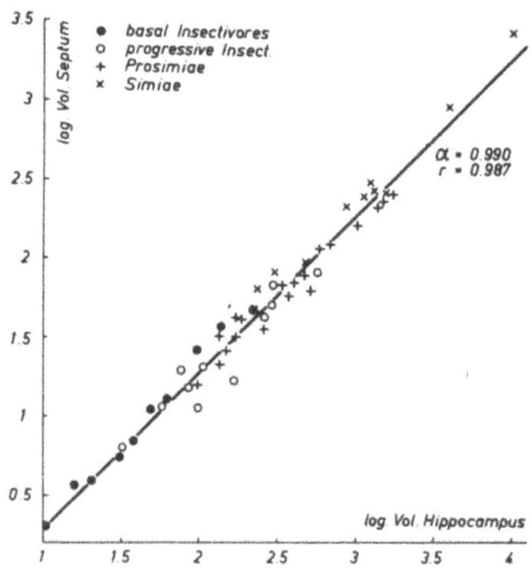

Figure 19. Septum plotted against hippocampus. Note the very good size correlation (r = 0.987) between these two structures.

primates in relation to the basal forms. This nucleus is the only one which reveals a clear cut reduction in the phylogenetic development of the primate septum.

Nucleus septalis dorsalis. Based on the 0.60 regression line, the size increase is 1.7 times for Galago, 1.6 times for cerecopithecus, and 1.5 times for colobus (Fig. 24).

Nucleus septalis medialis. The growth rates for Galago are 1.6 times, for cerecopithecus 1.4 times, and for colobus 1.1 times the basal size (Fig. 25). In spite of the fact that these growth rates are greater than one, the growth rates of this particular nucleus do not exceed that of the total septum.

Nucleus septalis fimbrialis. The growth rates are 1.6 times for Galago, 1.2 times for cerecopithecus, and 0.7 times for colobus (Fig. 27). Although colobus reveals an absolute reduction, in general there seems to be no marked size alteration of this nucleus in phylogeny. However, it definitely does not show the growth rate of the entire septum.

Bed nucleus of the stria terminalis. This structure has a clear and absolute increase in size through phylogeny. In higher monkeys the increase of this nucleus is greater than that for the septum as a whole. Consequently, the proportionate size of this nucleus in contrast to the rest of the septum is also larger. The growth rate in relation to the basal insectivores is 1.7 times for Galago, 2.7 times for cerecopithecus, and 2.1 times for colobus (Fig. 28).

Fimbria fornix. In the Galago this complex is nearly 2.0 times that of the basal insectivores, and colobus 2.4 times and cerecopithecus 2.9 times (Fig. 29). These figures represent an obvious absolute increase in size in phylogeny.

Subfornical body. The subfornical body is not considered a structure belonging to the septum. However, since it is a special periventricular organ lying along the posterior-ventral surface of the septum, it appears justifiable to treat it as a separate component. Based on a regression line with a slope of 0.49, the growth rates are 1.8 times for Galago, 4.4 times for colobus, and 5.7 times for cerecopithecus (Fig. 30). It is of interest to note that this represents a very marked increased growth rate for the subfornical organ. It is especially marked when compared to the growth rates of the septum as a whole which is approximately two times among simians and prosimians.

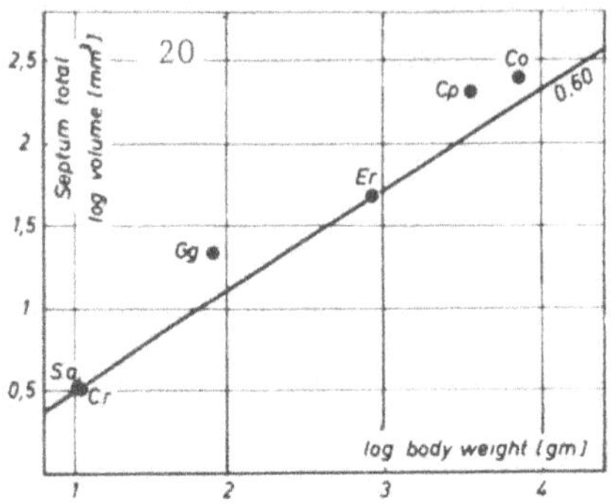
20
Septum total
log volume (mm³)
Co
Cp
0,60
Er
Gg
Sa
Cr
log body weight (gm)

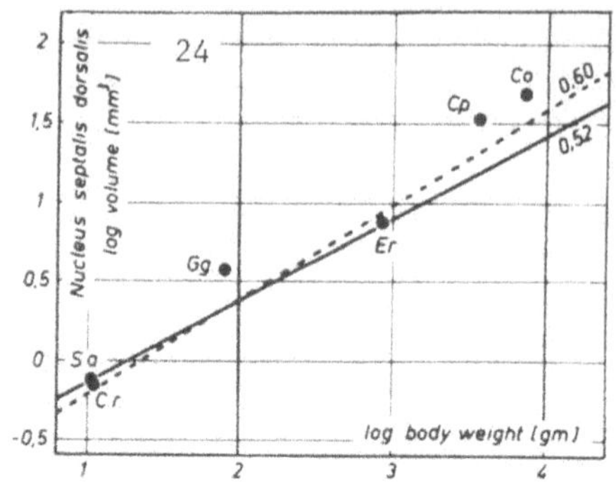
24
Nucleus septalis dorsalis
log volume (mm³)
Co
Cp
0,60
0,52
Er
Gg
Sa
Cr
log body weight (gm)

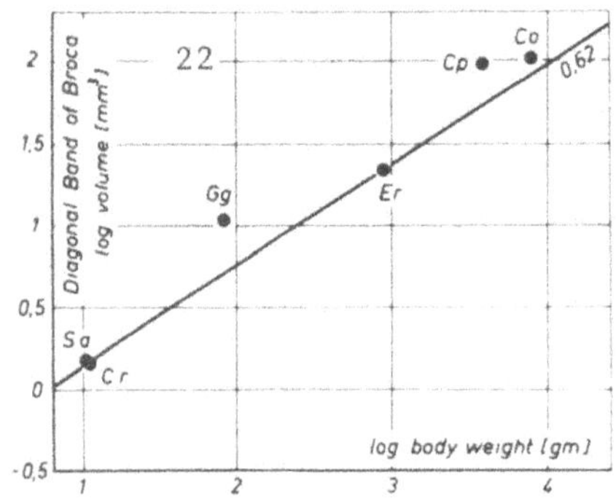
22
Diagonal Band of Broca
log volume (mm³)
Co
Cp
0,62
Er
Gg
Sa
Cr
log body weight (gm)

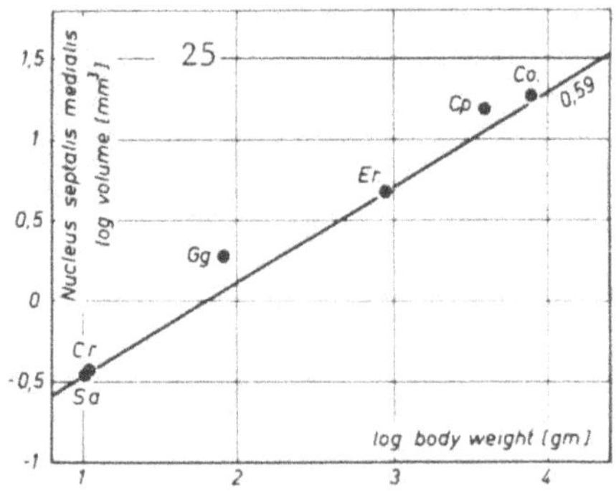
25
Nucleus septalis medialis
log volume (mm³)
Co.
Cp
0,59
Er
Gg
Cr
Sa
log body weight (gm)

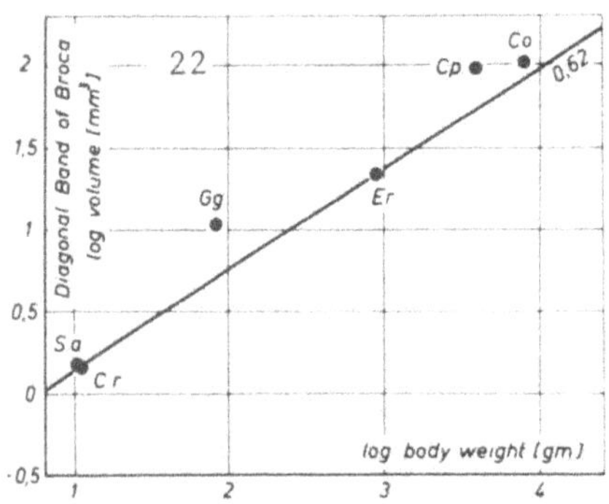
22
Diagonal Band of Broca
log volume (mm³)
Co
Cp
0,62
Er
Gg
Sa
Cr
log body weight (gm)

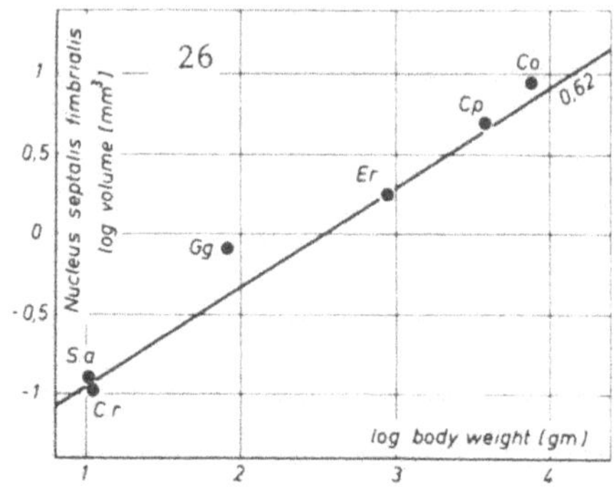
26
Nucleus septalis fimbrialis
log volume (mm³)
Co
Cp
0,62
Er
Gg
Sa
Cr
log body weight (gm)

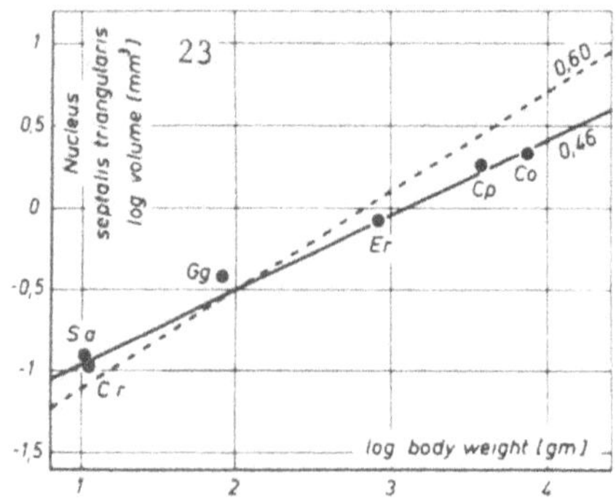
23
Nucleus
septalis triangularis
log volume (mm³)
0,60
0,46
Cp
Co
Er
Gg
Sa
Cr
log body weight (gm)

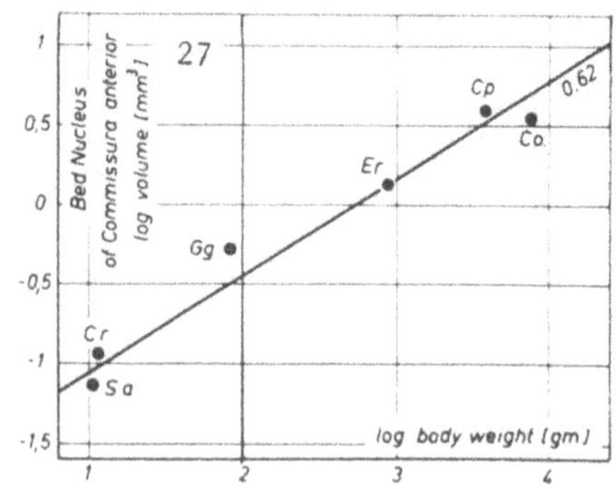
27
Bed Nucleus
of Commissura anterior
log volume (mm³)
Cp
0.62
Co.
Er
Gg
Cr
Sa
log body weight (gm)

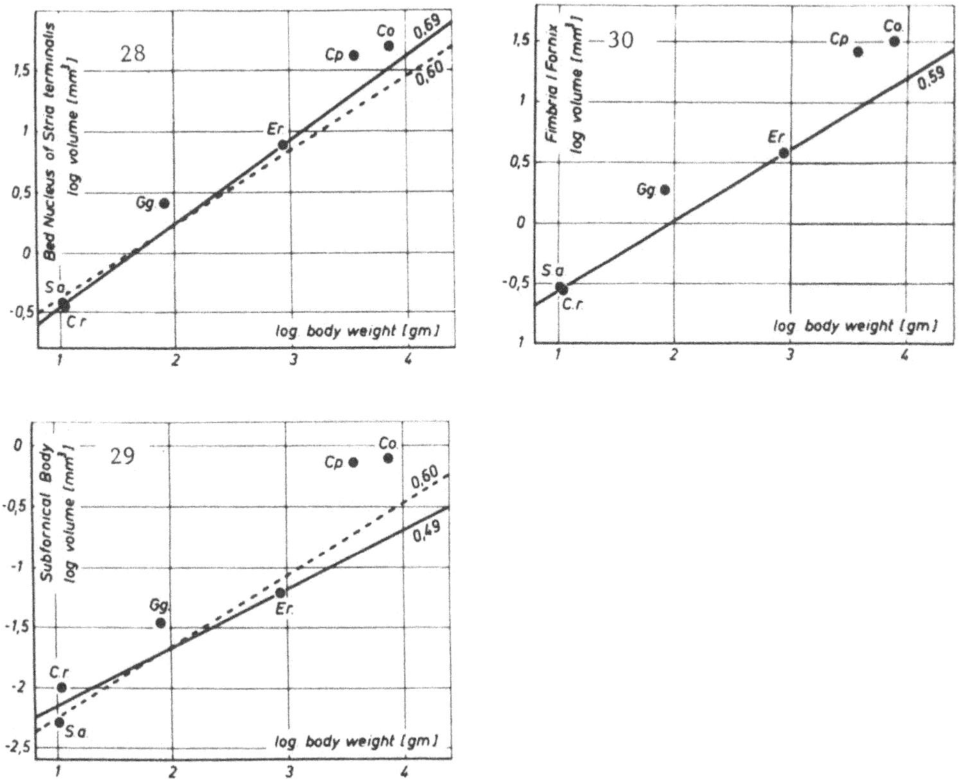

Figures 20-30. Regression lines established in basal insectivores by correlating body weights with septum as a whole (Fig. 20) and individual septal nuclear or fiber structures (Figs. 21-30, solid lines). The interrupted lines in Figures 21, 23, 24, 28, and 30 are constructed from the slope of the regression line obtained for the whole septum (0.60, Fig. 20) and properly adjusted to the values of specific septal nuclei in the basal forms. The species utilized were for the basal insectivores: Crocidura russula (Cr.), Sorex araneus (Sa.), and Erinaceus europaeus (Er.) and for the primates: Galago demidovii (Gg.), Cerecopithecus ascanius (Cp.), and Colobus badius (Co.). Taken from Andy and Stephan (3).

DISCUSSION

The septum has commonly been considered to contain a medial and lateral nucleus. In the above studies, it has been evident that the septal nuclei can be separated into four major groups which are readily identified from one another, especially in lower forms. Most of the nuclei described have been previously observed and have been considered as components of a preconceived medial and lateral septal nucleus. For example, the dorsal and ventral nuclear groups which we differentiate were described as components of the lateral septal nucleus by Loo (17) in the opossum, Humphrey (12) in the bat, Lauer (16) in the Panda, Jeserich (13) in the mink, Lauer (15) in the macaque and Brockhaus (6) in the human. The round cells described by Johnson (14) in the guinea pig and mole most likely represent the ventral group. Our nucleus dorsalis interna corresponds to the nucleus septalis hippocampalis as described by Loo (17) in the opossum, Humphrey (12) in the bat, Jeserich (13) in the mink and Johnson (14) in the mole. The nucleus septalis medialis corresponds with the descriptions given for the nucleus septalis medialis by Loo (17) in the opossum, Humphrey (12) in the bat, Fox (9) in the cat, Jeserich (13) in the mink, Lauer (16) in the Panda, and Johnson (14) in the rabbit. The large cells located posteriorly in the medial nucleus are considered as the septal component of the diagonal Band of Broca. The ventral continuation of these large cells is identified as the tubercular component of the same nucleus in order to emphasize the apparently similar nuclear characteristics and their belonging to the same nucleus. Nucleus septalis fimbrialis has not been usually identified as a caudal continuation of the lateral nucleus from which it may not be differentiated, according to Herrick (11) in the reptile, Loo (17) in the opossum, Young (26) in the rabbit, Fox (9) in the cat, Laudr (16) in the Panda, Johnson (14) in the guinea pig, and in the mole. Nucleus triangularis has been identified as a separate entity by Lauer (15) in the Panda, Young (26) in the rabbit, Humphrey (12) in the bat, Ramon y Cajal (7) in the rat, and Loo (17) in the opossum. It should be emphasized that for the human septum, Brockhaus (6) did not describe a nucleus which would correspond to the nucleus septalis triangularis. This is in keeping with the fact that this nucleus is quite small in the human septum in contrast to lower forms. The bed nucleus of the anterior commissure is identified by Gurdjian (10) in the rat, Young (26) in the rabbit, Johnson (14) in the mole, and Humphrey (12) in the bat. In the macaque, Lauer (15) states that it is continuous across the midline. In the cat, Fox (9) finds that the bed nucleus of the anterior commissure and stria terminalis are not separable. This is at variance with the findings of Andy and Stephan who describe lines of demarcation for most of its borders. In the mink, Jeserich (13) divides it into a dorsal and ventral component. He further considers the stria terminalis as dividing the dorsal component of the bed nucleus of the anterior commissure into a medial and lateral part. Divisions of the bed nucleus of the stria terminalis into three component groups may be, in part, supported by the divisions of the bed nucleus described by Jeserich (13) in the mink. Descriptions by other investigators in other animals were not found to coincide with those evident in the insectivores and primates of this study.

It must be emphasized that the development of the septum is progressive throughout primate phylogeny and attains its greatest development in the human brain. It must also be emphasized that in view of its progressive development, it most likely is not a vital contributor to olfactory function, since the olfactory system undergoes marked regression in phylogeny of the primate (22). The only septal nucleus which undergoes regression in phylogeny of primates is the nucleus septalis triangularis.

REFERENCES

1. Andy, O. J., and Stephan, H. The nuclear configuration of the septum of Galago demidovii. J. Comp. Neur., 111 (1959) 503-545.

2. Andy, O. J., and Stephan, H. Septal nuclei in the soricidae (Insectivores), Cyto-Architectonic study. J. Comp. Neur., 117 (1961) 251-274.

3. Andy, O. J., and Stephan, J. Septal nuclei in primate phylogeny (A quantitative investigation). J. Comp. Neur., 126 (1966) 157-170.

4. Andy, O. J., and Stephan, H. The septum in the human brain. J. Comp. Neurol.,133 (1968) 383-409.

5. Ariens Kappers, C. U., Huber, G. C., and Crosby, E. C. The comparative anatomy of the nervous system in vertebrates, including man. The MacMillan Company, New York, N. Y. (1936).

6. Brockhaus, H. Zur feineren Anatomie des Septum und des Striatum. J. Psych. Neur.,51 (1942) 1-56.

7. Cajal, Ramon Y. Histologie du systeme nerveus. Teil 2 A. Maloine, Paris (1911).

8. Dubois, E. Uber die Abhängigkeit des Hirngewichts von der Körpergrösse bei den Säugetieren. Arch. Anthrop.,25 (1897) 1-28.

9. Fox, C. A. Certain basal telencephalic centers in the cat. J. Comp. Neur.,72 (1940) 1-62.

10. Gurdjian, K. S. Olfactory connections of the albine rat, with special reference to the stria medullaris and anterior commissure. J. Comp. Neur.,38 (1925) 127-163.

11. Herrick, C. J. The morphology of the forebrain in amphibia and reptilla. J. Comp. Neur.,20 (1910) 413-547.

12. Humphrey, T. The telecephalon of the bat. 1. The non-cortical nuclear masses and certain pertinent fiber connections. J. Comp. Neur.,65 (1936) 603-711.

13. Jeserich, M. W. The nuclear pattern and the fiber connections of certain non-cortical areas of the telencephalon of the mink (Mustela vison). J. Comp. Neur.,83 (1945) 173-211.

14. Johnson, T. N. Studies on the brain of the guinea pig. 1. The nuclear pattern of certain non-cortical areas of the telencephalon of the mink (Mustela vison). J. Comp. Neur., 83 (1957) 173-211.

15. Lauer, E. W. The nuclear pattern and fiber connections of certain basal telencephalic centers of the macaque. J. Comp. Neur.,82 (1945) 215-254.

16. Lauer, E. W. Certain olfactory centers of the forebrain of the giant panda (Ailuropoda melanoleuca). J. Comp. Neur., 90 (1949) 213-242.

17. Loo, Y. T. The forebrain of the opossum (didelphis virginiana), Part II. J. Comp. Neur.,52 (1931) 1-148.

18. Rose, M. Uber das histogenetische Prinzip der Einteilung der Grosshirnrinde. J. Psychol. Neur.,(Lpz.) 32 (1926) 97-160. - Ders., Der Allocortex bei Tier und Mensch. J. Psychol. Neur. (Lpz.) 34 (1927) 1-111.

19. Shimazono, J. Das septum pellucidum des menschen. Arch. Ant. u. Physiol. Anat. Abt.,(1921) 55-61.

20. Smith. G. E. Morphology of the true limbic lobe, corpus callosum, septum pellucidum and fornix. J. Anat. (London), 30 (1895a) 157-167 and 185-205. Ders., The connection between the olfactory bulb and the hippocampus. Anat. Anz., 10.

21. Snell, O. Die Abhängigkeit des Hirngewichts von dem Körpergewicht und den geistigen Fähigkeiten. Arch. Psychiatr., Berlin 23 (1892) 436-446.

22. Stephan, J., and Andy, O. J. The septum (A comparative study on its size in insectivores and primates). J. Hirnforsch., 5 (1962) 229-244.

23. Stephan, H. Methodische Studien über den quantitativen Vergleich architektonischer Struktureinheiten des Gehirns. A. wiss. Zool.,164 (1960) 143-172.

24. Stephan, H., and Andy, O. J. Vergleichend-anatomische Untersuchungen an Insektivorengehirnen. V. Die quantitative Zusammensetzung der Oberflächen des Allocortex. Acta Anat. (Basel),44 (1961) 12-59.

25. Stephan, H., and Andy, O. J. Cytoarchitectonics of the septal muclei in old world monkeys (Cerecopithecus and Colobus). J. Hirnforsch.,7 (1964) 1-23.

26. Young, M. W. The nuclear pattern and fiber connections of the rabbit (Lepus coniculus). J. Comp. Neur., 65 (1936) 295-401.

AUTORADIOGRAPHIC STUDIES OF THE DEVELOPMENT AND CONNECTIONS OF THE SEPTAL AREA IN THE RAT[1]

L. W. Swanson and W. M. Cowan

Departments of Biology, and Anatomy and Neurobiology
Washington University, St. Louis, Missouri 63130

INTRODUCTION

The septum, or parolfactory area as it used to be called, occupies a pivotal position within the forebrain, being interposed between the hippocampus and amygdala on the one hand, and the hypothalamus and habenula on the other. Because of its topographic and connectional relationships, considerations of septal function have generally centered on its role as a component of the so-called limbic system. Interest in the neurobiology of the septum was particularly stimulated by reports of a lesion-induced "septal rage syndrome" (8), and the phenomenon of septally mediated intracranial self-stimulation (48). Recent advances in our understanding of the functional significance of the septal area are amply reviewed in other contributions to this Volume; it is our purpose here to review briefly the results of recent neuroanatomical studies of the development and connections of the septum. This is particularly timely in view of the fact that the septum and hippocampus are increasingly coming to be regarded as a model system for the study of neuronal plasticity (9,45,52) and development (54,61).

---

[1]Supported in part by Grants NS-03777, NS-10943 and MH-24604 from the National Institutes of Health. Here and subsequently we use the general term hippocampal formation to refer collectively to the fields of Ammon's horn (CA1 through CA4), the dentate gyrus, the subicular region (including the subiculum, presubiculum, and parasubiculum), and the entorhinal cortex. In the rat, at least, all the fibers of the postcommissural fornix are derived from the subiculum and the pre- and parasubiculum (59,63).

RESULTS AND DISCUSSION

The Topography and Cytoarchitecture of the Rat Septum

As commonly defined (4,29,61) the septum of most placental mammals comprises a collection of nuclei situated between the anterior horns of the lateral ventricles, dorsal to (and partly surrounding) the decussation of the anterior commissure, ventral to the corpus callosum, rostral to the fimbria and the ventral hippocampal commissure, and caudal to the frontal and infralimbic cortex and the nucleus accumbens. Figs. 1A to 1H, which are a series of photomicrographs of coronal sections through the septal region of the rat, clearly illustrate these topographical relationships. On cytoarchitectonic grounds the rat's septum, like that of several mammals (21,28,41,61), can be subdivided into medial, lateral, posterior and ventral divisions (61).

The medial division consists of a rather broad mass of large neurons, which may conveniently be divided into a medial septal nucleus dorsally, and a nucleus of the diagonal band of Broca ventrally. Interposed between the large neurons (which in Golgi preparations have lengthy spine-free dendrites) are smaller cells, which are particularly numerous in more ventral parts of the division. The lateral division (or lateral septal nucleus) consists largely of medium-sized neurons which may be broadly divided on the basis of the size, depth of staining, and density of the cells into three components: dorsal, intermediate and ventral. The cells of all three components have radiating dendrites with many spines. The posterior group consists of the septofimbrial and the triangular septal nuclei. The cells of the former lie embedded within the precommissural fornix; in size and appearance they resemble the cells of the lateral septal nucleus. The cells of the triangular septal nucleus, on the other hand, are small, almost granule-like, and are densely packed within the rostroventral part of the ventral hippocampal commissure. Finally, the ventral division consists of a somewhat heterogeneous collection of cell groups which may be referred to collectively as the bed nucleus of the stria terminalis since the entire mass appears to be in receipt of afferents from the amygdala by way of the stria terminalis (18) (Krettek and Price, personal communication).

In addition to these major divisions, the septal area in the rat also contains a number of less well defined cell masses including: (a) a small but quite distinct septohippocampal nucleus close to the dorsal midline (21,31,67); (b) a small, ovoid group of compactly arranged neurons situated near the origin of the postcommissural fornix at the level of the anterior commissure which Gurdjian (25) has described as the bed nucleus of the anterior commissure. Whether the subfornical organ, which lies beneath

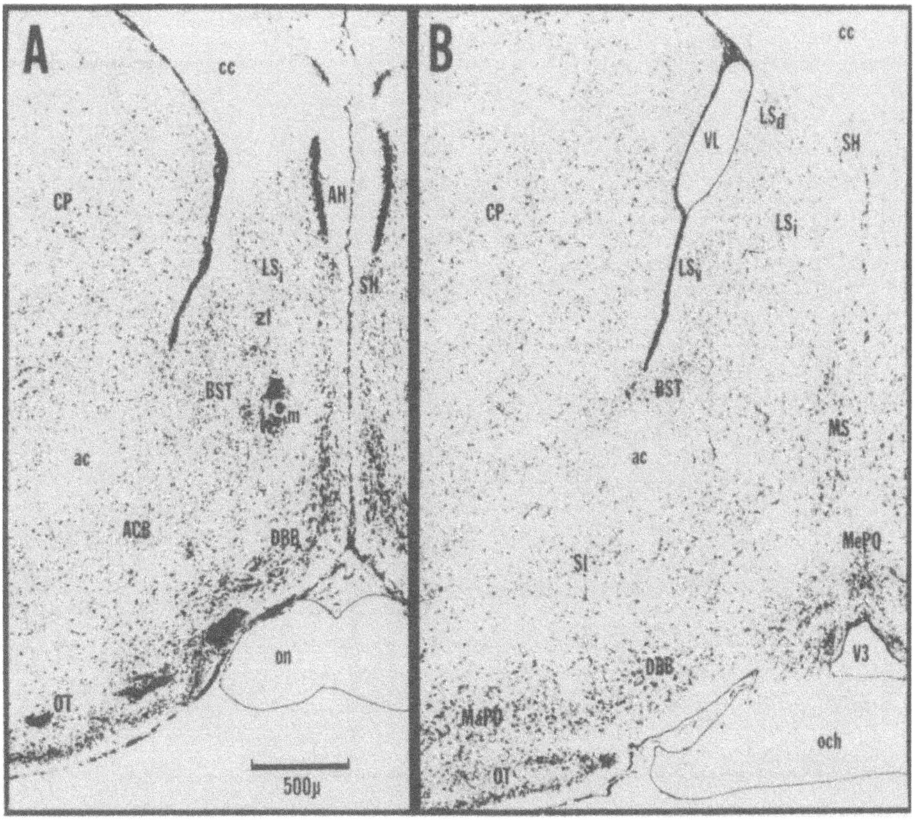

Figure 1, A,B. A series of photomicrographs to show the cytoarchitectonic subdivisions of the rat septal area and surrounding regions. Photomicrographs of representative frontal sections are arranged from anterior (A) to posterior (H). Nissl stain. From Swanson and Cowan (61).

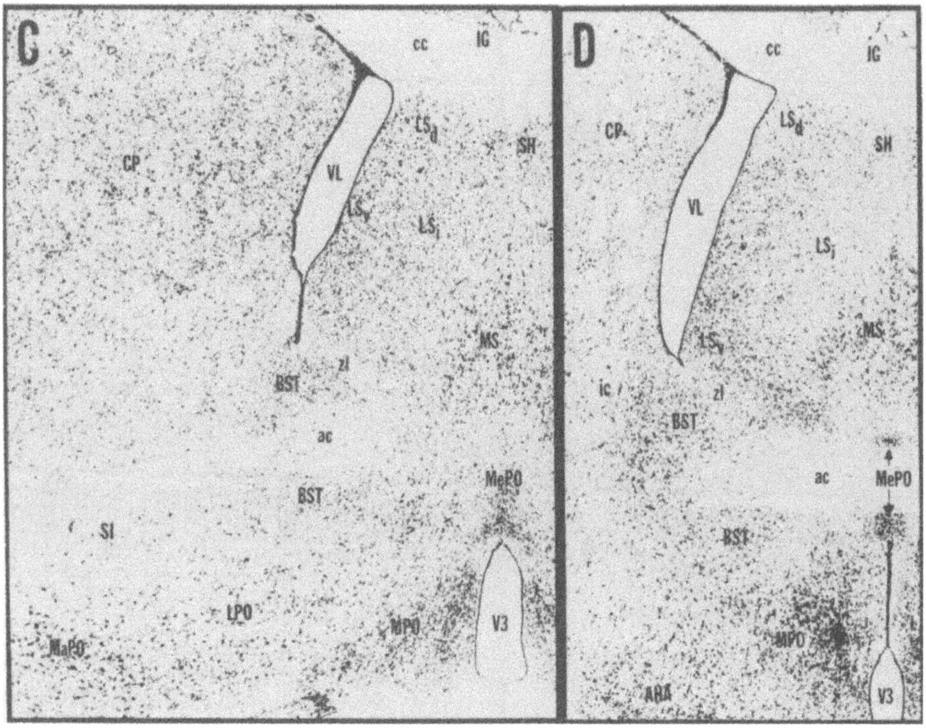

Figure 1, C,D. A series of photomicrographs to show the cytoarchitectonic subdivisions of the rat septal area and surrounding regions. Photomicrographs of representative frontal sections are arranged from anterior (A) to posterior (H). Nissl stain. From Swanson and Cowan (61).

the ventral psalterium, should be included within the septal complex, or be regarded as a separate entity, is debatable (1).

These subdivisions correspond fairly closely to those recognized by other workers (21,31,37,40,68) but the terminology adopted differs in many respects from that of certain other contemporary delineations, especially that of Andy and Stephan (4). From a careful examination of their published photomicrographs, it is clear that the differences are essentially terminological rather than substantive and that the mammalian septal region is remarkably uniform in its overall cytoarchitecture. Differences are, of course, to be found in different species but when the accompanying labels and arbitrarily drawn boundaries are ignored it is evident that these simply reflect differences in the relative development of individual nuclei, or divisions, rather than any basic difference in structural organization.

The Time of Neuron Origin in the Rat Septum

Several years ago Elliot Smith (20) put forward the view that the septum and hippocampus develop ontogenetically from a single cell mass, and as evidence for this pointed out that (a) even in adult animals, scattered neurons connect the two (21,32,47), and (b) in reptiles and acallosal mammals the two structures are contiguous (6). Unfortunately such topological considerations are misleading, and as a result of later work it has become clear that the septum and the hippocampal formation are formed from quite distinct telencephalic cell masses which, as discussed below, are derived from two widely separated areas of the neuroepithelium of the lateral ventricle (28,47,51,54). The actual development of the septum has not been the subject of much study. From the examination of normal embryonic material it has been suggested that the medial division (i.e., the medial septal-diagonal band complex) develops as a single cell mass (32,42), which is recognizable some time before the appearance of the lateral septum (28). The origin of the bed nucleus of the stria terminalis is unresolved; some consider it to be a telencephalic "olfactory" derivative (23), others that it is a general telencephalic or striatal derivative (42,49), and still others consider it to be a component of the hypothalamus (16,35,53).

Definitive evidence on this issue has had to await the development of ^{3}H-thymidine autoadiography. This technique not only enables the time of origin of particular neurons to be determined but also allows the migration of young neurons to be followed from the neuroepithelium in which they are generated to their definitive locations. Since the principles underlying this method are well known (56), it need only be mentioned that a single injection of ^{3}H-thymidine in a pregnant animal serves to label in-

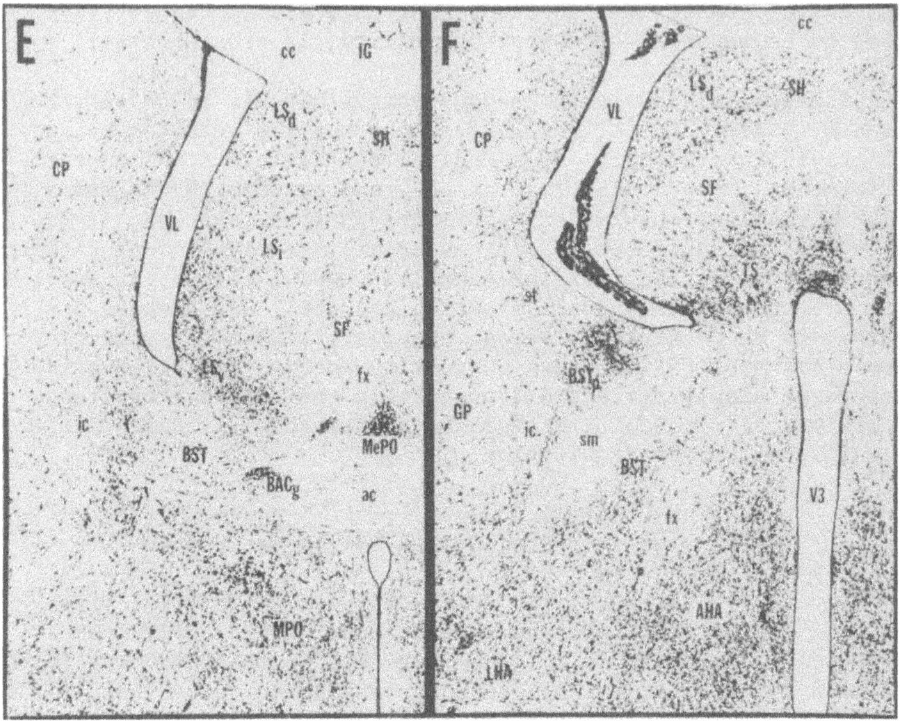

Figure 1, E,F. A series of photomicrographs to show the cytoarchitectonic subdivisions of the rat septal area and surrounding regions. Photomicrographs of representative frontal sections are arranged from anterior (A) to posterior (H). Nissl stain. From Swanson and Cowan (61).

tensely the nuclei of those neuroepithelial cells in the brain of the fetuses that are passing through their final round of DNA synthesis at, or shortly after, the time of the injection. On completing their last mitotic division the labeled young neurons and immature glial cells migrate away from the neuroepithelium and can be readily followed, in appropriately prepared material, to their final place of residence. We have recently used this approach to determine the time of neuron origin in the rat septum in a series of animals which had been pulse-labeled with ^{3}H-thymidine at successive ages from embryonic day 10 to term, and had been allowed to survive to young adulthood (61).

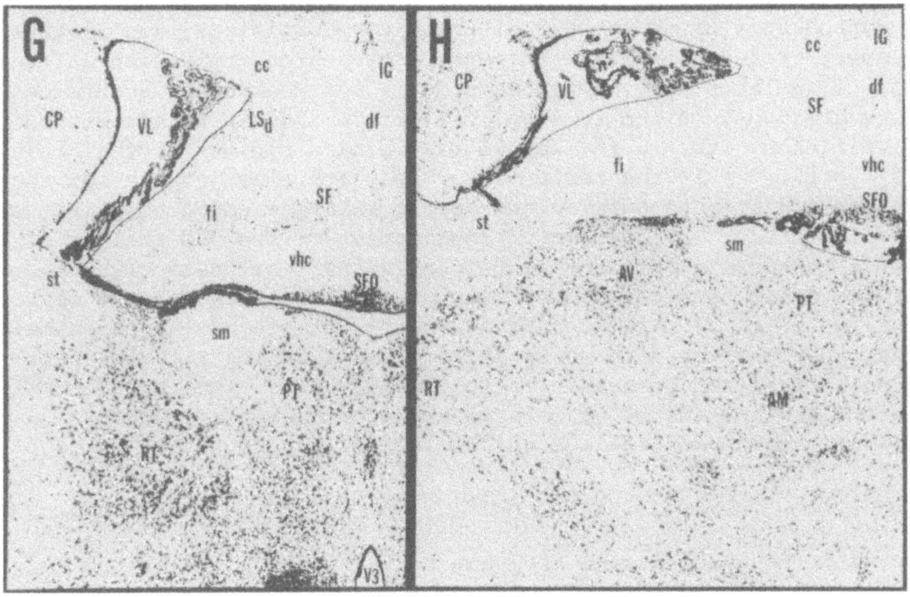

Figure 1, G,H. A series of photomicrographs to show the cytoarchitectonic subdivisions of the rat septal area and surrounding regions. Photomicrographs of representative frontal sections are arranged from anterior (A) to posterior (H). Nissl stain. From Swanson and Cowan (61).

The findings in these experiments (which have been, in general, confirmed independently in the mouse by Creps (15) are conveniently summarized by reference to one representative level through the rostro-caudal extent of the septum in Fig. 2. From this it may be seen that within each nuclear subdivision the cells are generated over a 5 to 6 day period although usually there is a distinct peak in the number of labeled cells on one particular day. Taken as a whole it is also clear that there is a simple "outside-in" sequence in the time of development of the cells in the medial, lateral and posterior divisions of the septum, such that the largest neurons (which are found in the medial division) arise first (from the neuroepithelium lining the medial wall of the anterior horns of the lateral ventricles) and are progressively displaced towards the midline by the later-formed cells. The last cells to be generated are the small neurons of the lateral septal nucleus which are found adjacent to the lateral ventricle. It is interesting that this pattern of neuron origin seems to occur quite independently of the cytoarchitectonic subdivisions, and in this respect it is not unlike the pattern of neurogenesis seen in the dorsal thalamus (5). At present it is not known what determines such gradients in the time of cell origin, but it is not without significance that in the septum (as in many regions of the brain (5, 28,57), it is the largest neurons which also project for the greatest distances outside the septum that are among the first to be generated. The pattern of development of the bed nucleus of the stria terminalis is rather more complex and the "date of birth" of its cells does not follow that of the adjoining lateral septal nucleus (Fig. 2). As we have only examined the brains of animals that survived for some weeks after pulse-labeling with ^{3}H-thymidine we have not been able to follow the route taken by the cells in the bed nucleus or to determine whether they are generated within the neuroepithelium of the lateral, or of the third, ventricles.

The Afferent and Efferent Connections of the Septum

Although the general organization of the afferent and efferent connections of the septum has been known for some time, many features of its connectivity until recently have been difficult to elucidate with conventional neuroanatomical methods. There are several reasons for this, although two are particularly cogent. The first is that the septum is traversed by a number of fiber tracts, primarily associated with the fornix system and the stria terminalis, whose presence greatly complicates the interpretation of normal neuroanatomical material, and whose interruption by non-specific lesions has confused the analysis of experimental material using the various degeneration methods. The second is that many of the fiber pathways of interest are of fine caliber and have proven difficult, and in some cases impossible, to demonstrate adequately with the usual techniques. To a large extent both of these difficulties can be obviated by the use of the recently developed autoradiographic method for tracing connections (11,36). This method is based on the selective uptake and incorporation of radioisotope-labeled amino acids into protein and glycoprotein precursors by neuronal somata and their subsequent distribution along the axons in one or other phase of axonal transport. It has been shown in other neural systems that this approach is not complicated by the involvement of fibers passing through the region into which the radioactive precursor is injected, and also to be significantly more sensitive than the commonly used degeneration methods in that it often reveals pathways and terminal projections which cannot be easily demonstrated by these other techniques (10,26, 63). Its principal limitation, like that of most other light microscopic methods, is that it does not permit the precise identification of the cells on which the afferent fibers terminate.

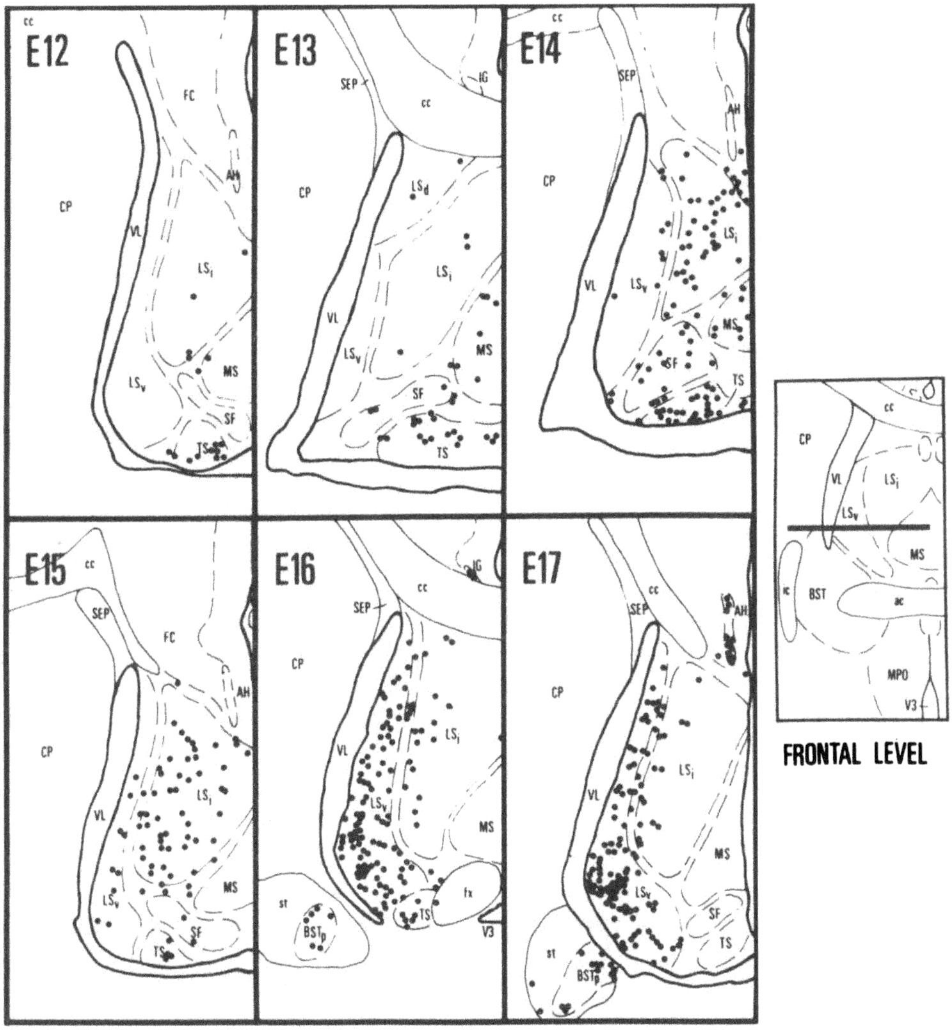

Figure 2. A series of projection drawings of horizontal sections through approximately the same frontal plane (see inset at right) in young adult rats individually pulse-labeled with ^{3}H-thymidine on embryonic days 12 through 17 (E12-E17) to show the distribution of heavily labeled neurons. A distinct "outside-in" sequence of development, largely independent of cytoarchitectonic subdivisions, is evident in the medial, lateral, and posterior septal nuclei. From Swanson and Cowan (61).

Electron microscopic autoradiography provides this additional refinement (19,27), but to date it has been applied to only a few neural systems.

The following account of the connections of the septum of the rat is based on an analysis of about 40 brains with very small injections of ^{3}H-proline in one or another of the septal nuclei, over 50 brains with similar injections in the hippocampal formation, and a further 50 brains with injections in various parts of the hypothalamus and preoptic region. Each subdivision of the septum has been labeled by an injection, without involvement of adjacent structures, in at least one experiment, and often in several brains (59,62).

The medial septal and diagonal band nuclei. That the medial septal-diagonal band nuclei project to the hippocampal formation is now well established (14,17,50), although the precise pattern of termination of the septal input to the hippocampus is still the subject of some debate, as is the question of whether or not the dorsal fornix is one of the contributing pathways to this projection as has been claimed by some workers (13,23,58) and denied by others (17,24,50). The findings in our autoradiographic material are in substantial agreement with the recent report by Mellgren and Srebro (44) that the medial septal-diagonal band complex projects through the fimbria and the dorsal fornix, and that this projection terminates in several regions of the hippocampal formation, in particular in fields CA2, 3, and 4 of the hippocampus itself, in the hilar region of the dentate gyrus, in the subiculum and parasubiculum, and in the entorhinal cortex. In view of the recent claims for sprouting of the "septal" input in the stratum moleculare of the dentate gyrus following the removal of the entorhinal afferents to this layer (9,46), it should be emphasized that we have been unable to find evidence for a significant projection from the medial septal-diagonal band complex to the stratum moleculare despite a repeated examination of several brains in which the other projections from the complex are readily identifiable. Patterns of silver grains indicative of terminal projections were particularly obvious over the stratum oriens and the stratum radiatum of field CA3, over the stratum polymorphe of the dentate gyrus, over the internal half of the stratum moleculare of the subiculum, and over all layers of the parasubiculum; a more diffuse and much larger projection was found to the medial and lateral entorhinal areas.

In general we have found that injections confined to the medial septal nucleus or to the nucleus of the diagonal band give similar projection patterns in the hippocampal formation, but as a rule the pathways from the diagonal band nucleus tended to be labeled more lightly. Furthermore, after injections involving the ventrolateral tip of the diagonal band nucleus, labeled fibers were followed anterodorsally over the genu of the corpus callosum, to a termination in layers I and V of the anterior limbic cortex. After larger injections, which were not strictly confined to the nucleus of the diagonal band, fibers could be followed even further caudally, around the splenium of the corpus callosum, to the dorsal part of the subiculum; at present it is not known whether these fibers arise in the diagonal band nucleus or from adjacent cells.

In addition to the projections to the hippocampal formation, the medial septal and diagonal band nuclei have both been found to project to the hypothalamus and the thalamus. The hypothalamic projection is to the pars medialis of the medial mammillary nucleus, with the fibers from the medial septal nucleus coursing through the medial preoptic and lateral hypothalamic areas, while those from the nucleus of the diagonal band are confined to the lateral preoptic and lateral hypothalamic areas. The thalamic projection is to the parataenial nucleus; this appears to arise only from the medial septal nucleus, since it has not been seen after injections confined to the nucleus of the diagonal band. Finally, it should be mentioned that there are probably intranuclear connections since small injections in any part of the medial septal or diagonal band nuclei always give rise to characteristic patterns of silver grains within the rest of the complex. This serves to confirm earlier suggestions by Fox (21) and Knook (24) that short connections exist between the medial septal and diagonal band nuclei.

The afferent connections of the medial septal and diagonal band nuclei have been well studied with the Nauta techniques; since they have been carefully reviewed by Raisman (50) we will only summarize them briefly and indicate those points on which our material has thrown some further light. Both the medial septal and diagonal band nuclei are said to be in receipt of fibers from the hippocampus (principally from field CA1), the medial forebrain bundle, the lateral septal nucleus, and a less well-defined areas which includes the basolateral group of amygdaloid nuclei, the piriform cortex and the olfactory tubercule. We have had occasion to reexamine the hippocampal input in some detail (59,63), and to a lesser extent that from the medial forebrain bundle, and the olfactory cortex (Swanson and Cowan, in preparation). From our autoradiographs it would appear that there are only a small number of hippocampal fibers to the circumscribed area containing the large cell bodies of the medial septal and diagonal band nuclei. Instead most of the hippocampal projection seems to end in the lateral septal nucleus, especially in its medial, relatively call-sparse zone where it adjoins the medial septal nucleus. However, since the dendrites of many medial septal neurons project into this region (61) the possibility remains that the hippocampal afferents may form axo-dendritic contacts with the cells in the medial septal and diagonal band nuclei, but definitive proof for this must await electron microscopic studies. On the other hand, our material fully supports Raisman's (50) conclusion that there is a substantial input to the medial septal and diagonal band nuclei from all parts of the lateral septum.

From an analysis of 11 injections involving the lateral preoptic area and the lateral hypothalamic area at various levels, as far caudally as the fields of Forel (Swanson, Saper, and Cowan, in preparation), it would appear that there is a restricted portion of the lateral preoptic-lateral hypothalamic area, at about the level of the supraoptic nucleus, which contributes fibers to the medial septal and diagonal band nuclei. Since after injections of ^{3}H-proline into this area labeled fibers could not be traced into the fimbria or dorsal fornix it seems likely that they terminate in the medial septal region. In addition, injections restricted to the magnocellular preoptic nucleus (Loo, '31; and others) labeled a number of fibers which appeared to end in the medial septal-diagonal band complex. The brains of six rats with injections involving either the olfactory tubercle or the piriform cortex have been examined and in none of these have we been able to identify afferents to the septal region.

Experimental evidence suggests that ascending catecholaminergic fibers from the brain stem pass through, and to some extent terminate in, the medial septal-diagonal band complex (2,39,45,66). At least some of these fibers contain the enzyme dopamine-β-hydroxylase (DBH) and are thus presumably adrenergic (noradrenaline- or adrenaline-containing) (65). Most of these DBH containing fibers appear to continue anterodorsally over the genu of the corpus callosum to innervate cortical areas.

The lateral septal nucleus. The efferent connections of the lateral septal nucleus have, with one exception, been difficult to define by experiments using degeneration methods (50); the exception is the massive projections into the medial septal and diagonal band nuclei discussed above. However, with the autoradiographic technique two further projections have been unequivocally demonstrated to the mammillary complex and to certain of the midline nuclei of the thalamus (62). In addition, a distinct organization has been found in the connections of this complex region, with the cells in each subdivision of the lateral septal nucleus projecting in a characteristic pattern. The scattered neurons in the dorsal part of the lateral septal complex appear to project only into the medial septal and diagonal band nuclei, whereas many fibers from the intermediate division continue caudally, through the medial forebrain bundle, to a termination in the vicinity of (but not within) the lateral mammillary nucleus of the same side. The projection of the more densely packed, smaller-celled, ventral division overlaps that of the intermediate division but, in addition, appears to give off fibers to the medial preoptic and anterior hypothalamic areas, and to the supramammillary region, and has a distinct termination in the anterior periventricular and parataenial nuclei of the thalamus.

The best known source of afferents to the lateral septal nucleus is the hippocampus, which, from the studies of Raisman et al. (50) appears to arise from all fields to Ammon's horn (i.e., fields CA1 through CA4). From our own autoradiographic material it would in fact appear that the lateral septal nucleus constitutes the sole subcortical projection field of Ammon's horn and of all but the ventral portion of the subiculum, which also gives rise to the medial corticohypothalamic tract, and a group of fibers that continue through the septum to the nucleus accumbens, anterior olfactory nucleus, taenia tecta, and adjacent infralimbic cortex (59). We have found that all parts of field CA1 and the subiculum project ipsilaterally to the lateral septal complex, while fields CA2 and CA3 project bilaterally upon the lateral septum. Like most of the recently investigated hippocampal connections (3,7,29,30,55), the efferents from the hippocampus to the lateral septal nucleus display a high degree of topographic organization. In general terms this organization may be summarized as follows (Fig. 3): the anterodorsally located neurons in fields CA1 and CA3, together with the dorsal parts of the subiculum, project exclusively to the dorsal aspect of the lateral septal complex; the more posteroventrally situated parts of these fields in turn project to the more ventral portions of the lateral septal nucleus. In the rostral part of the hippocampal formation one can clearly identify, in frontal sections, a medial to lateral organization in the arrangement of the various hippocampal fields, and it is also possible to demonstrate a comparable medial to lateral organization in its projection to the lateral septal nucleus. It is important to point out that even though small hippocampal injections of labeled amino acids into the hippocampus always give rise to quite restricted terminal projections in the septum (Fig. 3), there is always some overlap in the projection fields of adjacent regions of the hippocampus, but since we have not yet examined this issue with the electron microscope, we do not know whether axons from adjoining hippocampal fields terminate upon the same, or upon adjacent, septal neurons. No other part of the hippocampal formation has been found to project upon the lateral septal nucleus, and we have found little evidence for projections into the region after isotope injections involving the lateral hypothalamic and preoptic areas, the olfactory tubercle, or the piriform cortex (Swanson and Cowan, in preparation). For the sake of completeness we should add that a projection into the lateral septal nucleus from the proreal gyrus has recently been described in the cat (67);

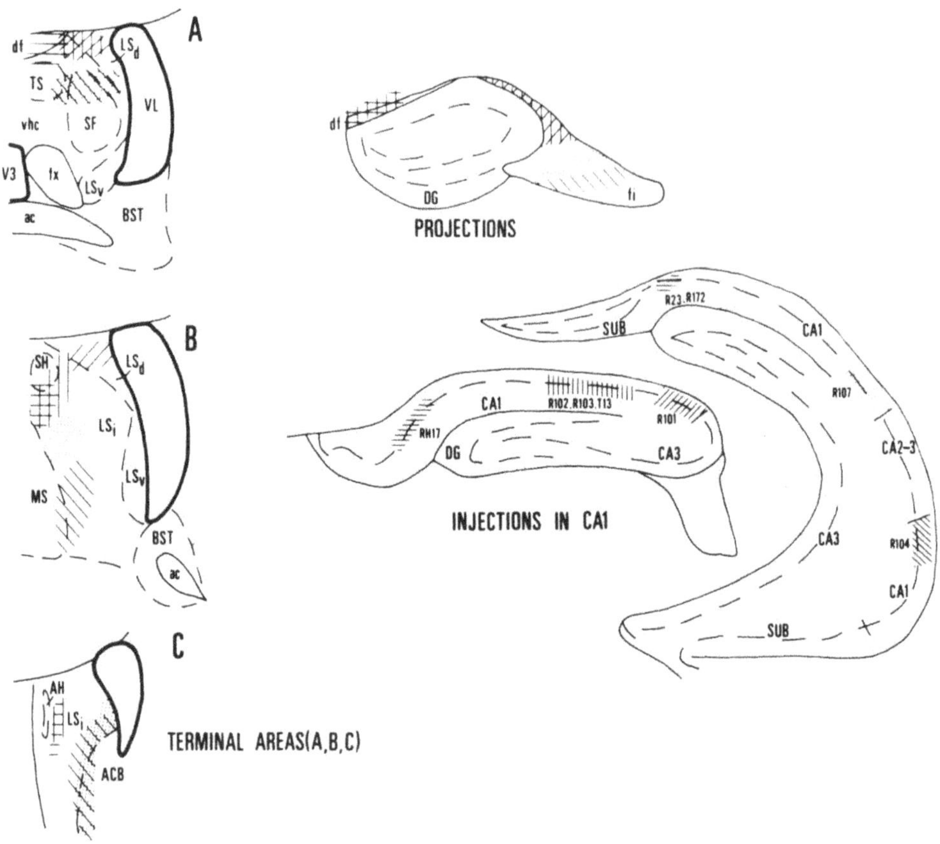

Figure 3. A summary of the topographic organization of hippocampo-septal projections in the rat. Anterodorsally in the hippocampus, neurons in field CA 1 are arranged from medial to lateral; their descending connections end in a corresponding medial to lateral sequence in the dorsal part of the lateral septum after traveling through the alveus and dorsal fornix. Caudally, the cells of field CA1 are arranged more vertically, and corresponding dorsoventral gradients are established between the cells of origin and terminal fields in the lateral septum; these fibers course primarily in the fimbria. The septal input from fields CA2 and CA3 is similarly arranged although it is bilateral. From Swanson and Cowan (63).

we have not had occasion to examine this pathway which is said to pass caudally over the corpus callosum and around the splenium, into the dorsal fornix, to reach the lateral septum. In addition, a few fibers from the amygdala may end in the ventral lateral septum (18) (Krettek and Price, personal communication).

Ascending catecholaminergic inputs to the lateral septum have been described (22,45). In DBH immunofluorescence-stained material (65) a few scattered adrenergic fibers were seen in the lateral septum, although most of the fibers were confined to a restricted bundle which emerged from the ventral bed nucleus of the stria terminalis and ascended in the lateral aspect of the lateral septum to course over the genu of the corpus callosum into cortical areas. A dopaminergic input to the lateral septum has also recently been described (39).

The triangular septal and septofimbrial nuclei. The connections of the posterior group of septal nuclei are poorly understood. Collectively they are said to receive an input from the fimbria (50) and the septofimbrial nucleus, at least, is said to project to the habenular complex (12,21,41,43,50). In our autoradiographic material we have been able to confirm a septo-habenular projection in the rat (62) and to show something of its organization. As Raisman (50) indicated it arises from the septofimbrial nucleus and terminates ipsilaterally within the medial habenular nucleus. Furthermore, we have shown that it is organized topographically in the sense that the cells in the anterodorsal part of the septofimbrial nucleus project to the medial part of the medial habenular nucleus, and as the isotope injections involve progressively more ventrolaterally situated neurons so the terminal projection field is found to occupy successively more lateral regions of the medial habenular nucleus. The triangular septal nucleus has been found to project bilaterally to both the medial and the lateral habenular nuclei; it also appears to send some fibers through the ipsilateral fasciculus retroflexus to the interpeduncular nucleus of the same side.

Degeneration has been described in the posterior part of the septum following hippocampal lesions of fimbrial section by Raisman *et al.* (51), and by others, but since these nuclei lie directly in the path of the hippocampal fibers to the lateral septum it is not possible to determine with certainty which if any of the fields of Ammon's horn or the subiculum project to it. Unfortunately the same difficulty applies to the autoradiographic method, so that while numerous silver grains are observed over the septofimbrial and triangular septal nuclei following ^{3}H-proline injections in fields CA1 and CA3, and the subiculum, it is now known whether these really constitute a terminal projection field or whether they are due simply to the labeling of fibers of passage. For the same reason we cannot rule out the possibility that the efferents of the medial septal and diagonal band nuclei which pass through the posterior nuclei on their way to the hippocampus, do not also give off collaterals which may terminate in the septofimbrial or triangular nuclei.

The bed nucleus of the stria terminalis. Because of the close relationship between the fibers of the stria terminalis and the cells of its bed nucleus, experimental degeneration methods have thus far completely failed to elucidate the efferent connections of the cells in the nucleus. However, minute injections of ^{3}H-proline, wholly confined within the boundaries of the nucleus, have revealed that it has a remarkably extensive pattern of projections to the amygdala, thalamus, and to successive populations of neurons associated with the medial forebrain bundle as far caudally as the locus coeruleus (62).

Following injections confined to the principal (or anterior) portion of the bed nucleus, projections have been found to course (a) through the stria terminalis to the central nucleus of the amygdala, (b) in association with the stria medullaris to the periventricular and parataenial nuclei of the thalamus of the same side and to the medial habenular nucleus of both sides (the fibers crossing in the habenular commissure), and (c) through the medial forebrain bundle to several regions including the nucleus accumbens, the substantia innominata, much of the preoptic area and the hypothalamus (with the notable exception of the core of the ventromedial nucleus, and the medial and lateral mammillary nuclei), to the ventral tegmental area, the central tegmental reticular field, the periaqueductal gray, the vicinity of the dorsal tegmental nucleus, and finally, to the locus coeruleus. Injections involving the small, cell-dense, posterior part of the bed nucleus displayed a somewhat different group of projections to: (a) the medial nucleus of the amygdala, (b) the periventricular, parataenial, anterodorsal, and anteroventral thalamic nuclei of the same side, (c) the medial habenular nucleus of both sides, and (d) the premammillary nuclei. The fibers to the premammillary nuclei pass through, and probably contribute collaterals to, the medial preoptic and the anterior and lateral hypothalamic areas.

It may be mentioned, for the sake of completeness, that the nearby nucleus accumbens, occasionally regarded as a component of the septum, appears to project exclusively to the globus pallidus and substantia nigra in the same way as the adjacent caudate-putamen (60).

The major source of afferents to the bed nucleus of the stria terminalis is the amygdala (18). This projection has recently been shown autoradiographically (Krettek and Price, personal communication) to be topographically organized such that fibers from the central, lateral, and basolateral nuclei end medially in the bed nucleus while those from the medial and posterior cortical nuclei terminate more laterally, and the projection from basomedial nucleus ends in an intermediate position. We have also been able to show (59) that the ventral part of the subiculum projects heavily to the bed of the nucleus, as does an area of the tuberal region of the hypothalamus, including the ventromedial nucleus (Saper, Swanson and Cowan, in preparation). There may well be other inputs to the nucleus, but with the exception of the unusually dense noradrenergic input to its ventral portion from the brainstem (22,65) these remain to be discovered. Interestingly, the small bed nucleus of the anterior commissure (of Gurdjian) receives a very dense, circumscribed noradrenergic terminal field (Fig. 4; also Swanson and Hartman 65).

SUMMARY

Perhaps the most obvious conclusion to be drawn from the observations we have presented is that the septum is an extremely complex region, both in terms of its cytoarchitecture and its connections. Almost certainly the view we now have of its organization is a rather gross oversimplification of its morphology, since we know virtually nothing of thesynaptic relationships between the various inputs and the neurons in the different septal nuclei. The fact that the dendrites of many of the cells of the medial septal

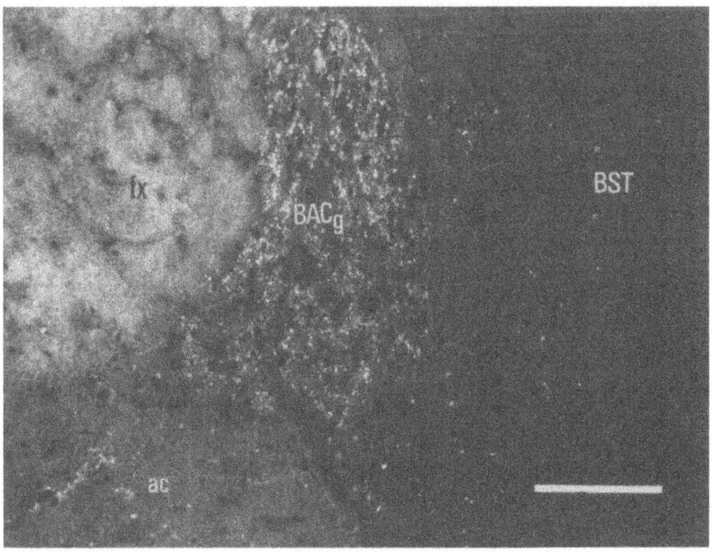

Figure 4. A photomicrograph of a section through the anterior commissure and descending column of the fornix showing a dense, circumscribed adrenergic input to Gurdjian's bed nucleus of the anterior commissure. A number of discrete pericellular varicosities surround most cells in this nucleus. DBH-immunofluorescence stain. Scale: 50 μm.

and diagonal band nuclei extend for considerable distances beyond the cytoarchitectonic boundaries of the nuclei has already been alluded to; this means that afferents which are described as ending in an adjoining nucleus (for example, the intermediate division of the lateral septal nucleus) may well influence directly the activity of neurons in the medial septal and diagonal band nuclei, and also indirectly the neural centers to which these latter nuclei project. The dendrites of the cells in other nuclei tend to be more restricted, but potentially they present the same problem especially when they adjoin the medial septal and diagonal band nuclei. Of equal importance is the fact that we know very little about the physiological nature of the various connections we have described, for example, whether they are excitatory or inhibitory, what neurotransmitters they utilize, and at a more significant level, with what behavioral mechanisms they are implicated. Lastly, it should be emphasized that in describing the various septal nuclei as receiving, or contributing to, particular projections we are in fact treating them as discrete entities composed of homogeneous neuronal populations. At this time this is a necessary but perhaps misleading representation. It is known from Golgi preparations, and even from the study of Nissl-stained sections, that most if not all of the septal nuclei consist of more than one class of neuron. The medial septal nucleus, for instance, contains large neurons which we may assume from the work of Daitz and Powell (17) to be the main source of the septo-hippocampal projections since they undergo retrograde degeneration following interruption of the fimbria. However, small and medium-sized cells which are unaffected by fimbrial section are also found in the medial eptal-diagonal band complex. Do these different classes of neurons have different inputs? Do they have quite different efferent projections? Are some of them interneurons? These and other questions can only be answered with a great deal more work, using the neuroanatomical methods currently available.

With these reservations in mind it may be helpful to summarize briefly the major connections of the various subdivisions of the septum (insofar as we know them from the work of the past two decades, and more particularly, from our recent autoradiographic studies) and to place them in the wider context of such neural systems as the limbic system, the medial forebrain bundle, and the hypothalamus.

The septum is reciprocally linked to the hippocampal formation through the now well established projection of the medial septal-diagonal band complex to fields CA1-CA3 of Ammon's horn, to the polymorphic zone of the dentate gyrus, and to the subicular region and entorhinal cortex. The return projection to the septum is provided by the extensive and well organized input of the precommissural fornix (from the major hippocampal fields and the subiculum) to the various subdivisions of the lateral septal nucleus, which project in turn upon the medial septal-diagonal band complex. A similar pattern of reciprocal connections exists between the septum and the amygdala: the bed nucleus of the stria terminalis both receiving a substantial projection from most of the nuclei of the amygdala, and projecting back to the central and medial amygdaloid nuclei. In a similar manner the septum (especially the medial septal-diagonal band complex and the bed nucleus of the stria terminalis) projects extensively upon the preoptic region and hypothalamus (and even beyond this into the brainstem) and receives a fairly substantial input from a restricted portion of the lateral preoptic and hypothalamic areas (to the medial septal and diagonal band nuclei) and from the ventromedial hypothalamic nucleus (to the bed nucleus of the stria terminalis).

The principal interest which attaches to these patterns of connectivity is that they provide for the integration of neural activity in anatomically (and probably functionally) quite diverse neural systems. Thus the hippocampus (using the term in the strictest sense to include only the fields of Ammon's horn), which appears to have no direct input to the hypothalamus (3), can, through its projection to the lateral septal nucleus, have a fairly direct influence upon hypothalamic activity, both through the direct lateral septal input into the medial forebrain bundle, and through more indirect routes, including a relay through the medial septal-diagonal band complex which also projects to the hypothalamus. To cite another example, the amygdala has no direct connections to the caudal hypothalamus. It can, however, through its projection upon the bed nucleus of the stria terminalis potentially play a significant role in regulating the activity of the caudal hypothalamus including the mammillary complex, and beyond this many centers in the brainstem, as far caudally as the widely projecting local coeruleus. Finally, the massive topographically organized input from the septum to the mammillary complex should be re-emphasized (Fig. 5), especially in light of the unexpectedly extensive input to the mammillary complex now known to come from other forebrain regions, including the subicular cortex (59), and most of the preoptic area and hypothalamus (Saper, Swanson and Cowan, in prepartaion).

Although much less is known about the projection of the septofimbrial and triangular septal nuclei upon the habenular complex, this is also likely to be of significance since potentially at least it provides for a fairly direct link between the hippocampus and hypothalamus on the one hand, and the habenular-interpeduncular nuclear complex on the other. The same is likely to be true for the septal projections upon the anterior and rostral midline nuclei of the thalamus for, through such connections, the amygdala and the hippocampus are again brought into relation with the

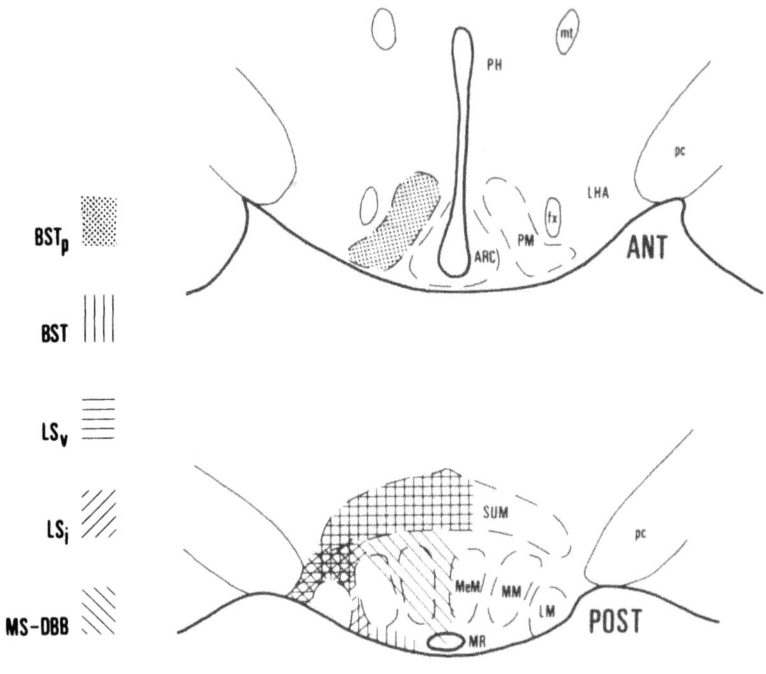

Figure 5. A summary diagram of the topographical organization of the septal input to the mammillary complex in the rat as shown autoradiographically. All of the septum, except the posterior nuclei, projects to the mammillary complex. From Swanson and Cowan (62).

anterior limbic and cingulate areas, and with the nucleus accumbens of the striatal complex.

Since the time of Elliot Smith (20) it has been fashionable to think of the septum as simply a "relay station" between the hippocampal formation and the hypothalamus. Clearly it does serve this function (especially if we are careful to leave the term "relay station" undefined) but is role in the forebrain must now be conceived of much more broadly than this. Morphologically, it is a highly organized complex of nuclei, with a remarkable variety of connections to and from some of the least understood regions of the brain. Functionally, it is likely to prove at least as complex, and until we know more about the types of information it receives from such structures as the amygdala, hippocampus, and brainstem, we shall continue to be frustrated in our efforts to assign a function to the system as a whole or to any of its anatomical subdivisions.

LIST OF ABBREVIATIONS

ACB	nucleus accumbens
AH	anterior hippocampal rudiment
AHA	anterior hypothalamic area
AM	anteromedial nucleus (thalamus)
ARC	arcuate nucleus (hypothalamus)
AV	anteroventral nucleus
BAC_g	bed nucleus of the anterior commissure (Gurdjian)
BST	bed nucleus of the stria terminalis
CER	cerebellum
DP	caudate-putamen
DBB	nucleus of the diagonal band (Broca)
DG	dentate gyrus
FC	frontal cortex
GP	globus pallidus
IC_m	medial island of Calleja
IG	induseum griseum
LHA	lateral hypothalamic area
LM	lateral mammillary nucleus
LPO	lateral preoptic area
$LS_{d,i,v}$	lateral septal nucleus (p. dorsalis, intermedialis, ventralis)
MaPO	magnocellular preoptic nucleus
MeM	medial mammillary nucleus, p. medialis
MM	medial mammillary nucleus, p. lateralis
MPO	medial preoptic area
MR	mammillary recess
MS	medial septal nucleus
OB	olfactory bulb
OT	olfactory tubercle
PH	posterior hypothalamic area
PM	premammillary nucleus
PT	parataenial nucleus
RT	reticular nucleus (thalamus)
SEP	subependymal plate
SF	septofimbrial nucleus
SFO	subfornical organ
SH	septohippocampal nucleus
SI	substantia innominata
SUB	subiculum
SUM	supramammillary area
TS	triangular nucleus
VL	lateral ventricle
V3	third ventricle
ac	anterior commissure
cc	corpus callosum
df	dorsal fornix

fi	fimbria
fx	fornix
ic	internal capsule
mt	mammillothalamic tract
och	optic chiasm
on	optic nerve
pc	cerebral penduncle
sm	stria medullaris
st	stria terminalis
vhc	ventral hippocampal commissure
zl	zona limitans

REFERENCES

1. Akert, K., H. D. Potter, and J. W. Anderson. The subfornical organ in mammals. I. Comparative and topographic anatomy. J. Comp. Neur., 116 (1961) 1-13.

2. Anden, N. E., A. Dahlstrom, K. Fuxe, K. Larsson, L. Olson, and U. Ungerstedt. Ascending monoamine neurons to the telencephalon and diencephalon. Acta Physiol Scand., 67 (1966) 313-326.

3. Andersen, P., T. U. P. Bliss, and K. K. Skrede. Lamellar organization of hippocampal excitatory pathways. Exp. Brain Res., 13 (1971) 222-238.

4. Andy, O. J., and H. Stephan. The Septum of the Cat. C. C. Thomas, Springfield, 1964.

5. Angevine, J. B. Time of neuron origin in the diencephalon of the mouse. An autoradiographic study, J. Comp. Neur., 139 (1970) 129-188.

6. Ariens Kappers, C. U., G. C. Huber, and E. C. Crosby. The Comparative Anatomy of the Nervous System of Vertebrates, Including Man. New York: Hafner Publishing Co., 1936.

7. Blackstad, T. W., K. Brink, J. Hem, and B. Jeune. Distribution of hippocampal mossy fibers in the rat. An experimental study with silver impregnation methods. J. Comp. Neur., 138 (1970) 433-450.

8. Brady, N. V., and W. J. H. Nauta. Subcortical mechanisms in emotional behavior; affective changes following septal forebrain lesions in the albino rat. J. Comp. Physiol. Psychol., 56 (1953) 339-346.

9. Cotman, C. W., D. A. Matthews, D. Taylor, and G. Lynch. Synaptic rearrangement in the dentate gyrus: histochemical evidence of adjustments after lesions in immature and adult rats. Proc. Nat. Acad. Sci. USA, 70 (1973) 3473-3477.

10. Cowan, W. M., and M. Cuénod. The use of axonal transport for the study of neural connections; A retrospective survey. In: The Use of Axonal Transport for Studies of Neuronal Connectivity. Internat. Symposium, Switzerland. M. Cuénod and W. M. Cowan, eds., Elsevier, Amsterdam, in press.

11. Cowan, W. M., D. I. Gottlieb, A. E. Hendrickson, J. L. Price, and T. A. Woolsey. The autoradiographic demonstration of axonal connections in the central nervous system. Brain Res., 37 (1972) 21-51.

12. Cragg, B. G. The connections of the habenula in the rabbit. Exp. Neurol., 3 (1961a) 388-409.

13. __________ Olfactory and other afferent connections of the hippocampus in the rabbit, rat, and cat. Exp. Neurol, 3 (1961b) 588-600.

14. Cragg, B. G., and L. H. Hamlyn. Some commissural and septal connexions of the hippocampus in the rabbit. J. Physiol. (Lond.), 135 (1957) 460-485.

15. Creps, E. S. Time of neuron origin in preoptic and septal areas of the mouse: an autoradiographic study. J. Comp. Neur., 157 (1974) 161-243.

16. Crosby, E. C., and R. T. Woodbourne. The comparative anatomy of the preoptic area and the hypothalamus. A. Res. Nerv. Ment. Dis., Proc., 20 (1940) 52-169.

17. Daitz, H. M., and T. P. S. Powell. Studies of the connexions of the fornix system. J. Neurol. Neurosurg. Psychiat., 17 (1954) 75-82.

18. de Olmos, J. S., and W. R. Ingram. The projection field of the stria terminalis in the rat brain. An experimental study. J. Comp. Neur., 146 (1972) 303-334.

19. Droz, B., H. L. Koenig, and L. Di Giamberardino. Axonal migration of protein and glycoprotein to nerve endings. I. Radioautographic analysis of the renewal of protein in nerve endings of chicken ciliary ganglion after intracerebral injection of [^{3}H] lysine. Brain Res., 60 (1973) 93-127.

20. Elliot Smith, G. Some problems relating to the evolution of the brain. Arris and Gale Lectures. Lancet, 1 (1910) 106, 147-221.

21. Fox, C. A. Certain basal telencephalic centers in the cat. J. Comp. Neur., 72 (1940) 1-62.

22. Fuxe, K. Evidence for the existence of monomine-containing neurons in the central nervous system. IV. The distribution of monoamine terminals in the central nervous system. Acta Physiol. Scand., Suppl. 247, 1965.

23. Gerebtzoff, M. A. Sur quelques voies d'association de l'écorce cérébrale (recherches anatomo-expérimentales). J. Belg. Neurol., 39 (1939) 205-221.

24. Guillery, R. W. Degeneration in the hypothalamic connexions of the albino rat. J. Anat., 91 (1957) 91-115.

25. Gurdjian, E. S. Olfactory connections in the albino rat, with special reference to the stria medullaris and the anterior commissure. J. Comp. Neur., 38 (1925) 128-163.

26. Hendrickson, A. Electron microscopic radioautography: identification of origin of synaptic terminals in normal nervous tissue. Science, 165 (1969) 194-196.

27. Hendrickson, A. E., N. Wagoner, and W. M. Dowan. An autoradiographic and electron microscopic study of retino-hypothalamic connections. Z. Zellforsch., 135 (1972) 1-26.

28. Hines, M. Studies on the growth and differentiation of the telencephalon in man. The fissure hippocampi. J. Comp. Neur., 34 (1922) 73-171.

29. Hjorth-Simonsen, A. Some intrinsic connections of the hippocampus in the rat: an experimental analysis. J. Comp. Neur., 147 (1973) 145-162.

30. Hjorth-Simonsen, A., and B. Jeune. Origin and termination of the hippocampal perforant path in the rat studies by silver impregnation. J. Comp. Neur. 144 (1972) 215-231.

31. Humphrey, T. The telencephalon of the bat. I. The non-cortical nuclear masses and certain pertinent fiber connections. J. Comp. Neur., 65 (1936) 603-711.

32. Johnston, J. B. The morphology of the septum, hippocampus and pallial commissures in reptiles and mammals. J. Comp. Neur., 23 (1913) 371-478.

33. ________. Further contributions to the study of the evolution of the forebrain. J. Comp. Neur., 35 (1923) 337-481.

34. Knook, H. L. The Fibre-Connections of the Forebrain. F. A. Davis, Co., Philadelphia, 1966.

35. Kuhlenbeck, H., and W. Haymaker. The derivatives of the hypothalamus in the human brain; their relation to the extrapyramidal and autonomic systems. Mil. Surg., 105 (1949) 26-52.

36. Lasek, R., B. S. Joseph, and D. G. Whitlock. Evaluation of a radioautographic neuroanatomical tracing method. Brain Res., 8 (1968) 931-939.

37. Lauer, E. W. The nuclear pattern and fiber connections of certain basal telencephalic centers in the macaque. J. Comp. Neur., 82 (1945) 215-254.

38. La Vail, J., and W. M. Cowan. The development of the chick optic tectum. II. Autoradiographic studies. Brain Res., 28 (1971) 421-441.

39. Lindvall, O. Mesencephalic dopaminergic afferents to the lateral septal nucleus of the rat. Brain Res., 87 (1975) 89-95.

40. Lindvall, O., and A. Björklund. The organization of the ascending catecholamine neuron systems in the rat brain, as revealed by the glyoxylic acid fluorescence method. Acta Physiol. Scand., Suppl. 412, 1974.

41. Loo, Y. T. The forebrain of the opossum, Didelphis virginiana. J. Comp. Neur. 52 (1931) 1-148.

42. Macchi, G. The ontogenetic development of the olfactory telencephalon in man. J. Comp. Neur., 95 (1951) 245-305.

43. McLardy, T. Observations on the fornix of the monkey. I. Fiber studies. J. Comp. Neur., 103 (1955) 327-343.

44. Mellgren, S. I., and B. Srebro. Changes in acetylcholinesterase and distribution of degeneration fibers in the hippocampal region after septal lesions in the rat. Brain Res., 52 (1973) 19-36.

45. Moore, R. Y., A. Björklund, and U. Stenevi. Plastic changes in the adrenergic innervation of the rat septal area in response to denervation. Brain Res., 33 (1971) 13-35.

46. Nadler, J. V., C. W. Cotman, and G. S. Lynch. Altered distribution of choline acetyltransferase and acetylcholinesterase activities in the developing rat dentate gyrus following entorhinal lesion. Brain Res., 63 (1973) 215-230.

47. Nauta, W. J. H., and W. Haymaker. Hypothalamic nuclei and fiber connections. In: The Hypothalamus. W. Haymaker, E. Anderson, and W. J. H. Nauta, eds., C. C. Thomas, Springfield, pp. 136-209, 1969.

48. Olds, J., and P. Milner. Positive reinforcement produced by electrical stimulation of septal areas and other regions of the rat brain. J. Comp. Physiol. Psychol, 47 (1954) 419-427.

49. Papez, J. W. The embryological development of the hypothalamic area in mammals. A. Res. Nerv. Ment. Dis., Proc., 20 (1940) 31-51.

50. Raisman, G. The connexions of the septum. Brain, 89 (1966) 317-348.

51. Raisman, G., W. M. Cowan, and T. P. S. Powell. An experimental analysis of the efferent projection of the hippocampus. Brain, 89 (1966) 83-108.

52. Raisman, G., and P. M. Field. A quantitative investigation of the development of collateral reinnervation after partial deafferentation of the septal nuclei. Brain Res., 50 (1973) 241-264.

53. Rose, J. E. The ontogenetic development of the rabbit's diencephalon. J. Comp. Neur., 77 (1942) 61-129.

54. Schlessinger, A. R., W. M. Cowan, and D. I. Gottlieb. An autoradiographic study of the time of origin and the pattern of granule cell migration in the dentate gyrus of the rat. J. Comp. Neur., 159 (1975) 149-176.

55. Shipley, M. T. The topographical and laminar organization of the presubiculum's projection to the ipsi- and contralateral entorhinal cortex in the guinea pig. J. Comp. Neur., 160 (1975) 127-145.

56. Sidman, R. L. Autoradiographic methods and principles for study of the nervous system with thymidine-H^3. In: Contemporary Research Methods in Neuroanatomy. W. J. H. Nauta and S. O. E. Ebbesson, eds., Springer-Verlag, New York, pp. 252-274, 1970.

57. Sidman, R. L. and P. Rakic. Neuronal migration, with special reference to developing human brain: a review. Brain Res., 62 (1973) 1-35.

58. Sprague, J. M., and M. Meyer. An experimental study of the fornix in the rabbit. J. Anat., 84 (1950) 354-368.

59. Swanson, L. W., and W. M. Cowan. Hippocampo-hypothalamic connections: origin in subicular cortex, not Ammon's horn. Science, 189 (1975) 303-304.

60. Swanson, L. W., and W. M. Cowan. A note on the connections and development of the nucleus accumbens. Brain Res., 92 (1975) 324-330.

61. Swanson, L. W., and W. M. Cowan. The morphology and time of neuron origin of the rat septal area. J. Comp. Neur., (in preparation), 1975.

62. Swanson, L. W., and W. M. Cowan. An autoradiographic study of the efferent connections of the rat septal area. J. Comp. Neur., (in preparation), 1975.

63. Swanson, L. W., and W. M. Cowan. The extrinsic connections of the hippocampus. An autoradiographic study in the rat. J. Comp. Neur., (in preparation), 1975.

64. Swanson, L. W., W. M. Cowan, and E. G. Jones. An autoradiographic study of the efferent connections of the ventral lateral geniculate nucleus in the albino rat and the cat. J. Comp. Neur., 156 (1974) 143-164.

65. Swanson, L. W., and B. K. Hartman. The central adrenergic system. An immunofluorescence study of the location of cell bodies and their efferent connections in the rat utilizing dopamine-β-hydroxylase as a marker. J. Comp. Neur., (in press), 1975.

66. Ungerstedt, U. Stereotaxic mapping of the monoamine pathways in the rat brain. Acta Physiol. Scand., Suppl. 367, 1971.

67. Voneida, T. J., and G. J. Royce. Ipsilateral connections of the gyrus proreus in the cat. Brain Res., 76 (1974) 393-400.

68. Young, M. W. The nuclear pattern and fiber connections of the non-cortical centers of the telencephalon of the rabbit (Lepus cuniculus). J. Comp. Neur., 65 (1936) 295-401.

ACETYLCHOLINESTERASE HISTOCHEMISTRY OF THE SEPTAL REGION IN THE RAT

B. Srebro, S. I. Mellgren and W. Harkmark

Institute of Physiology and Institute of Anatomy
University of Bergen, 5000 Bergen, Norway

INTRODUCTION

Based on the normal and post-lesion distribution of AChE-contining cell bodies and fibres, Lewis and Shute (13) have described the existence of several "cholinergic systems" in the rat brain. One of the most prominent forebrain systems was defined as "groups of AChE-containing neurons which are intimately related to the hippocampal formation and its projection pathways in the fore-brain and mid-brain" (13). The main part of this system consists of a cholinergic septo-hippocampal projection, the so-called "septal radiation", which originates in the AChE-rich cell bodies of the medial septal and diagonal band nuclei. Subsequently, the cholinergic nature of these cells in the medial septum received a variety of indirect evidence from the experimental histochemical and quantitative neurochemical analyses (3,5,12,14,19,20).

According to Krnjevic and Silver (11) the entire medial wall of the developing cerebral hemisphere in the cat is innervated by AChE-containing fibres arising in the medial septal and diagonal band nuclei. AChE-preparations from the septal region of mature brains (10,13) reveal conspicuous granular material in the neuropil, cell bodies, and in strongly reactive fibre bundles. This suggests that the cholinergic projection arises there and leaves the septum primarily in a caudal, but also in a dorsal and a rostral direction.

We have confirmed these findings in material obtained from the developing rat brain. The AChE activity in the perikarya of the medial septum was discernible on the first day of postnatal life (23). Enzyme reaction in these cells and in the surrounding neuropil increased rapidly during the first postnatal week with a clear gradient of intensity along the septo-hippocampal axis. The hippocampal region of the newborn rat exhibited enzyme activity in the neuropil and in scattered perikarya on the second day after birth (16). A rapid increase of AChE activity in all areas of the hippocampal region was observed during the second and third postnatal week.

The septal origin of the major portion of the hippocampal AChE has been suggested in several studies (3,14,17,18,19). We have been able to demonstrate that the septal projection is more extensive than previously described (17) as seen in the Fink-Heimer impregnated material and AChE-preparations following medial septal lesions. Although an exact correspondence between degeneration pattern and changes in AChE activity could not be established, there was a marked effect of medial septal lesions on enzyme activity in all parts of the hippocampal region, including the entorhinal area, para-, and presubiculum (Fig. 1).

We have previously observed (17) early changes in the AChE activity of the neuropil in the dorsal septum following small unilateral lesions in the medial septum. These changes preceded any detectable loss of enzyme activity in the hippocampal region and were correlated well with the distribution of degenerating terminals on and around cells of the dorsal septum. On the other hand, recent electrophysiological studies (4,15) suggest that the septum contains inhibitory interneurons of which some may be activated locally by recurrent collaterals (4). Septal cells also respond to microiontophoretically administered acetylcholine (ACh) (21).

These findings prompted us to re-examine the distribution of AChE in the septal region of the rat and to determine histochemically the localization of a possible cholinergic innervation of septal neurons.

The distribution of the AChE activity in the septal region was only briefly commented on in our previous work (17). Recently a more detailed description together with the cytoarchitectonics of this region has been published by Biaołwas and Narkiewicz (2).

METHODOLOGY

The present study was based on material obtained from the adult male rats (Möll-Wistar). The animals were decapitated under light ether anesthesia. The brains were removed and immediately frozen with carbon dioxide. Sections 20 μm thick were cut either frontally, horizontally, or sagittally, and dried at room termperature. A modified Koelle copper thiocholine method (9) for frozen

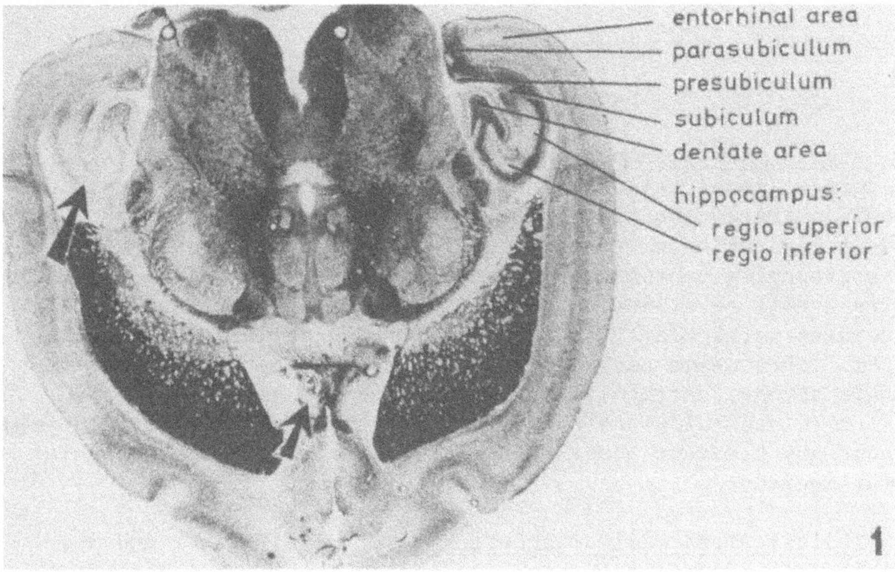

Figure 1. Horizontal AChE stained section from adult rat 3 days after unilateral medial septal lesion (lower arrow). All areas and fields of the ipsilateral hippocampal region show a considerable loss of enzyme activity (upper arrow). 6.5 X. (Reduced 15% for reproduction).

sections (6) was used. In brief, the sections were incubated at 37°C in a medium containing 4 mM acetylthiocholine iodine, 2 mM copper sulfate, 10 mM acetylthiocholine iodide, 2 mM copper sulfate, 10 mM glycine, and 50 mM acetate buffer (pH 5.0). To inhibit nonspecific cholinesterase, ethopropazine in a final concentration of 2.0 mM was used. Incubation times of 1, 2, and 3 hrs. were employed. Parallel sections after short incubation times were counterstained with thionin.

In the present study we have applied a modification of Andy and Stephan's classification of the septal nuclei in the cat brain (1).

RESULTS

Medial Septal Structures

Anterior hippocampus (H) (Figs. 2 and 3). Most rostrally under the corpus (genu) callosum the anterior continuation of the hippocampus dominates the midline. In AChE-preparations this nucleus shows a lamination where the granular cells are AChE-negative while the surrounding neuropil is moderately to heavily stained. It is mostly a diffuse stain, but there are also highly reactive fibre structures particularly in the periphery bordering the cingulate cortex. The enzyme activity is more intense ventrally, and some of this strong reaction is confined to medium sized perikarya. Still more ventrally and caudally this strongly reactive part joins without any clearcut border the most rostral level of the medial septal nucleus.

Medial septal nucleus anterior part (MA) (Figs. 2 and 3), exhibits an increasing enzyme reaction in the ventral direction, but there is a sharp border consisting of a pale zone towards the intensely stained tuberculum olfactorium (OT, Fig. 2). The AChE activity is confined to neuropil as diffusely scattered granules as well as paramedially running fibre structures. However, the strongest staining intensity is seen in rather large perikarya. Other cell bodies are more moderately stained and might be identical with the small, vertically arranged cells described by Białowas and Narkiewicz (2).

Medial septal nucleus posterior part (MP). According to Andy and Stephan (1) this is a paried structure which lies a little more caudal than MA, and dorsal to the nucleus of the diagonal band of Broca (BD), and extends laterally (in the cat and man) to the subependymal layer of the lateral ventricle. Białowas and Narkiewicz (2) point out that the MA nucleus spreads out in the caudal direction as a thin layer among various structures of the

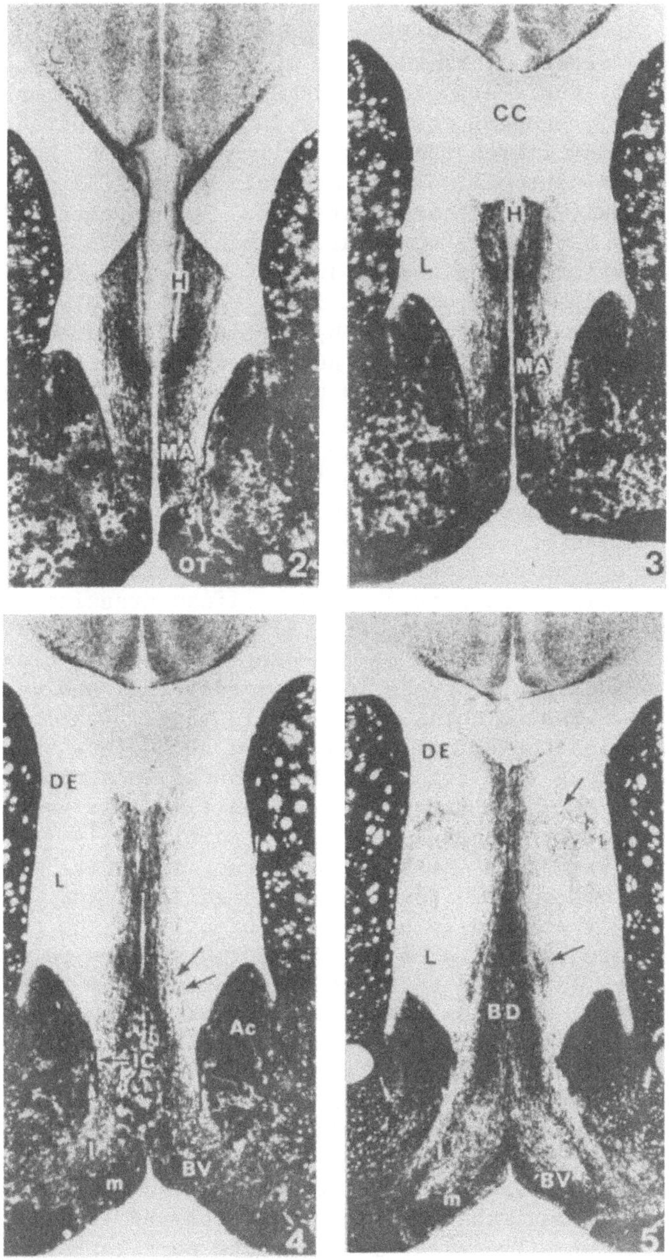

Figures 2-5. Frontal AChE stained sections (each separated by approximately 200 μM) through the septal region from rostral (Fig. 2) to caudal (Fig. 9) levels. For abbreviations see <u>Abbreviations</u> at the end of this section. 16 X. (Reduced 30% for reproduction).

septum and shows no AChE activity. They state that "sometimes this region is called nucleus medialis posterior (MP)", to which they apply the label "transitional area" as a better descriptive term. In our material, immediately lateral to the BD nucleus at rostral levels, there are stnads of AChE-positive fibres (upper arrows, Fig. 4), running principally in a vertical direction. In addition to these fibres there is a slight amount of background activity in the neuropil. There are only a few AChE positive perikarya, weakly to moderately stained. In the caudal direction, however, these paramedial structures exhibit an AChE reaction which form distinctive paired islands of highly stained neuropil (lower arrow, Fig. 5; arrow, Fig. 11). Medially this structure is separated from the DB nucleus by a pale intermediate zone (Figs. 5 and 11). More caudally the MP nucleus becomes slightly broader (lower arrow, Fig. 6) and appears to connect with the AChE positive structures (upper arrow, Fig. 6) of the dorsal septum (see below). Thus, on the basis of the present material, it seems that only part of the MP contains both AChE-reactive perikarya and neuropil (lower arrows, Figs. 5 and 6; arrow, Fig. 11).

Nucleus of the diagonal band of Broca; dorsal part (BD) and ventral part (BV). There is a rather diffuse transition rostrally between the medial septal (MA, Fig. 3) and diagonal band (BV, Fig. 4) nuclei. The nucleus of the diagonal band is clearly subdividable in AChE preparations into a dorsal (BD) and ventral (BV) limb (Fig. 5). The BD thins out dorsally (Figs. 4 and 5) and on the midline joins the dorsal septum. The BD nucleus shows a moderate to strong AChE reaction, distributed in a manner quite similar to the MA. Its strongly AChE-positive cell bodies are arranged vertically and parasagittally, particularly in the rostrodorsal part. The BV is organized into two, a lateral and a medial AChE-reactive component. Rostrally the medial strand (m, Fig. 4) is very strongly stained in both the neuropil and the perikarya. In between these two parts of the BV there is a low to moderate enzyme activity localized to fibre structures. The lateral part (1, Fig. 4 and 5) contains AChE precipitates in the neuropil and fibres which are only a little denser than in the intermediate zone. In addition, it has many strongly stained perikarya. In sections incubated for 1 hr. it appears that both the cytoplasm and proximal processes react heavily, while the nuclei are more or less pale. More caudally the background activity of the medial strand is somewhat reduced near the midline and the overall staining intensity is much the same in the medial and lateral strands (m and 1, Fig. 6). The paler intermediate part (arrow, Fig. 12) forms a ring-like structure surrounded by more strongly reacting elements, dominated by very darkly stained cell bodies. Most laterally, adjacent to the tuberculum olfactorium (OT, Fig. 12), there is a strong neuropil activity in addition to a heavy perikaryal stain. Dorsal to this ring structure where the BV meets the BD (Fig. 12) the two-bladed configuration fades. More caudally the medial septum is characterized by a broadening of the BD nucleus (Figs. 7 and 8). The latter seems to lose contact (Fig. 7) with the AChE stained ring structure described above. Ventrally the BD nucleus is characterized by strongly stained cell bodies which are not organized in any particular orientation, whereas dorsally the perikarya are arranged in a longitudinal fashion. From the dorsal part a large number of AChE stained fibres can be seen running into the fimbria (arrows, Fig. 8). Only a small part of the nucleus of the diagonal band (BD, Fig. 9) is present at the level of the anterior commissure just below the columns of the fornix (Fx, Fig. 9). The fornix itself is traversed by strongly AChE-reactive fibres.

Triangular septal nucleus (A). Above the columns of fornix an intensely stained structure corresponding to nucleus triangularis septi is observed (A, Fig. 9). The reaction product here is concentrated for the most part in small dots which have the same intensity as the strongly positive cell bodies of the nucleus of the diagonal band. However, clear perikarya-like structures, such as those found in the nucleus of the diagonal band, are difficult to recognize because of the diffuse transition between the highly stained dots and their surroundings. Only a careful examination of sections incubated for 1 hr. confirms that the aforementioned dark structures represent medium sized cell bodies with stained proximal processes. The remaining tissue components show a diffuse stain in the neuropil and few AChE positive fibres.

Lateral Septal Structures

Nucleus lateralis septi (L). This nucleus (Figs. 3-8) is essentially negative in AChE preparations. All of its perikarya are devoid of enzyme activity and the neuropil shows only traces of diffuse stain, especially in the medial direction. The border between the lateral septal nucleus and the nucleus accumbens (Ac, Fig. 4) is sharply demarcated because of the very high intensity of the enzyme reaction in the latter structure. Rostrally, the border between the lateral and dorsal septal nucleus (see below) is diffuse in AChE preparations, while more caudally (Figs. 6-8) the appearance of AChE positive tissue components helps to delineate both nuclei.

Nucleus accumbens (Ac). This nucleus (Figs. 4-6) and the insulae of Calleja (IC, Fig. 4) on its medial wall are among the most strongly AChE-stained structures in the brain. The rostral portion of the nucleus accumbens has an especially dark medial part in which the precipitate seems to be localized in both the neuronal cell bodies and the neuropil. The intensity of the stain,

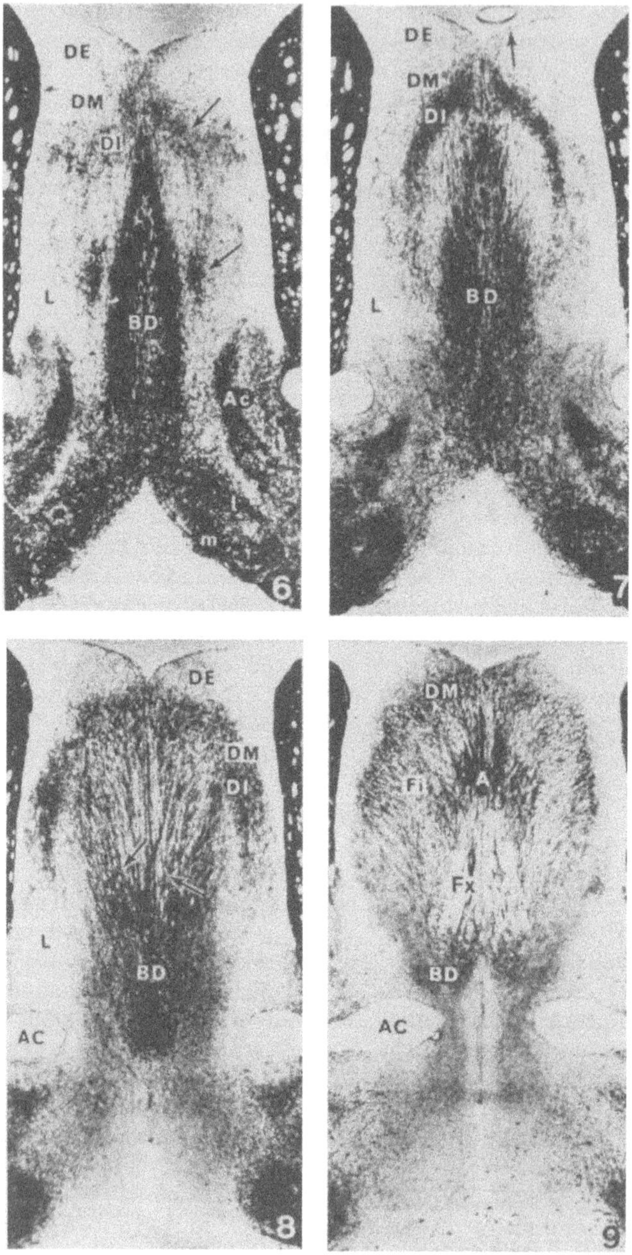

Figures 6-9. Frontal AChE stained sections (each separated by approximately 200 μm) through the septal region from rostral (Fig. 2) to caudal (Fig. 9) levels. For abbreviations see Abbreviations at the end of this section). 16X. (Reduced 30% for reproduction).

however, makes it difficult to detect single perikarya and their processes. On the other hand, at more caudal levels, where the reaction is less intense (Ac, Figs. 6 and 12) such perikarya can be observed readily.

Dorsal Septal Structures

Fiber Components. Close to the midline and at almost all rostrocaudal levels, the dorsal aspect of the septum is characterized by moderately stained fibres. Rostrally, these fibres run from the lateral part of the anterior hippocampus rostral and dorsal to the corpus callosum (Figs. 2 and 3). More caudally there is an increasing number of sub-callosal fibres (arrow, Fig. 7) some of which seem to course through the corpus callosum. The supracallosal fibres can be observed through the rosto-caudal extent of the septum, and they are more strongly stained than the subcallosal fibres.

Dorsal septal nucleus. The AChE-containing components of the lateral part of the dorsal septum appear at midrostro-caudal levels as a low intensity staining diffuse neuropil (upper arrow, Fig. 5). The external part of the dorsal septal nucleus, equivalent to Andy and Stephan's "pars externa" of the cat septum (DE, Figs. 4-8) is essentially AChE-negative, like the lateral septal nucleus. The AChE-reactive part of the dorsal septal nucleus, on the other hand, seems to correspond to the "pars intermedia" (DM, Figs. 6-8) and the "pars interna" (DI, Figs. 6-8) in the same terminology. More caudally, there is an increase in enzyme activity of the dorsal septum staining most intensely ventrally (DI, Fig. 7), and giving a characteristic "arrow-head" appearance. Still more caudally, the ventral reactive part (DI, Fig. 8) is accompanied by an equally stained external layer (DM) with a paler zone in between, thus forming a characteristic bilayer structure (Fig. 10). It appears that the enzyme reaction is localized mainly in the neuropil, although there are accumulations of stained particles on and around some of the perikarya. However, these perikarya are not as strongly reactive as those in the diagonal band nucleus, and it is not possible to observe their initial processes. Thus, the degree of staining in these cells is probably insufficient to warrant the suggestion that they might be cholinergic. The bilayer appearance of enzyme stain in the dorsal septum disappears at the most caudal levels (DM, Fig. 9)

Septofimbrial nucleus (Fi). This nucleus (Fig. 9) is composed of neurons interstitial to the fibres of the fimbria. This area is dominated by the strongly AChE-stained fimbria fibres, but examination of AChE preparations counterstained with thionin, reveals a few scattered septofimbrial perikarya with moderate AChE stain.

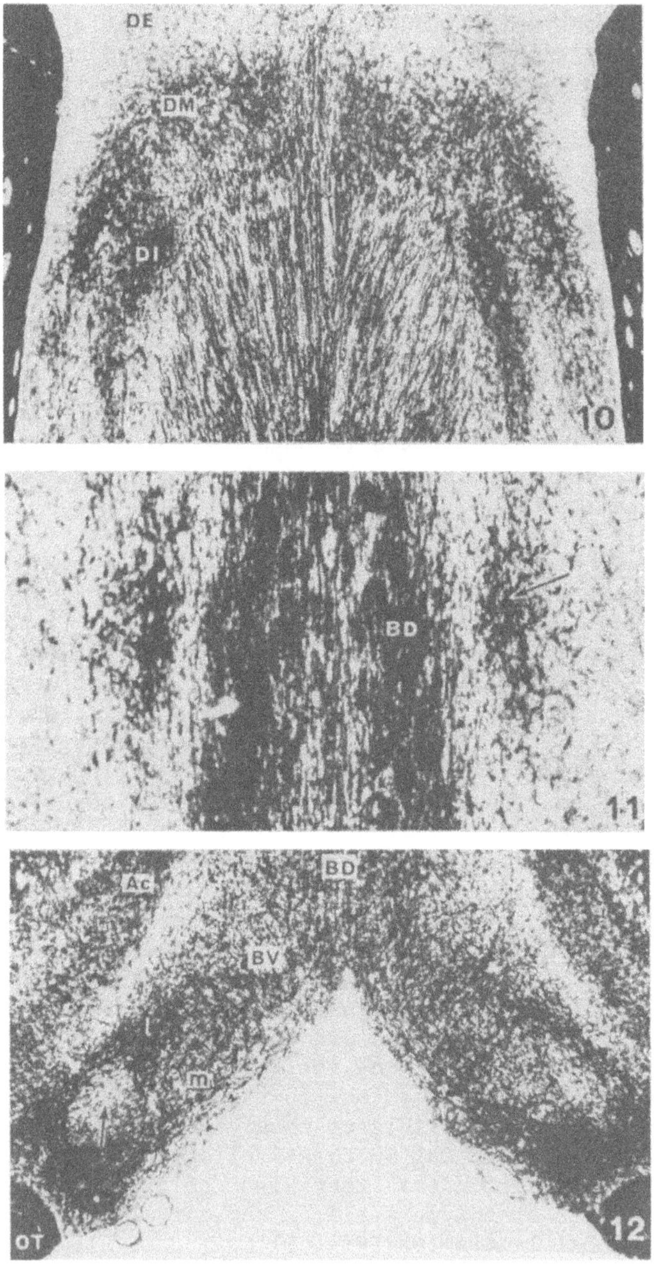

Figures 10-12. Higher magnifications of the caudal part of the dorsal septum (40 X, equivalent to Fig. 8) and of the dorsal (64 X) and ventral (40X) limb of the diagonal band nucleus (equivalent to Fig. 6). (Reduced 30% for reproduction).

DISCUSSION

In agreement with previous findings (2,8,13) the present report has confirmed that the main part of the histochemically demonstrable AChE in the spetal region is confined in the neuronal cell bodies, fibre systems, and neuropil of the medial septum. However, a detailed analysis of the pattern of AChE distribution and its correlation with the cytoarchitectonic organization of the individual nuclei leads us to the conclusion that a relatively small portion of the enzyme activity is found in the medial septal nucleus. Thus, only the most rostral part of this nucleus (pars anterior; MA) and the isolated paramedial islands (pars posterior; MP) at the midrostrocaudal levels contain AChE-positive perikarya and fibres.

On the other hand, the AChE activity in the nucleus of the diagonal band (and its two components, BV and BD) showed a complex pattern of distribution, which changed its organization and intensity of reaction along the rostro-caudal axis. Most of enzyme activity in this nucleus was observed in strongly reactive perikarya and in a dense plexus of fibre structures. The intensity of the neuropil stain was more variable than that of the cells and fibres.

The second most AChE-reactive part of the septal region was found in the caudal part of the dorsal septum. Here the enzyme activity was localized primarily in the neuropil and in the fibres of the area which form an intensely stained bilayer structure. In agreement with the cytoarchitecture of this region the most external and dorsal quadrant (DE) was devoid of enzyme activity, suggesting a differentiation in the dorsal septum similar to that described in the cat brain (1).

The non-perikaryal nature of the enzyme activity in the dorsal septum is supported by our experimental findings (7) of a loss of AChE stain in this area following ipsilateral medial septal lesions. These latter results, furthermore, suggest that the AChE-reactive part of the dorsal septum is innervated by a number of cholinergic fibres terminating in its neuropil. The origin of these AChE-containing fibres is unknown at present. One possibility is that they are axon collaterals from the cholinergic septo-hippocampal projection, they may originate, on the other hand, in intrinsic AChE-positive septal neurons, or in extraseptal structures.

The present results suggest that the caudal part of the dorsal septum may be a major "cholinoceptive" area of the septal region in the rat, while the nucleus of the diagonal band may be "cholinergic" in nature. The presence of histochemically demonstrable AChE in the central nervous system must be interpreted with a caution, however, and can be regarded only as an indicator (not a proof) of a possible presence of cholinergic neurons and synapses (22). Additional neurochemical and electrophysiological evidence is required to confirm the "cholinoceptive" and the "cholinergic" nature of the parts of the septal region described in the present paper.

ABBREVIATIONS

A = triangular septal nucleus
AC = anterior commissure
Ac = nucleus accumbens
BD = nucleus of the diagonal band of Broca, dorsal part
BV = nucleus of the diagonal band of Broca, ventral part
CC = corpus callosum
DE = dorsal septal nucleus, external part
DI = dorsal septal nucleus, internal part
DM = dorsal septal nucleus, intermediate part
Fi = septofimbrial nucleus and fimbria
Fx = fornix
H = anterior continuation of the hippocampus
IC = Islands of Calleja
L = lateral septal nucleus
MA = medial septal nucleus, anterior part
MP = medial septal nucleus, posterior part
OT = olfactory tubercle
l and m = lateral and medial blades of the diagonal band nucleus, ventral part.

REFERENCES

1. Andy, O. J. and Stephan, H. The Septum of the Cat. Thomas, Springfield, Ill., 1963, pp. 84.

2. Białowas, J. and Narkiewicz, O. Acetylcholinesterase activity in the septum of the rat: A histochemical topography, Acta Neurobiol. Exp., 34 (1974) 573-582.

3. Chippendale, T. J., Cotman, C. W., Kozar, M. D. and Lynch, G. S. Analysis of acetylcholinesterase synthesis and transport in the rat hippocampus: Recovery of acetylcholinesterase activity in the septum and hippocampus after administration of diisopropylfluorophosphate, Brain Research, 81 (1974) 485-496.

4. DeFrance, J. F., Kitai, S. T. and Shimono, T. Electrophysiological analysis of the hippocampal-septal projections: II Functional characteristics, Exp. Brain Res., 17 (1974) 485-496.

5. Fonnum, F. Application of microchemical analysis and subcellular fractionation techniques to the study of neurotransmitters in discrete brain areas, Adv. Biochem. Psychopharmacol., 6 (1972) 75-88.

6. Genesler-Jensen, F. A. and Blackstad, T. Distribution of acetylcholine esterase in the hippocampal region of the guinea pig. Enthorhinal area, parasubiculum, and presubiculum, Z. Zellforsch., 114 (1971) 460-481.

7. Harkmark, W., Mellgren, S. I. and Srebro, B. Acetylcholinesterase histochemistry of the septal region in rat and human: Distribution of enzyme activity, Brain Research, 1975 (in press).

8. Jacobowitz, D. M. and Palkovits, M. Topographic atlas of catecholamine and acetylcholinesterase-containing neurons in the rat brain. I Forebrain (Telencephalon, Diencephalon), J. Comp. Neurol., 157 (1974) 13-28.

9. Koelle, G. B. The histochemical localization of cholinesterases in the central nervous sytem of the rat, J. Comp. Neurol., 100 (1954) 211-228.

10. Krnjevic, K., and Silver, A. A histochemical study of cholinergic fibres in the cerebral cortex, J. Anat. (Lond.): 99 (1965) 711-759.

11. Krnjevic, K. and Silver, A. Acetylcholine esterase in the developing forebrain, J. Anat. (Lond.), 100 (1966) 63-89.

12. Kuhar, M. J., Sethy, V. H., Roth, R. H. and Aghajanian, G. K. Choline: Selective accumulation by central cholinergic neurons. J. Neurochem., 20 (1973) 581-593.

13. Lewis, P. R. and Shute, C. C. D. The cholinergic limbic system: Projections to hippocampal formation, medial cortex, nuclei of the ascending cholinergic reticular system, and the subfornical organ and the supra-optic crest, Brain, 90 (1967) 521-540.

14. Lewis, P. R., Shute, C. C. D. and Silver, A. Confirmation from choline acetylase analyses of a massive cholinergic innervation to the rat hippocampus, J. Physiol. (Lond.), 191 (1967) 215-224.

15. McLennan, H. and Miller, J. J. The hippocampal control of neuronal discharges in the septum of the rat, J. Physiol. (Lond.), 237 (1974) 607-624.

16. Mellgren, S. I. Distribution of acetylcholinesterase in the hippocampal region of the rat during postnatal development. Z. Zellforsch., 141 (1973) 375-400.

17. Mellgren, S. I. and Srebro, B. Changes in acetylcholinesterase and the distribution of degenerating fibres in the hippocampal region after septal lesions in the rat, Brain Research, 52 (1973) 19-36.

18. Mosko, S., Lynch, G. and Cotman, C. W. The distribution of septal projections to the hippocampus of the rat, J. Comp. Neurol., 152 (1973) 163-174.

19. Oderfeld-Nowak, B., Narkiewicz, O., Białowas, J., Dabrowska, J., Wieraszko, A. and Gradkowska, M. The influence of septal nuclei lesions on activity of acetylcholinesterase and choline acetyltransferase in the hippocampus of the rat, Acta Neurobiol. Exp., 34 (1974) 583-601.

20. Palkovits, M., Saavedra, J. M., Kobayashi, R. M. and Brownstein, M. Choline acetyltransferase content of limbic nuclei of the rat, Brain Research, 79 (1974) 443-450.

21. Segal, M. Responses of septal nuclei neurons to microiontrophoretica-ly administered putative neurotransmitters, Life Sciences, 14 (1971) 1345-1351.

22. Silver, A. The Biology of Cholinesterases, North-Holland Publishing Co., Amsterdam, 1974, pp. 596.

23. Srebro, B. and Mellgren, S. I. Changes in postnatal development of acetylcholinesterase in the hippocampal region after early septal lesions in the rat, Brain Research, 79 (1974) 119-131.

ORGANIZATION OF THE HIPPOCAMPAL-SEPTAL AXIS[1]

Allan Siegel and Henry Edinger

Departments of Anatomy, Physiology, and Neuroscience
New Jersey Medical School, Newark, New Jersey

The series of studies described below and in the following chapter (8) represent an attempt to determine the nature of the neural relationships that exist between the hippocampus and septum of the mammal. It represents a multidisciplinary approach of brain stimulation, neuroanatomy, and neurophysiology.

BRAIN STIMULATION STUDIES

Hippocampus

Our interest in the nature of the relationship of the hippocampus to the septum evolved from a study of the role of the hippocampus in the regulation of aggressive behavior. Previous neurophysiological and behavioral studies of the hippocampus have indicated that: (1) the hippocampus may participate in the regulation of emotional behavior (4,10,16,18,21,25), and (2) regional functional differences between dorsal and ventral hippocampus may exist (1,3,9,13). Therefore, Siegel and Flynn(28) attempted to provide direct evidence for a modulatory role of the hippocampus in one form of emotional behavior--predatory aggression elicited by electrical stimulation of the hypothalamus in the cat, and to

---

[1]Supported by NIH Grant NS 07941-04 and by the Benevolent Foundation of Scottish Rite Freemasonry, Northern Jurisdiction, U.S.A. The authors acknowledge the technical assistance of Miss Ann Yoanides for her skillful histological preparation of the slides and Dr. William Zeman for his valuable assistance in the care and preparation of the animals.

determine whether involvement of different regions of the hippocampus would produce equivalent or differential effects upon aggressive behavior.

Electrodes for stimulation and recording were aseptically implanted into the dorsal and ventral hippocampus of 30 adult cats which did not spontaneously attack anesthetized rats. In addition, electrodes were also implanted into those regions of hypothalamus from which quiet biting attack behaviour could be elicited. The paradigm for testing the influence of the hippocampus upon the hypothalamic attack behavior was as follows: the hypothalamus was stimulated at an intensity that reliably produced a directed attack upon a rat within 30 sec. On the next trial, the hypothalamus and hippocampus were stimulated concurrently. The current intensity passing through the hypothalamic electrode during dual stimulation was always identical to the stimulating current during single stimulation. Trials were arranged in pairs; i.e., dual timulation always preceded or followed stimulation of the hypothalamus alone, the exact order derived from a table of random numbers. Latency scores on trials in which afterdischarges occurred were not considered in the statistical treatment of the data.

The differences in latency scores for paired trials of single and dual stimulation were analyzed separately for each hippocampal electrode site by a t-test for paired observations. Fig. 1 describes the mean attack latencies for hypothalamic stimulation alone, and in combination with hippocampal stimulation. Eight pairs of electrode tips located in the dorsal hippocampus suppressed attack behavior elicited from stimulation of the hypothalamus. Two pairs of electrodes in which stimulation failed to produce suppression of attack were located in the subcallosal gyrus above the hippocampus. Significant decreases in attack latencies were obtained from stimulation of the ventral hippocampus as well as from regions adjacent to it. Electrode sites which produced suppression and facilitation are depicted in Fig. 2. Fig. 3 shows two pairs of electrode tracks, one associated with suppression and the other associated with facilitation.

An attempt was also made to determine the effects of selected lesions of the dorsal or ventral hippocampus upon hypothalamically-elicited attack behavior. In this phase of the experiment attack latencies were compared on a series of pre- and postlesion sessions. Radio frequency lesions were placed bilaterally following a series of 50 trials extending over three sessions. Following the lesions, an additional 50 trials were administered. Lesions of the ventral hippocampus, in 7 of 9 cats, resulted in a marked suppression of attack latencies. Fig. 4 describes the pre- and postlesion attack latencies of an individual cat. The height of each dot represents the attack latency from hypothalamic stimulation. The

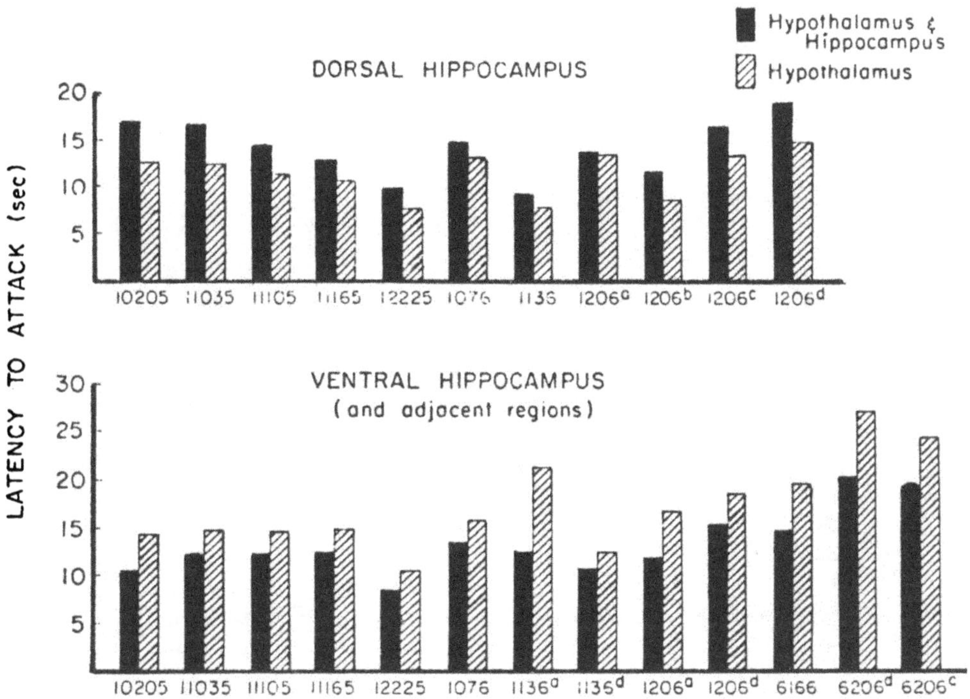

Figure 1. Bar graph indicates that concurrent stimulation of dorsal hippocampus and hypothalamus increases latency to attack whereas concurrent stimulation of ventral hippocampus (and adjacent regions) and hypothalamus decreases attack latency.

a) Stimulation of ipsilateral hypothalamus and hippocampus (right side);
b) current intensity applied through electrodes aimed for right hippocampus was raised to 0.30 mA from 0.20 mA, the intensity customarily used;
c) stimulation of hypothalamus and contralateral hippocampus;
d) stimulation of ipsilateral hypothalamus and hippocampus (left side).

first 10 postlesion trials were conducted several hours after the lesion was made while the last 20 trials occurred one month after lesion. Lesions involving the dorsal hippocampus did not reveal quite as consistent a pattern of results as compared to the results described following lesions of the ventral hippocampus. Three cats showed significant facilitation, one cat showed suppression, and little change was noted in two others.

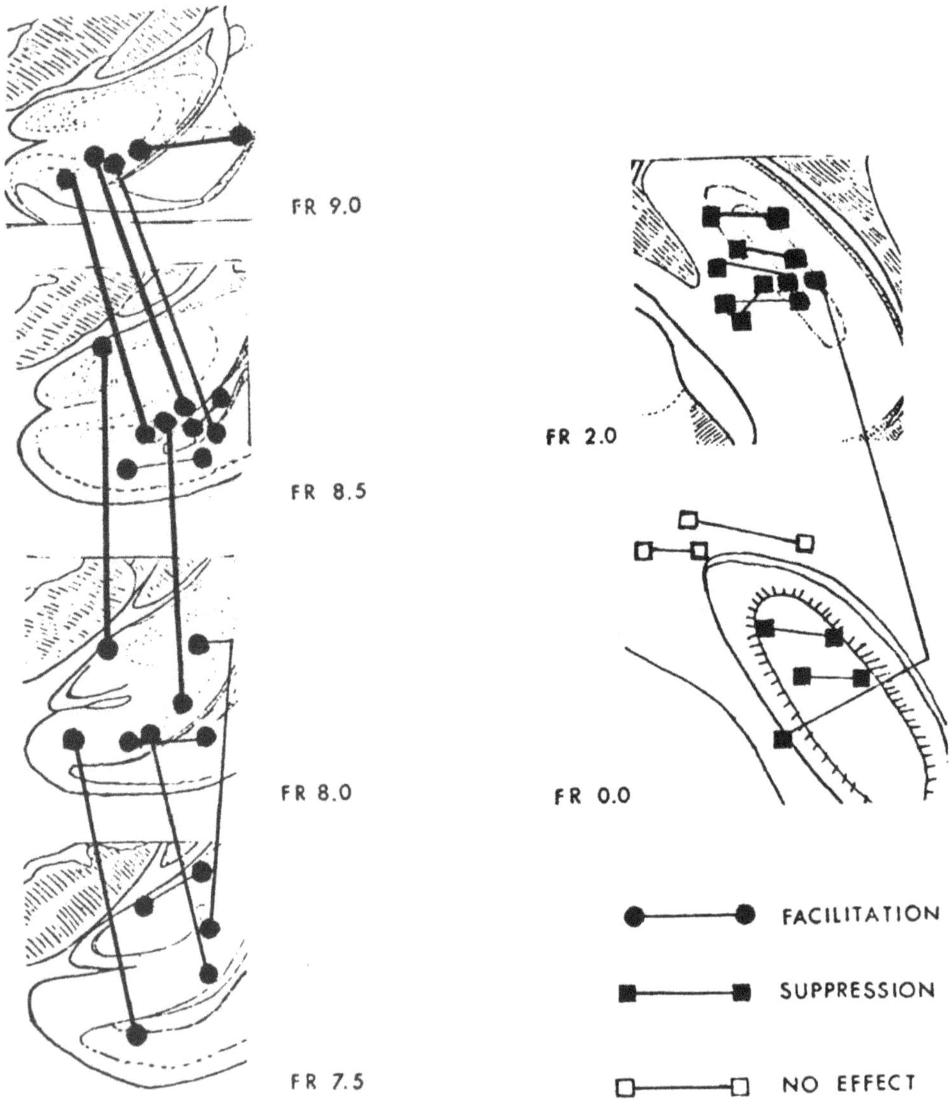

Figure 2. Sites of tips of electrode pairs from which stimulation produced suppression or facilitation of attack behavior. Stimulation of the hypothalamus was bipolar. The lines join the two points stimulated. For purposes of illustration all tips have been placed on the right side. Frontal sections are taken from the Atlas of Jasper and Ajmone Marsan (1954).

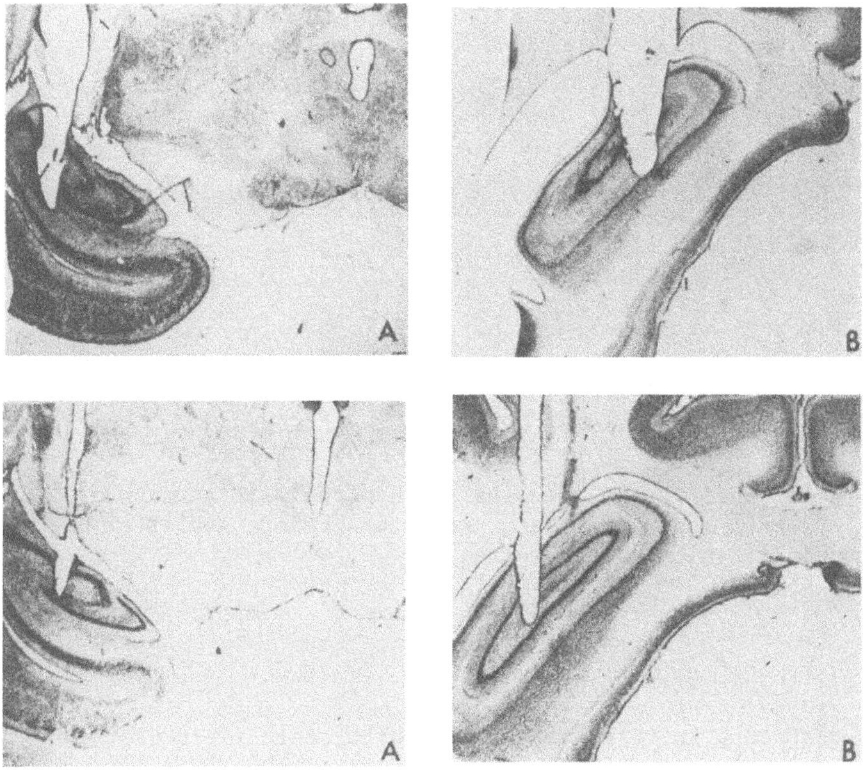

Figure 3. Photographs show sites of tips of electrode pairs in the hippocampus of two cats. Stimulation between pair A (ventral hippocampus) resulted in facilitation of attack; suppression of attack resulted from stimulation of pair B (dorsal hippocampus).

The results of this experiment have led us to draw the following conclusions regarding hippocampal function. Firstly, these findings provide direct evidence of a modulatory role of the hippocampus upon aggressive behavior. Secondly, this structure cannot be considered to be functionally uniform in its influence upon specific behavioral processes. Instead, there exists a functional differentiation of the hippocampus with respect to aggressive behavior. These findings suggest that the ventral hippocampus exerts a tonic excitatory influence, either directly or indirectly, upon neurons in the hypothalamus which partake in the elaboration of attack behavior. In contrast, it appears that the dorsal hippocampus exerts an inhibitory influence upon the attack response. Thirdly, the widespread distribution of effective hippocampal electrode points does not suggest a functional organization of

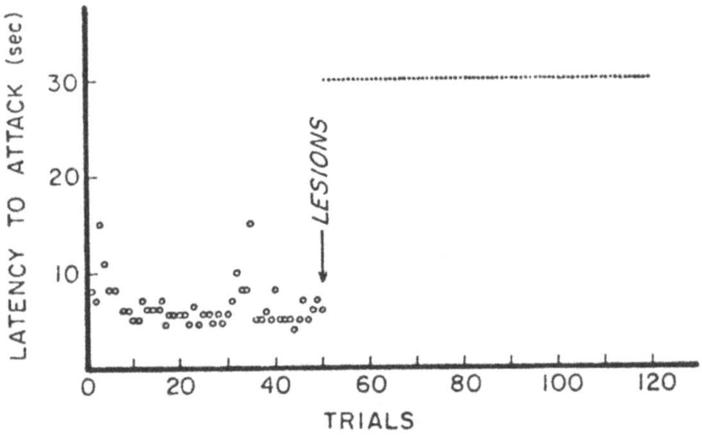

Figure 4. Individual latencies for hypothalamic attack on all trials for an individual cat. Responses before and after ventral hippocampal lesions are compared.

this structure in terms of a cytoarchitectonic field distribution.

Septal Area

A number of studies have indicated that lesions of the septal area result in an increase in aggressive behavior (2,5,6,17,34). However, other studies have revealed either an increase in placidity (7) or little change (11) following septal lesions. Cholinergic stimulation has been reported to elicit aggressive behavior in the cat (12) while electrical stimulation of the septum of the monkey inhibited aggression (26). The experiment described below (30 was attmepted in order to clarify the role of the septal area in the regulation of aggressive behavior. The paradigm adopted was identical to that used in our study of the hippocampus and aggression described above. Twenty-eight male and female cats that did not spontaneously attack rats were utilized in this experiment. The effects of septal stimulation upon three different types of hypothalamic responses were explored. In addition to quiet biting attack, we observed the effects of septal stimulation upon affective display (i.e., pronounced hissing, pupillary dilatation, arching of the back, retraction of the ears) elicited from the ventromedial nucleus and upon flight behavior (i.e., vigorous attempts to escape from the cage) elicited from wide regions of the hypothalamus. A map of hypothalamic electrode points producing these effects is indicated in Fig. 5.

The effects of electrical stimulation of the septal area upon mean latencies to attack are shown in Fig. 6. The results indicate that 14 of 21 electrode sites identified in the septal area produced a statistically significant suppression effect upon hypothalamically elicited attack behavior. Two points produced facilitation of hypothalamic attack and five other points did not significantly alter the response. An analysis of the electrode sites in the septal area that modulated attack behavior indicates that the suppression effect was not restricted to any specific region or nuclear group (see Fig. 7).

The effects of electrical stimulation of the septal area upon mean latencies to escape and affective display (hissing onset) are shown in Fig. 8. In contrast to the effects upon quiet biting attack, septal stimulation showed some evidence of facilitation of flight and hissing. Stimulation of six sites facilitated flight or hissing, two sites inhibited these responses, and from one additional site no change was produced.

These findings indicate that the septal area, exerts a selective inhibitory influence upon one hypothalamic process tested--quiet biting attack. The fact that the septal area does not inhibit other hypothalamic responses is of interest. This finding demonstrates that the septum exercises a selected control of hypothalamic processes. Further, this finding also indicates that the septal inhibition of attack behavior cannot be interpreted in terms of a general inhibition of all motor responses.

NEUROANATOMICAL STUDIES

The experiments of Siegel and Flynn (28) described above demonstrated a functional differentiation of the dorsal and ventral hippocampus in the cat. In order to determine the anatomical mechanisms by which modulation of attack behavior is achieved Siegel and Tassoni (81) compared the efferent projections of the dorsal and ventral hippocampus in the cat. The efferent pathways of the dorsal hippocampus have been described in a variety of species (15,19,20,22,23,24). On the other hand, information regarding efferent projections of the ventral hippocampus had been lacking in the literature.

Radio frequency lesions were placed into either the dorsal or ventral hippocampus of 25 adult cats. Animals were allowed to survive for four to eight days. They were then perfused with 0.9% saline and 10% formalin, stored in formalin for at least one week,

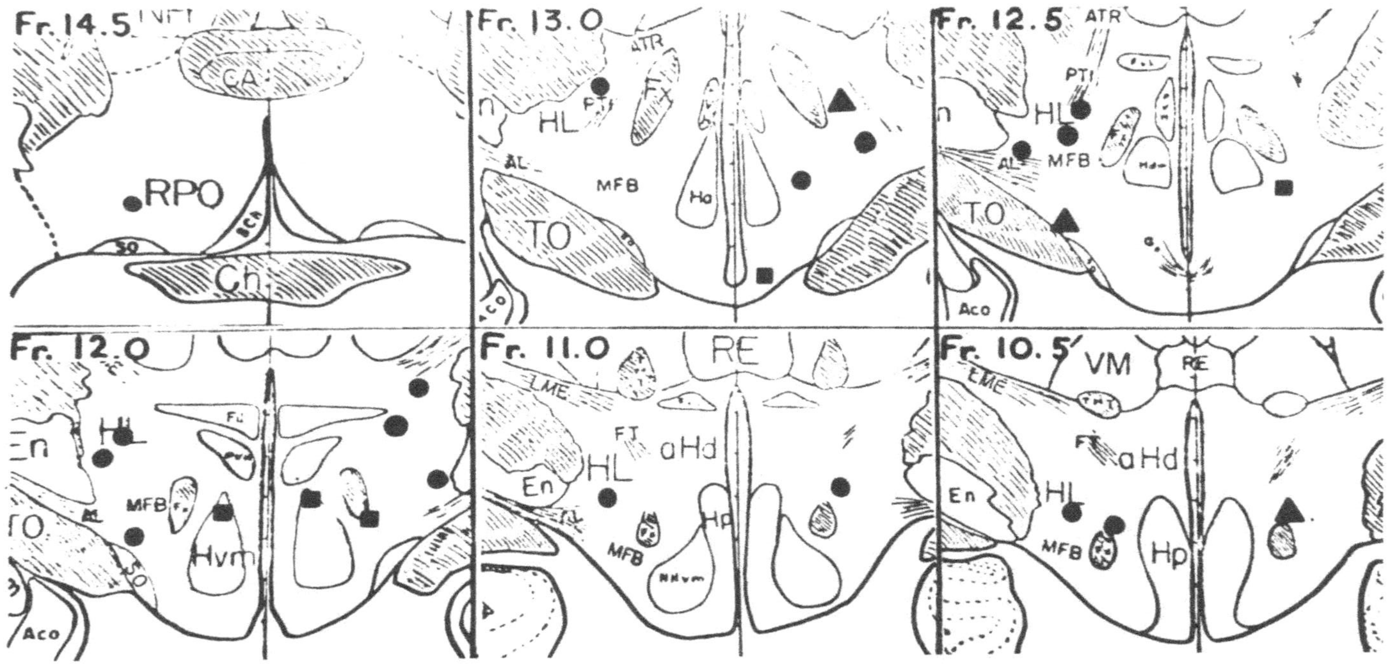

Figure 5. Plot of hypothalamic points producing attack, hissing, and escape. Attack is indicated by squares, and escape by triangles. Sections are taken from the Atlas of Jasper and Ajmone Marsan (1954).

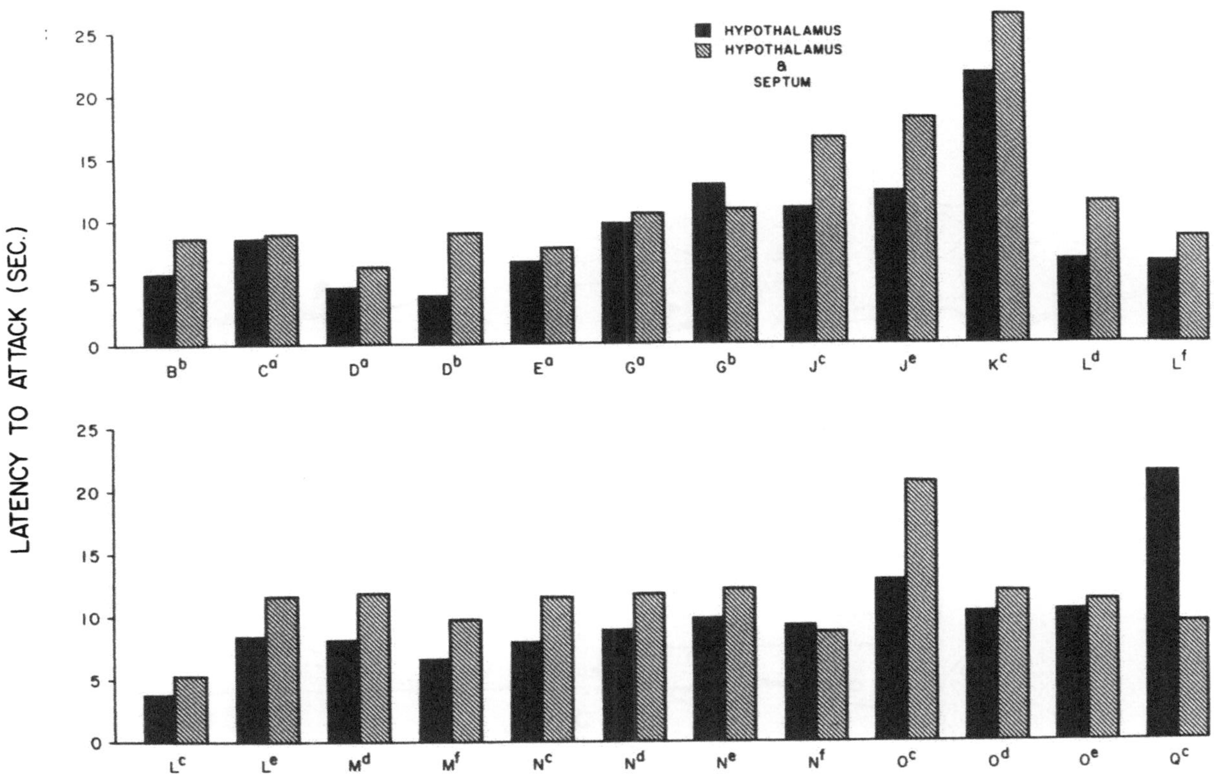

Figure 6. Bar graph indicates that concomitant stimulation of hypothalamus and septum increases latencies to attack. a) bipolar stimulation on right side; b) bipolar stimulation on left side; c) monopolar stimulation from anterior electrode on left side; d) monopolar stimulation from anterior electrode on right side; e) monopolar stimulation from posterior electrode on left side; f) monopolar stimulation from posterior electrode on right side.

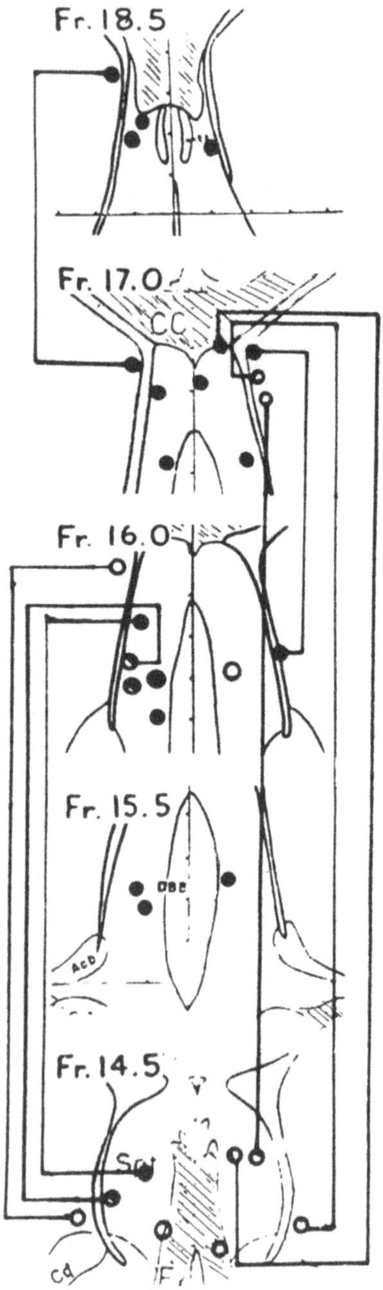
Fr. 18.5
Fr. 17.0
CC
Fr. 16.0
Fr. 15.5
DBB
Acb
Fr. 14.5
Spt
cd

A

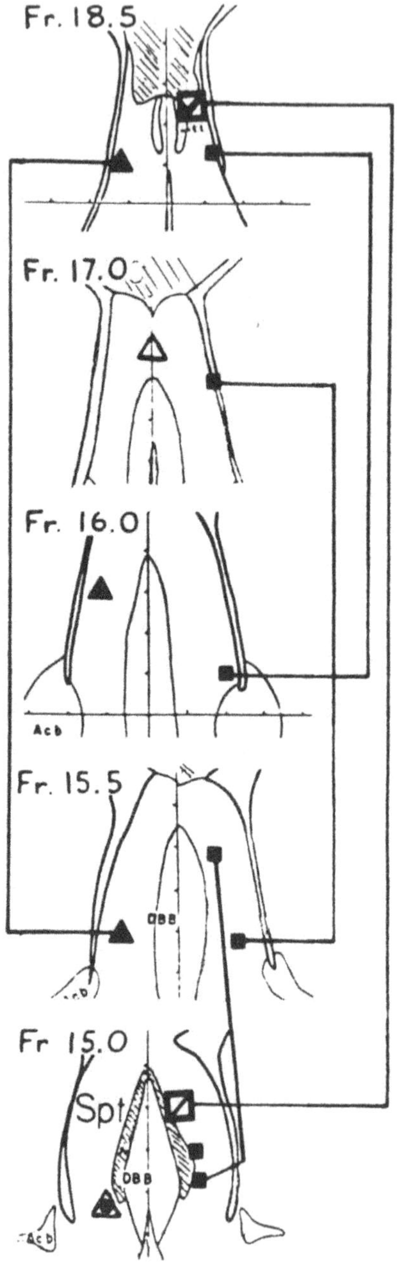
Fr. 18.5
Fr. 17.0
Fr. 16.0
Acb
Fr. 15.5
DBB
Fr 15.0
Spt
DBB
Acb

B

Figure 7. a) Electrode sites in septum modulating attack behavior. b) Electrode sites modulating hissing and escape. Lines connecting two points indicate bipolar stimulation. Isolated points indicate monopolar stimulation. Closed circles = suppression of attack; half-opened circles = facilitation of attack; open circles = no effect; closed triangles = facilitation of escape; open triangle = no effect; triangle with closed circle inside = suppression of escape; closed squares = facilitation of hissing; diagonal within squares = suppression of hissing.

and then sectioned on a freezing microtome at 30 microns. Sections were stained principally by the Fink Heimer 1 method for identifying degenerating axons and their terminals and by cresylechtviolet for identification of relevant cell groups.

Thirteen cats received lesions of the ventral hippocampus and eight cats sustained lesions of the dorsal hippocampus. Fig. 9 displays photographs of the main focus of lesions placed in either the dorsal or ventral hippocampus in four cats. An analysis of the pattern of tissue destruction of the different cytoarchitectonic fields within the ventral or dorsal hippocampus indicated that there were no systematic differences in frequency of destruction of CA fields between those two regions of the hippocampus. The results of this study indicated two major findings. One significant observation was the presence of massive quantities of terminal degeneration in the septal area in contrast to the much smaller

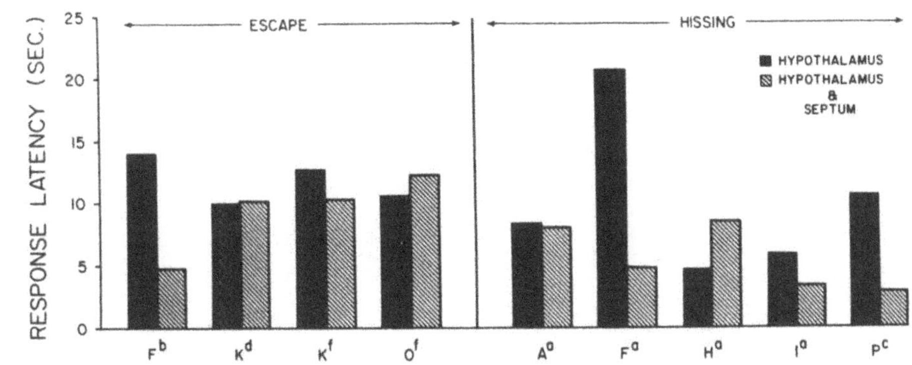

Figure 8. Bar graph indicates that concomitant stimulation of hypothalamus and septum tends to facilitate hissing and escape behavior. Superscripts are identical to those described in Fig. 6.

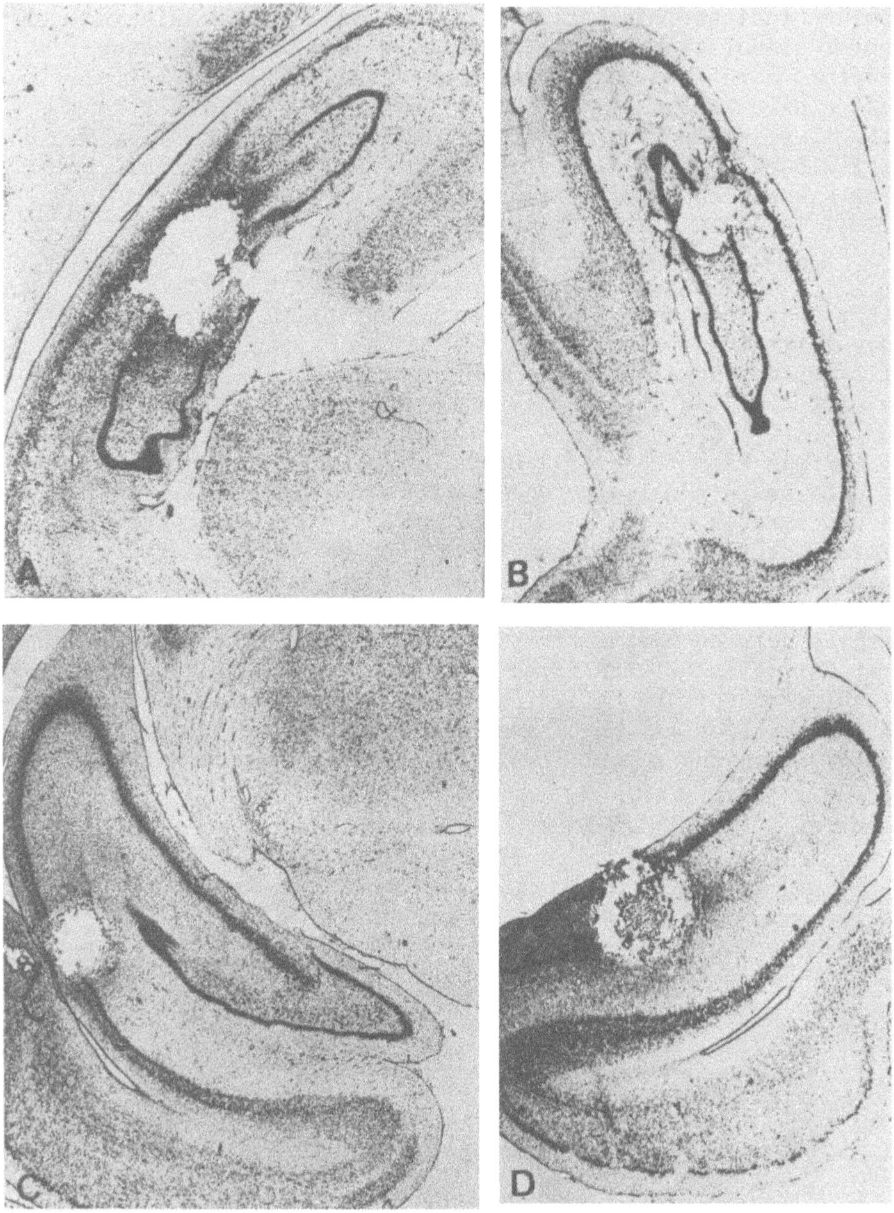

Figure 9. Photographs indicate the main focus of dorsal (A and B) and ventral (C and D) hippocampal lesions. A lateral approach was employed to produce the lesions shown in C. A dorsal approach was used to make lesions shown in A, B, and C.

numbers of degenerating fibers that reached the hypothalamus following lesions of the hippocampus. The second important observation was the presence of a clearly organized topographical distribution of fibers from the ventral and dorsal hippocampus to the septal area in the cat. Specifically, fibers from the ventral hippocampus pass through the fimbria and lateral aspect of the fornix to terminate in massive quantities through the fimbria and lateral aspect of the fornix to terminate in massive quantities throughout the lateral septal area and nucleus accumbens. In contrast, fibers from the dorsal hippocampus pass through the medial aspect of the fornix and terminate in somewhat less massive quantities in the medial septal area. Both regions, however, project similarly to the region of the nucleus of the diagonal band olfactory tubercle. Both regions of hippocampus appear to project also to the anteroventral nucleus of thalamus, to the hypothalamus, and mammillary bodies. It was our further observation that fibers from the ventral hippocampus appeared to terminate in greater amounts at a more anterior level of lateral hypothalamus while fibers from the dorsal hippocampus appeared to terminate in greater quantities at more posterior levels of lateral hypothalamus. In both cases, the small quantities of degeneration indicated only a very limited distribution of fibers to the lateral hypothalamus from both regions of hippocampus. The efferent distribution of fibers from the ventral and dorsal hippocampus are shown in Fig. 10 and Fig. 11.

The conclusions drawn from this study appear to be in conflict with the conclusions reached by Raisman (23) following his study of the efferent projections of the hippocampus in the rat. One possible interpretation of the differences between the conclusion of Raisman and the results of studies performed in our laboratory is that a species variation exists between rat and cat. Therefore, in order to attempt to resolve this controversy, Siegel, et al. (27) reinvestigated the organization of the hippocampal projection to the septum in a wider spectrum of mammalian species--gerbil, rat, rabbit, and cat.

Four gerbils, 18 rats, 8 rabbits, and 7 cats were used in this study. The procedures employed were similar to the procedures employed by Siegel and Tassoni (31) described above. Briefly, lesions of the dorsal or ventral hippocampus were made by surgical aspiration or by radio frequency currents. Survival times of 2-5 days were adopted for all animals in this study. Following perfusion of animals, brains were removed and stored in neutral buffered 10% formalin for at least one to two weeks prior to sectioning. Brains were then transferred into a 30% sucrose formalin solution several days prior to sectioning in order to reduce freeze-thawing artifacts in the tissue and subsequently embedded in albumin gelatin. Sections were again stained by the Fink-Heimer 1 method and by cresylechtviolet.

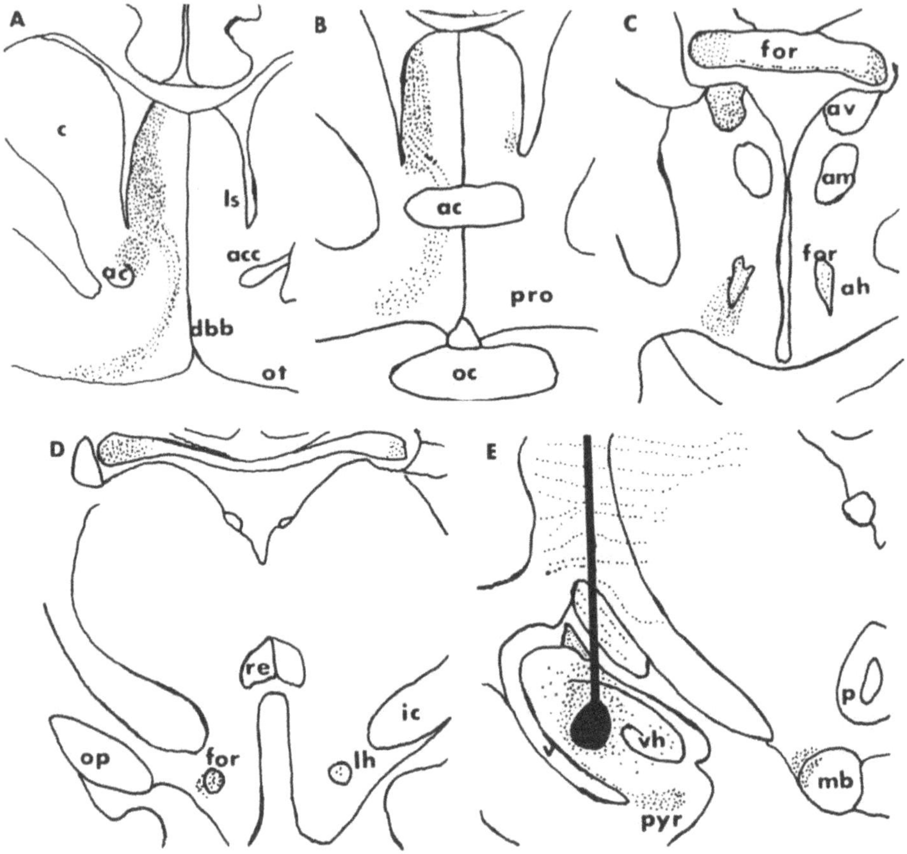

Figure 10. Course of degeneration resulting from a lesion placed in the ventral hippocampus. (Abbreviation used in Fig. 10 - 11 and 25-26: ac, anterior commissure; acc, nucleus accumbens; ah, anterior hypothalamus; am, anterior medial nucleus; av, anterior ventral nucleus; c, caudate nucleus; d, nucleus of Darkschewitsch; dbb, diagonal band of Broca; dh, dorsal hippocampus; for, fornix; ic, internal capsule; lh lateral hypothalamus; Is, lateral septum; mb, mammillary bodies; ms, medial septum, ndb, ventral nucleus of diagonal band of broca; oc, optic chiasm; op, optic tract; ot, olfactory tubercle; p, posterior hypothalamic nucleus; pa, paraventricular nucleus; pro, preoptic area; pry, pyriform cortex; re, nucleus reuniens; v; inferior horn of lateral ventricle; vh, ventral hippocampus).

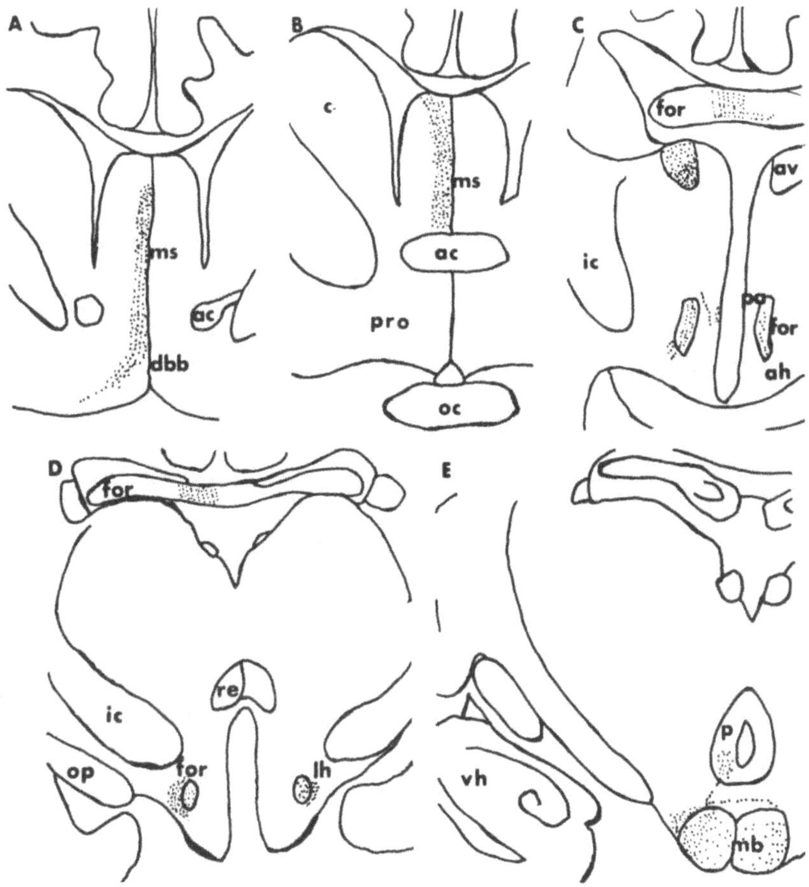

Figure 11. Course of degeneration resulting from a lesion placed in the dorsal hippocampus.

LESIONS OF THE GERBIL HIPPOCAMPUS

Restricted lesions were placed systematically at different levels of the hippocampus in 4 gerbils. Cases G1 and G2 sustained lesions of the anterior and posterior portions of the dorsal hippocampus. Case G3 included a lesion of the posterior hippocampus at a position midway between the dorsal and ventral hippocampus. The lesion in case G4 was restricted to a narrow region of the ventral hippocampus. A map of the extent of these lesions is shown in Fig. 12.

Following lesions of the dorsal hippocampus at either anterior (G1) or posterior (G2) levels degenerating axons were

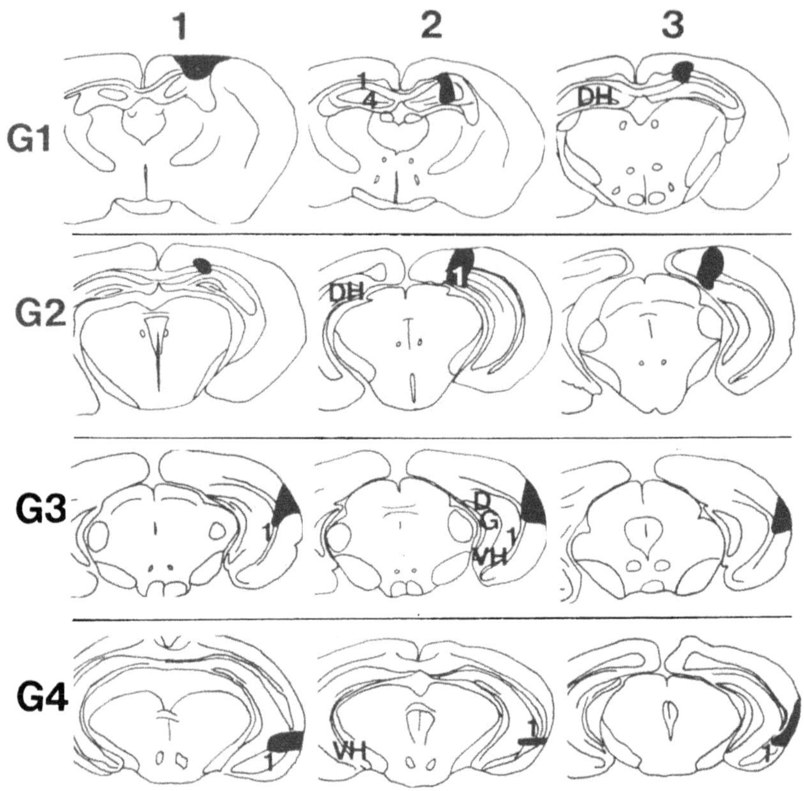

Figure 12. Outline of lesions of gerbil cases G1 - G4 plotted at their rostral (column 1), middle (column 2), and caudal (column 3) positions. (Abbreviations used in Fig. 12-24: a, amygdala; ac, anterior commissure; acc, nucleus acumbens; c, caudate nucleus; cc, corpus callosum; dg, dentate gyrus; f, fornix; fi, fimbria; oc, optic chiasm, olfactory tubercle; s, septum; 1-4, CA1-CA4).

distributed to both ventral and dorsal aspects of the medial septal area where they were observed to terminate in dense quantities. There was no terminal degeneration present in the lateral septal area. Additional fibers of passage were traced to the nucleus of the diagonal band and adjacent olfactory tubercle. The projection from case G3 passed to an intermediolateral portion of the septum where terminal degeneration was observed in dense quantities. In this case, terminal degeneration was also noted in the nucleus accumbens, nucleus of the diagonal band, and olfactory tubercle. In the case of G4 massive quantities of terminal degeneration were observed in the lateral septal area and nucleus

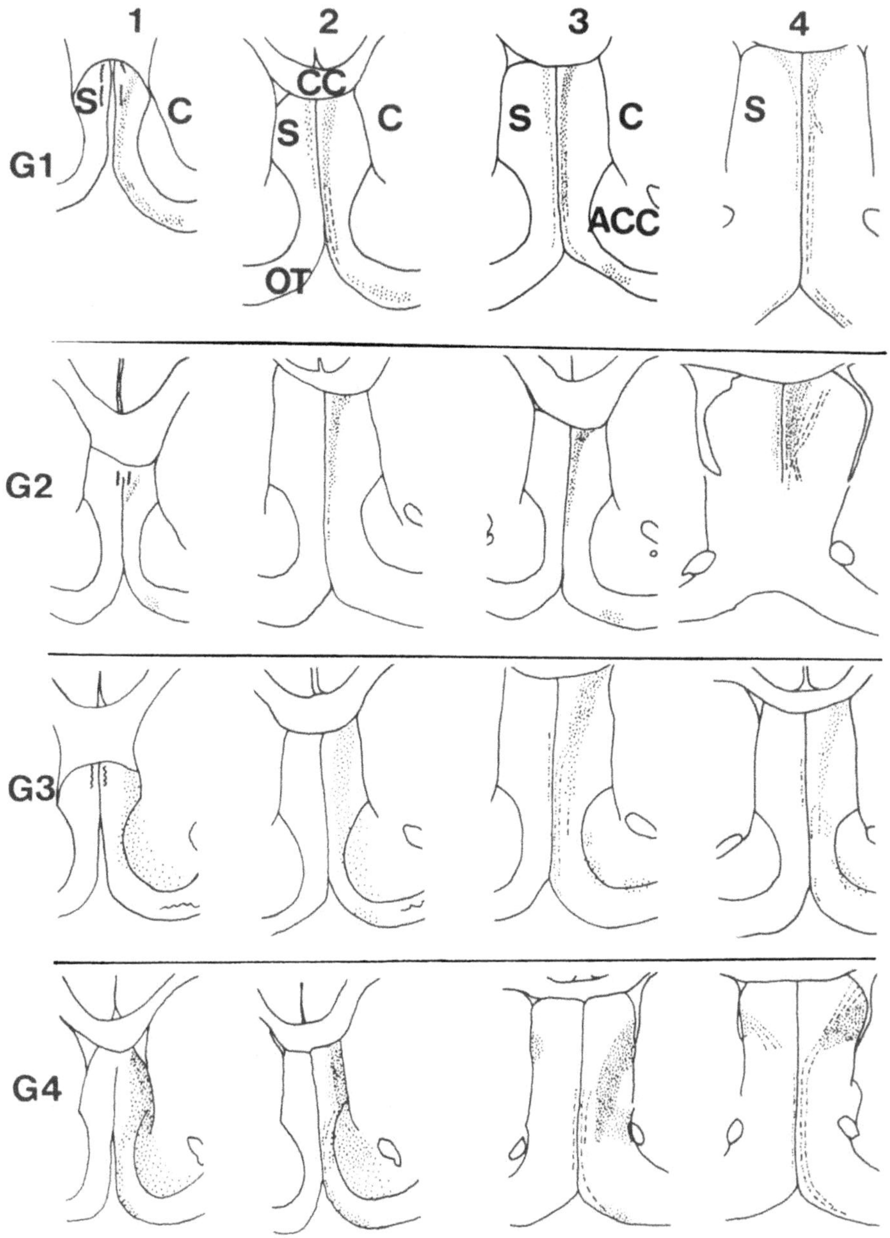

Figure 13. Course of degeneration in the septal area after lesions of different portions of the hippocampus in gerbils G1-G4.

accumbens following a lesion of the ventral hippocampus. Fibers of passage were observed in the medial septal area but no terminal degeneration was noted. These fibers of passage ultimately terminated in the nucleus of the diagonal band and olfactory tubercle.

LESIONS OF THE DORSAL HIPPOCAMPUS OF THE RAT

Lesions of the anterior hippocampus were placed in rats R1 - R3. The lesions in each of these cases produced damage that included the regio inferior (CA3 and/or CA4). A description of these lesions is shown in Fig. 14. Fig. 15 and Fig. 16 illustrate lesions of the posterior portion of the dorsal hippocampus. It is of interest to note that in case R4 a massive lesion included destruction of all CA fields and the dentate gyrus of the middle third of the hippocampus. Fig. 17 indicates that the dorsal hippocampal projection to the septal area is invariant in spite of variations in the placement of the lesions with respect to the positions occupied by the CA fields.

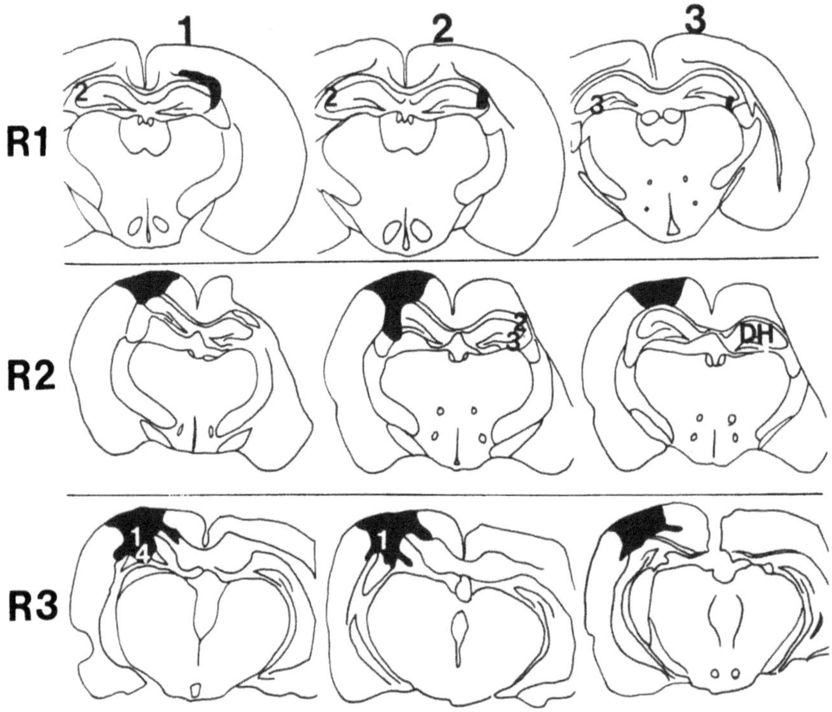

Figure 14. Outlines of lesions of the anterior dorsal hippocampus of rat cases R1-R3 plotted at their rostral, middle, and caudal positions.

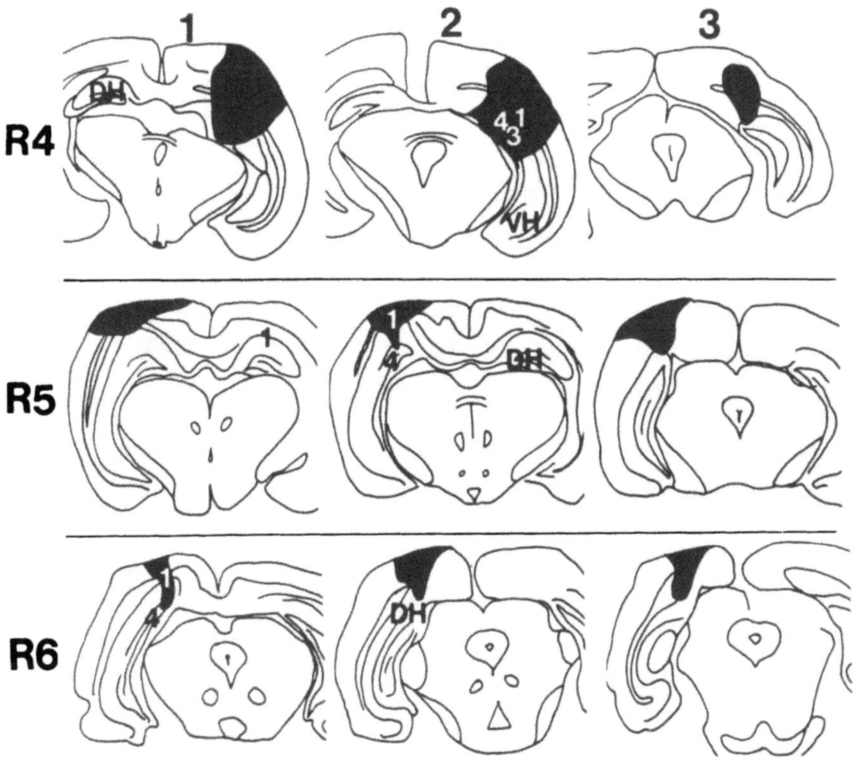

Figure 15. Outlines of lesions of the posterior dorsal hippocampus of rat cases R4-R6 plotted at their rostral, middle, and caudal positions.

LESIONS OF THE VENTRAL HIPPOCAMPUS OF THE RAT

Lesions placed in cases R10-R12 were restricted to the ventral hippocampus and shown in Fig. 18. It should be noted here that in each of these cases damage to the region of field CA1 was evident.

The pattern of degeneration following lesions of the ventral hippocampus of cases R10-R12 were highly similar to one another and also to the pattern described for gerbil case G4 which also received a lesion of the ventral hippocampus. In each of these cases degenerating axons were traced directly into the fimbria from the site of the ventral hippocampal lesion. No degenerating fibers were observed in the fornix--the region containing degenerating axons following lesions of the dorsal hippocampus. In each of the cases sustaining ventral hippocampal lesions,

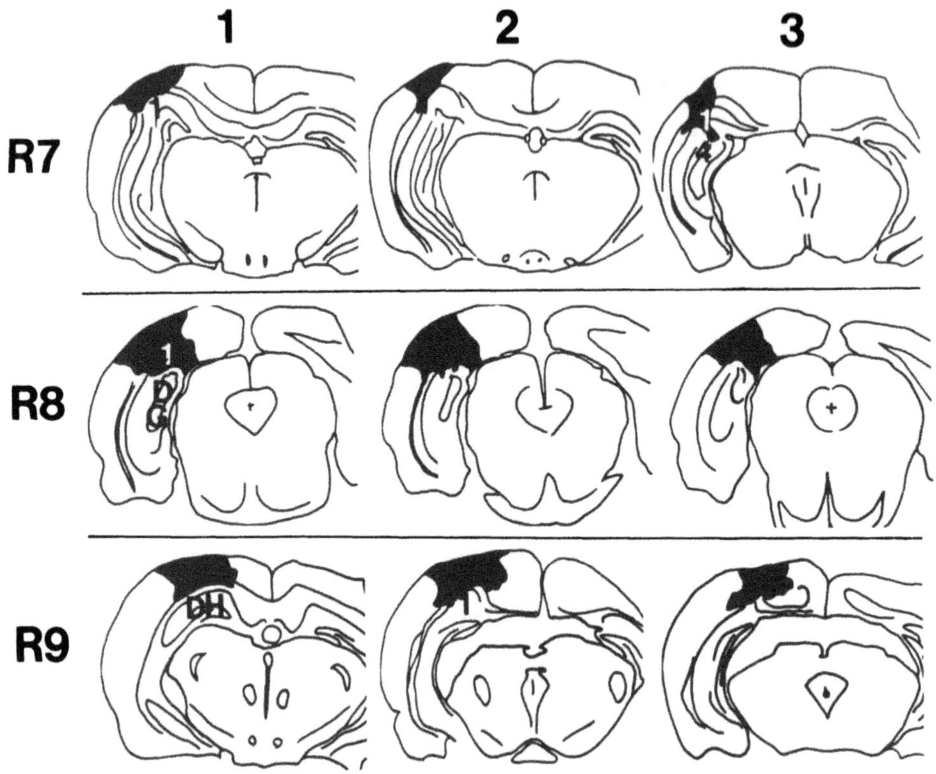

Figure 16. Outlines of lesions of the posterior dorsal hippocampus of rat cases R7-R9 plotted at their rostral, middle, and caudal positions.

degenerating fibers terminated in massive quantities throughout the rostro-caudal extent of the lateral septum and nucleus accumbens. No terminal degeneration was observed in the medial aspect of the septum. Fibers of passage from the lateral septum penetrated through the medial septum into the diagonal band and were observed to terminate specifically in the nucleus of the diagonal band and olfactory tubercle. This pattern of degeneration is described in Fig. 19.

OTHER LESIONS OF THE HIPPOCAMPUS OF THE RAT

Three cases involved combined lesions of the dorsal hippocampus and the fimbria. Following the placement of these lesions massive quantities of terminal degeneration were observed throughout

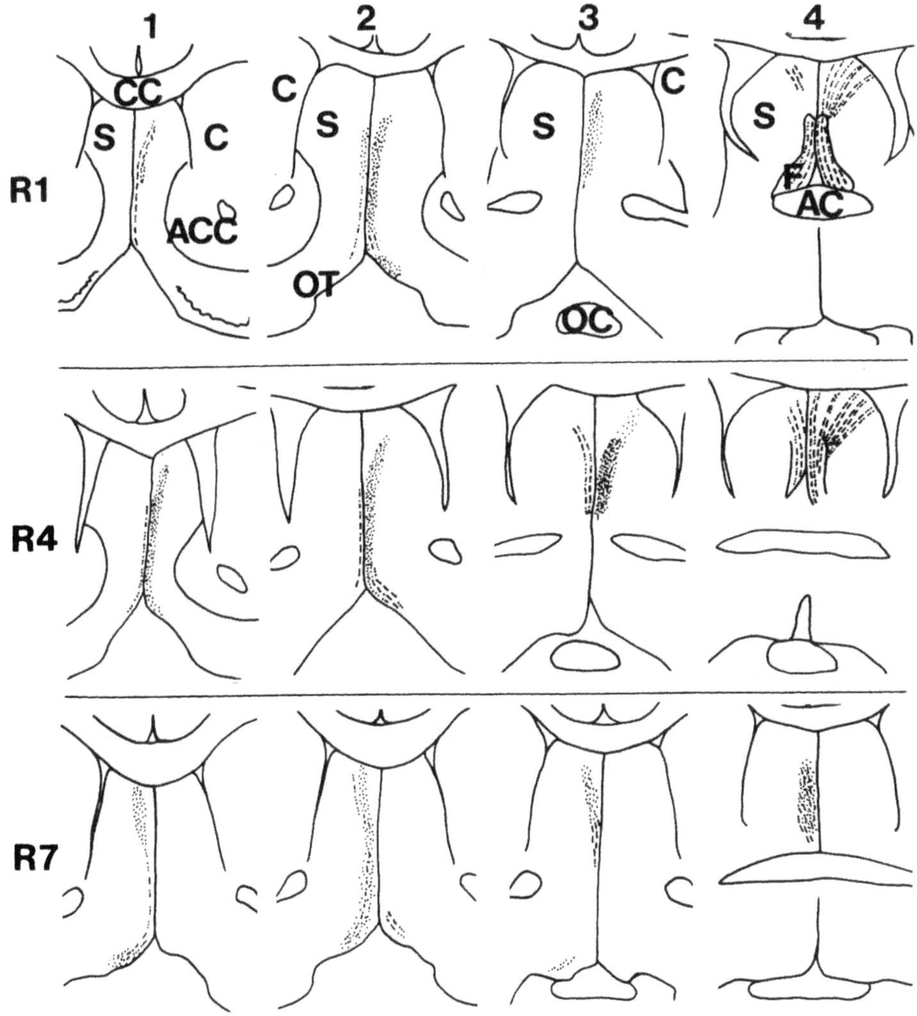

Figure 17. Course of degeneration in the septal area after lesions of different levels of dorsal hippocampus in rat cases R1, R4, and R7.

the total extent of both medial and lateral septum and nucleus accumbens. In addition, degenerating axons were traced through the medial septal area and diagonal band to the nucleus of the diagonal band and adjacent olfactory tubercle.

In one case a large lesion produced destruction of fields CA1 and CA2 of both dorsal and ventral hippocampus. The pattern of

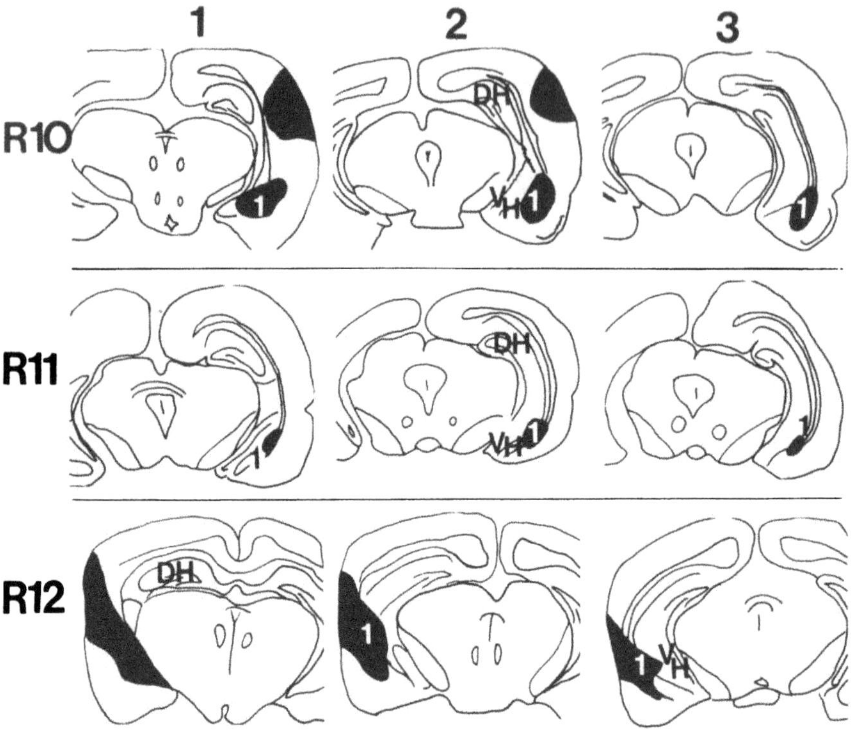

Figure 18. Outlines of lesions of the ventral hippocampus of rat cases R10-R12 plotted at their rostral, middle, and caudal positions.

degeneration in the septum following this lesion replicated the pattern described immediately above for those cases sustaining combined lesions of the dorsal hippocampus and fimbria.

In another case a superficial lesion was placed in the alveus at a position midway between the dorsal and ventral hippocampus. This lesion resembled quite closely the one placed in case G3 of the gerbil hippocampus. Terminal degeneration was similarly limited to an intermediolateral position within the septum. The far lateral and far medial aspects of the septal area showed no evidence of terminal degeneration.

LESIONS OF THE HIPPOCAMPUS OF THE RABBIT

Suction lesions were placed in different portions of the hippocampus in 8 rabbits. The results of three cases will be described briefly. In case Rb1 a superficial lesion was placed in field CA1 of the dorsal hippocampus. In addition, damage to

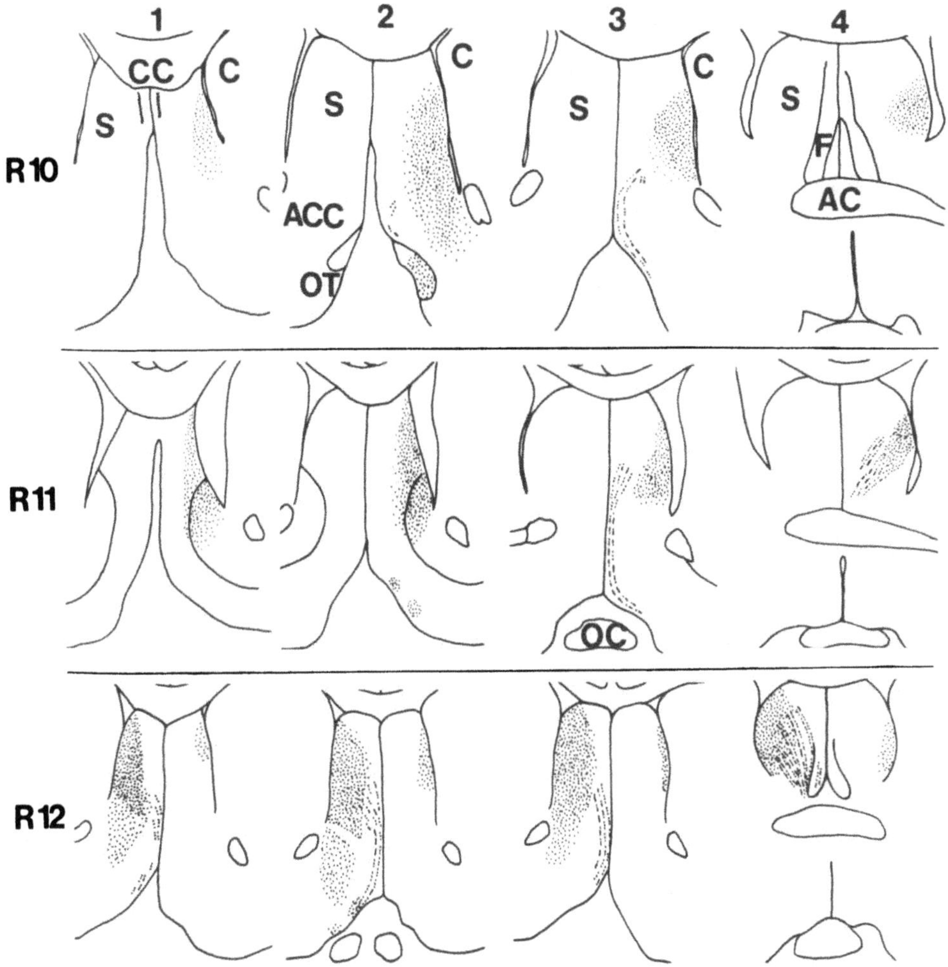

Figure 19. Course of degeneration in the septal area after lesions of the ventral hippocampus in rat cases R10-R12.

adjacent portions of the subiculum was noted as well. The distribution of fibers to the septum following this lesion is similar to the results obtained with lesions of the dorsal hippocampus of the rat and gerbil. Again, fibers from the dorsal hippocampus terminate entirely within the medial septal area and region of olfactory tubercle. Case Rb5 received a combined lesion of the dorsal hippocampus and fimbria. We observed that terminal degeneration was present in massive quantities throughout the entire septum and nucleus accumbens. This observation is identical to the pattern of degeneration observed in the septum following a combined lesion in the rat. Case Rb7 received a lesion that destroyed posterior levels of the ventral hippocampus. This animal showed extensive destruction of all CA fields and the dentate gyrus of the ventral posterior hippocampus. Further destruction of the neighboring entorhinal cortex and subiculum was noted as well. Following this lesion massive quantities of terminal degeneration were present in the lateral two thirds of the rostro-caudal extent of the septum. No terminal degeneration was present in the medial septal area. Instead, fibers of passage were followed in a ventral direction through the medial septal area to the diagonal band and olfactory tubercle. The patterns of terminal degeneration within the septum for these three cases are illustrated in Fig. 20.

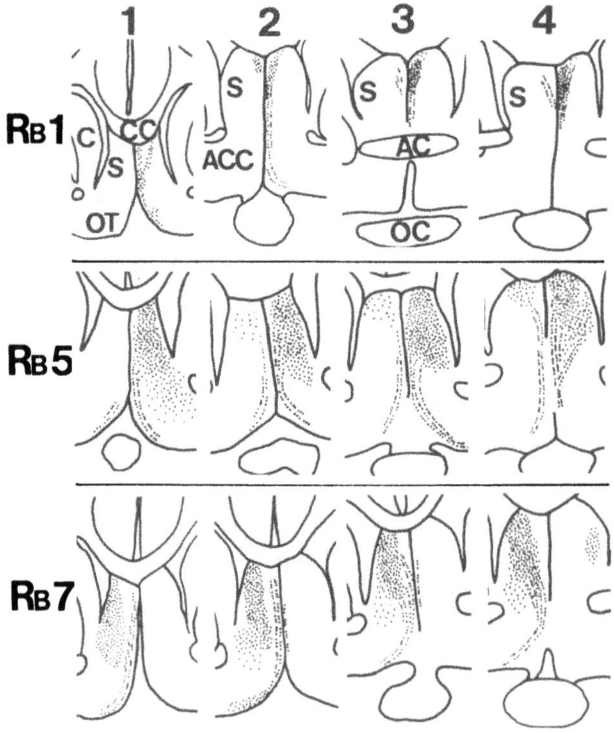

Figure 20. Course of degeneration in the septal area after lesions of the hippocampus in rabbit cases Rb1, Rb5, and Rb7.

LESIONS OF THE HIPPOCAMPUS OF THE CAT

This phase of the experiment represented an attempt to replicate the findings of Siegel and Tassoni (31). The results of three cases will be briefly described. Case C2 contained a discrete lesion that damaged field CA1 and portions of the dentate gyrus. From the site of this lesion degenrating axons were traced through the medial aspect of the fornix and terminal degeneration was observed throughout the caudal two thirds of the medial septum. Other degenerating fibers were followed through the diagonal band to the olfactory tubercle. No terminal degeneration was present in either the lateral septal area or nucleus accumbens. Case C3 sustained a combined lesion of the dorsal hippocampus and fimbria. The pattern of degeneration observed in this animal was identical to the results following combined lesions of the dorsal hippocampus and fimbria seen in the rat and rabbit. Again, axonal degeneration was traced through both medial and lateral aspects of the fornix to the septum. A stereotaxic lesion was placed in the ventral hippocampus in case C5. At anterior levels of the lesion damage to the CA1-CA2 region was noted, and at posterior levels, considerable destruction of both fields CA1 and CA3 as well as the dentate gyrus was noted. In this case degenerating axons of passage were followed through the lateral limb of the fornix and terminated throughout the rostro-caudal extent of the lateral septal area and nucleus accumbens. Although degenerating fibers of passage reached the medial septal area in route to the nucleus of the diagonal band and olfactory tubercle, no terminal degeneration was observed in this region of the septum. Thus, this aspect of the study directly replicated the findings observed initially by Siegel and Tassoni (31).

The results of our experiments on a series of subprimate mammalian species have confirmed our earlier ideas regarding the organization of the efferent projections of the hippocampus. In brief, our findings now clearly demonstate that the notion of a differential projection from different cytoarchitectonic fields of the hippocampus to the septum is an incorrect one. Instead, the organization of the topographical hippocampal projection to the septal area can be understood in terms of the position of cells lying in a dorsal-ventral continuum within the hippocampus. It should be noted that in every case and for all species considered a lesion placed in the dorsal hippocampus produced terminal degeneration restricted to the medial septum. In a similar manner, every case which involved damage to the ventral hippocampus resulted in terminal degeneration restricted to the lateral septum. The studies described above also demonstate that these topographical projections to the septum from the hippocampus remain invariant throughout different phylogenetic levels of the mammalian nervous system.

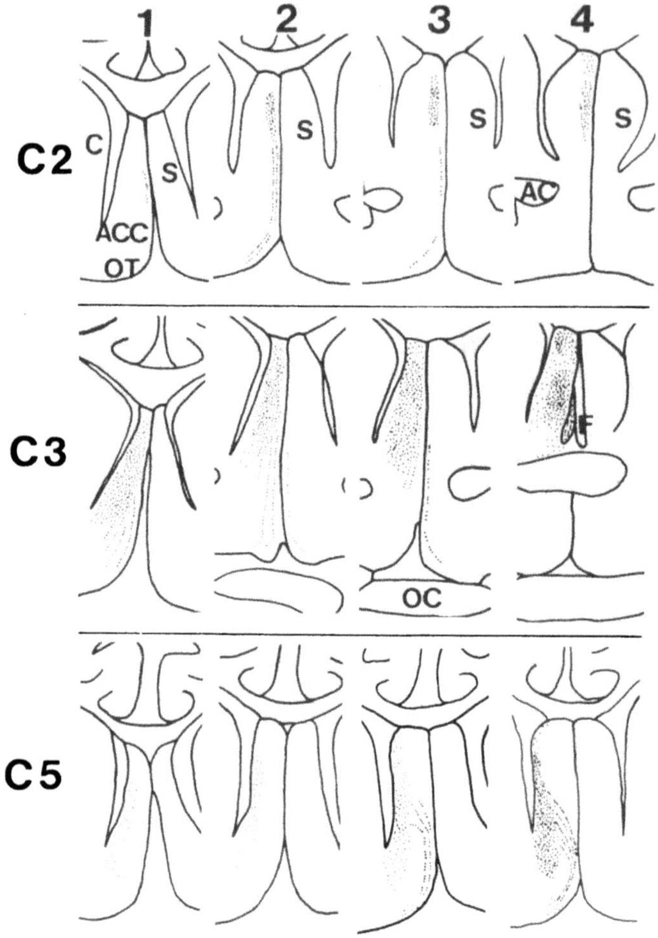

Figure 21. Course of degeneration in the septal area after lesions of different levels of the hippocampus in rat cases C2, C3, and C5.

The question of relating these anatomical findings to our knowledge of the role of the hippocampus and septal area in aggressive behavior has been a difficult one at best. One conclusion that we have been able to reach from these studies is that the hippocampus modulates hypothalamic attack behavior indirectly via its connections with the septum. Said otherwise, neurons in the lateral and medial septum represent the initial linkages in the pathway underlying modulation of the hypothalamus by the ventral and dorsal hippocampus, respectively. Further speculation regarding the functional organization of the hippocampus and septum is considered in this book (8).

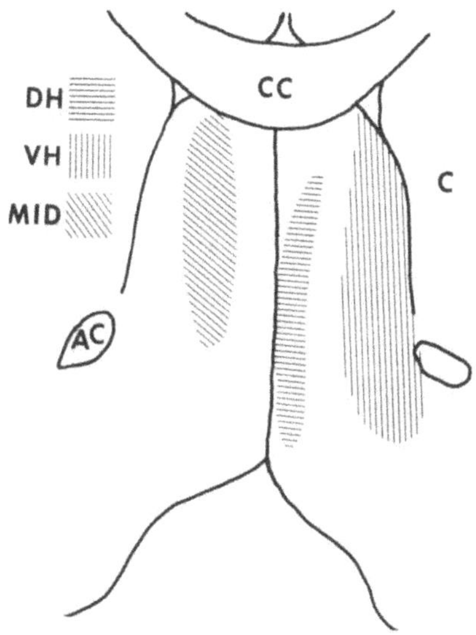

Figure 22. Schematic diagram summarizing the topographical projections from the dorsal hippocampus (DH), ventral hippocampus (VH), and posterior crus of the hippocampus (MID) to the medial, lateral, and intermedilateral portions of the septal area, respectively.

HIPPOCAMPAL PROJECTIONS TO THE SEPTUM IN THE MONKEY

The studies described above indicate the topographical organization of the hippocampal projection to the septum in the non primate mammalian nervous system. The primate hippocampus differs from other mammalian forms because there exists no dorsal hippocampus. Instead, the hippocampus lies buried deep within the temporal lobe throughout its anterior-posterior extent. Thus, Siegel, et al. (29) asked the following questions: What is the nature of the projection of the hippocampus to the septum in the primate brain? and Is this projection topographically organized? Twenty-two adult male squirrel monkey have been analyzed thus far by the Fink Heimer 1 method for identifying degenerating axons and their terminals. Diagrams of the placement of a lesion in the anterior, portion of the hippocampus and the projections from this lesion to the septum is shown in Fig. 23. A lesion in the posterior hippocampus and its projections to the septum is shown in Fig. 24.

Following a lesion of the anterior portion of the hippocampus (i.e., the area of the hippocampus lying in the anterior portion

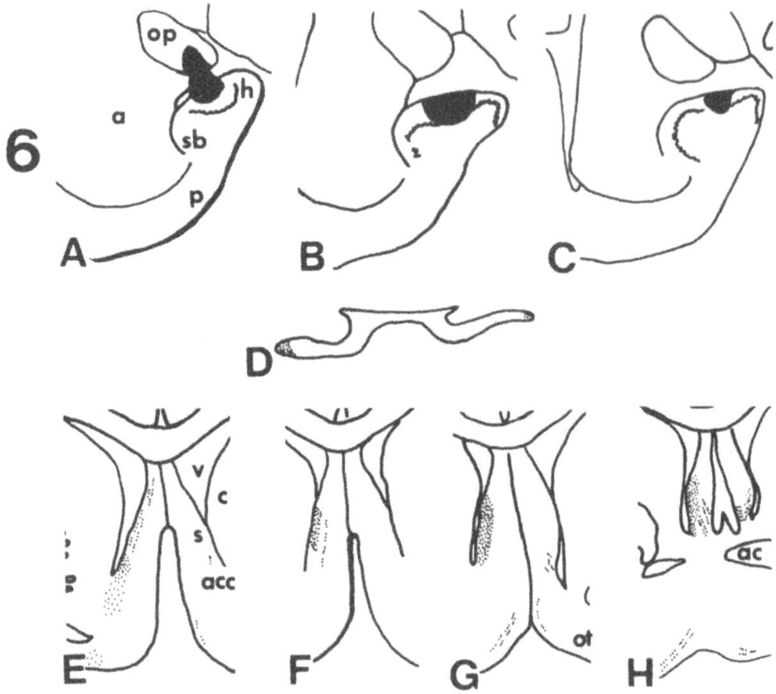

Figure 23. Outline of a lesion of the anterior hippocampus in the squirrel monkey plotted at its rostral, middle, and caudal positions and course of degeneration in the septal area following this lesion.

of the temporal lobe near the amygdala) degenerating axons are observed passing through the fimbrial into the lateral margin of the fornix. These fibers are then traced rostrally into the lateral aspect of the septum where they are observed to terminate. In contrast, following a lesion of the posterior portion of the hippocampus, (i.e., the area of the hippocampus lying in a posterior position in the temporal lobe), enter the medial aspect of the fornix and pass in a rostral direction to the medial septal area where they terminate. Following a lesion of the middle third of the hippocampus, degenerating axons pass through the posterior portion of the fimbria and enter the fornix where they are situated in an intermediolateral position. These fibers are then observed to terminate in an intermediolateral position within the septal area. Fibers from all regions of hippocampus appear to pass through the diagonal band and terminate in the region of the nucleus of the diagonal band and olfactory tubercle.

16

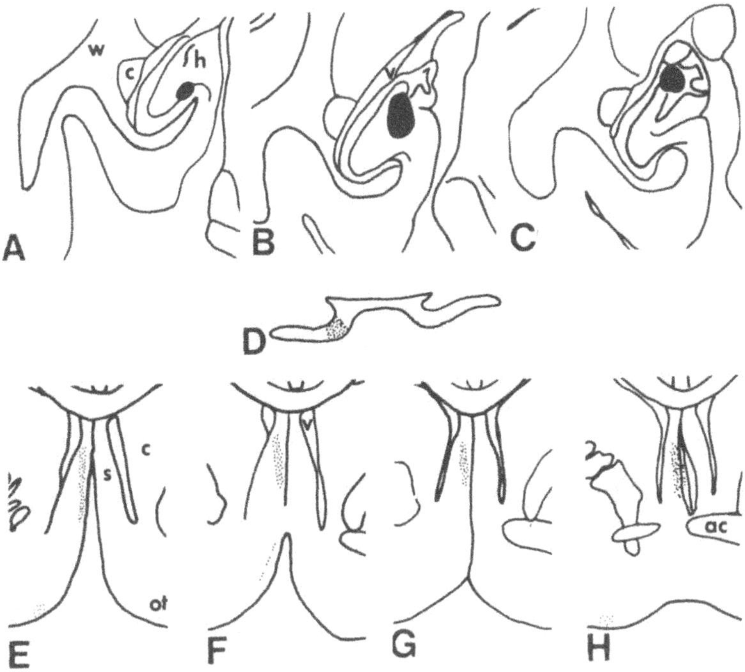

Figure 24. Outline of a lesion of the posterior hippocampus in the squirrel monkey plotted at its rostral middle, and caudal positions and course of degeneration in the septal area following this lesion.

These observations therefore lead us to conclude that the primate hippocampus is, indeed, topographically organized with regard to its projections to the septal area. It is reasonable then to suggest that, in the primate, the anterior portion of the hippocampus is homologous to the ventral hippocampus of lower mammalian forms while the posterior hippocampus of the primate is homologous to the dorsal hippocampus.

PROJECTIONS FROM THE SEPTUM IN THE CAT

It is appropriate here to indicate the results of one additional neuroanatomical study which adds to our knowledge of the organization of the hippocampal-septal axis. Siegel and Tassoni (32) attempted to determine the efferent projections of the lateral and medial portions of the septum in the cat. This study was conducted on 15 adult animals of both sexes. Lesions were placed in

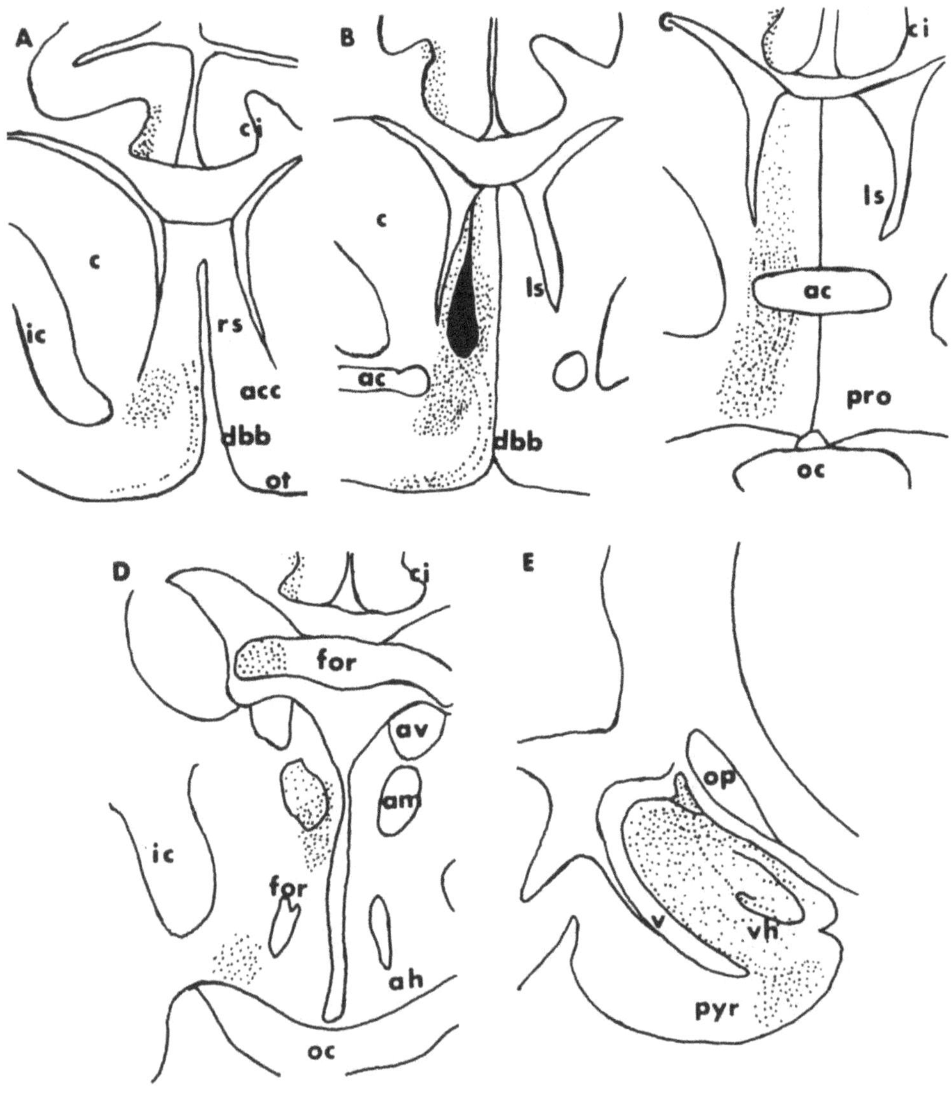

Figure 25. Course of degeneration resulting from a lesion placed in the cat's lateral septum.

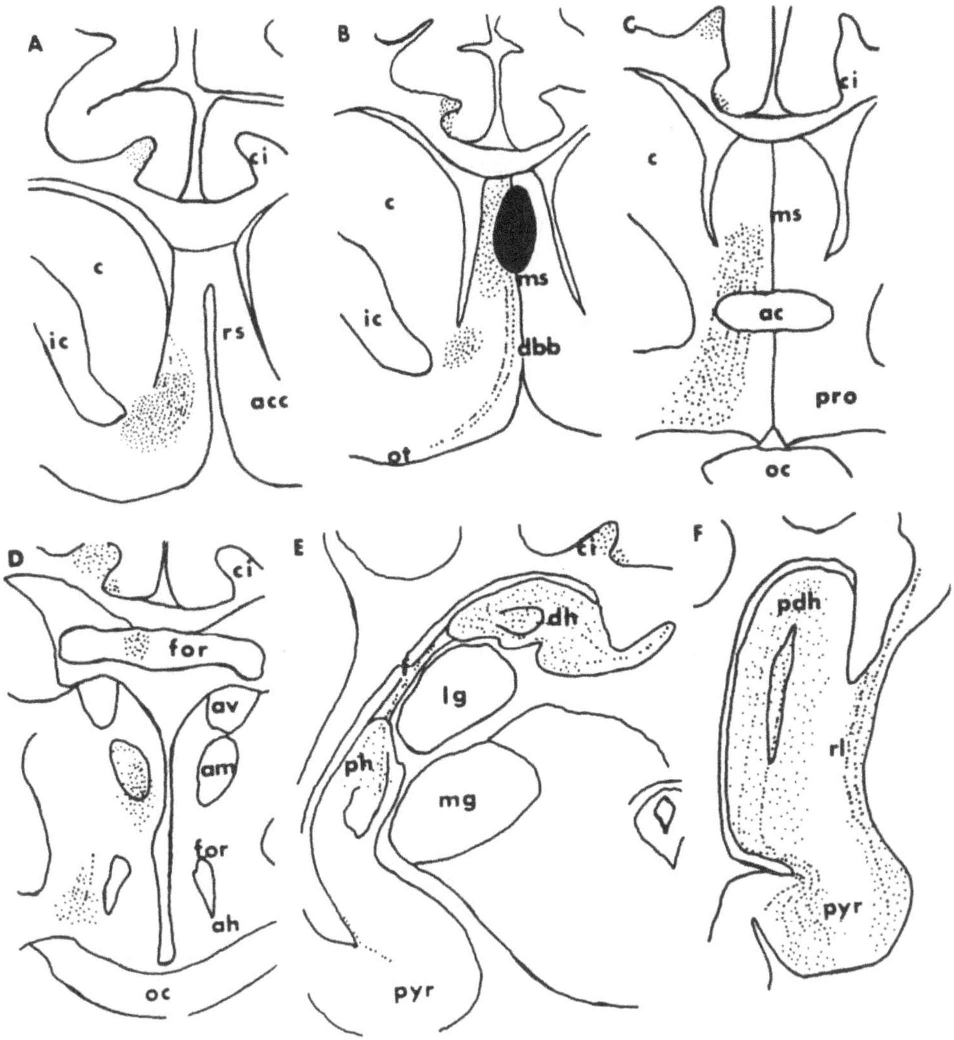

Figure 26. Course of degeneration resulting from a lesion placed in the cat's medial septum.

either the lateral or medial portions of the septum. In general, the lesions were placed within the anterior half of the septum in order to avoid damage to the fornix or its descending columns. Briefly stated, the results of this study indicated that there exists a differential projection from each of these regions to the hippocampus. Specifically, fibers from the lateral aspect of the septum project to the ventral hippocampus while fibers from the medial septum project back to the dorsal hippocampus. The projections to each of the regions of the hippocampus appears to be non specific in that fibers destined for the dorsal hippocampus appear to be distributed to all CA fields and, likewise, fibers destined for the ventral hippocampus are similarly distributed to all CA fields. This finding indicates that the dorsal hippocampus medial septum and ventral hippocampus-lateral septum comprise independent reciprocal pathways. One additional finding should also be noted. Fibers from each of the regions of the septum appear to project in massive quantities to the hypothalamus. The largest component of this projection also appears to terminate in the lateral preoptic area and adjacent portions of the anterior lateral hypothalamus. Some fibers do, however, project in a caudal direction and reach as far downstream as the posterior end of the hypothalamus. This pathway thus constitutes an anatomical mechanism by which the septal area can modulate hypothalamic processes such as aggression.

REFERENCES

1. Adey, W. R., Dunlop, C. W. and Hendrix, C. E. Hippocampal slow waves, Arch. Neurol. (Chic.) 3 (1960) 74-90.

2. Ahmed, S. S. and Harvey, J. A. Long-term effects of septal lesions and social experience on shock-elicited fighting in rats, J. Comp. Physiol. Psychol., 66 (1968) 596-602.

3. Andy, O. J., Webster, C. L., Mukawa, J. and Bonn, P. Electrophysiological comparisons of the dorsal and ventral hippocampus. In. p. Passouant (Ed.), Physiologie de L'Hippocampe, C.N.R.S., Paris, 1961, p. 411.

4. Bard, P. and Mountcastle, V. B. Some forebrain mechanisms involved in expression of rage with special reference to suppression of angry behavior, Ass. Res. Nerv. Ment. Dis., 27 (1948) 362-404.

5. Blanchard, R. J. and Blanchard, D. C. Limbic lesions and reflexive fighting, J. Comp Physiol. Psychol, 66 (1968) 603-605.

6. Brady, J. V. and Nauta, W. J. H. Subcortical mechanisms in emotional behavior: affective changes following septal forebrain lesions in the albino rat, J. Comp. Physiol. Psychol., 46 (1953) 339-346.

7. Clody, D. E. and Carlton, D. L. Behavioral effects of lesions of the medial septum of rats, J. Comp. Physiol. Psychol., 67 (1969) 344-351.

8. Edinger, H. M. and Siegel, A. Functional aspects of the hippocampal-septal axis. This volume.

9. Elul, R. Regional differences in the hippocampus of the cat, Electroenceph. Clin. Neurophysiol., 16 (1964) 470-502.

10. Green, J. D., Clemente, C. D. and DeGroot, J. Experimentally induced epilepsy with injury to Ammon's horn, Arch. Neurol. Psychiat. (Chic.), 78 (1957) 259-263.

11. Harrison, J. M. and Lyon, M. The role of the septal nuclei and components of the fornix in the behavior of the rat, J. Comp. Neurol. 108 (1957) 121-137.

12. Hernandez-Peon, R., Cevez-Ibarra, G., Morgane, P. J. and Timolaria, C. Limbic cholinergic pathways involved in sleep and emotional behavior, Exp. Neurol, 8 (1963) 93-111.

13. Hughes, K. R. Dorsal and ventral hippocampal lesions and maze learning: influence of preoperative environment, Canad. J. Psychol., 19 (1965) 325-332.

14. Jasper, H. H. and Ajmone Marsan, C. A. Stereotaxic Atlas of the Diencephalon of the cat, Nat. Res. Council Canada: Ottawa, 1954.

15. Johnson, T. N. An experimental study of the fornix and hypothalamotegmental tracts in the cat. J. Comp. Neurol., 125 (1965) 29-40.

16. Kaada, B. R., Jansen, J. Jr. and Andersen, P. Stimulation of the hippocampus and medial cortical areas in unanesthetized cats, Neurology (Minneap.), 3 (1953) 844-847.

17. King, F. A. Effects of septal and amygdaloid lesions on emotional behavior and conditioned avoidance responses in the rat, J. Nerv. Ment. Dis., 126 (1958) 57-63.

18. Maclean, P. D. and Delgado, J. M. R. Electrical and chemical stimulation of fronto-temporal portion of limbic system in the waking animal, Electroenceph. clin. Neurophysiol., 5 (1953) 91-100.

19. Nauta, W. J. H. An experimental study of the fornix system in the rat, J. Comp. Neurol., 104 (1956) 247-271.

20. Nauta, W. J. H. Hippocampal projections and related neural pathways to the midbrain in the cat. Brain, 81 (1958) 319-340.

21. Papez, J. W. A proposed mechanism of emotion, Arch. Neurol. Psychiat. (Chic.), 28 (1937) 725-743.

22. Powell, T. P. S. and Cowan, W. M. An experimental study of the efferent connections of the hippocampus, Brain, 78 (1955) 115-132.

23. Raisman, G. The connections of the septum. Brain, 89 (1966) 317-348.

24. Raisman, G., Cowan, W. M. and Powell, T. P. S. An experimental analysis of the efferent projection of the hippocampus, Brain, 89 (1966) 83-108.

25. Rothfield, L. and Harmar, P. J. On the relation of the hippocampal-fornix system to the control of rage responses in cats, J. Comp. Neurol., 101 (1954) 265-282.

26. Rubinstein, E. H. and Delgado, J. M. R. Inhibition induced by forebrain stimulation in the monkey, Amer. J. Physiol., 205 (1963) 941-948.

27. Siegel, A., Edinger, H. M. and Ohgami, S. The topographical organization of the hippocampal projection to the septal area: a comparative neuroanatomical analysis in the gerbil, rat, rabbit and cat, J. Comp. Neurol., (1974) (In Press).

28. Siegel, A., and Flynn, J. P. Differential effects of electrical stimulation and lesions of the hippocampus and adjacent regions upon attack behavior in cats, Brain Res., 7 (1968) 252-267.

29. Siegel, A., Ohgami, S. and Edinger, H. M. Projections of the hippocampus in the squirrel monkey, Society for Neuroscience: St. Louis, Mo., October, (1974) (abs.).

30. Siegel, A., and Skog, D. Effects of electrical stimulation of the septum upon attack behavior elicited from the hypothalamus in the cat. Brain Research, 23 (1970) 371-380.

31. Siegel, A. and Tassoni, J. P. Differential dfferent projections from the ventral and dorsal hippocampus of the cat, Brain Behav. Evol., 4 (1971) 185-200.

32. Diegel, A. and Tassoni, J. P. Differential efferent projections of the lateral and medial septal nuclei to the hippocampus in the cat. Brain Behav. Evol. 4 (1971) 201-219.

33. Valenstein, E. S. and Nauta, W. S. H. A comparison of the distribution of the fornix system in the rat, guinea pig, cat and monkey. J. Comp. Neurol., 113 (1959) 337-363.

34. Yutzey, D. A., Meyer, D. R. and Meyer, P. M. Effects of simultaneous septal and neo-or limbic-cortical lesions upon emotionality in the rat, Brain Research, 5 (1967) 452-458.

THE SEPTO-HIPPOCAMPAL SYSTEM: SIGNIFICANCE OF THE SUBICULUM[1]

R. B. Chronister, R. W. Sikes and L. E. White, Jr.

Department of Neurobiology, Division of Neuroscience
University of South Alabama, Mobile, Alabama 36688

INTRODUCTION

The hippocampal formation has been a favorite area of anatomic investigation for a number of years. Excellent descriptions of its morphology can easily be found in the classical anatomical literature. For example, Meynert (19,20,21) divided the area into a fascia dentata, a cornu ammonis (synonomous with ammonshorn or hippocampus proper), and a subiculum or sigmoid process (this latter area included more than what is now referred to as subiculum). As the concept of a hippocampal lobule or hippocampal gyrus became elaborated, considerable confusion remained in the definition of subiculum. Some statements can be found that appear to equate subiculum with the whole so-called hippocampal gyrus. (This confusion can still be found in a modern medical dictionary (11). However, close examination of this literature reveals that the subiculum as used by many authors, referred to the areas now called subiculum, presubiculum and parasubiculum.

Obersteiner (23) pointed out that the portion of the hippocampal gyrus adjacent to the hippocampus - i.e. the subiculum - is different than the portion of the hippocampal gyrus near the tempero-occipital cortex. A distinction was even made between a medial and a lateral subiculum (38). Brodmann (3) clarified the nature of the subiculum by differentiating the presubiculum (area

[1]Expert histological assistance was provided by B. Kimball. The technical help of L. Gardner and R. Murray is appreciated. This project was supported by NIH Grant No. NS-10809.

27) and parasubiculum (area 49) from the subiculum proper. Following in Brodmann's tradition, Filimonoff (12) termed the presubiculum and parasubiculum periallocortex in distinction to the subiculum which he included in allocortex.

Slightly different terminology arose from the students of Ramon y Cajal. Lorento de No' (17,18) in his analysis of the hippocampal formation distinguished a prosubiculum, subiculum, presubiculum and parasubiculum. Prosubiculum was introduced originally to represent the transition between hippocampus and subiculum (34). Lorente de No' (18) however, gave the prosubiculum major prominence.

Numerous criteria have been used for differentiating hippocampus from subiculum. These have included the termination of stratum radiatum (1,18) and the appearance of light-staining cells (30) in the subiculum. Parcellations are dependent upon method and consequently some of these differing criteria are due to the differing methodologies employed. Confusion, similar to the problem of subiculum and related to methodology, has also been reported with respect to the hippocampal efferent fibers.

The extrinsic fiber connections of the hippocampus have been examined in numerous studies (see historic reviews (1,6,14,22,25, 36). In spite of disagreement with regard to the termination pattern (25,32), general areas of fiber termination have been reported. These connections include fibers to the septal area, to the nucleus of the diagonal band, to the nucleus accumbens septi, to the anterior thalamus, to the mamillary bodies, and to the regions immediately adjacent to the anterior portions of the third ventricle. With the exception of the last-mentioned tract, the medial corticohypothalamic tract, all of these fibers have a hippocampus proper origin.

The medial corticohypothalamic pathway has been reported to emanate from the subicular component of the hippocampus (25). Recent morphologic studies (32) have defined no additional efferent fiber systems of a subicular origin. However, electrophysiological studies (5) have demonstrated that the subiculum is very important to normal hippocampal function. This finding is apparently incompatible with such a minor contribution to extra-hippocampal relationships. Preliminary analyses (4) of the fiber projections of the subiculum have failed to show degeneration in the area of the third ventricle. Instead, degenerating fibers were traced into the region of the nucleus accumbens septi. Because of the importance of the subiculum the present study was carried out to systematically map the fiber connections of the subiculum. In order to avoid possible ambiguities concerning the present results, the areas of the hippocampus under consideration will be defined in general detail.

In general, the hippocampal formation can be considered to be S-shaped in its normal plane (Fig. 1A). Approximately one half of the S is allocortex and is characterized usually by an emphasis of one cell type. The other half is periallocortex which is characterized by a laminated structure consisting of more than one type of neuron but lacking the six-layered scheme of isocortex. In addition, periallocortex possesses the distinctive cell free layer of lamina dissecans separating a lamina principalis externa from a lamina principalis interna.

The allocortical hippocampal formation is compromised of the dentate fascia (fascia dentata or dentate gyrus) with granular

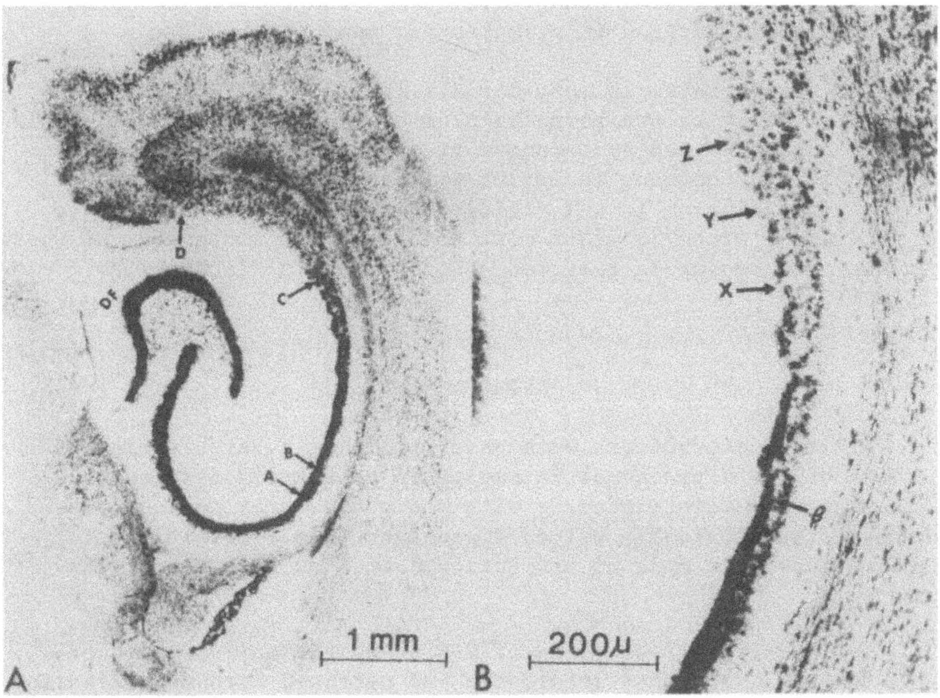

Figure 1. A. a 20μ horizontal section through the hippocampal formation of the rat. Abbreviations: A = border between CA3 and CA2; B = border between CA2 and CA1; C = border between CA1 and subiculum; D = border between subiculum and presubiculum; cresylviolet. B. High power of transition from CA1 to subiculum. Same section as 1A. Abbreviations: X = border between CA1 and prosubiculum A; Y = border between prosubiculum A and prosubiculum B (corresponds to C in 1A); Z = beginning of subiculum proper; β - cell free sublayer of CA1. Cresylviolet.

appearing cells, CA3 or regio-inferior with large pyramidal cells, CA1 or regio superior with small pyramidal cells and the subiculum. CA1 is further characterized by the appearance of sublayers within stratum pyramidale (Fig. 1B). This sublayer (β) abruptly ends at the boundary between CA1 and subiculum (Fig. 1B) and marks the beginning of the subicular type of cortex (by this definition, a portion of prosubiculum (7) is included as a part of subiculum). The subiculum continues until the sudden onset of the lamina principalis externa of the periallocortex. The development of this external lamina marks the beginning of presubiculum (Brodmann's area 27). Some dorso ventral levels of the hippocampsl formation, the granular appearing lamina principalis externa of the presubiculum overlays the cells of the subiculum (Fig. 1). This portion of presubiculum is for purposes of clarity referred to as presubiculum α to emphasize the obvious transitional character between subiculum and presubiculum proper. Presubiculum α is highly exaggerated in primates and man.

The presubiculum in most mammals (but not all) (26) is joined laterally by the parasubiculum or Brodmann's area 49. This area is characterized by a random or diffuse cellular appearance particularly pronounced in lamina principalis externa. Lateral to the parasubiculum is the rigidly laminated entorhinal region or Brodmann's area 28. (For more detail on the morphology of these regions, the reader is referred to other papers) (1,3,6,7,26,27, 28,29,31,33).

METHODOLOGY

All the observations were made utilizing a large series of lesions of the hippocampal formation in Long Evans derived male rats. Lesions were produced, with radio frequency current passed through stereotoxically placed microelectrodes. Several large lesions were also made by conventional methods with large electrodes.

Following survival times of 2-6 days, the animals were exsanguinated under ether anesthesia and perfused with normal saline followed by 4% paraformaldehyde dissolved in Millonig's buffer. The brains were removed and stored in the fixative from 1 day to several weeks. The brains were sectioned at 20μ in the coronal or horizontal planes depending on the dorso-ventral placement of the lesion. Sections were processed in several reduced silver methods including Original Nauta (35), both varieties of Fink-Heimer (13) and the cupric silver method (9). Alternate series were also processed with cresylviolet for cytoarchitectural verification of the areas under consideration.

RESULTS

As can be seen in Fig. 2, lesions were clearly placed in the subiculum. Additional lesions were placed in the presubiculum/parasubiculum region, in CA3, in the posterior periallocortex and proisocortex, in the entorhinal cortex, in the dorsal fornix, and in the fimbria. Subicular lesions were available in the dorso-posterior subiculum, in the ventro-posterior subiculum, and in the middle levels of the subiculum. These targets were chosen because of the possibility of obtaining lesions without significant demage to other hippocampal subfields (with the exception of the dentate fascia).

Following a subicular lesion, considerable degeneration could be found in the perforant path destined for CA3 and the dentate fascia. The distribution of terminal degeneration was restricted to regions known to receive fibers from entorhinal cortex (2,15, 16). Occasional degeneration fragments could be seen in the commissural zone of the dentate fascia. Degeneration was also followed into the lateral alveus and from there into the fornix-fimbrial system. Due possibly to electrode passage through the dorsal psalterium, degenerating axons were also seen in the posterior regions in the dorsal fornix.

After a mid-subicular lesion (on the dorso-ventral axis) degenerating fibers could be traced through the lateral portions of the ipsilateral fornix-fimbrial system into the region of the septal area. Very scant amounts of terminal degeneration could be seen in septal area itself with the exception of small amounts in the ipsilateral lateral septal region. There are, however, two major components of the degenerating axons. The precommissural component courses slightly lateral to the medial septal regions (in horizontal section) and arches into the area of ipsilateral nucleus accumbens septi. In coronal section, these fibers can be seen coursing lateral to the anterior hippocampus and descending near the border of the lateral ventricle. Near the inferior aspect of the ventricle, the degenerating fibers bifurcate with some turning lateral and being distributed to the nucleus accumbens septi (Fig. 3A). The others continue to descend and enter the ipsilateral nucleus of the diagonal band (Fig. 3B) and show extensive terminal degeneration.

The second major component of the degenerating axons following a subicular lesion travel in the postcommissural fornix. These fibers enter the region of the ipsilateral anterior thalamus with dense terminal degeneration (Fig. 4A) seen in nucleus anteroventralis thalami (AV) and smaller amounts in nucleus anteromedialis thalami (AM). Little or no terminal degeneration was present in anterodorsalis. A considerable number of degenerating axons

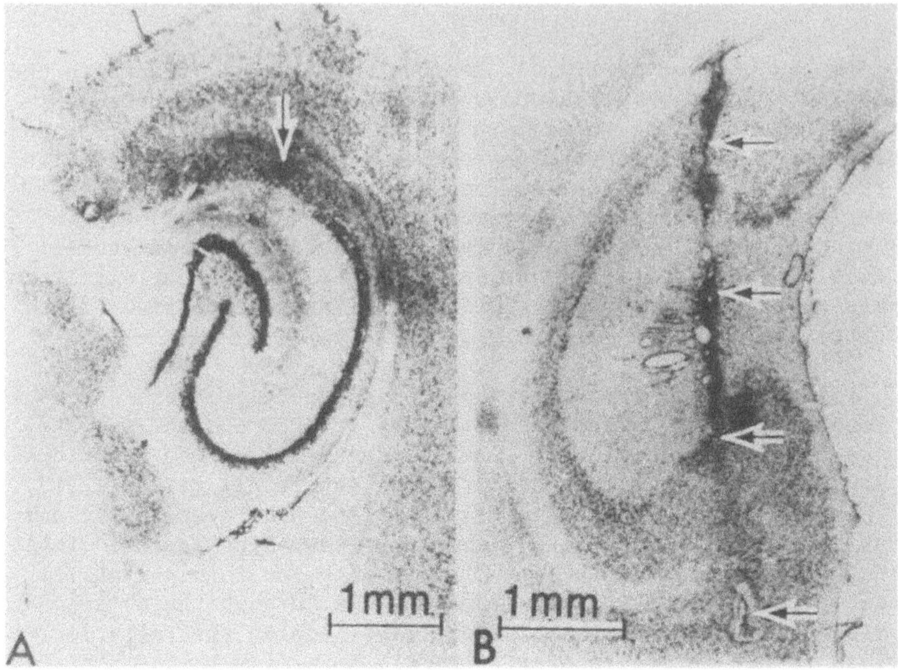

Figure 2. A. Horizontal section through rat hippocampal formation showing a typical subicular lesion (at arrow tip). Cresylviolet. B. Coronal section showing an electrode path lesion of posterior subiculum (both dorsal and ventral) and terminating in periallocortex. Cresylviolet.

continue in the ipsilateral column of the fornix. These fibers can be traced to the region of the medial mamillary nucleus (Fig. 4B) with few degenerating terminals in the lateral mamillary nucleus.

Lesions in the other regions of the subiculum produced similar degeneration patterns. In these situations (i.e. dorso posterior subiculum, ventroposterior subiculum) there was no degeneration seen in the medial corticohypothalamic pathway destined for midline hypothalamus. Based on the present data, the source of this pathway cannot be confirmed.

The path of the degenerating fibers from the subiculum lesions seemed to course exclusively in the fornix-fimbrial system and not

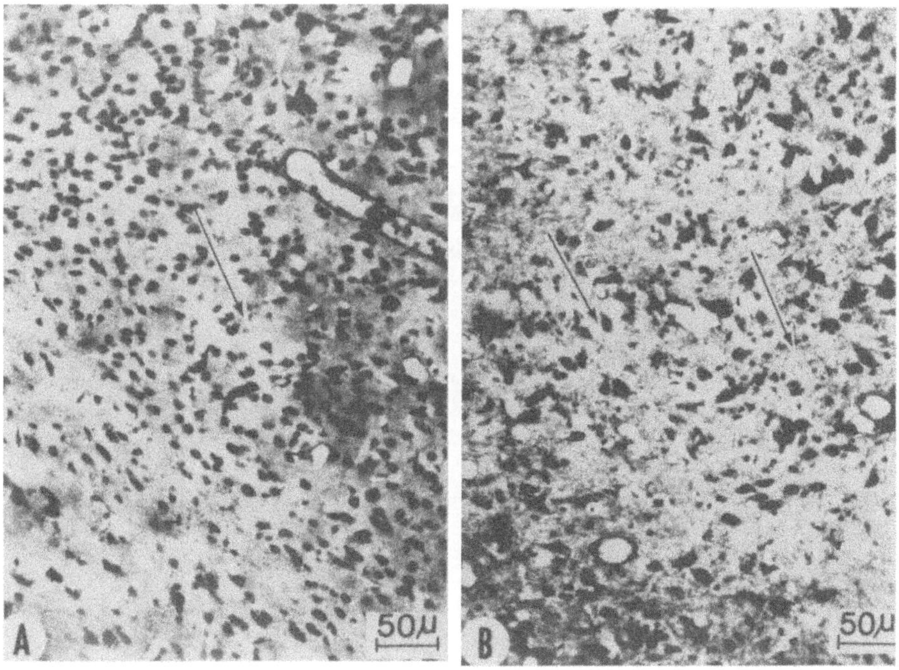

Figure 3. A. Coronal section demonstrating terminal degeneration in nucleus accumbens septi (see arrow) of rat. Modified Fink-Heimer. B. Coronal section showing terminal degeneration in diagonal band (see arrows). Modified Fink-Heimer.

in the dorsal fornix. To get some insight into the distributions of these major fiber systems, lesions were produced in the dorsal fornix, in the fimbrial complex anterior to hippocampus, and in the fimbria emanating from the middle levels of hippocampus.

A lesion of the anterior fimbrial complex near its lateral border produces extensive degeneration in both the midline and the lateral portions of the septal area. The midline degeneration appears to be mostly in the columns of the fornix. Precommissural components could be traced to nucleus accumbens septi and the nucleus of the diagonal band. Intense terminal degeneration can be seen as a cap on the lateral portion of the mid-to-dorso-lateral septal region. This can be followed for the entire antero-posterior extent of the septal complex (Fig. 5). Degeneration is also seen in the contralateral lateral septal region. Interestingly, the densest contralateral septal degeneration is nearly homotopic in position in relation to the ipsilateral position as described earlier (8).

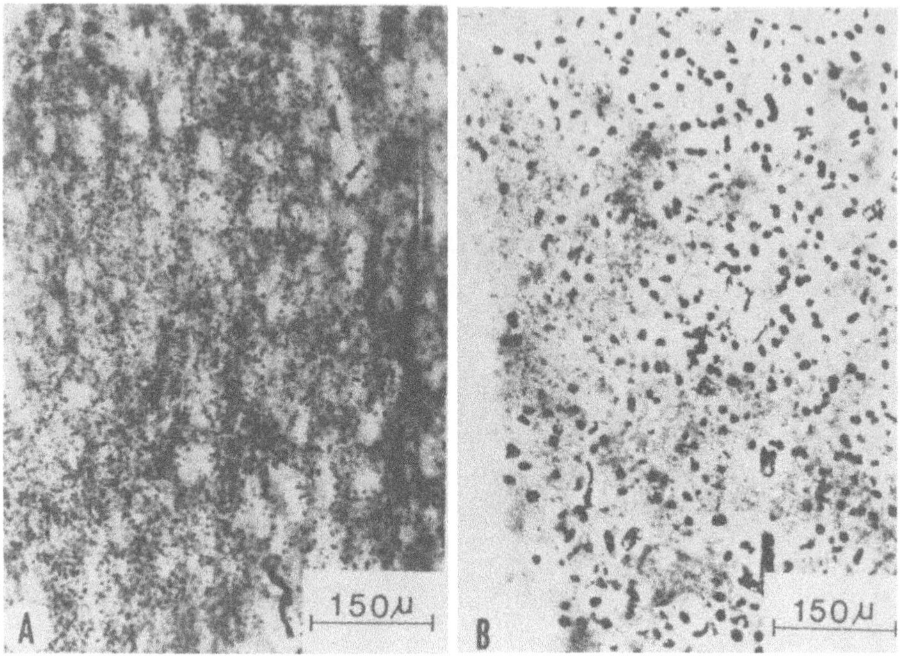

Figure 4. A. Terminal degeneration in anterior thalamus (AV) following a subicular lesion. Fink-Heimer II. B. Terminal degeneration in medial mamillary region following a subicular lesion. Fink-Heimer II.

Degenerating fibers can be traced to the anterior thalamus with terminal degeneration in AV and AM. There is intense terminal degeneration in the mamillary bodies with the primary degeneration seen in the medial portions. Small amounts of degeneration could be followed to the regions of the third ventricle. This pathway probably corresponds to the medial corticohypothalamic tract.

A lesion of the middle portion of the fimbria produces a similar pattern of degeneration. In the septal area, the intense terinal degeneration is seen slightly dorsal and more medial than that following the anterior fimbrial lesion. Degeneration could also be seen contralaterally with the densest terminal degeneration again in the near-homotopic focus.

Degenerating fibers of the precommissural fornix can be seen in nucleus accumbens septi and nucleus of the diagonal band. The accumbens degeneration could be traced to the extreme rostral por-

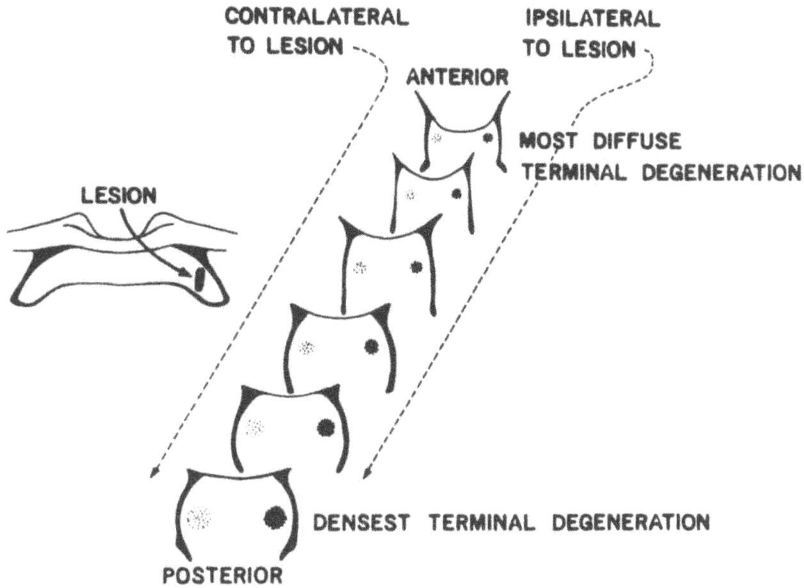

Figure 5. Schematic diagram of densest terminal degeneration in the total anterior-posterior extent of the lateral septal area. Note the horizontal columnar organization of the degeneration both ipsilateral and contralateral to the lesion.

tion of this nucleus. Here, the fibers become very fine and impossible to follow. The degeneration in the nucleus of the diagonal band is very intense and can be followed to its region lateral to the preoptic area and ventral to the anterior commissure. (This lesion also encroached upon the stria terminalis and this degeneration pattern could reflect interruption of this bundle.) The degeneration into the nucleus of the diagonal band which could be followed directly from the precommissural fornix terminated in the more dorsal portions of this nucleus. As in the anterior fimbrial lesion, little terminal degeneration could be seen in the medial septal regions. For example, the degeneration in the lateral septal areas can be followed as far rostral as the region of the anterior hippocampus. At this level, numerous fibers can be seen descending into the region of the nucleus of the diagonal band. However, consistent terminal degeneration does not appear in the medial septal nucleus. Thalamic degeneration in the post-commissural system, following the middle fimbrial lesion, is most massive in anteroventralis with some also in anteromedialis and very little in anterodorsalis. Fibers continue and terminate extensively in the medial mamillary nucleus. Degenerating fibers can also be seen coursing out of the postcommissural fornix and entering the areas adjacent to the third ventricle. Most of these fibers appeared to terminate in the region of the arcuate nucleus but some fibers could be seen more laterally but without distinct termination patterns.

Interrupting the dorsal fornix results in extensive degeneration in the columns of the fornix. Little degeneration is present in the septal area itself but some fibers could be seen coursing in the precommissural system to the region of the nucleus accumbens septi and perhaps the nucleus of the diagonal band. This precommissural component, however, is very small. Postcommissural degeneration is triking. The ipsilateral column of the fornix shows heavy amounts of degenerating fibers. Most of this degeneration is distributed to the medial nucleus of the mamillary body. Smaller amounts of terminal degeneration are present in the anterior thalamus (AV and AM).

On the bases of these various lesions of the major fiber systems of the hippocampus, it can be concluded that the fiber systems degenerating following a subiculum lesion (at least the levels studied in the present experiment) course in the fornix-fimbrial system and not the dorsal fornix. (The source of the dorsal fornix must be in the dorsal hippocampus and according to Swanson's chapter in this book arises from dorsal subiculum.)

One of the problems with lesion studies is in the interruption of fibers of passage. To circumvent this problem, lesions of the periallocortex hippocampal formation were examined for their fiber relations. Larse lesions of the entorhinal region, with small encroachment into the depths of presubiculum and parasubiculum result in very small, and almost negligible, degeneration in anterior thalamus and an even smaller amount in the mamillary bodies. On the other hand- lesions of the presubiculum and parasubiculum (Fig. 6A) result in extensive degeneration in the anterior thalamic regions (Fig. 6B). (One such case showed degeneration in the anterodorsal nucleus as well as in the anteroventral and anteromedial nuclei. Histologic analysis of the lesion site showed extensive brainstem damage in this lesion. Since this was the only example of degeneration in the anterodorsal thalamic nucleus, it seems reasonable to conclude that degeneration in AD is due to fibers of irigin in the brainstem.) The terminal degeneration is dense in AV and AM. However, the degeneration is the columns of the fornix is not very heavy and only a small amount could be seen entering the region of the mamillary bodies (Fig. 6C). Since a portion of the periallocortex overlies a portion of the subicular layers (presubiculum α), this small amount of mamillary degeneration could be a result of lesions of this area. One

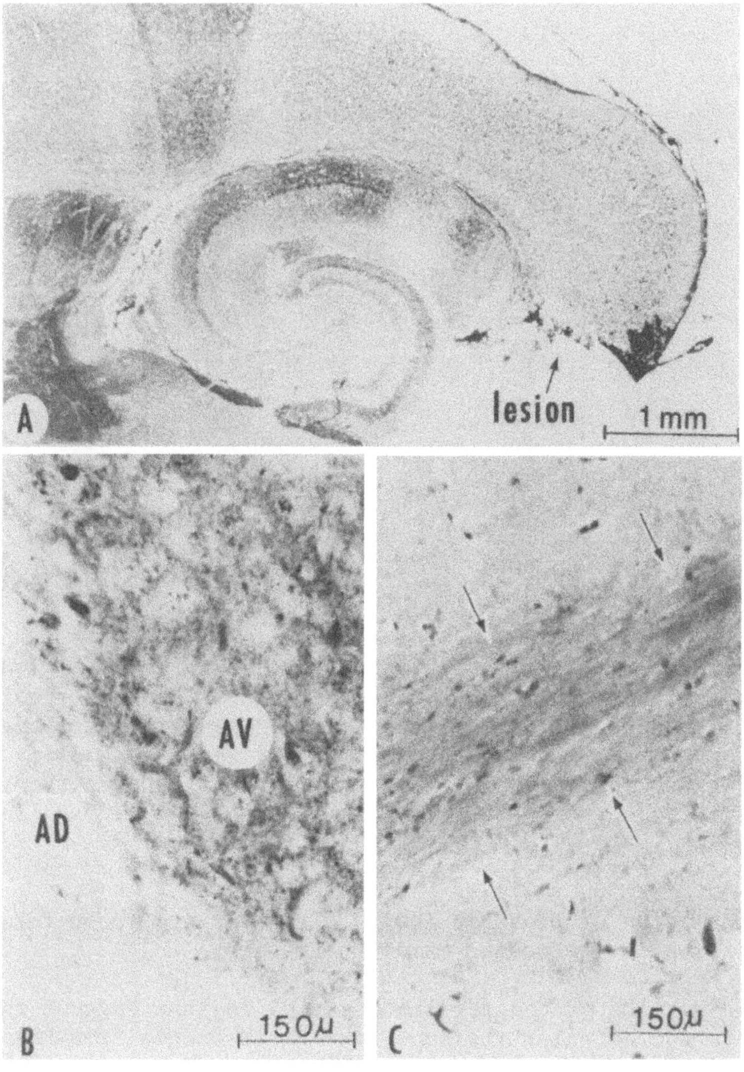

Figure 6. A. A horizontal section showing complete destruction of the pre- and parasubicular areas. Fink-Heimer I. B. Terminal degeneration in anterior thalamus following the lesion in 6A. Abbreviations - AV = anteroventral nucleus; AD = anterodorsal nucleus. Fink-Heimer I. C. The near absence of degenerating fibers in the columns of fornix rostral to the mamillary bodies. The fornix is outlined by arrows. Fink-Heimer I.

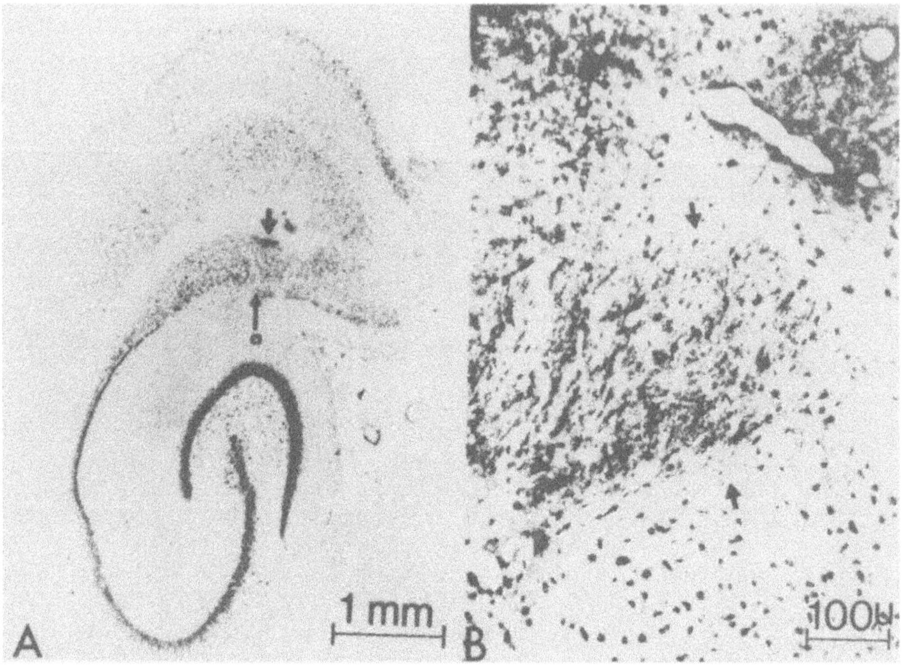

Figure 7. A. An isolated lesion of presubjculum α (at tip of large arrow), a = border between presubiculum α and subiculum. Horizontal section. Creysylviolet. B. Degenerating fibers in the column of the fornix and destined for medial mamillary nucleus following the lesion in 7A. Fornix column is between the arrows. Fink-Heimer.

isolated lesion of presubiculum α (Fig. 7) resulted in terminal degeneration in the medial mamillary region.

The results of the periallocortical lesions suggest that the fibers to the anterior thalamus take their origin from the periallocortical presubiculum and parasubiculum whereas the fibers to the mamillary bodies arise primarily in the subiculum. Lesions of the dorsal periallocortex of the cingulate region do not show resultant degeneration in the dorsal fornix or fornix-fimbrial system.

DISCUSSION

The data of the present experiment necessitate a change in the conventional view of the fiber connections of the hippocampus.

Most reference sources suggest that all areas of the hippocampus contribute fibers to the post commissural fornix destined for the anterior thalamus and mamillary bodies. This does not appear to be the case. The present results can be summarized as follows and are diagrammed additionally in Fig. 8.

(1) Lesions of the entorhinal area result in no terminal degeneration in septum, anterior thalamus, or mamillary bodies.

(2) Lesions of presubiculum and parasubiculum result in copious terminal degeneration in anterior thalamus (AV and AM) and small amounts in mamillary body.

(3) Lesions of subiculum result in considerable amounts of degeneration (of presubicular-parasubicular origin?) in anterior thalamus (AV and AM) and medial mamillary lesions.

These results suggest a possible separation of the fornix system into discrete compartments. Fibers from the periallocortical presubiculum and parasubiculum course into the alveus and through the subiculum on their way to the anterior thalamus. Subicular axons enter the alveus and course to the mamillary bodies. In the hippocampus, both of these groups of fibers are joined by fibers from the CA3-CA2-CA1 complexes. These fibers from the hippocampus proper travel in the fornix system to the

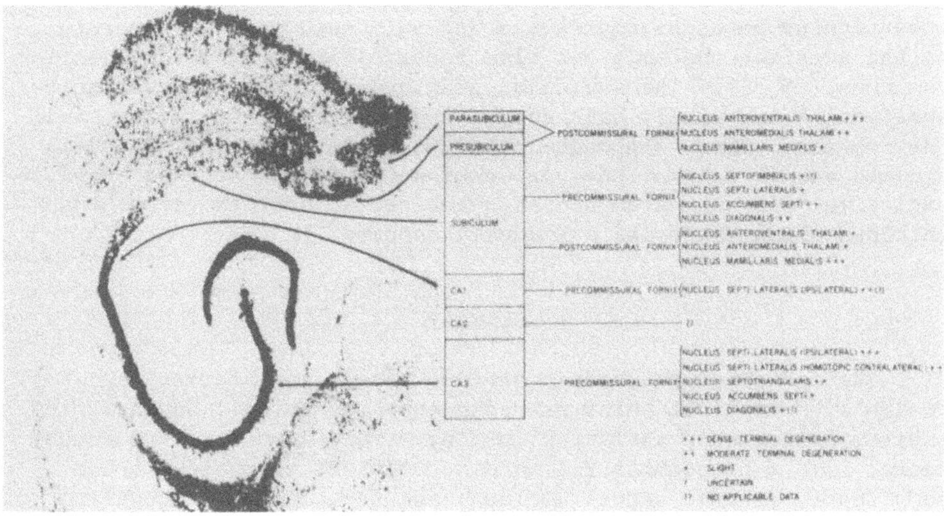

Figure 8. A summary of the various projections of the hippocampal subfields and their sites of termination.

septal area. The CA3 fibers terminate bilaterally in homotopic regions of the lateral septal area. A lesion of the fimbria, thus interrupting all of the fiber systems from the hippocampal formation, appears to result in a major increase in ipsilateral terminal degeneration. This suggests that the fibers of CA1 origin are distributed ipsilaterally only (as Swanson, see chapter in this book, also found).

In no instance was there clear-cut terminal degeneration in medial septum. However, the medial septum is known to receive a considerable amount of hippocampal input. The organization of the neurons and dendritic processes of the medial septum is extremely complex (unpublished observations) and clearly invades the descending fornix fibers on their way to the diagonal band region.

No consistent degeneration was observed in the medial corticohypothalamic pathway in the present study. This system, in normal fibers stains, can be seen in sagittal section to arch posterior to the anterior commissure, descend, and enter the periventricular region near the nucleus suprachiasmaticus. Whether this area represents the terminus of the medial corticohypothalamic pathway cannot be demonstrated in normal tissue. In the present study, no lesions were placed in the special subicular origin of this pathway.

In conclusion, the hippocampus proper has rather exclusively massive fiber connection with the septal region. However, as suggested previously (24), there is a major contribution to the fornix system from periallocortical structures. Furthermore, the presubicular-parasubicular areas not only send fibers to terminate in the anterior thalamus but also receive fibers from the anterior thalamus (10,37). The subicular origin axons, which appear to be substantial in number, have very strong terminations in the mamillary region. Thus, although the precommissural fornix system is a mixed one in origin, the postcommissural fornix appears to arise exclusively in subiculum and adjacent periallocortex rather than in hippocampus proper as previously assumed.

SUMMARY

Lesions were produced in various hippocampal formation subfields with emphasis being placed on the subiculum. Degenerating fibers, following a subicular lesion, were traced into two components: 1) A precommissural fornix component to nucleus accumbens septi, lateral septal nucleus, and the nucleus diagonalis; 2) A massive postcommissural fornix component to the anterior thalamus and the medial mamillary bodies. Further investigation suggested that the anterior thalamic component arises in the presubiculum-parasubiculum region while the mamillary component was of subicular origin. The remaining subfields of the hippocampus appear to project solely into the region of the septal complex. These data suggest a dissociation of efferent relationships within the hippocampal formation.

REFERENCES

1. Blackstad, T. W. Commissural connections of the hippocampal region in the rat, with special reference to their mode of termination. J. Comp. Neurol., 105 (1956) 417-537.

2. Blackstad, T. W. On the termination of some afferents to the hippocampus and fascia dentata. Acta Anat., 35 (1958) 202-214.

3. Brodmann, K. Vergleichend Lokalisationslehre der Grosshirnrinde. In Ihren Prinzipien Dargestellt auf Grund des Zellenbaues. Leipzig, Verlag Von Johann Ambrosius Barth, (1925) 1-324.

4. Chronister, R. B. and Zornetzer, S. F. Fiber connections of allocortex: The subiculum hippocampi. Anat. Rec., 175 (1973) 293.

5. Chronister, R. B., Zornetzer, S. F., Berstein, J. J., and White, L. E., Jr. Hippocampal theta rhythm: Intra-hippocampal formation contributions. Brain Research, 65 (-974) 13-28.

6. Chronister, R. B. and White, L. E., Jr. Fiberarchitecture of the hippocampal formation: Anatomy, projections and structural significance. In R. Isaacson and K. Pribram (Eds) The hippocampus: A comprehensive treatise. New York: Plenum, In press 1975a.

7. Chronister, R. B. and White, L. E., Jr. Comparative analyses of allocortex. J. Comp. Neurol., In press 1975b.

8. DeFrance, J. F., Shimono, T., and Kitai, S. T. Anatomical distribution of the hippocampal fibers afferent to the lateral septal nucleus. Brain Research, 34 (1971) 176-180.

9. DeOlmos, J. S. and Ingram, W. R. An improved cupric-silver method for impregnation of axonal and terminal degeneration. Brain Research, 33 (1971) 523-529.

10. Domesick, V. B. The fasciculus cinguli in the rat. Brain Research, 20 (1970) 19-32.

11. Dorland's illustrated medical dictionary, 24th Ed. Philadelphia: Saunders, 1965.

12. Filimonoff, I. N. A rational subdivision of the cerebral cortex. Arch. Neurol. Psychiat. 58 (1947) 296-311.

13. Fink, R. P. and Heimer, L. Two methods for selective silver impregnation of degenerating axons and their synaptic endings in the central nervous system. Brain Research, 4 (1967) 369-374.

14. Guillery, R. W. Degeneration in the postcommissural fornix and the mamillary peduncle of the rat. J. Anat., 90 (1956) 350-371.

15. Hjorth-Simonsen, A. Projection of the lateral part of the entorhinal area to the hippocampus and fascia dentata. J. Comp. Neurol., 146 (1972) 219-232.

16. Hjorth-Simonsen, A. and Jeune, B. Origin and termination of the hippocampal perforant path in the rat studied by silver impregnation. J. Comp. Neurol., 144 (1972) 225-232.

17. Lorente de No', R. Studies on the structure of the cerebral cortex. J. Psychol. Neurol., 46 (1933) 381-438.

18. Lorente de No', R. Studies on the structure of the cerebral cortex. II. Continuation of the study of the ammonic system. J. Psychol. Neurol., 46 (1934) 113-177.

19. Meynert, Th. Der Bau der Grosshirnrinde und seine örtlichen Verschiedenheiten nebst einem pathologischanatomischen Corollarium. Vierteljahr. Psychiat., 1 (1867) 198-217.

20. Meynert, Th. Der Bau der Grosshirnrinde und seine örtlichen Verschiedenheiten, nebst einem pathologischanatomischen Corollarium. Vierteljahr. Psychiat., 2 (1868) 88-113.

21. Meynert, Th. The brains of mammals. In S. Stricker (ed) Manual of human and comparative histology. London: New Sydenham Society, 1872, 367-537.

22. Nauta, W. J. H. An experimental study of the fornix system in the rat. J. Comp. Neurol., 104 (1956) 247-271.

23. Obersteiner, H. The anatomy of the central nervous organs. London: Charles Griffin, 1890.

24. Pilleri, G. Uber die Faserdegeneration der Fimbria hippocampi nach Ausfall der Ammonshornrinde und der benachbarten Gebiete. Arch. Psychiat. Zeit, ges. Neurol., 198 (1959) 287-302.

25. Raisman, G., Cowan, W. M. and Powell, T. P. S. An experimental analysis of the efferent projection of the hippocampus. Brain, 89 (1966) 83-108.

26. Rose, M. Histologische Lokalisation der Grosshirnrinde bei kleinen Säugetieren (Rodentia, Insectivora, Chiroptera). J. Psychol. Neurol., 19 (1912) 119-207.

27. Rose, M. Der Allocortex bei Tier und Mensch. I. Teil. J. Psychol. Neurol., 34 (1926) 1-111.

28. Rose, M. Die sog. Riechrinde beim Menschen und beim Affen. II. Teil des "Allocortex bei Tier und Mensch". J. Psychol. Neurol., 34 (1927) 261-401.

29. Rose, M. Cytoarchitektonischer Atlas der Grosshirnrinde der Maus. J. Psychol. Neurol., 40 (1929) 1-51.

30. Rose, M. Cytoarchitektonic und Myeloarchitektonic der Grosshirnrinde. In Bumke, O. and O. Foerster. Handbuch der Neurologie, Bd.1, Berlin: Springer, 1935, 588-778.

31. Sgonina, K. Zur vergleichenden Anatomie der Entorhinal und Prasubikularregion. J. Psychol. Neurol., 48 (1937) 56-163.

32. Siegel, A. and Tassoni, J. P. Differential efferent projections from the ventral and dorsal hippocampus of the cat. Brain Behav. Evol., 4 (1971) 185-200.

33. Smith, R. W. and White, L. E., Jr. The fiberarchitectonics of the cat hippocampsl formation. J. Comp. Neurol., 123 (1964) 11-28.

34. Vogt, C. and Vogt, O. Allgemeinere Ergebnisse unserer Hirnforschung. J. Psychol. Neurol., 25 (1919) 279-462.

35. White, L. E., Jr. Enhanced reliability in silver impregnation of terminal axonal degeneration - Original Nauta method. Stain Tech., 35 (1960) 5-9.

36. White, L. E., Jr. A morphologic concept of the limbic lobe. Int. Rev. Neurobiol., 8 (1965) 1-34.

37. White, L. E., Jr., Nelson, W. M. and Foltz, E. L. Cingulum fasciculus study by evoked potentials. Exp. Neurol., 2 (1960) 406-421.

38. Zuckerkandl, E. Über Das Riechcentrum. Eine Vergleichend-Anatomische Studie. Stuttgart: Ferdinand Enke, 1887, 1-123.

ARTERIAL AND VENOUS VESSEL PATTERNS IN RAT SEPTAL STRUCTURES[1]

Peter Coyle

Department of Anatomy, The University of Michigan

Ann Arbor, Michigan

INTRODUCTION

Previous studies of the rat cerebral vasculature were directed in large part towards investigation of the circle of Willis and regional distributions of the major cerebral vessels. Green (8) identified the three major cerebral arteries and illustrated the superficial veins. The developing arteries were investigated by Moffat (10,11,12) and patterns of variation in the circle of Willis were studied by Brown (1). Accounts of some smaller arterial branches including the choroidal arteries were given in Craigie's Neuroanatomy of the Rat (17) and by Moffat (12). Capillary densities for different strata in the hippocampus and fascia dentata were reported by Craigie (3,4,5). The rat subfornical artery, a branch of the anterior cerebral artery, was described by Spoerri (14). Accounts of vessel patterns in rat septal structures were not found in the English literature. No information was located on distributions of septal venous vessels for this commonly utilized laboratory animal. The current study was undertaken to identify and chart the course of vessels supplying and draining septal structures. Hopefully, the data can be applied to improve experimental approaches in investigating septal structures and provide

[1]The author wishes to express his appreciation for encouragement extended by J. B. Ranck and J. F. DeFrance to pursue this project. Partial project support came from NIH Grant NS-10970 and NSF GB-26184, J. B. Ranck investigator. Gratitude is also expressed to W. L. Brudon for his assistance in preparing the figures and to E. W. Lauer and R. T. Woodburne for proofreading the manuscript.

background material for interpretation of findings following procedures which could lead to vascular insults.

MATERIALS AND METHODS

Fifty-five adult white rats unselected with regard to sex were deeply anesthetized with ether. Following a ventral approach to open the thoracic cavity, heparin was injected into the left ventricle to reduce blood coagulation. The right atrium was snipped off in forty-five animals perfused via the arterial system. These animals were divided into five groups of nine rats. Each group was perfused transcardially through the left ventricle with either Ward's vinyl acetate, Ward's latex, Ward's bioplastic combined with India ink, India ink or General Electric Silicone RTV Rubber. The last named substance was combined with RTV Diluent and in some cases India ink, or mercuric oxide to facilitate visualization of the vessels. The materials were placed in a syringe and injected under hand pressure. No attempt was made to regulate the perfusion pressure. The silicone rubber and India ink specimens were the best for study.

In the remaining ten animals, saline was perfused through the arterial system or in a retrograde direction through the venous system. Following saline treatment, the ascending aorta was transected, a hypodermic needle was inserted into the right atrium via the right ventricle and either Ward's bioplastic or the silicone rubber was injected in a retrograde direction through the venous system.

Brains perfused via the arterial system were removed after at least one hour following the perfusion and stored in 10% formalin. For animals perfused via the venous system, decapitation was performed and the whole head was immersed in the fixation fluid for a period of one to three days. All specimens were studied for general distribution of surface vessels before clearing procedures or sectioning in transverse, horizontal or sagittal planes or dissections were undertaken. Sections between 1/2 and 2 mm in thickness were cut with a hand held razor blade. Some whole brains and half brains cut in the sagittal plane were macerated in 2-4% potassium hydroxide for 3 to 4 days, dehydrated in alcohol and cleared in glycerin. Other specimens, following either sectioning or dissections, were cleared in glycerin without maceration.

Unmacerated materials retain key landmarks necessary for accurate topographic descriptions. India ink perfused specimens are best left unmacerated. The ink does not provide a rigid framework and consequently macerated tissues collapse and the ink diffuses. The silicone rubber is flexible and this property leads to distortion of the smaller vessels in macerated preparations. The bioplastic and acetate specimens were good for studying casts; distortion is minimal but landmarks are lost following total maceration. The observations were made under dissecting and light microscopes. Cleared thick sections were transilluminated whereas dissected materials were lighted similarly or directly. A Model ED-10 Polaroid Camera was utilized to document the findings.

RESULTS

Specimens perfused via the arterial system with vinyl acetate, latex, or the silicone rubber served for investigation of the arterial system. These materials do not pass through the capillaries as does India ink. Specimens perfused with the bioplastic or rubber injected into the venous system and the ink preparations were utilized for studying venous patterns.

Arterial Supply to the Septum

In rats, the circle of Willis receives blood from the paired internal carotid artery and the basilar arter (Fig. 1). The anterior cerebral artery courses superior to the optic nerve. Upon reaching the longitudinal fissure the vessel of each hemisphere usually joins its counterpart to form an azygos portion of the anterior cerebral artery. The vessel passes forward as the inferior limb of an arc to reach the rostral septal region (Fig. 2). The rostral extent of the azygos portion is usually at the departure point of a large paired olfactory artery. The olfactory vessel supplies branches to the frontal cortex and medial, dorsal, and dorsal-lateral regions of the olfactory bulb. Distal to this artery, the anterior cerebral artery changes direction to form the superior limb of the arc and follows a course over the corpus callosum to supply the dorsal-medial cortex before extending onto the dorsal surface of the hemisphere. Distal branches of the anterior cerebral artery form extensive anastomoses with the middle cerebral artery (Fig. 3) and to a much less extent anastomose with branches of the posterior cerebral artery.

The larger and longer arterial branches to the septum arise mostly from the inferior limb of the anterior cerebral artery (Fig. 2). In following the course of the azygos vessel, septal arteries to the two hemispheres arise more or less in an alternating pattern; a branch to one hemisphere, the next branch to the other and so forth. The long septal vessels course from a rostral inferior to caudal superior position within the septum (Fig. 2). The vessels tend to parallel each other and are located in the medullary substance but do not course at right angles to

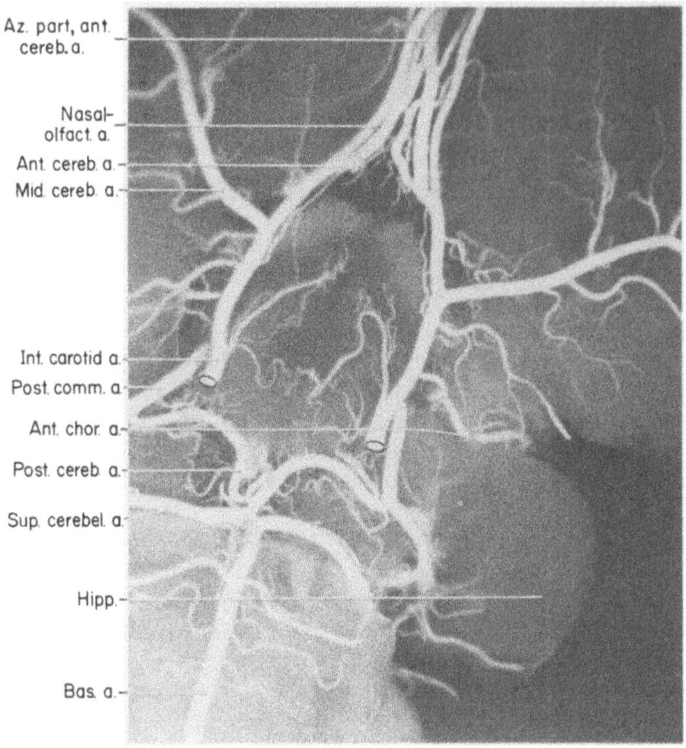

Figure 1. Basal view of brain to demonstrate major arteries of the circle of Willis. Cortex removed to expose temporal end of hippocampus and initial course of anterior chroidal artery. Silicone rubber perfusion.

the surface as observed for neocortical vessesl (Fig. 4). Medial precommissural septal structures in the course of the long vessels include the small and large cell divisions of the medial septal nucleus and more caudally the septal portion of the nucleus and tract of the diagonal band of Broca and the precommissural fornix. Caudal to the anterior commissure, the nucleus triangularis septi, fibers of the descending column of the fornix, and the nucleus septofimbrialis receive blood from branches coursing in the septal gray substance superior to the anterior commissure. Branches of vessels in the medial septal region extend to supply lateral septal structures (Fig. 4), including portions of the lateral septal nucleus, medial regions of the nucleus accumbens septi and bed nucleus of the stria terminalis. The long vessels branch most profusely rostral to the descending column of the fornix into which smaller branches penetrate and could be followed into the ventral hippocampal commissure.

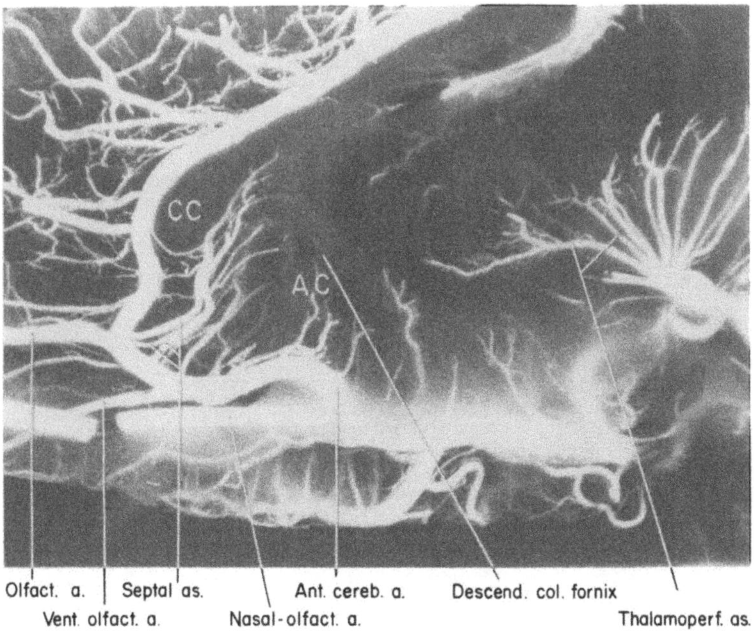

Figure 2. Parasagittal view of brain. Note course of anterior cerebral artery and its septal branches. Vessels injected with silicone rubber and tissue cleared in glycerin.

A smaller number of vessels arise from the paired portion of the anterior cerebral artery to supply the most inferior septal regions. No branches of medial striate arteries were traced to septal structures other than the bed nucleus of the stria terminalis. Short, smaller diameter vessels arise from the superior limb of the anterior cerebral artery in the area rostral to the body of the corpus callosum. Occasionally, vessels arise from the supracallosal portion of the anterior cerebral artery and course around the rostral extent of the corpus callosum (Fig. 2). These smaller vessels supply the superior portion of the anterior continuation of the hippocampus, structures associated with the precallosal induseum griseum and the rostral extent of the lateral septal nucleus.

The number and sites of origin of vessels supplying the caudal-superior septal region varied (Fig. 5). In some specimens, vessels to this region were observed arising from the superior limb of the cerebral artery as well as from the inferior limb. Vessels supplying the more caudal septal structures entered the medullary substance at levels caudal to those supplying rostral regions.

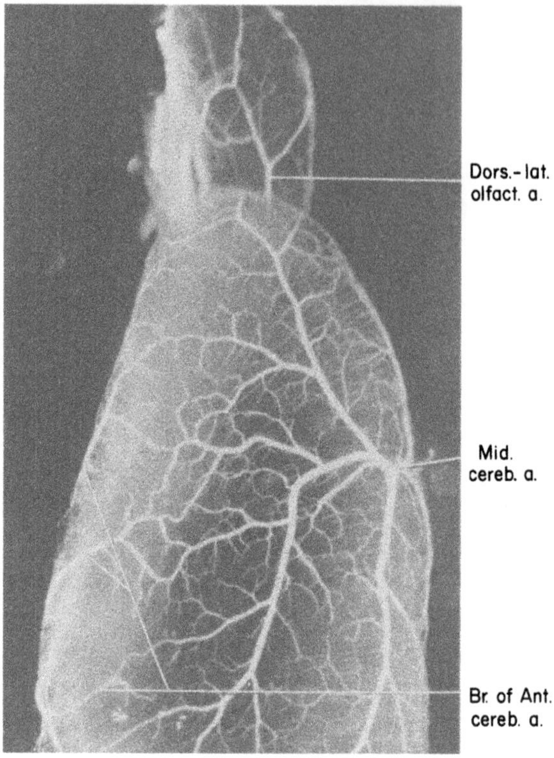

Figure 3. Dorsal view of cerebral hemisphere. Note anastomoses of anterior and middle cerebral arterial branches. Arteries injected with silicone rubber, tissue partially macerated in potassium hydroxide and cleared in glycerin.

Nearly all vessels were intramedullary rostral to the anterior commissure. Branch patterns in the rostral septum did not appear to follow any consistent configuration and capillary density appeared to be greater in medial than in lateral septal structures although a quantitative analysis was not undertaken. Occasionally before entering the descending column of the fornix, the stem vessels divide into ascending and descending branches which follow a course parallel to the fornix with penetrating branches entering at an angle normal to the long axis of the fiber bundle (Fig. 5).

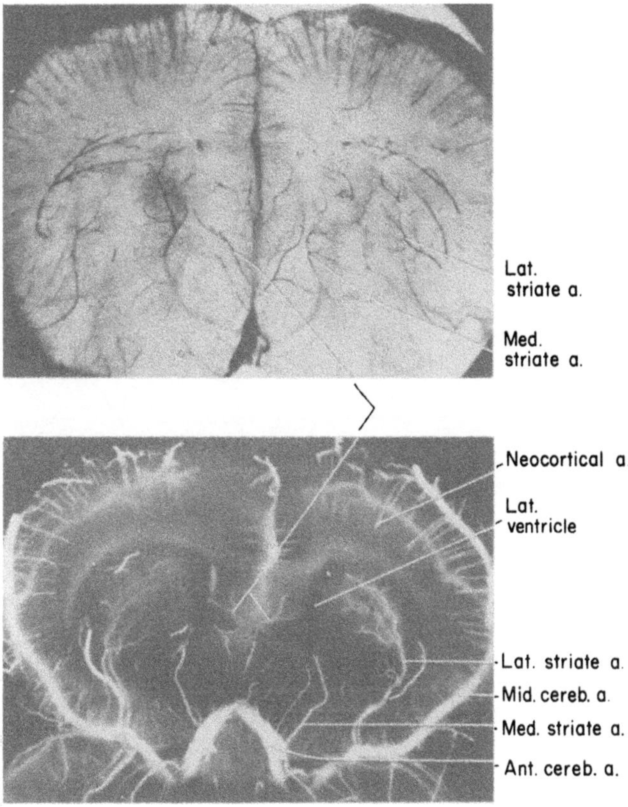

Figure 4. Thick frontal sections at precallosal level (upper) and at level of anterior commissure (lower). Note medial to lateral course of septal arteries. Upper specimen perfused with bioplastic-ink mixture. Lower specimen injected with silicone rubber. Both specimens partially macerated in potassium hydroxide and cleared in glycerin.

Venous Drainage of the Septal Region

The venous system is more complexly arranged than the arterial supply. The medial septum was observed to have three venous fields; a rostral, a middle, and a caudal field. Each field drains in a different direction.

The anterior vein of the corpus callosum courses caudal to the azygos portion of the anterior cerebral artery and takes a straighter course than the artery from the base of the brain to the corpus callosum (Fig. 6). The vein forms the posterior

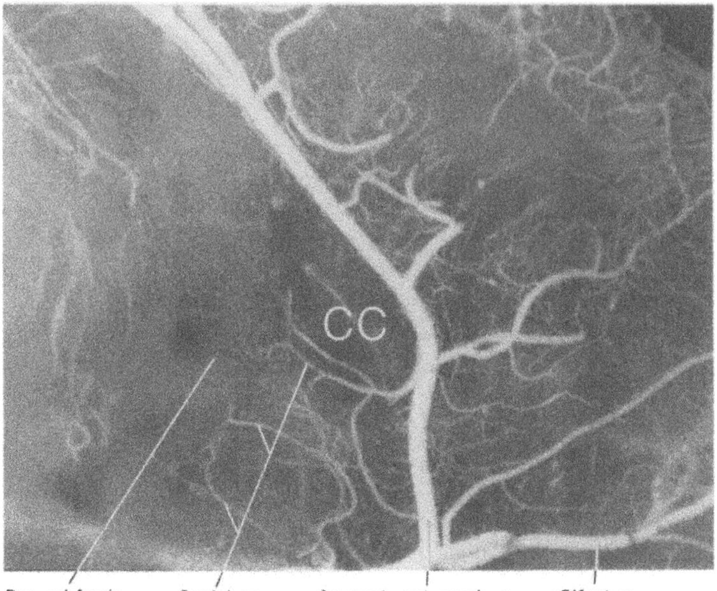

Figure 5. Parasagittal section. Note branching of septal arteries rostral to the fornix. Specimen perfused with silicone rubber, macerated in potassium hydroxide, cleared in glycerin.

boundary of the rostral venous field and receives tributaries drining rostral septal structures located between the vein and the anterior cerebral artery. The middle field drains into the same vein but the flow is in a rostral direction (Fig. 6). This field extends from the anterior vein of the corpus callosum to approximately the level of the anterior commissure. Drainage from the rostral and middle septal fields is from a lateral to medial direction. The anterior vein of the corpus callosum is joined by several other smaller veins draining basal structures including the medial aspect of the tuberculum olfactorium, horizontal limb of the diagonal band of Broca and the lamina terminalis.

Drainage from the caudal field including septal structures superior as well as caudal to the anterior commissure and from the descending column of the fornix, ventral hippocampal commissure, and portions of the septal end of the hippocampus is to the internal cerebral vein (Fig. 7). Venous tributaries are particularly evident in the septum immediately superior to the anterior commissure (Figs. 6, 7). Tributaries from the septal region join others draining the thalamus and other non-septal structures at the internal cerebral vein. Drainage from caudal-lateral septal structures is into the internal cerebral vein. The

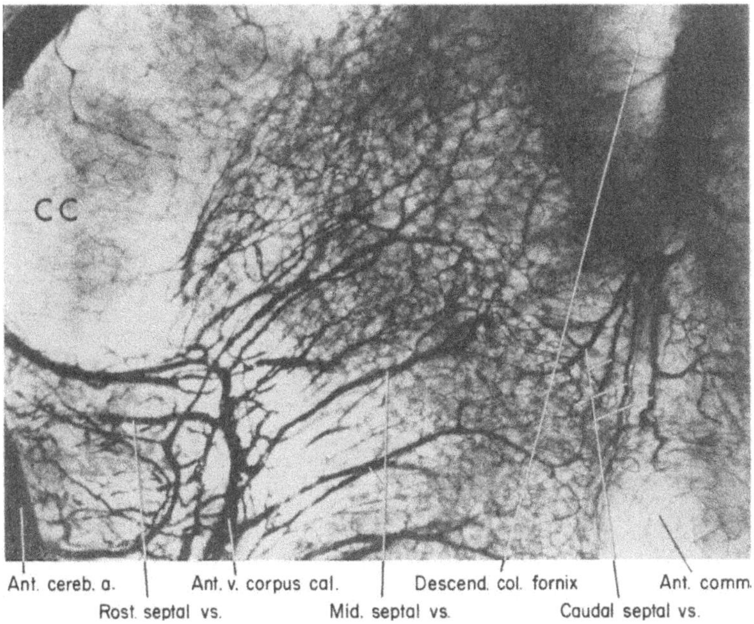

Figure 6. Thick parasagittal section. Note different sets of septal veins. Specimen perfused with India ink and cleared in glycerin. Tissue transilluminated.

extent of the fields in the lateral septal region could not be unequivocally determined in ink preparations since small vessels could not be traced to larger known veins.

DISCUSSION

Rat septal structures receive blood from vessels arising in largest part from the azygos portion of the anterior cerebral artery. The artery is supplied by the internal carotid arteries and to a lesser extent from the vertebral-basilar system. Since the septal portion of the anterior cerebral artery is unpaired, occlusion of, or injections into, either internal carotid artery would be expected to result in bilateral perfusion effects. The larger and longer septal vessels arising from the cerebral artery course from a rostral inferior to caudal superior position in the septum. Objects such as electrodes, cannulae, thermistors, and the like inserted into medial septal structures, or a lesion in the rostral medial septum may interrupt the anterior cerebral artery or vessels supplying more caudal and lateral septal structures.

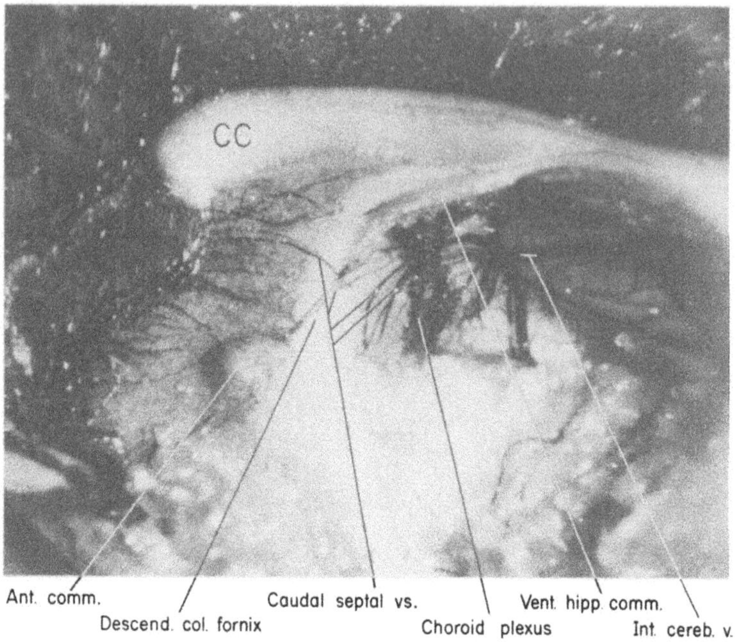

Figure 7. Sagittal view of India ink perfused specimen, dissected to illustrate vascular structures in the caudal septal region.

Several variables determine the extent and degree of tissue involvement following vascular insult. Gross anatomical vessel variations, vessel spasms, collateral supplies through anastomosing vessels, and tissue metabolic requirements vary from individual to individual and provide for diversity in lesion sizes in a population. Anastomoses between arteries and capillary fields are not structurally uniform throughout the nervous system (17). Circulation under vasomotor and neural transmitter influences (6,9,16) and secondary effects due to changes in acid base status or cerebral metabolism cannot be excluded as variables. The extent of the real vascular bed can be demonstrated following placement of a lesion or object implant by utilizing thioflavin S (7) as a fluorescent material to demonstrate patent vessels at time of death.

Septal vascular patterns in rat are comparable to those observed for rabbit by Prolo and Stilwell (13). The rostral inferior to caudal superior course of arteries in the septum is the same in both animals. In rabbit, the septal nuclei stand high in the list of vascularity receiving a considerable supply from the anterior cerebral artery with many branches entering the anterior perforated substance. In rat, the total supply to the septum comes from the anterior cerebral artery and no septal vessels were observed to enter via the anterior perforated substance. Unlike rabbit, no striate vessels from the middle cerebral artery could be demonstrated to supply septal structures in rat. Septal arteries were traced into the ventral hippocampal commissure and septal portion of the descending column of the fornix. A similar pattern may exist for rabbit but was not described by Prolo and Stilwell (13). Vascular supply to the fimbria, alveus, and dorsal hippocampal commissure is from vessels other than those passing through the septum. Consequently, a septal lesion would not interrupt supplies to these structures.

Unfortunately not much data is available for comparing venous patterns in different forms. In all rats examined, rostral, middle, and caudal septal fields were observed. As noted for veins in other species, there was considerable variation in the number and course of the branches. Variations in the smaller branches contributed to the difficulty of establishing the exact extent of the different fields. The line of demarcation separating the middle and caudal fields appeared to be the lamina rostralis. As in rat, short intrinsic septal veins in human run medial to the column of the fornix (2) to join the internal cerebral vein independent of the vein of the septum pellucidum. The vein of the septum pellucidum in man takes a subependymal course on the ventricular side, drains a large portion of the frontal pole of the cerebral hemisphere and runs lateral to the column of the fornix (2,15). The latter named vessel was not investigated in rat, but may drain structures in the lateral septum.

SUMMARY

Patterns of arterial and venous vessels in adult rat septal structures were investigated to provide the septal and hippocampal experimentalists with a better anatomical basis for data interpretation following object implants. The longer septal arteries arise from the proximal portion of the azygos part of the anterior cerebral artery to course from a rostral inferior to caudal superior position within the septum. The most caudal structures receiving supply from these vessels passing superior to the anterior commissure include the ventral hippocampal commissure, septal portion of the descending column of the fornix, and caudal septal nuclei. These vessels and shorter septal arteries arising from the paired anterior cerebral artery and its azygos part supply more rostral regions including the medial and lateral septal nuclei and portions of the diagonal band nucleus. Both vessel types enter the septal medullary substance on its medial aspect and follow a lateral course. No branches from the middle cerebral artery could be traced to septal structures. Objects implanted into the septum may interrupt arteries supplying more caudal superior lateral septal structures but could spare direct involvement to arteries supplying the dorsal hippocampal commissure, fimbria, alveus, hippocampus and fascia dentata.

Three venous fields were observed in the medial septum. The rostral venous field is located between the anterior cerebral artery and the anterior vein of the corpus callosum, the middle field from this vein to approximately the level of the anterior commissure. Behind the commissure the caudal field includes the caudal septal nuclei, the septal portion of the descending column of the fornix, and the ventral hippocampal commissure. Drainage from the rostral and middle fields is to the anterior vein of the corpus callosum while tributaries from the caudal field anastomose with the internal cerebral vein. Objects or chemicals implanted in the caudal superior septum could damage septal veins joining venous tributaries of the dorsal fornix, dorsal hippocampal commissure and septal end of the hippocampus.

ABBREVIATIONS

Ant. cereb. a.	anterior cerebral artery
Ant. chor. a.	anterior choroidal artery
Ant. comm.	anterior commissure
Ant. v. corpus cal.	anterior vein of the corpus callosum
Az. part, ant. cereb. a.	azygos part of the anterior cerebral artery
Bas. a.	basilar artery
Br. of ant. cereb. a.	branches of the anterior cerebral artery
CC	corpus callosum
Caudal septal vs	caudal septal veins
Des. col. fornix	descending column of the fornix
Descend. col. fornix	descending column of the fornix
Dors.-lat. olfact. a.	dorsal-lateral olfactory artery
Hipp.	hippocampus
Int. carotid a.	internal carotid artery
Int. cereb. v.	internal cerebral vein
Lat. striate a.	lateral striate artery
Lat. ventricle	lateral ventricle
Nasal-olfact. a.	nasal-olfactory artery
Neocortical a	neocortical artery
Med. striate a.	medial striate artery
Mid. cereb. a.	middle cerebral artery
Mid. septal vs.	middle septal veins

Olfact. a.	olfactory artery
Post. cereb. a.	posterior cerebral artery
Post. comm. a.	posterior communicating artery
Rost. septal vs.	rostral septal veins
Septal as.	septal arteries
Sup. cerebel. a.	superior cerebellar artery
Thalamoperf, as.	thalamoperforating arteries
Vent. olfact. a.	ventral olfactory artery
Vent. hipp. comm.	ventral hippocampal commissure

REFERENCES

1. Brown, J. O., The morphology of circulus arteriosus cerebri in rats. Anat. Rec., 156 (1966) 99-106.

2. Browning, W., The Veins of the Brain and its Envelopes. F. B. O'Connor, Jr., Brooklyn, N. Y., (1884), pp. 1-79.

3. Craigie, E. H., The vascular supply of the archicortex of the rat. I. The albino rat (Mus norvegicus albinus). J. Comp. Neurol., 51 (1930) 1-11.

4. Craigie, E. H., The vascular supply of the archicortex of the rat. II. The albino rat at birth. J. Comp. Neurol., 52 (1931) 349-364.

5. Craigie, E. H., The vascular supply of the archicortex of the rat. III. The wild norway rat (Mus norvegicus) in comparison with the albino. J. Comp. Neurol., 52 (1931) 359-364.

6. Forbes, H. S., Wolff, H. G., Cerebral circulation. III. The vasomotor control of cerebral vessels. Arch. Neurol. Psychiat. (Chic.) 19 (1928) 1057-1086.

7. Fourman, J., and Moffat, D. B., Fluorescence techniques in the study of microcirculation of the kidney. From: In Vivo Techniques in Histology, ed. by G. H. Bourne, Williams and Wilkins Company, Baltimore, (1967), pp. 239-259.

8. Green, E. C., Anatomy of the rat. Trans. Amer. Phil. Soc., N. S., 27 (1935) 1-370.

9. Kuschinsky, W., Wahl, M., and Neiss, A., Evidence for cholinergic dilatatory receptors in pial arteries of cats. Pflügers Arch., 347 (1974) 199-208.

10. Moffat, D. B., The development of the hindbrain arteries in the rat. J. Anat., (Lond) 91 (1957) 25-39.

11. Moffat, D. B., The development of the anterior cerebral artery and its related vessels in the rat. Amer. J. Anat. 108 (1961) 17-29.

12. Moffat, D. G., The development of the posterior cerebral artery. J. Anat. (Lond) 95 (1961) 485-494.

13. Prolo, D. J., and Stilwell, D. L., Jr., Arterial supply of the diencephalon and some associated areas of the rabbit brain. J. Comp. Neurol., 119 (1962) 229-254.

14. Spoerri, V. O., Über die Gefässversorgung des Subfornikalorgans der Ratte. Acta anat., 54 (1963) 333-348.

15. Wolf, B. S., and Huang, Y. P., The subependymal veins of the lateral ventricles. Am. J. Roentgenol., 91 (1964) 406-426.

16. Wolff, H. G., The cerebral circulation. XIa. The action of acetylcholine. Arch. Neurol. Psychiat. (Chic.), 22 (1929) 686-690.

17. Zeman, W., and Innes, J. R. M., Craigie's Neuroanatomy of the Rat, Academic Press, New York, (1963), pp. 30-41.

Part II
Physiology and Pathology

SEPTAL-HYPOTHALAMIC RELATIONSHIPS

G. J. Mogenson[1]

University of Western Ontario

The contributions of the septum to physiological and behavioral responses have been frequently linked with those of the hypothalamus. The early observations of changes in blood pressure, heart rate and respiration from stimulation of the septum (19,20,41) were considered in terms of effects on the hypothalamus and lower brain stem. In the classical study of Kaada (18) the autonomic and somatomotor responses elicited by stimulation of the septum and other limbic forebrain structures were attributed to effects on the hypothalamus. The demonstrations of the rage syndrome (6) and of polydipsia (17) following lesions of the septum were considered to be the result of removing inhibitory effects on neural systems represented in the hypothalamus for emotional expression (3) and drinking (43), respectively. It is not surprising, therefore, that a presentation on septal-hypothalamic relationships has been included in this symposium devoted to the septum.

---

[1]The author would like to express his sincere appreciation to H. Abdelaal, F. R. Calaresu, J. Ciriello, S. Black, C. Gentil, P. McGarry, J. J. Miller, W. Sibole, D. Tegart and T. Wishart, colleagues with whom he worked on the septum and especially to J. J. Miller who gave permission to reproduce previously unpublished material from his Ph.D. dissertation. Helpful suggestions about the manuscript were provided by Blanche Box, John Kucharczyk and Mark Evered. The illustrations were prepared by Rebecca Woodside and the manuscript typed by Mrs. Marianne Jeffery. The author's research presented in this article was supported by grants from the Medical Research Council of Canada and the National Research Council of Canada.

Although the title has the virtue of being brief there is the danger that it may promise too much. It is appropriate, therefore, to indicate at the outset that the purpose of this article is to discuss attempts in my laboratory during the last few years, to investigate the influences of the septum on the hypothalamus and on some of the functions to which it contributes. These investigations utilized physiological, electrophysiological and behavioral techniques.

THE SEPTUM AND RESPONSE INHIBITION

The earlier evidence suggested that the septum exerts primarily inhibitory effects on physiological and behavioral responses. Ranson and his colleagues reported that electrical stimulation of the septal area produced a decrease in blood pressure, heart rate, and respiration in anesthetized animals. In unanesthetized animals Malmo (23) observed a decrease in heart rate following septal stimulation. In the first studies of the effects of the septum on endocrine function, Endroczi, Schreiberg and Lissak (14) demonstrated that stimulation of the septum reduced adrenocortical secretion. The demonstration of a passive avoidance deficit following septal lesions was considered evidence that the septum is involved in response inhibition, the inhibitory effects being transmitted caudally to the hypothalamus and midbrain (25,26). The rage syndrome and polydipsia following septal lesions, as indicated above, were attributed by many investigators to the removal of inhibitory effects on hypothalamic systems that subserve emotional expression and water intake, respectively. The first electrophysiological studies also suggested that stimulation of the septum primarily inhibits neurons of the hypothalamus (38).

The Septum and Behavioral Arrest

The first experiments in my laboratory were to demonstrate inhibitory effects of the septum which, according to the model shown in Fig. 1A, are exerted caudally on the hypothalamus and brainstem. It was decided to see whether stimulating the septum would produce behavioral arrest. Rats were trained to run along a straight alley to obtain food. When their speed of running had become stable, half-second trains of stimulation to the septum were triggered by the interruption of a photo beam as the animal passed the middle of the straight alley. In most of the animals the septal stimulation produced a dramatic behavioral arrest lasting 1-2 sec. Locomotion ceased suddenly, the animal remained in a fixed posture, and when the stimulation terminated it continued running along the alley as if nothing had happened.

The Septum in the Control of Water and Salt Intakes

We next confirmed the observations of Harvey and Hunt; lesions of the septum increased water intake in a large percentage of rats (76). We also demonstrated that electrical stimulation of the septum reduced water intake without altering food intake (47). As shown in Fig. 2B the lap rate of rats drinking after a period of water deprivation gradually decreased during stimulation of the septum. Since stimulation was observed subsequently to attenuate drinking elicited by electrical stimulation of the lateral hypothalamus (42), it appeared that the septum inhibited "drinking systems" represented in the lateral hypothalamus. Either the septum exerted tonic inhibitory effects on these systems or in some way was concerned with satiety (analogous to the contribution of the ventromedial hypothalamus to the satiation of food intake).

Lubar and co-workers (21,22) raised the possibility that increased water intake following septal lesions was secondary to water diuresis. It was suggested that septal lesions altered the release of antidiuretic hormone. We investigated this possibility by measuring sodium levels in rats with septal lesions (4). Septal lesioned rats whose water intake was increased following surgery had serum sodium and osmolality levels similar to those of sham operated animals (see Table 1). When the animals were water deprived for 24 hours these values were still the same for lesioned and control animals. If septal lesions disrupted mechanisms that control the output of antidiuretic hormone, serum sodium and

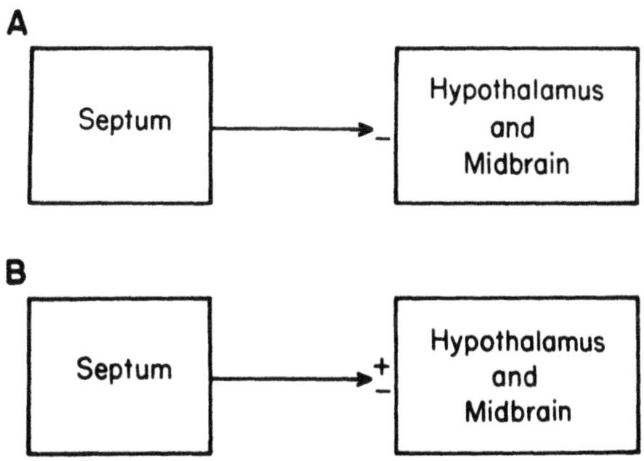

Figure 1. Models that have influenced studies of the mechanisms by which the septum influences physiological and behavioral responses. A. Initially it appeared that the septum had inhibitory effects on the hypothalamus and midbrain. B. Subsequently it was shown that the septum exerted both facilitatory and inhibitory (modulatory) effects on the hypothalamus and midbrain.

serum osmolality should have increased following water deprivation. Urine osmolality of rats with septal lesions also did not differ from sham operated controls (Fig. 2). It may be concluded that the increased water intake of rats with septal lesions is not secondary to systemic dehydration due to the disruption of mechanisms that conserve body water. The results shown in Table 1 also indicate that the hyperdipsia is not due to the removal of an inhibitory effect on hypovolemic thirst, as hypothesized by Blass and Hanson (5).

Another series of experiments also supported the view that the septum exerts inhibitory effects on the hypothalamus (see Fig. 1a). These experiments involved investigating effects of stimulating the septum on the intake of sodium chloride solution. Several investigators have reported that lesions of the septum increased the intake of isotonic and hypertonic saline solutions (8,15,16). This is not due to the lesions causing a sodium deficit since serum sodium levels are normal (Table 1). Since lesions of the septum do not influence saline intake if the hypothalamus has been lesioned previously it appears that the inhibitory effect on saline intake is exerted on the hypothalamus (2).

The effects of electrical stimulation of the septum on the intake of sodium chloride solution was investigated by Gentil, Mogenson and Stevenson (15) using a two-bottle preference test. Stimulation of both the medial septum and the lateral septum reduced the intake of either 0.9% or 1.5% sodium chloride solution. The effect was specific to saline solutions since the intakes of 5% glucose or of 1% potassium chloride were not affected by septal stimulation in the two-bottle preference test (Table II). The stimulation had no effect on the volume or specific gravity of urine excreted following a stomach load of 12 ml of water. When rats were placed on a salt-free diet the attenuating effect of septal stimulation on saline intake disappeared after three or four days (Fig. 4). By this time the output of sodium in the urine had dropped from 1.162 ± 0.04 mEq/1/24 hr to 0.945 ± 0.18 mEq/1/24 hr. Gentil, Mogenson and Stevenson (15) also stimulated the lateral hypothalamus when rats were tested in the two-bottle preference test. The hypothalamic stimulation increased the intake of both sodium chloride solution and water.

From the lesion and stimulation studies just reviewed it appears that the septum exerts an inhibitory effect on the mechanisms that control salt intake. Whether this inhibition is because the septum receives and integrates inputs that signal sodium concentration and sodium needs in the body, as suggested by Antunes-Rodrigues, Gentil, Negro-Vilar and Covian (1), or because the septum is involved in taste and motivational processes cannot be decided without further investigation.

TABLE I: EFFECTS OF LESIONS OF THE MEDIAL SEPTUM ON ARTERIAL PRESSURE IN NORMOTENSIVE, GENETICALLY HYPERTENSIVE AND DOCA TREATED RATS

	Week 1	Week 3	Week 7
Normotensive			
Sham (n = 12)	111 ± 4.0*	113 ± 3	148 ± 4.0
Lesion (n = 7)	123 ± 5.0*	115 ± 2	144 ± 7.0
Genetically hypertensive			
Sham (n = 5)	157 ± 7.0	155 ± 4	218 ± 6.0
Lesion (n = 7)	151 ± 6.0	154 ± 4	201 ± 8.0
DOCA			
Sham (n = 6)	152 ± 7	164 ± 8	198 ± 7
Lesion (n = 5)	153 ± 8	168 ± 12	225 ± 18

The values are arterial pressure (mean ± SEM mmHg). The indirect, tail cuff method was used for weeks 1 and 3 in restrained, unanesthetized rats. For week 7 arterial pressure was measured from the cannulated femoral artery in rats anesthetized with sodium pentobarbital.

(* difference significant at $p < 0.05$, using one-tailed t test).

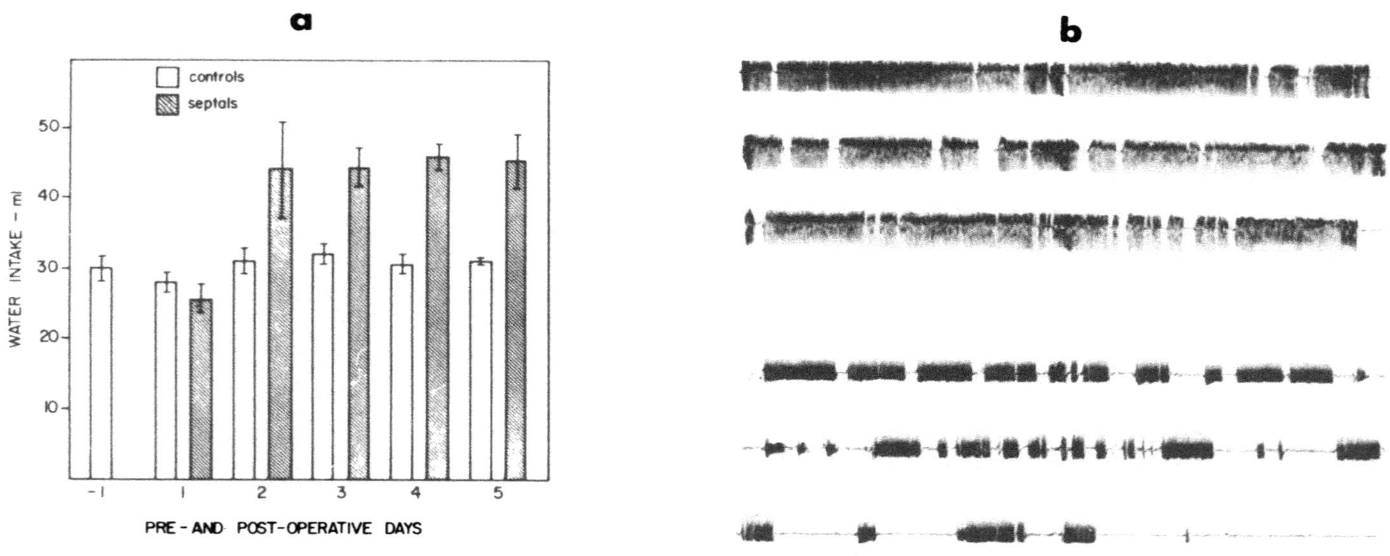

Figure 2. (a) Average daily water intake in eight rats with large septal lesions (cross-hatched bars) compared to the daily intake of eight sham-operated rats (open bars). Both the medial septum and lateral septum were damaged extensively. (b) Lap rates of drinking by a rat after being water-deprived for 23 hours (top) and under similar conditions (bottom) when the septum was being stimulated. For further details, see Wishart and Mogenson (47).

TABLE II: SERUM VALUES DURING DEPRIVATION OF FOOD AND WATER AND DURING HYPOVOLEMIA

Serum Solute	Group	n	Treatment: Ad lib	Treatment: Total Deprivation	Treatment: Hypovolemia
Sodium mEq/1	Sham operated	12	146 ± 0.5*	148.3 ± 0.4	144.9 ± 0.7
	Septal drinker	8	148.3 ± 0.6	148.9 ± 0.4	145.2 ± 1.4
	Septal non-drinker	12	147.8 ± 0.3	148.0 ± 0.3	146.2 ± 0.9
	Vehicle injected	10	-	-	147.0 ± 0.4
	Sham operated	12	298.0 ± 1.9	300.3 ± 1.9	315.3 ± 3.6
	Septal drinker	8	295.9 ± 2.4	303.1 ± 2.1	316.0 ± 4.3
	Septal non-drinker	12	297.7 ± 2.0	301.7 ± 1.3	314.5 ± 2.0
	Vehicle injected	10	-	-	302.6 1.6

* Standard error of the mean

(Modified from Black and Mogenson, 1973)

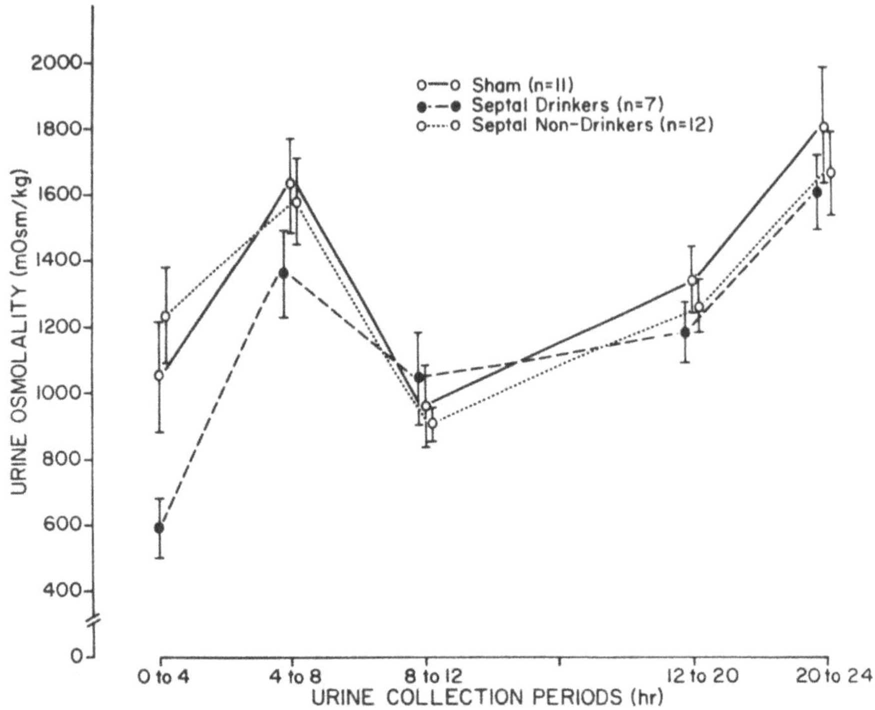

Figure 3. Mean values of urine osmolality for sham-operated, septal drinker and septal non-drinker groups of rats during water deprivation. From Black and Mogenson (4).

The Septum and Experimental Hypertension

Recently Marwood, Gan and Lockett (24) reported that female rats of an inbred hypertensive strain housed in a sound-attenuated room (semi-anechoic room with a sound level of less than 45 decibels) developed hypertension within a few weeks. These investigators suggested that the reduced auditory input might influence the interplay of the reticular activating system and limbic forebrain structures resulting in a change in secretions by the pituitary gland. They investigated this hypothesis by making lesions of the septum and measuring the effects on arterial pressure. Lesions of the medial septum increased arterial pressure of the genetically hypertensive rats by a magnitude similar to that observed in rats exposed to a reduced ambient noise level.

The effect of medial septal lesions on arterial pressure was investigated in two series of experiments in normotensive rats, in normotensive rats treated with deoxycorticosterone to

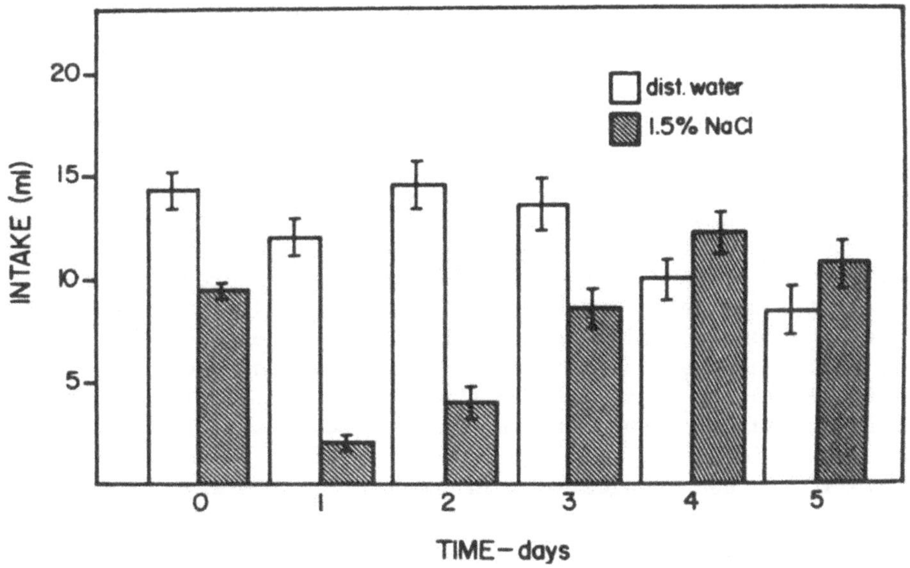

Figure 4. Effects of electrical stimulation of septum on intakes of NaCl solution and water in rats on a salt-free diet. Day 0 was a control condition with no septal stimulation. Following this non-stimulation test on day 0 rats were placed on salt-free diet. Septal stimulation was presented on days 1 to 5. Note gradual increase in intake of NaCl solution from day 1 to day 5 despite septal stimulation. From Gentil, Mogenson and Stevenson (15).

elevate their blood pressure and in genetically hypertensive rats (McGarry and Mogenson, 1974, unpublished results). Blood pressure was determined at weekly intervals for 4 to 6 weeks in unanesthetized, restrained animals using a tail blood pressure cuff and at the end of this period by cannulating the carotid artery with the animals anesthetized with nembutal. Lesions of the medial septum increased arterial pressure in a few of the animals of each group, particularly during the first week or so following surgery. However, for most of the comparisons the differences between the lesioned groups and sham operated groups was not statistically significant (Table 2).

Further experiments are needed to investigate the effects of septal lesions on arterial pressure and to evaluate the hypothesis of Marwood, Gan and Lockett that lesioning the septum by removing an inhibitory input to the hypothalamus increased anterior pituitary-adrenal cortical activity. More discrete lesions should be used in some of the experiments in order to test the possibility that the hypertension is related to damage to a particular region of the septum and of damage to certain efferent projections.

THE SEPTUM AND RESPONSE MODULATION

Although for some responses or functions the septum may have exclusively inhibitory effects, as implied in the previous section, there is a growing body of evidence that the septum can also exert facilitatory effects. Several investigators have reported that septal stimulation increases blood pressure, heart rate, and rate of respiration as well as decreasing these responses (7,10,11,12, 23). Both tachycardia and bradycardia have been observed from stimulation of the same site in the septum (44). Stimulation of the septum has been shown to increase as well as decrease drinking responses elicited by stimulation of the lateral hypothalamus (42). It has been suggested by a number of investigators that the septum can either facilitate or inhibit hypothalamic and brainstem systems, as shown in Fig. 1B, and in this way exert modulatory effects on a variety of responses (9,29,30,42).

The Effects of Stimulation of the Septum on Arterial Pressure

A detailed mapping study of cardiovascular responses to electrical stimulation of the septum in rats was carried out in collaboration with F. R. Calaresu (7). Both hypertensive and hypotensive responses were observed and these responses depended on the site of stimulation and the nature of the anesthetic. With stimulation of the lateral septum arterial pressure was decreased in rats anesthetized with chloralose and increased in animals under urethan (Fig. 5, A and D). On the other hand, with stimulation of the medial septum arterial pressure was increased in animals under chloralose and decreased in animals under urethan (Fig. 5, C and E). When the border between the medial and lateral septum was stimulated in rats under chloralose there was an initial hypertensive response followed by hypotension (Fig. 5B). These elicited changes were also observed following the administration of d-tubocurarine indicating that they were not secondary to respiratory responses or to muscular activity induced by the stimulation. Under chloralose the baseline arterial pressure was 15-20 mmHg higher than under urethan suggesting a higher level of sympathetic activity in chloralosed animals.

These results indicate quite clearly that the septum can exert modulatory, and not only inhibitory, effects on arterial pressure. However, it is difficult to explain on the basis of existing knowledge the observation that the elicited arterial pressure changes are in opposite directions under the two anesthetics. One possibility is that each anesthetic selectively eliminates certain synaptic pathways so that the mode of functioning of the system is completely different for the two conditions. Another possibility is that the differential effects of stimulation are related to the different baseline values of arterial pressure under the two anesthetic agents so that the system is operating in different regions of its functional range.

Under urethan a rhythmic oscillation of arterial pressure was freuqnelty observed during the hypertensive response elicited by electrical stimulation of the septum (Fig. 5F). The consistent frequency of oscillation (24.6 ± 0.2/min) suggested the possibility of "the existence in the septum of a generator of rhythmic discharges controlling the level of arterial pressure"(7).

Recently the effects of electrical stimulation in rats anesthetized with sodium pentobarbital have been investigated (Ciriello, Calaresu and Mogenson, 1974, unpublished observations). Stimulation of the lateral septum caused a decrease in arterial pressure, stimulation of the medial septum an increase, and stimulation of the border between these two regions an initial increase followed by a decrease (Fig. 6). The results are similar to those obtained when the animals were anesthetized with chloralose.

The Effects of Stimulation of the Septum on the Activity of Hypothalamic Neurons

Microelectrode extra-cellular recordings have been made from neurons of the ventromedial hypothalamus (33,38,45), the posterior hypothalamus (13), the lateral hypothalamus (28), and the paraventricular nucleus of the hypothalamus (36) while stimulating the septum. In all cases the stimulation has been observed to facilitate as well as inhibit the discharge of hypothalamic neurons. From such results it has been tempting to speculate that the septum is able to exert modulatory effects on functions associated with the hypothalamus.

Miller and Mogenson (28) have reported that septal stimulation facilitated or inhibited the discharge of many neurons of the lateral hypothalamus depending on their baseline firing rate. If the hypothalamic neuron was discharging slowly septal stimulation usually speeded it up (Fig. 7b) whereas if it was discharging rapidly the same septal stimulation decreased its firing rate (Fig. 7c). Taking advantage of convergence of inputs to hypothalamic neurons from the septum and from the periphery Miller and Mogenson investigated the effects of stimulating the septum while using stimulation of the sciatic nerve or olfactory bulb to

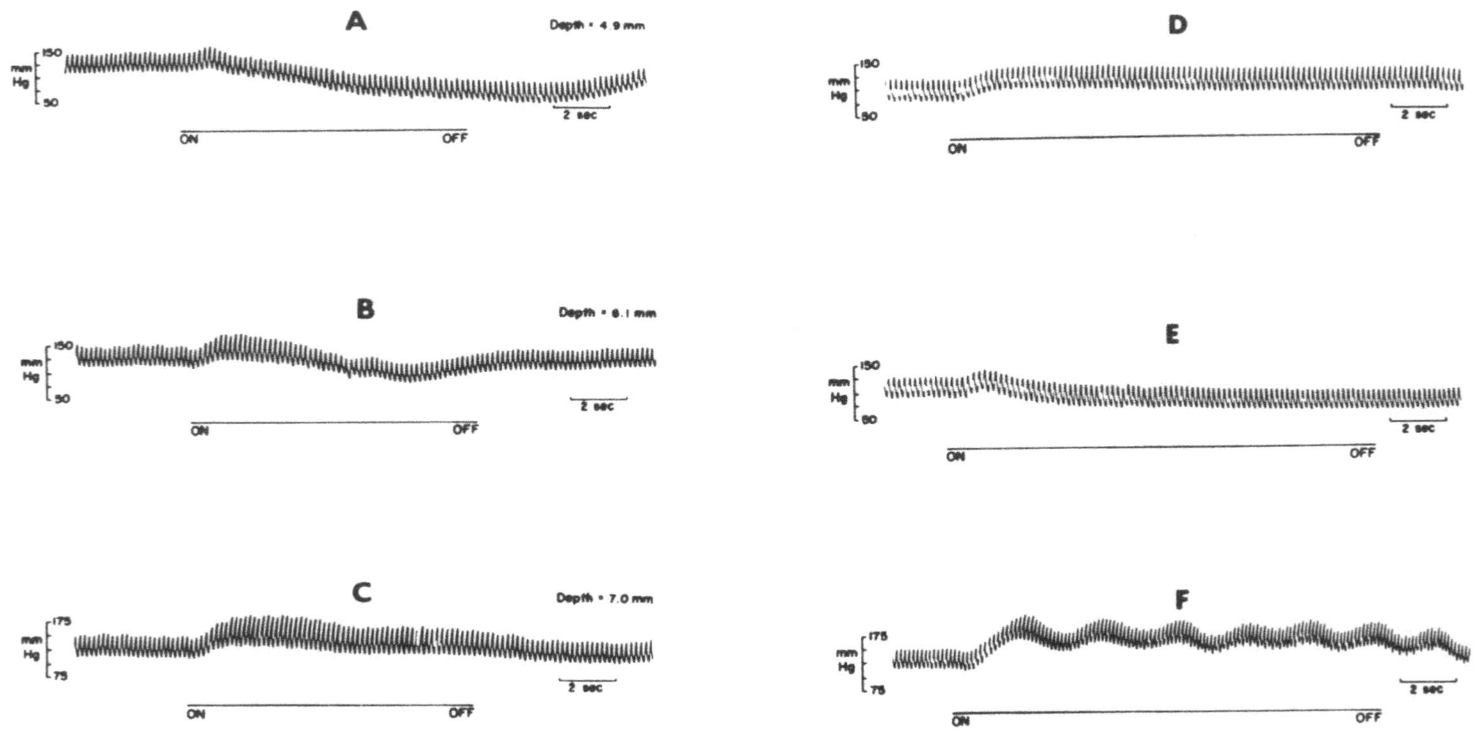

Figure 5. Typical arterial pressure responses elicited by electrical stimulation (80 Hz, 2 msec pulse duration, 10 volts) of the septum. A: stimulation of the lateral septum under chloralose. B: stimulation at border between medial and lateral septum under chloralose. C: stimulation of medial septum under chloralose. D: stimulation of lateral septum under urethan. E: stimulation of medial septum under urethan. F: oscillatory response during stimulation of lateral septum under urethan. Modified from Calaresu and Mogenson (17).

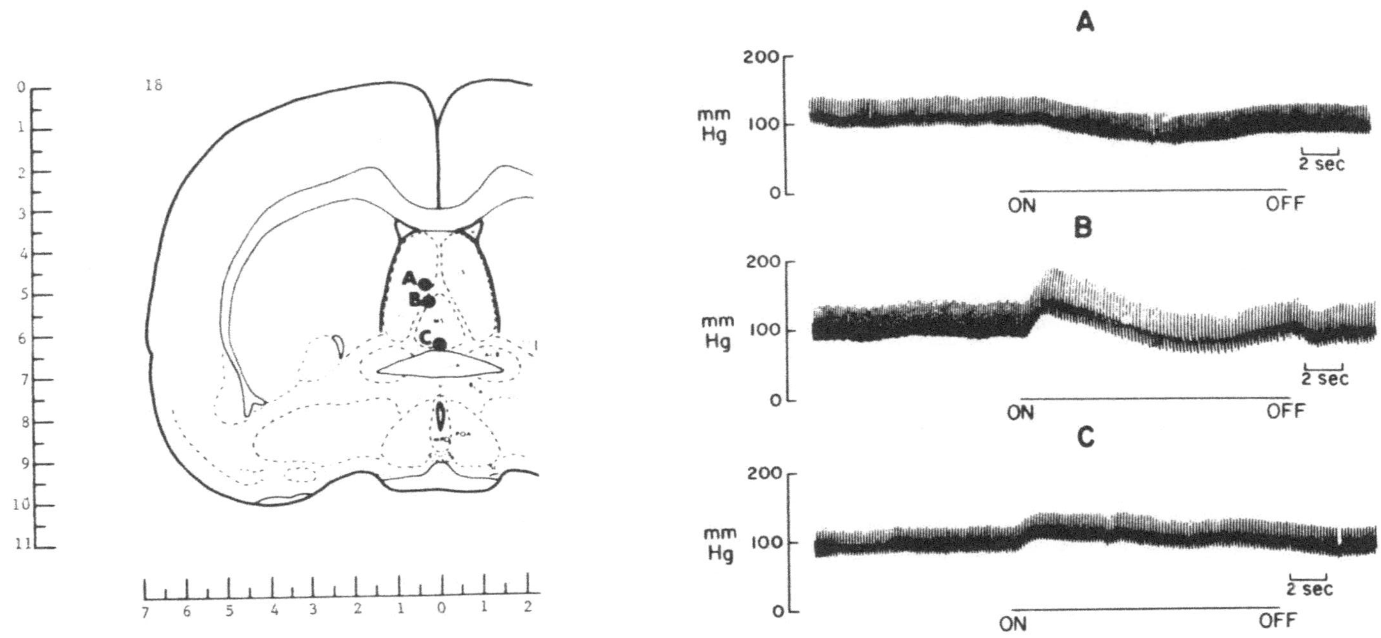

Figure 6. Typical arterial pressure responses elicited by electrical stimulation (80 Hz, 2 msec pulse duration, 400-600 μA) in rats anesthetized with sodium pentobarbital. A. Stimulation of the lateral septum elicited a hypotensive response. B. Stimulation at the border of lateral and medial septum elicited initial hypertension followed by hypotension. C. Stimulation of the medial septum elicited a hypertensive response (Cirielo, Calaresu and Mogenson, 1974, unpublished observations).

manipulate the firing rate of the hypothalamic neuron. The effect of septal stimulation on a hypothalamic neuron whether, excitatory or inhibitory, depended on the discharge rate of that neuron. These results suggest that the septum exerts modulatory effects on hypothalamic neurons and that whether the modulatory effect is excitatory or inhibitory depends on the background sensory input and other factors that determine the activity of hypothalamic neurons at a particular time.

The hypothalamic neuron was more likely to be facilitated with stimulation of the mediodorsal septum and more likely to be inhibited with stimulation in the bed nucleus of the stria terminalis. In later sections evidence is presented which suggests that the fornix transmits primarily excitatory effects and the stria terminalis inhibitory effects. However, since wide regions of the septum contribute fibers to both fornix and stria terminalis for most sites of stimulation the baseline discharge rate of the hypothalamic neuron was a critical factor.

Effects of Stimulation of the Septum on Self-Stimulation of the Lateral Hypothalamus

Miller and Mogenson (29) also investigated the modulatory effects of septal stimulation on the rate of self-stimulation of the lateral hypothalamus. In view of the earlier observation that the occurrence of excitation or inhibition of hypothalamic neurons depended on the baseline level of discharge a comparison was made of the effects of septal stimulation when the intensity of hypothalamic stimulation was at threshold and when it was suprathreshold. This was to investigate the possibility that septal stimulation would enhance self-stimulation when the stimulating current and the rate of self-stimulation were low and would suppress self-stimulation when the stimulating current and the rate of self-stimulation were higher.

Bipolar electrodes were implanted stereotaxically into the septum and into the lateral hypothalamus under nembutal anesthesia. When the hypothalamic electrode had been positioned in the region of the lateral hypothalamus from which self-stimulation has been reliably elicited (22) the septal electrode was lowered and single square wave stimuli were administered while recording evoked field potentials from the hypothalamic electrode. The field potentials had two main components (see Fig. 9: number 1, star shows large 10-14 msec component; number 3, star shows large 18-23 msec component, and Fig. 15), which we will see later, represent activity being transmitted via the precommissural fornix and the stria terminalis.

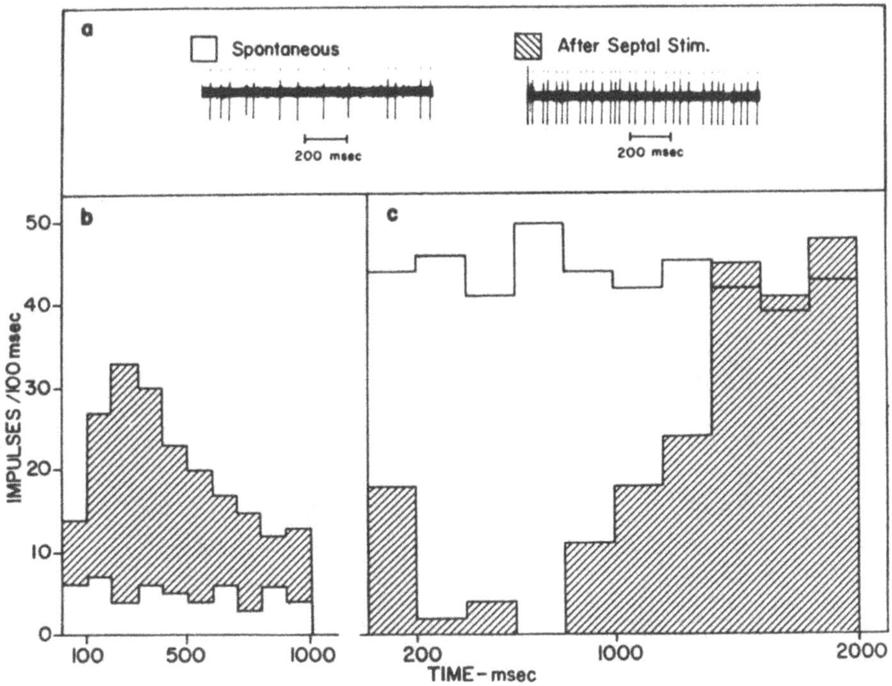

Figure 7. (a) Action potentials recorded extracellularly from a neuron in the lateral hypothalamus of the rat. Following electrical stimulation of the septum (single pulse, duration 0.5 msec, current intensity 0.25 mA) this neuron discharged at a faster rate (right). (b) Frequency histograms showing facilitatory effects of stimulating the septum when the baseline firing rate of the hypothalamic neuron is relatively slow (spontaneous rate: blank, septal stimulation, cross hatched). The ordinate is the total number of impulses per 100 msec in each address of the histogram for 15 trials. (c) Frequency histograms showing inhibitory effects of stimulating the septum when the baseline firing rate of the hypothalamic neuron is relatively fast. For (b) and (c) the stimulation parameters were 20 Hz, 0.5 msec pulses, 0.30 mA. From Mogenson, 1973, as modified from Miller and Mogenson (28).

The results for the self-stimulation test beginning one week later are shown in Table IV for the 9 rats tested with both low current and high current levels. The animals were trained to press a lever to obtain 0.5 sec trains (50 Hz, 0.5 msec pulse duration) of stimulation to the lateral hypothalamus. A pulse of stimulation presented to the septum 5 msec before each pulse to the lateral hypothalamus increased the rate of self-stimulation when

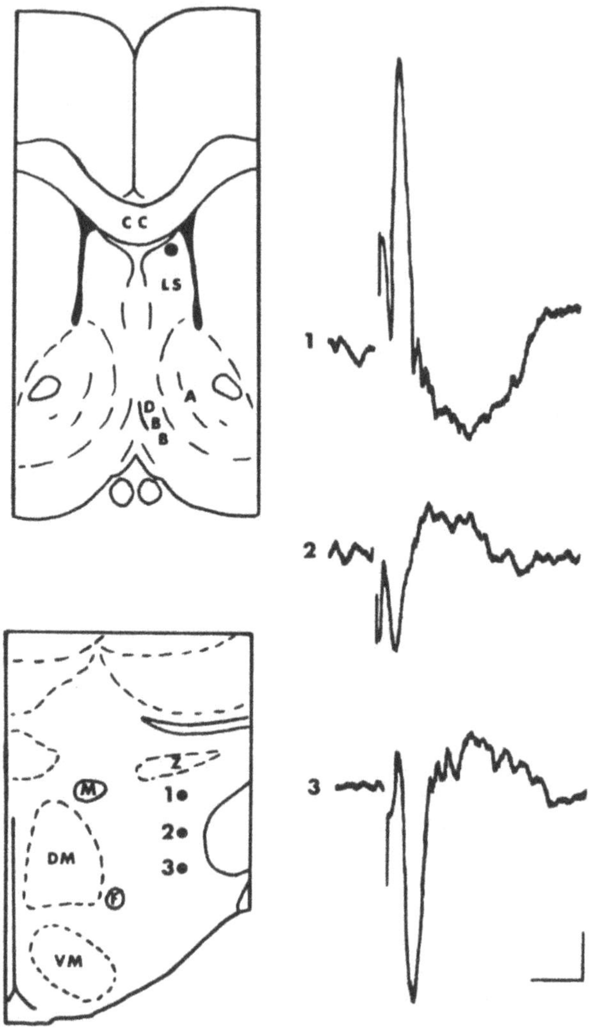

Figure 8. Location of a stimulating electrode site (top) in the dorsolateral septum and LH recording sites 1, 2, 3 (bottom) from which phase reversals in the 3-6 msec component were observed. A, nucleus accumbens; AC, anterior commissure; CC, corpus callosum; DBB, diagonal band of Broca; DM, dorsomedial hypothalamus; F, fornix; FH, fimbrial hippocampus; H, hippocampus; LS, lateral septum; MS, medial septum; M, mamillothalamic tract; ST, stria terminalis; VM, ventromedial hypothalamus; Z, zone incerta. Calibration, 10 msec, 10 μv. From Miller and Mogenson (30).

TABLE III: EFFECT OF SEPTAL STIMULATION ON PREFERENCE OF RATS FOR DISTILLED WATER OR GLUCOSE, DISTILLED WATER OR SODIUM CHLORIDE, AND DISTILLED WATER OR POTASSIUM CHLORIDE

	Water, ml	Solution, ml	Total intake, ml
A) 5% Glucose			
No stimulation	4.81 ± 1.13	15.00 ± 2.00	19.81 ± 1.19
Stimulation	4.90 ± 1.28	14.90 ± 1.95	19.80 ± 1.11
B) 1.5% Sodium chloride			
No stimulation	9.81 ± 0.24	10.50 ± 1.46	20.31 ± 1.51
Stimulation	15.00 ± 0.50	4.37 ± 1.00	19.37 ± 1.32
C) 1% Potassium chloride			
No stimulation	11.81 ± 0.77	5.12 ± 0.71	16.81 ± 0.72
Stimulation	12.37 ± 1.03	5.48 ± 0.80	17.86 ± 1.09

Results expressed as mean ± SE for 2 days, n = 8.

From Gentil, Mogenson and Stevenson (1971)

the low current level was presented to the hypothalamus. When the intensity of the hypothalamic stimulation was higher the pulses of stimulation to the septum reduced the rate of self-stimulation. It appears that the modulatory effects exerted by the septum on hypothalamic neurons concerned with self-stimulation depend on the activity of these neurons.

PATHWAYS SUBSERVING THE MODULATORY EFFECTS OF THE SEPTUM ON THE LATERAL HYPOTHALAMUS

Field potentials recorded from the lateral hypothalamus in response to electrical stimulation of the septum were studied extensively in an attempt to obtain some indication of the pathways by which the septum exerts its modulatory effects on the hypothalamus (30). Following single pulse stimulation (duration 0.1-0.5 msec, 0.05 - 0.4 mA current intensity) potentials were observed that consisted of three components. These components had peak latencies of 3-6, 10-14 and 18-23 msec and onset latencies of 0.5 - 2.0, 5.0 - 8.0, and 12.0 - 16.5 msec respectively.

The short latency component (3-6 msec), recorded from the dorsal region of the lateral hypothalamus, was most prominent with stimulation of the dorsolateral septum (Fig. 8). When the stimulating electrode was lowered the 3-6 msec component decreased in amplitude and was masked by the larger 10-14 msec component.

The 10-14 msec and 18-23 msec components were always present but their amplitude was related to the site of stimulation. As shown in Fig. 9 the 10-14 msec component (responses 1 and 2) was of larger amplitude when the dorsomedial septum was stimulated whereas the 18-23 msec component (responses 3 and 4) was more prominent when the ventrolateral region was stimulated. By recording at different depths in the lateral hypothalamus it was observed that phase reversal of the 10-14 msec component occurred at the level of the dorsal region of the lateral hypothalamus whereas phase reversal of the 18-23 msec component occurred somewhat lower at the level of the dorsal aspect of the ventromedial hypothalamus.

The relationship of the amplitudes of the 10-14 msec and 18-23 msec components of the field potentials to sites of stimulation suggested that they are related to activity being transmitted along different pathways. Stimulation of the dorsomedial septum which resulted in the large amplitude 10-14 msec component was in the region of the precommissural fornix fibers. On the other hand stimulation of the ventrolateral septum was in the region of the bed nucleus of the stria terminalis. A direct test of this

TABLE IV: EFFECTS OF SEPTAL STIMULATION ON LH SELF-STIMULATION RATES[a]

Rat No.	Low current		High Current	
	Control	5 msec	Control	5 msec
1	30± 3.9	124± 4.4*	177± 6.6	158± 8.8
5	37± 5.5	60± 3.8*	143±12.9	130±12.9*
6	25± 3.0	96± 4.6*	190± 6.0	117± 4.3*
7	73±14.9	104± 9.8*	159± 4.2	155± 3.0
8	49± 7.7	74±12.3*	201± 2.7	210± 7.2
10	97±16.3	129±15.2*	198± 2.2	185± 9.1
11	64±20.3	101±15.0*	176± 8.8	155± 8.3*
12	59± 7.4	93± 4.7*	117± 4.8	92± 2.5*
20	79± 3.4	137± 6.2*	130± 3.6	72± 4.3*

a Mean lever presses (x±SEM) for 3-min test periods;

* $p < .05$;

Modified from Miller and Mogenson (1972).

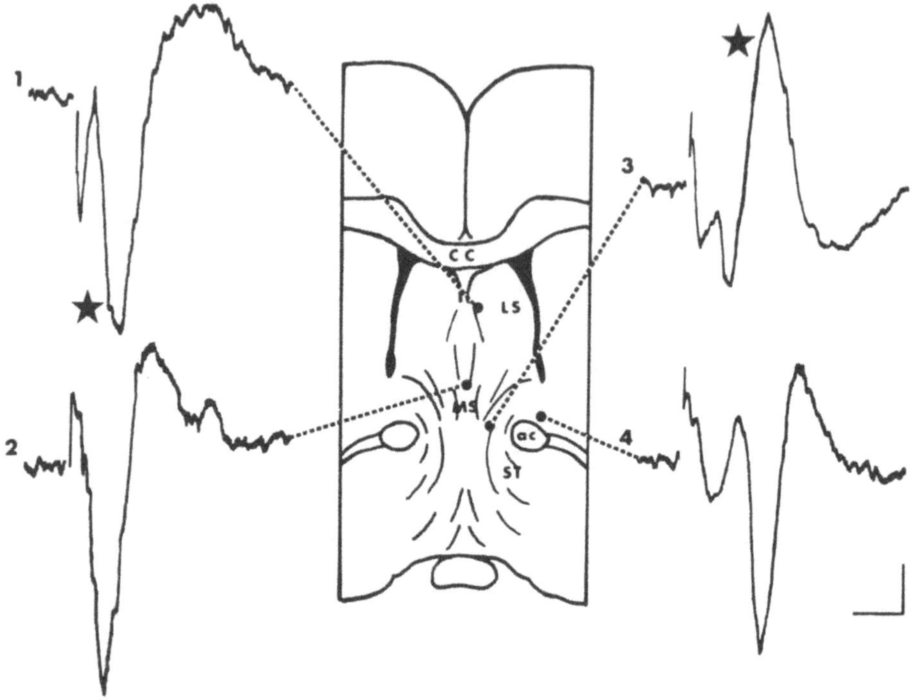

Figure 9. Lateral hypothalamic (LH) evoked responses produced by stimulation of the dorsomedial and midline region and the ventrolateral region of the septum; 1 and 2 are examples of the 10-14 msec component, 3 and 4 are examples of the 18-23 msec component. Calibration, 10 msec, 50 μv. From Miller and Mogenson (30).

possibility, by sectioning the fornix and the stria terminalis between the septal stimulating sites and the lateral hypothalamic recording sites was not possible. However, with stimulation of the hippocampus (Fig. 10A) a field potential with peak latency of 11-17 msec was recorded from the hypothalamus (Fig. 10B). This evoked response was similar to the 10-14 msec component recorded to stimulation of the dorsomedial septum except the latency was somewhat longer. When the dorsomedial septum was lesioned (Fig. 10C) the field potential elicited by stimulation of the hippocampus was markedly attenuated (Fig. 10D). With stimulation of the stria terminalis (Fig. 11B) a field potential similar to the 18-23 msec component, but with a peak latency of 32-38 msec was recorded (Fig. 11A). This evoked potential was attenuated (Fig. 11C) by lesions of the posterior portion of the bed nucleus of the stria terminalis (Fig. 11D).

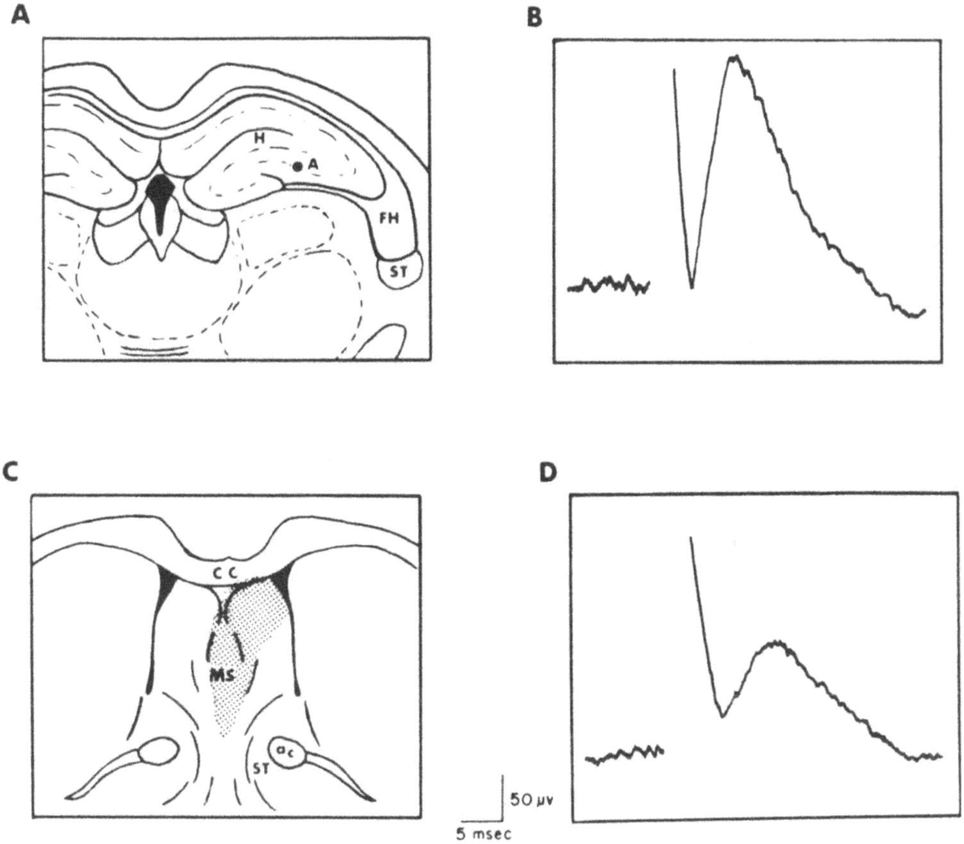

Figure 10. Evoked response recorded (B) in the dorsal region of the lateral hypothalamus to stimulation of the hippocampus (A, point marked A). After a lesion of the dorsal-midline region of the septum (C) the evoked response was attenuated (D). Calibration, 5 msec, 50 µv. Modified from Miller and Mogenson (30).

Extra-cellular microelectrode recordings were made from 71 neurons in the lateral hypothalamus and single pulse stimulation was administered to the septal sites corresponding to the pre-commissural fornix dorsomedial region (PCF) and to the region of the bed nucleus of the stria terminalis (ST). Approximately 70% of the units were influenced by stimulation of the PCF site and about the same percentage by stimulation of the ST site. Three times as many neurons were activated by stimulation of the PCF site as were activated by stimulation of the ST site (Table V). The discharge pattern and post stimulus time histogram for a neuron activated by stimulation of the PCF site are shown in Fig. 12a. Fig. 13

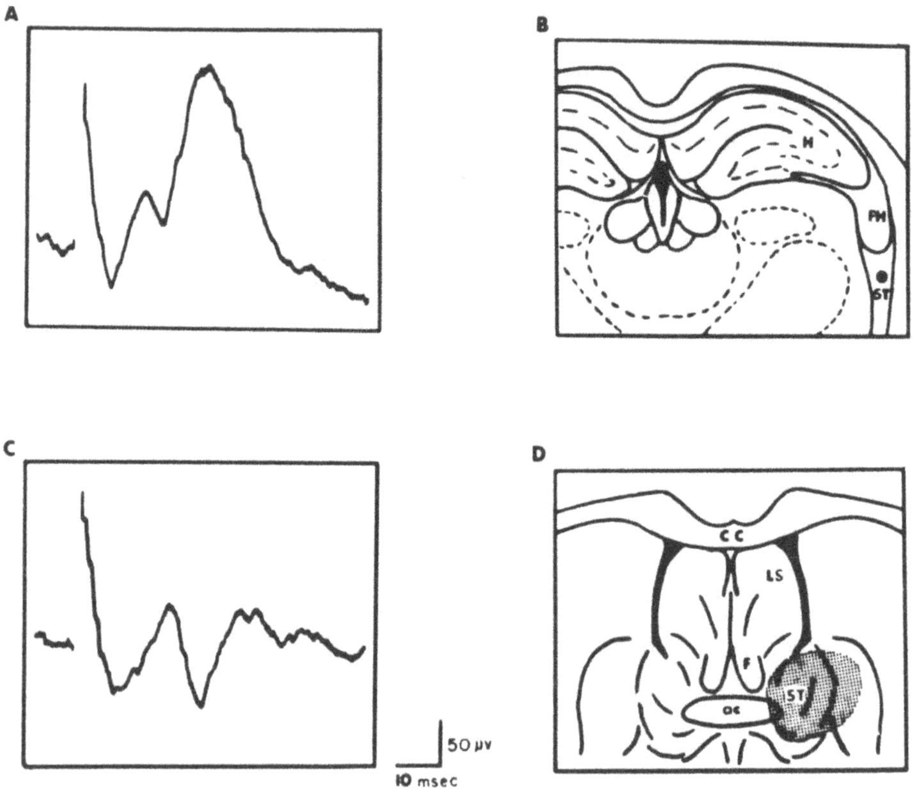

Figure 11. Evoked potential (A) recorded in the hypothalamus to stimulation of the stria terminalis (B, point marked ST) are similar to those obtained with stimulation of the septum in the region of the bed nucleus of the stria terminalis (for comparison, see Miller and Mogenson [30], Fig. 3-3), except they have a somewhat longer latency. This evoked potential is attenuated (C) by lesions of the septum (D) and the attenuation is greater when the lesion is placed in the more posterior part of the septum destroying the bed nucleus of the stria terminalis. These observations suggest that the fibers mediating the inhibitory effect on elicited drinking might not be of septal origin but "fibers of passage" from the amygdala to the hypothalamus. Modified from Miller and Mogenson (30).

demonstrates the relationship between the increased discharge of lateral hypothalamic neurons elicited by PCF stimulation and the 10-14 msec component of the field potential. Two and one-half times as many neurons were inhibited by stimulation of the ST sites as were inhibited by stimulation of the PCF site (Table V). The discharge pattern and post-stimulus time histogram for a neuron inhibited by ST stimulation are shown in Fig. 12b. The onset of the inhibition corresponded with the onset of the 18-23 msec component of the field potentials. When the region between the PCF site and the ST site was stimulated the hypothalamic neurons were first activated and then inhibited (activation-inhibition sequence). Population histograms showing the activation-inhibition sequence for 27 lateral hypothalamic neurons are shown in Fig. 14 together with a typical superimposed field potential.

It is concluded from these electrophysiological experiments that the 10-14 msec component of the field potential elicited by stimulation of the septum is transmitted via the precommissural fornix fibers and that it reflects excitatory effects on neurons of the lateral hypothalamus. The 18-23 msec component is transmitted via the stria terminalis and reflects inhibitory effects on neurons of the lateral hypothalamus. Because of the latency of these components and the attenuation of their amplitude at high frequency repetitive stimulation (>50 Hz) it appears that both are multi-synaptic pathways.

The 3-6 msec component of the evoked field potentials has not been studied in detail because it was frequently masked by the larger 10-14 msec component. It did appear to be mediated by direct fiber projections, however, since the latency was relatively short and it was not attenuated when high frequency repetitive stimuli (>100 Hz) were administered to the dorsal septum. Miller (27) has suggested that the 3-6 msec component may represent activity transmitted by a pathway that has been traced by fiber degeneration techniques from the dorsal region of the septum ventrally through the septum and along the medial forebrain bundle to the lateral hypothalamus (34,35,39,40). From preliminary evidence Miller suggests that the 3-6 msec component represents inhibitory effects on hypothalamic neurons and that these inhibitory effects may account for the inhibition of some hypothalamic neurons when the PCF site was stimulated (see Table V, 19%).

TABLE V: RESPONSES OF NEURONS OF THE LATERAL HYPOTHALAMUS TO STIMULATION OF TWO REGIONS OF THE SEPTUM

Stimulation Site	Nature of Response		
	+	±	-
Dorsomedial (PCF)	31	50	19
Ventrolateral (ST)	10	38	52

Scores are the percentage of neurons displaying the following responses: + facilitation, ± facilitation-inhibition sequence, - inhibition From Miller (1972).

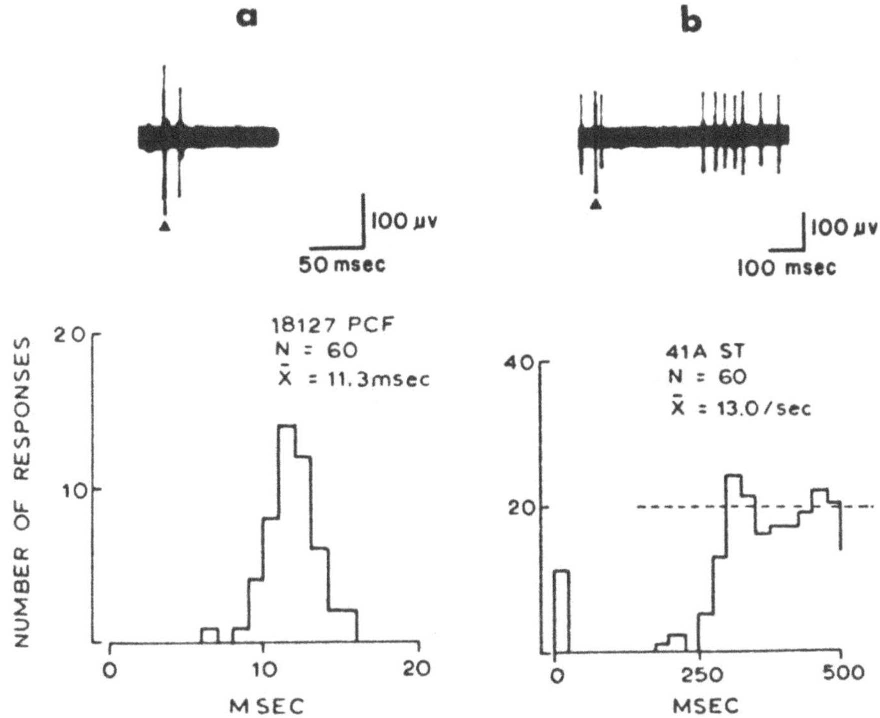

Figure 12. (a) Action potentials of a neuron of the lateral hypothalamus (top) and the corresponding post-stimulus time histogram (bottom) activated by single pulse stimulation of the precommissural fornix (PCF) region of the septum. (b) Action potentials of a neuron of the lateral hypothalamus (top) and the corresponding post-stimulus time histogram (bottom) inhibited by single pulse stimulation of the bed nucleus of the stria terminalis (ST). N - number of stimulation trials. From Miller (27).

THE EFFECTS OF STIMULATION OF THE SEPTUM ON DRINKING RESPONSES ELICITED BY STIMULATION OF THE LATERAL HYPOTHALAMUS

It was indicated earlier that electrical stimulation of the septum reduced water intake in rats (see Fig. 23). The results of the electrophysiological experiments discussed in the previous section suggested that the septum may facilitate as well as inhibit drinking. Experiments that investigated this possibility will now be described.

A bipolar electrode was implanted in the lateral hypothalamus and those rats in which drinking was elicited by electrical stimulation (50 Hz, 0.2 msec pulses, 10 sec trains) were selected. A

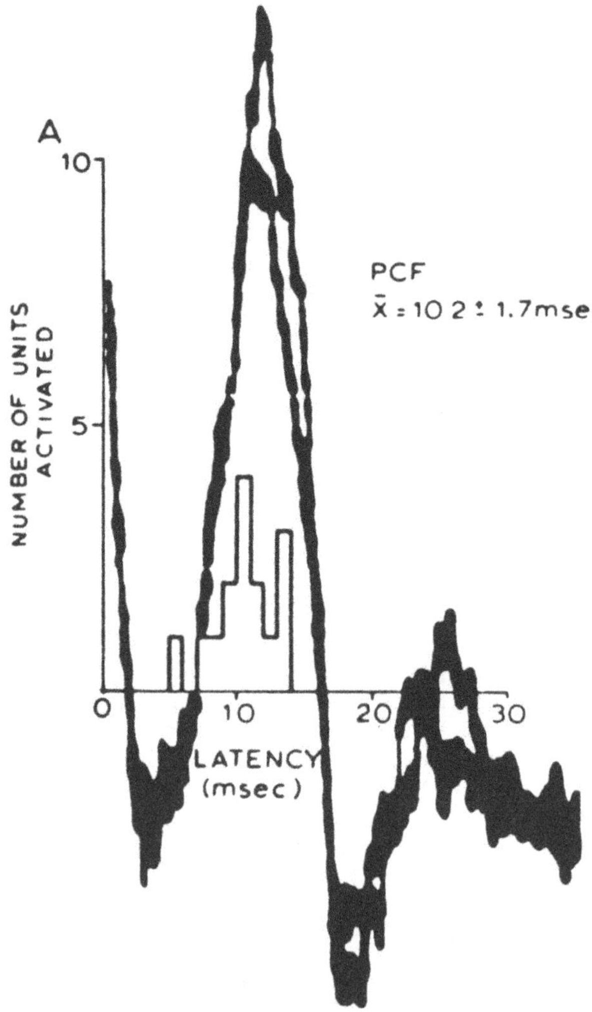

Figure 13. Population histogram of the discharges of 15 neurons of the lateral hypothalamus activated by stimulation of the pre-commissural fornix region (PCF) with a superimposed field potential recorded from the same site in the lateral hypothalamus. Mean latency of activation is 10.2 sec. NOte the relationship between the increased discharge rate of the neurons and the 10-14 msec component of the field potential. From Miller (30).

second bipolar electrode was then positioned in the septum. During this procedure single pulse stimuli were administered to the septum and field potentials recorded from the lateral hypothalamic electrode (Fig. 15A, Electrode E_3). When the electrode was positioned in the dorsomedial septum (Electrode E_1) the 10-14 msec component was prominent in the evoked field potentials and when the electrode was positioned in the bed nucleus of the stria terminalis (Electrode E_2) the 18-23 msec component was prominent. In some animals electrodes were placed in both the mediodorsal and ventrolateral sites. A week later behavioral testing began. Drinking was again elicited by 10 sec trains of hypothalamic stimulation (50 Hz, 0.2 sec pulses). During some of the stimulation periods a pulse of stimulation to the septum preceded by 5 msec each pulse to the lateral hypothalamus (for further details see Sibile, Miller and Mogenson [30]).

Septal stimulation both facilitated and inhibited drinking elicited by electrical stimulation of the lateral hypothalamus, depending on the site of stimulation (Fig. 16). With stimulation of the mediodorsal septum, in the region of the precommissural fornix fibers, the elicited drinking was facilitated (rats 1-5). This is the region from which the large amplitude 10-14 msec component of the field potential was elicited. On the other hand with stimulation of the region of the bed nucleus of the stria terminalis the elicited drinking was reduced (rats 10,11). This is the region from which the large amplitude 18-23 msec component of the field potential was recorded. Stimulation of sites between the dorsomedial region and the ventrolateral region had variable effects on drinking elicited by hypothalamic stimulation with no overall change (rats 6-9).

Fig. 17 shows the results for one of the animals in which one stimulating electrode was implanted in the dorsomedial region and another in the ventrolateral region (Fig. 17a). The drinking elicited by stimulation of the hypothalamus was enhanced when the dorsomedial septum was stimulated and reduced when the ventrolateral septum was stimulated (Fig. 17b).

It is clear that stimulation of the septum both facilitates and inhibits elicited drinking. Histological and electrophysiological evidence suggest that these modulatory effects are mediated by the fornix and the stria terminalis (Fig. 15B).

The effects of stimulating the septum on spontaneous drinking following 23 hr water deprivation was investigated in some of the animals. The septal stimulation did not change the water intake of these animals. Presumably this is because drinking elicited by stimulation of the lateral hypothalamus is not due to the

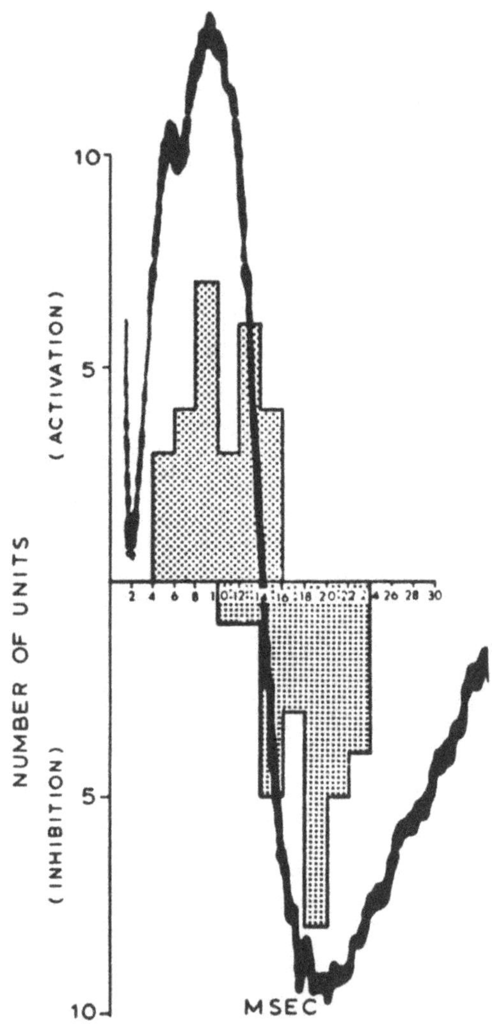

Figure 14. Population histogram for 27 neurons of the lateral hypothalamus showing an activation-inhibition sequence following single pulse stimulation to sites of the septum between the dorsomedial precommissural fornix region and the ventrolateral region (bed nucleus of the stria terminalis). The superimposed evoked response shows the relationship of the activation-inhibition sequence to the 10-14 and 18-23 msec components of the field potentials. From Miller (27).

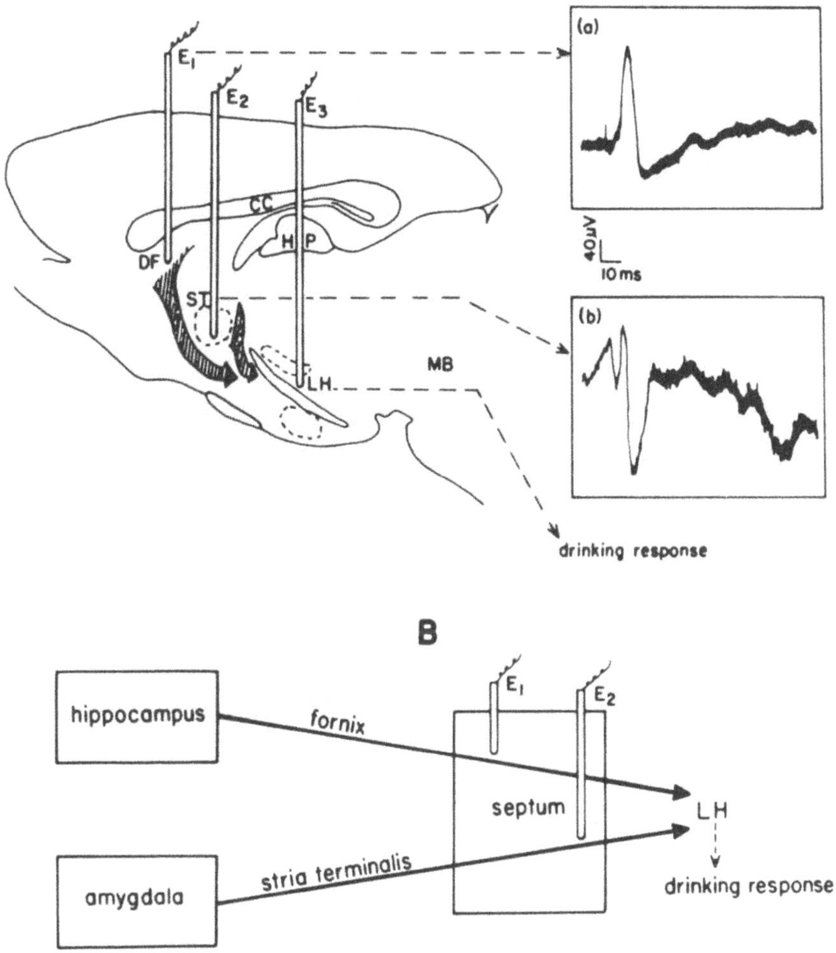

Figure 15. A. Evoked field potentials recorded from an electrode in the lateral hypothalamus (E3) from which drinking had been elicited previously by chronic electrical stimulation. Stimulation of the dorsomedial septum, in the region of the precommissural fornix fibers (Electrode E1), elicited the 10-14 msec component of the field potential (a). Stimulation of the ventrolateral septum, in the region of the bed nucleus of the stria terminalis (Electrode E2), elicited the 18-23 msec component of the field potential. B. From electrophysiological and lesioning studies it appears that precommissural fornix fibers mediate the facilitatory effect on elicited drinking and fibers of the stria terminalis the inhibitory effects, Miller and Mogenson (30).

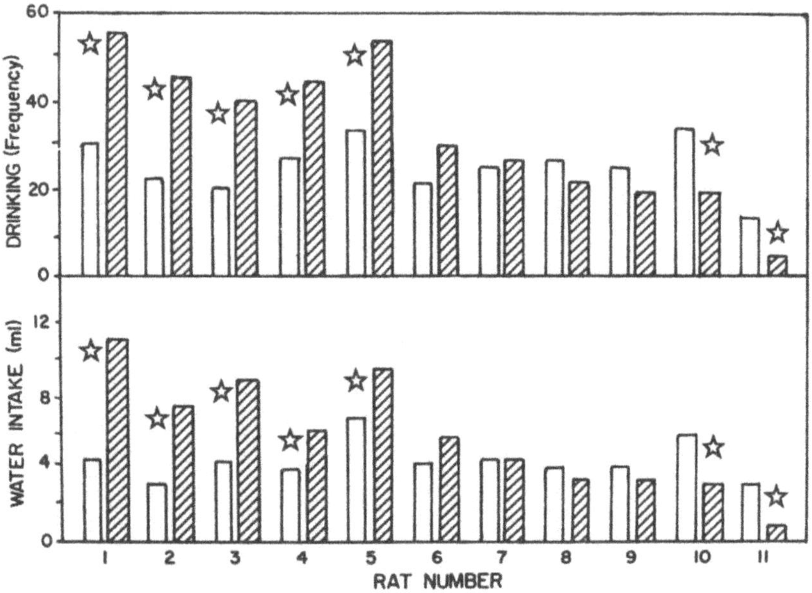

Figure 16. Average frequency of drinking and average water intake for 11 rats for three test sessions. Open columns indicate hypothalamic stimulation alone, cross hatching indicates when a pulse of stimulation was presented to the septum 5 msec before each pulse to the hypothalamus. Stars designate differences that are statistically significant ($p < .01$). From Sibole, Miller and Mogenson (42).

activation of neural systems associated with deficit signals but rather with incentive signals (31).

SUMMARY

The functions of the septum are not very well understood and there has been a tendency to consider its contributions to physiological and behavioral responses in relation to those associated with the hypothalamus. The septum, like the other structures of the limbic system, has prominent connections with the hypothalamus and midbrain. In the earlier investigations there was a tendency to consider that the septum was concerned with response inhibition, exerting inhibitory effects on such hypothalamus systems as those concerned with emotional expression and drinking behavior. Subsequent electrophysiological studies have shown that stimulation of the septum activates as well as inhibits hypothalamic neurons

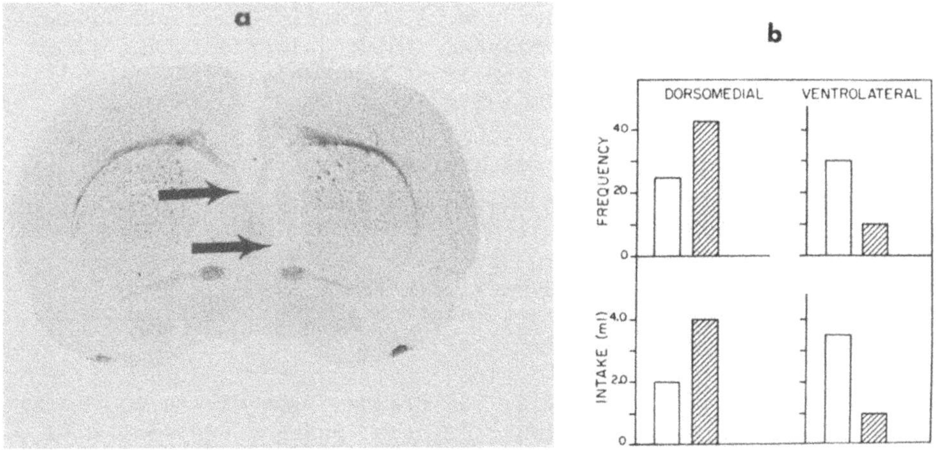

Figure 17. Two sites of stimulation of the septum in the same animal are shown by arrows. The upper arrow indicates the electrode placement in the dorsomedial septum from which there was facilitation of the drinking elicited by hypothalamic stimulation. The lower arrow indicates the electrode placement in the ventrolateral septum from which there was suppression of the elicited drinking. From Mogenson (1973).

and it is considered that the septum exerts modulatory effects on hypothalamic and midbrain systems. The modulatory effects of septal stimulation on cardiovascular responses, on self-stimulation of the lateral hypothalamus were discussed. Electrophysiological and behavioral evidence was presented which suggests that precommissural fornix fibers transmit primarily facilitatory effects on hypothalamic neurons and stria terminalis fibers transmit primarily inhibitory effects. One or other of these pathways may be dominant depending in part on the discharge rate of the hypothalamic neuron being monitored, which is related to the pattern and magnitude of inputs to the system.

REFERENCES

1. Antunes-Rodrigues, J., Gentil, C. G., Negro-Vilar, A. and Covian, M. R. Role of adrenals in the changes of sodium chloride intake following lesions in the central nervous system. Physiol. Behav., 5 (1970) 89-93.

2. Antunes-Rodrigues, J., Saad, W. A., Gentil, C. G. and Covian, M. R. Mechanism of decreased sodium chloride intake after hypothalamic lesions: effect of hydrochlorothiazide. Physiol. Behav., 5 (1970) 1183-1185.

3. Bard, P. A diencephalic mechanism for the expression of rage with special reference to the sympathetic nervous system. Amer. J. Physiol., 84 (1928) 490-515.

4. Black, S. L., and Mogenson, G. J. The regulation of serum sodium in septal lesioned rats: a test of two hypotheses. Physiol. Behav., 10 (1973) 379-384.

5. Blass, E. M. and Hanson, D. G. Primary hyperdipsia in the rat following septal lesions. J. Comp. Physiol. Psychol., 70 (1970) 87-93.

6. Brady, J. V. and Nauta, W. J. H. Subcortical mechanisms in emotional behavior: Affective changes fol-owing septal forebrain lesions in the albino rat. J. Comp. Physiol. Psychol., 46 (1953) 339-346.

7. Calaresu, F. R., and Mogenson, G. J. Cardiovascular responses to electrical stimulation of the septum in the rat. Amer. J. Physiol., 223 (1972) 777-782.

8. Chiaraviglio, E. Effect of lesions in the septal area and olfactory bulbs on sodium chloride intake. Physiol. Behav., 4 (1969) 693-697.

9. Covian, M. R. Studies on the neurovegetative and behavioral functions of the brain septal area. In: W. R. Adey and T. Tokizane (Eds), Structure and Function of the Limbic System. Progress in Brain Research, Vol. 27. Elsevier, Amsterdam, 1967, 189-217.

10. Covian, M. R., Antunes-Rodrigues, J. and O'Flaherty, J. J. Effects of stimulation of the septal area upon blood pressure and respiration in the cat. J. Neurophysiol., 27 (1964) 394-407.

11. Covian, M. R., Lico, M. C. and Antunes-Rodrigues, J. Blood pressure and respiratory changes by stimulation of brain septal area in rats. Acta Physiol. Latino Amer., 16 (1966) 99-105.

12. Covian, M. R. and Timo-Iaria, C. Decreased blood pressure due to brain septal stimulation: parameters of stimulation, bradycardia, baroreceptor reflex. Physiol. Behav., 1 (1966) 37-43.

13. Dafny, N. and Feldman, S. Effects of stimulating reticular formation hippocampus and septum on single cells in the posterior hypothalamus. Electroenceph. clin. Neurophysiol. 26 (1969) 570-587.

14. Endröczi, E., Schreiberg, G. and Lissák, K. The role of central nervous activating and inhibitory structures in the control of pituitary-adrenocortical function. Effects of intracerebral cholinergic and adrenergic stimulation. Acta Physiol. Acad. Sci. Hung., 24 (1963) 211-221.

15. Gentil, C. G., Mogenson, G. J. and Stevenson, J. A. F. Electrical stimulation of septum, hypothalamus, and amygdala and saline preference. Amer. J. Physiol., 220 (1971) 1172-1177.

16. Grace, J. E. Central nervous system lesions and saline intake in the rat. Physiol. Behav., 3 (1968) 387-393.

17. Harvey, J. A. and Hunt, H. F. Effect of septal lesions on thirst in the rat as indicated by water consumption and operant responding for water reward. J. Comp. Physiol. Psychol., 59 (1965) 49-56.

18. Kaada, B. R. Somatomotor, autonomic and electrocorticographic responses to electrical stimulation of 'rhinecephalic' and other structures in primates, cat and dog. A study of response from the limbic subcallosal, orbital insula, pyriform and temporal cortex, hippocampus, fornix and amygdala. Acta physiol. scand., 24 (1951) 1-285.

19. Kabat, H. Electrical stimulation of points in the forebrain and midbrain. The resultant alterations in respiration. J. Comp. Neurol., 64 (1936) 187-211.

20. Kabat, H., Magoun, H. W. and Ranson, S. W. Reaction of the bladder to stimulation of points in the forebrain and midbrain. J. Comp. Neurol., 63 (1936) 211-239.

21. Lubar, J. F., Boyce, B. A. and Schaefer, C. F. Etiology of polydipsia and polyuria in rats with septal lesions. Physiol. Behav. 3 (1968) 289-292.

22. Lubar, J. F., Schaefer, C. F. and Wells, D. G. The role of the septal area in the regulation of water intake and associated motivational behavior. Ann. N. Y. Acad. Scis., 157 (1969) 875-893.

23. Malmo, R. B. Slowing of heart rate after septal self-stimulation in rats. Science, 133 (1961) 1128-1130.

24. Marwood, J. F., Gan, E. K. and Lockett, M. F. Lesions in the septal nuclei of the rat raise mean systemic arterial pres-sure and prevent the development of sound-withdrawal hypertension. J. Pharm. Pharmacol., 25 (1973) 614-620.

25. McCleary, R. A. Response specificity in the behavioral effects of limbic system lesions in the cat. J. Comp. Physiol. Psychol., 55 (1961) 605-613.

26. McCleary, R. A. Response-modulating functions of the limbic system; initiation and suppression. In: E. Stellar and J. M. Sprague (Eds), Progress in Physiological Psychology, Vol. 1. Academic Press, New York, 1966, 209-272.

27. Miller, J. J. Modulatory influences of the septum on hypothalamic functions: an electrophysiological and behavioral analysis. Ph.D. Thesis, University of Western Ontario, 1972.

28. Miller, J. J. and Mogenson, G. J. Effect of septal stimulation on lateral hypothalamic unit activity in the rat. Brain Res., 32 (1971) 125-142.

29. Miller, J. J. and Mogenson, G. J. Modulatory influences of the septum on lateral hypothalamic self-stimulation in the rat. Exp. Neurol., 33 (1971) 671-683.

30. Miller, J. J. and Mogenson, G. J. Projections of the septum to the lateral hypothalamus. Exp. Neurol., 34 (1972) 229-243.

31. Mogenson, G. J. and Phillips, A. G. Motivation: A psychological construct in search of a physiological substrate. In: Progress in Psychobiology and Psychological Psychology, Vol. 6. 1975 (In Press).

32. Mogenson, G. J. and Stevenson, J. A. F. Drinking and self-stimulation with electrical stimulation of the lateral hypothalamus. Physiol. Behav., 1 (1966) 251-254.

33. Murphy, J. T., Dreifuss, J. J. and Gloor, P. Topographical differences in the responses of single hypothalamic neurons to limbic stimulation. Amer. J. Physiol., 214 (1968) 1443-1453.

34. Nauta, W. J. H. An experimental study of the fornix in the rat. J. Comp. Neurol., 104 (1956) 247-272.

35. Nauta, W. J. H. Hippocampal projections and related neural pathways to the midbrain in the cat. Brain, 81 (1958) 319-340.

36. Negoro, H., Visessuwan, S. and Holland, R. C. Inhibition and excitation of units in paraventricular nucleus after stimulation of the septum, amygdala and neurohypophysis. Brain Res., 57 (1973) 479-483.

37. Negro-Vilar, A., Gentil, C. G. and Covian, M. R. Alterations in sodium chloride and water intake after septal lesions in the rat. Physiol. Behav., 2 (1967) 167-170.

38. Oomura, Y., Ooyama, H., Yamamoto, T., Naka, F., Kobayashi, N. and Ono, T. Neuronal mechanisms of feeding. In: W. R. Adey and T. Tokizane (eds), Structure and Function of the Limbic System. Progress in Brain Research, Vol. 27. Elsevier, Amsterdam, 1967, 1-33.

39. Powell, E. W. Septal efferents in the cat. Exp. Neurol. 14 (1966) 328-337.

40. Raisman, G. The connexion of the septum. Brain, 89 (1966) 317-348.

41. Ranson, S. W., Kabat, H. and Magoun, H. W. Autonomic responses to electrical stimulation of the hypothalamus, preoptic region and septum. Arch. Neurol. Psychiat., 33 (1935) 467-477.

42. Sibole, W., Miller, J. J. and Mogenson, G. J. Effects of septal stimulation on drinking elicited by electrical stimulation of the lateral hypothalamus. Exp. Neurol., 32 (1971) 466-477.

43. Stellar, E. The physiology of motivation. Psychol. Rev. 61 (1954) 5-22.

44. Stuart, D. G., Kawamura, Y. and Hemingway, A. Activation and suppression of shivering during septal and hypothalamic stimulation. Exp. Neurol., 4 (1961) 485-506.

45. Tsubokawa, T. and Sutin, J. Mesencephalic influence upon the hypothalamic ventromedial nucleus. Electroenceph. Clin. Neurophysiol., 15 (1963) 804-810.

46. Wishart, T. B. and Mogenson, G. J. Effects of food deprivation on water intake in rats with septal lesions. Physiol. Behav., 5 (1970) 1481-1486.

47. Wishart, T. B. and Mogenson, G. J. Reduction of water intake by electrical stimulation of the septal region of the rat brain. Physiol. Behav., 5 (1970) 1399-1404.

A FUNCTIONAL ANALYSIS OF THE SEPTAL NUCLEI[1]

J. F. DeFrance

Morin Memorial Laboratory, Department of Anatomy
Wayne State University, Detroit, Michigan

INTRODUCTION

At the end of the nineteenth century and into the early part of the twentieth, a number of investigators (22,23,25,32,44,54), emphasized the intimate relationship between the hippocampal formation and the septum provided by the fimbria-fornix system. The fimbria-fornix system was thought to arise exclusively from the hippocampal formation and subiculum. Cajal (44) epitomized the relationship by stating that the septum should be considered as a center associated with, or subordinate to, the hippocampus. Later, Crosby (12) suggested that there might be a reciprocal connection from the septum to the hippocampus within the fimbria-fornix pathway. This connection was subsequently confirmed (13,46).

Green and Arduini (27) and Petsche and co-workers (38,39) demonstrated that the close anatomical relationship between the two limbic system regions indeed represents a close functional relationship. These studies showed that the hippocampal activity depended upon the integrity of the septum (or septal nuclei).

Recently, the septal afferents coursing via the fimbria-fornix system has been the subject of study (14-18,21,36,40). Still,

---

[1]This study was supported by National Science Foundation Grant GB 35532 and U. S. Public Health Service Grants NB 00405, RR 5384. The authors wish to express their gratitude to H. Yoshihara, J. Kocsis, R. McCrea, and D. Agresta for their capable assistance and especially to S. T. Kitai for his support and advice throughout the study.

knowledge is lacking concerning possible functional differences associated with the various septal nuclei.

In this regard, it might be differences in the amount of available inhibition that is most interesting. Since the hippocampal input to septal neurons is excitatory, it may well be the patterns of inhibition, "sculpturing" out information, which make for functional differences between septal regions. The present study represents such a search within the septum of the cat.

The cyoarchitectural picture of the septum has lead most investigators (24,28,29,34,35,53) to the concept that it consists of two principal regions: a medial and a lateral septal nucleus. Nevertheless, the comparative studies of Andy and Stephan (5-8) suggest an alternate organizational scheme. The major septal divisions recognized in these studies are the medial, dorsal, lateral, and caudal groups. Anticipating the results, the present study finds physiological support for the organization suggested by Andy and Stephan (7).

METHODOLOGY

The results for this report were taken from a series of 260 acutely prepared cats, mature and of both sexes.

Surgical: the animals were anesthetized with a thiamylal sodium (30 mg/kg)-chloralose (70-80 mg/kg) combination or with sodium pentobarbital (40 mg/kg). The animals were pretreated with methyl atropine (0.1 - 0.2 mg/kg). A tracheotomy was performed for artificial respiration. Respiratory CO_2 was monitored and maintained between 3.5% and 4.5%. Femoral arterial blood pressure, EKG, and rectal temperature were monitored. A cephalic vein was connulated for drug and fluid delivery.

Following an extensive craniotomy, the cortex and corpus callosum overlying the septum and dorsal hippocampus were removed by suction. Paraffin oil at 38°C was used to cover the exposed brain tissue.

Stimulation: Stainless steel, bipolar stimulating electrodes (50-100 Kohms) were visually placed upon the lateral aspect of the ipsilateral fimbria (IFim [L]), medial aspect of the ipsilateral fimbria (IFim [M]), or the ipsilateral fornix (IFx). Bipolar electrodes (75-100 Kohms) were stereotaxically placed in the medial forebrain bundle (MFB) - rostral hypothalamus area to activate the ventral septal afferents (VSA) or to antidromically activate septal cells. For local stimulation (LOC) within the septum and subfield stimulation in the hippocampus, glass microelectrodes

(2MNaCl, 1-3 megohm) filled with fast green (52) were used. The magnitude of the microstimulating current was routinely monitored. Stimulus durations were varied between 0.1 and 3.0 msec. For most stimulus paradigms, however, 0.2 msec was the preferred duration.

Recording: Field potentials, extra - and intracellular unitary potentials were recorded in the septal nuclei following stimulation of related pathways and adjoining cellular regions. Glass microelectrodes were used for recording, filled with 2 MNaCl for field potentials (1-3 meghom) and extracellular units (5-10 megohm) and 2MKCl for intracellular recordings (10-20 megohm). In addition, the recording electrodes were filled with fast green to mark important recording spots.

The recording microelectrode was zeroed under a microscope with respect to known and consistent landmarks: the midline and the function between the septal gray and fimbria-fornix system (Fig. 1A ref. zero). The microelectrode was then moved known distances for recording.

Field potentials evoked by hippocampal stimulation were analyzed in the same manner as those described for the rat (16). There were no marked differences between the waveform of the evoked responses between the two species. Fig. 1B (lower trace) shows a test field response from a paired-stimulus testing (PST) paradigm with the test control superimposed. The various are components indicated. Briefly, the early negativity (N_0 or N_1) reflects either a compound fiber volley (N_0) or the antidromic activation of cells (N_1), or both in some instances. Extracellular synaptic currents associated with EPSPs make up the greater part of the N_2 component. N_3 is an envelope of action currents resulting from the discharge of the cells (i.e., principal cells) which project out of the immediate septal group. Therefore, the amplitude of N_3 serves as an index for the relative number of principal cells discharging from a given experimental manipulation. This strategy is similar to the population spike analysis in the hippocampus (2). The amplitude of the N_3 component was determined by subtraction of the N_2 component from the combined N_{2-3} negativity. The amplitude of the N_2 component was taken from the test response to stimuli at a 5 msec interstimulus interval (ISI). Generally, the N_2 amplitude remained unchanged for all intervals greater than 4 msec. The positivity, P_1, reflects the current flow leading to intracellular IPSPs.

Well-isolated extracellular unitary recordings were also obtained in the septum. Unitary discharges, determined to be of somatic origin by subsequent intracellular penetration, presented a positive-going spike followed by a smaller and more prolonged negativity (Fig. 1C, arrow). Responses of this kind are similar to

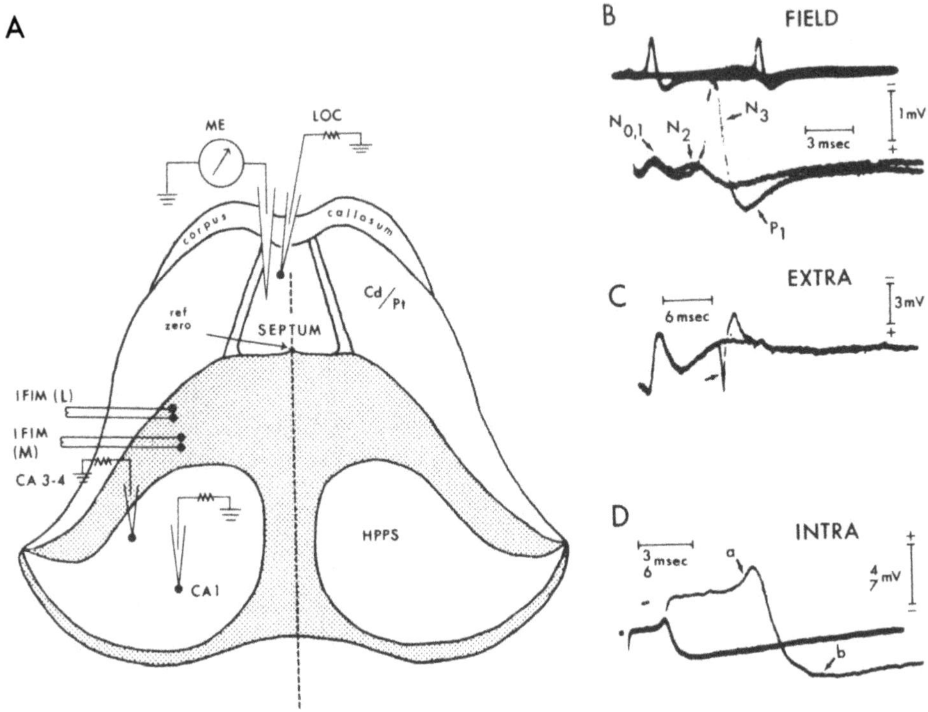

Figure 1. Experimental arrangement and identification of components. A. Schema of experimental arrangement. B. Field responses to paired stimuli with the identification of the $N_{0,1}$, N_2, N_3, and P_1 components. Upper trace shows both the conditioning and test response. The lower trace shows only the test response, with the control superimposed. C. Extracellularly recorded unitary discharge. Note that the spike is positive-going. D. Intracellularly recorded synaptic potentials from a principal cell in response to IFim stimulation. "a" indicated the EPSP component and "b" indicated the IPSP component.

giant extracellular spikes (26,51) and occur when the recording electrode is near or in contact with the neuronal membrane.

Intracellularly recorded potentials from principal cells following IFim stimulation, show a depolarizing-hyperpolarizing sequence (Fig. 1D). The depolarizing potentials may be EPSPs, action potentials, or both. The hyperpolarizing potentials are IPSPs (17). A minimum transmembrane potential of 30 mV was established as a criterion for intracellular recordings to be included in the analysis.

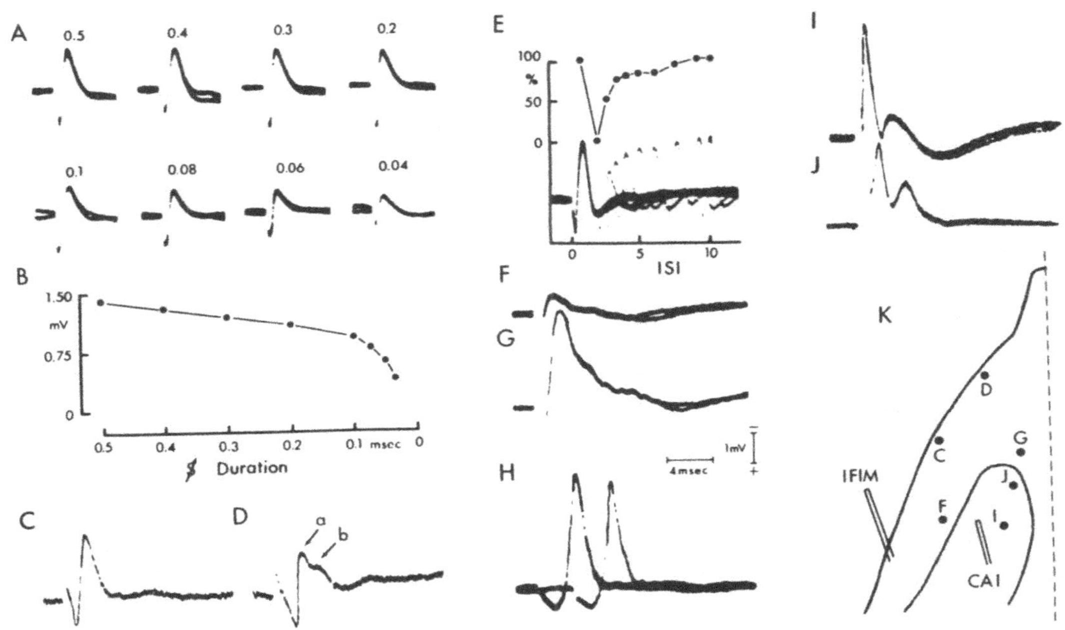

Figure 2. Response characteristics of the fimbria-fornix system and septal course of dorsal CA1 axons. A. Fimbria (IFim) fiber volleys at a fixed stimulus intensity, but varying stimulus duration. B. Plot of the changes in amplitude as a function of the stimulus duration for the responses shown in A. C, D. Fiber responses recorded at two different distances from the point of stimulation. E. PST of a IFim fiber volley. The peak amplitudes are plotted as a function of the ISI. F, G. Tandem of fiber responses following dorsal CA1 stimulation. Refere to K for stimulation and recording points. H. Antidromic response recorded in pyramidal cell layer of dorsal CA1 (same as stimulation site in F, G) following stimulation of point G, I, J. Fiber responses of dorsal CA1 following stimulation of CA1 as shown in K. K. Schema of stimulation and recording points for C–J.

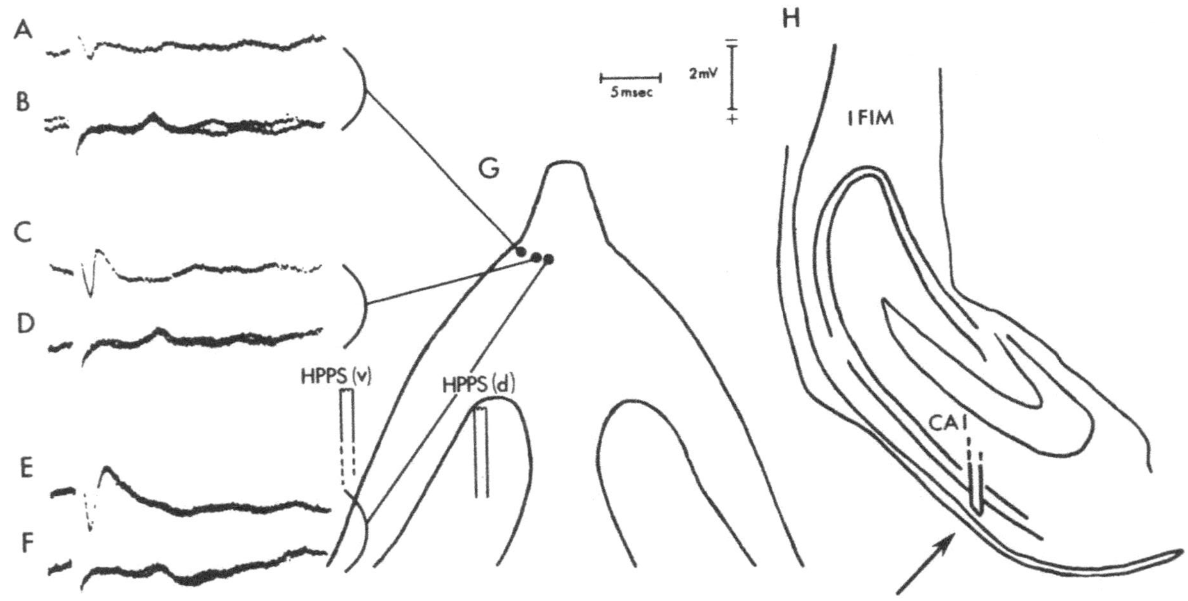

Figure 3. Comparison of fiber volleys evoked by dorsal hippocampal (HPPS [d]) and ventral hippocampal (HPPS [v]) stimulation. A, C, E. Compound fiber volley following HPPS (d) stimulation. B, D, F. Compound fiber volley following HPPS (v) stimulation. G. Schema showing recording points for A-F. H. Sketch drawn from the histological section locating the HPPS (v) stimulation electrode.

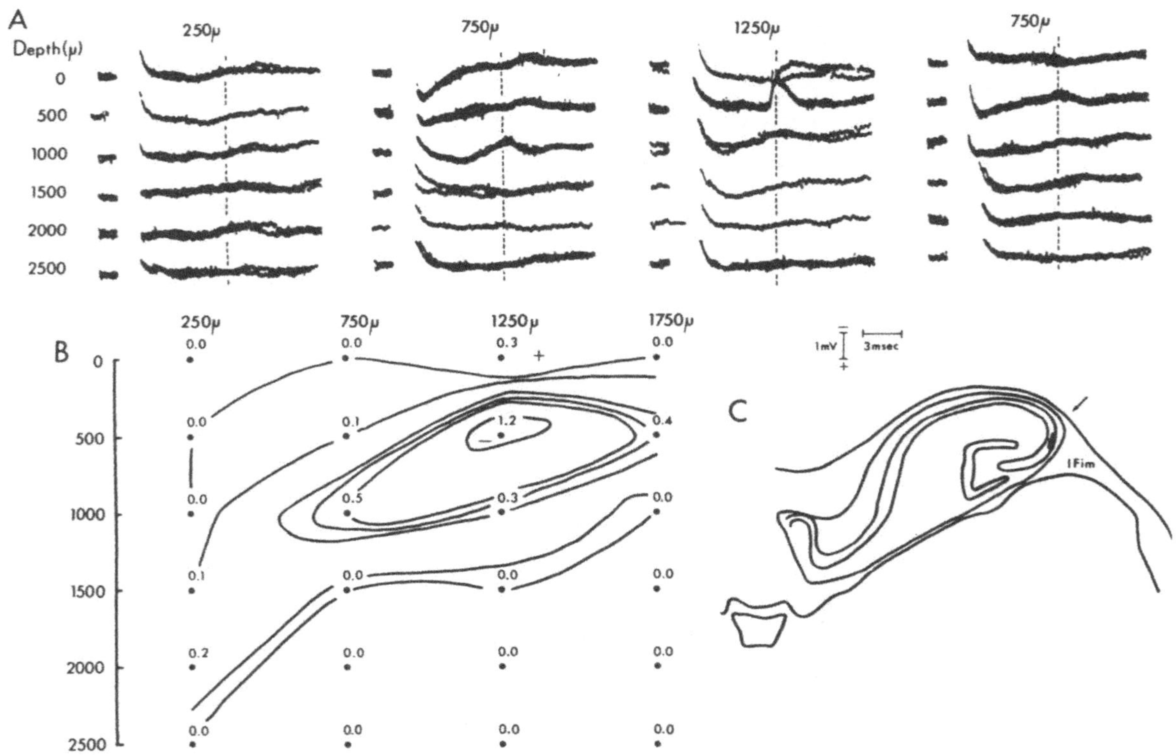

Figure 4. Mapping septal field potentials following microstimulation of CA3. A. Record specimens of evoked field responses for indicated depths and distance from the midline (250μ, 750μ, 1250μ, 1750μ). B. Isopotential plot of responses shown in A. C. Sketch of histological section locating tip of stimulation electrode. The stimulation electrode resistance was 1 megohm and the stimulation current was 75μA.

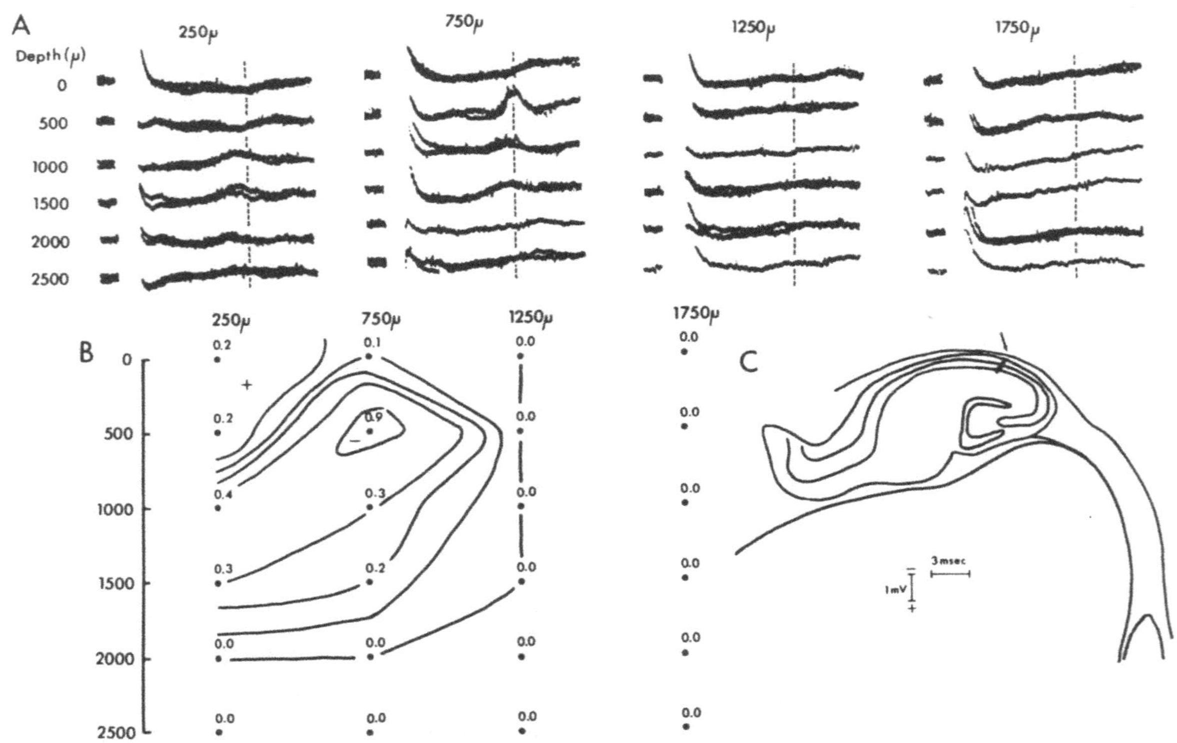

Figure 5. Mapping septal field potentials following microstimulation of CA1. A. Record specimens from the same location as shown in Fig. 4. B. Isopotential plot of responses shown in A. C. Sketch of histological section locating tip of stimulating electrode. Stimulation current was 75μA passed through a 1 megohm electrode.

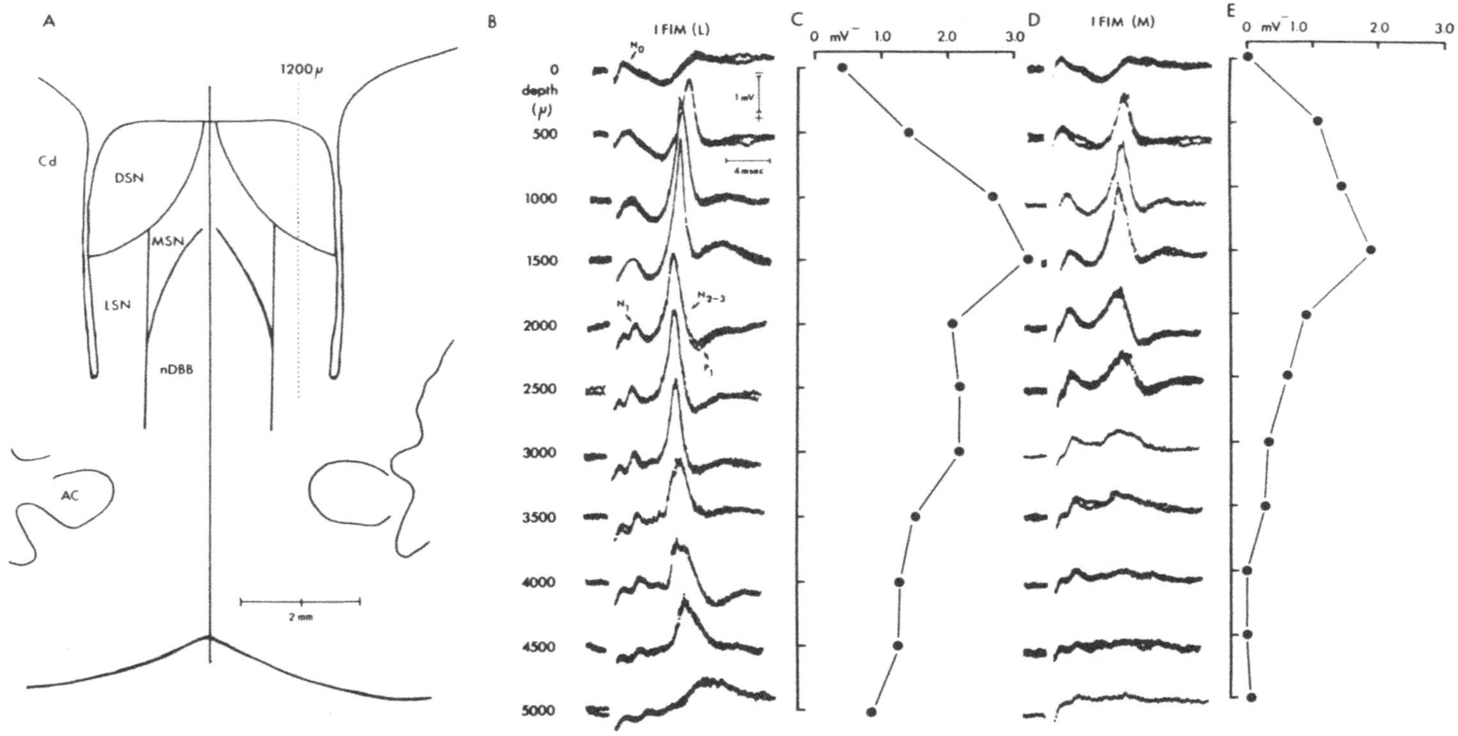

Figure 6. Field profiles in the lateral septal region (1200µ from the midline to IFim and VSA stimulation. A. Schema, including location of recording probe. B. Profile of field potentials in response to IFim stimulation. C. Plot of the peak amplitudes in B as a function of depth. D. Profile of field potentials in response to VSA stimulation. E. Plot of the peak amplitudes in D as a function of depth. Negative up in all instances. Onset of the stimulus corresponds to the break from the baseline.

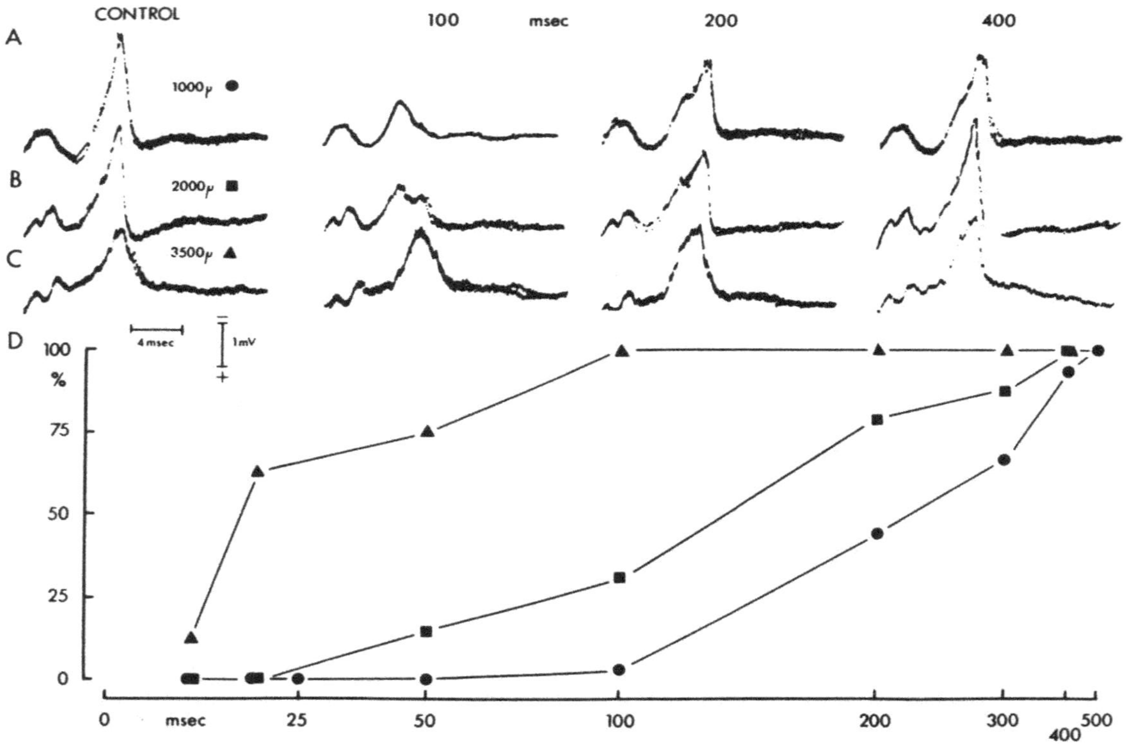

Figure 7. Effects of PST with an IFim-IFim combination upon field responses in the lateral septal region. A. Record specimens of test responses recorded at a depth of 1000μ. Control is the control test response. Test responses at 100 msec, 200 msec, and 400 msec ISIs are illustrated. B. Similar series as in A, except at a depth of 2000μ. C. Similar series as in A, B, except at a depth of 3500 . D. Plot of the N_3 amplitude with % of control as a function of ISI. The N_3 amplitude was determined by subtracting the N_2 amplitude determined from a 5 msec interval from the combined N_{2-3} amplitude.

Histology: Following the experiment, a DC current was passed through the stimulating electrodes to mark the site of the stimulated area. For the microstimulating electrodes filled with fast green, an anionic current (10-20 nA) was passed for approximately 20 minutes to deposit dye in the extracellular medium surrounding the tip (52). The same procedure was carried out to mark the recording sites.

RESULTS AND DISCUSSION

Fiber Characteristics and Hippocampal Fields of Origin

Before looking at septal responses to hippocampal or fimbria activation the characteristics of fimbria and fornix responses needed to be established. Fig. 2A-K shows the results of different studies of the fimbria-fornix system. Fig. 2K is a schema of experimental arrangement and recording points of responses shown in Fig. 2C, D, F-J.

The result of altering the stimulus duration upon compound fiber volleys of the fimbria is shown in Fig. 2A, B. The stimulus intensity is held constant at 2.0 times the threshold (2.0T). As the stimulus duration is reduced to 0.1 msec, there is a linear diminution in the response amplitude. However, with durations less than 0.1 msec there is a rapid, non-linear, reduction in response amplitude. Stimulus durations of 0.1 - 0.2 msec were selected therefore, for efficient stimulation of the fimbria pathways.

The conduction velocity of the hippocampal efferents within the fimbria ranged from 1.8 - 2.3 M/sec. The calculated conduction velocity for the responses in Fig. 2C, D was 2.1 M/sec, when calculated by the latency difference between the peak of the response in Fig. 2C and the "a" peak of the response in Fig. 2D. However, the conduction velocity of fibers giving rise to the 'b' peak is 1.9M/sec. The fimbria might not, then, consist of a single population of fibers.

Additional evidence for a "complex" fimbria comes from the PST paradigm for the fimbria fiber volley (Fig. 2E). Recovery of the test fiber volley commences at a 1.5 msec interstimulus interval (ISI). Recovery then steadily progresses until a 4 msec ISI, where the response in 75% of the control response amplitude. Recovery then does not commence to improve until a 7.5 msec ISI with full recovery being achieved in a 10 msec ISI. Recovery of the fiber volley is then a two step process. The reason for the unexpectedly prolonged reocvery times is not known. Complete surgical isolation of the fimbria had no effect upon the time course for the recovery. Also, without effect, was the application of tetraethylammonium (TEA) ions which blocks K^+ conduction (30). Hence, the response suppression was not due to spike after-hyperpolarization, which is K^+ dependent. However, the prolonged time-course for complete recovery of the fiber volley was not reflected by the refractory periods of single fibers recorded in the septum. Extracellularly recorded single units (fiber responses and antidromic action potential) within the septal nuclei typically recovered in an ISI of 0.7 - 4.0 msec. It must be borne in mind at the number of action potentials is not the only determinant of the amplitude of a compound fiber volley. Changes in the amplitude of the action potentials can result from changes in the transmembrane potential upon which the action potentials are superimposed (20). Extracellular potential changes have been shown in some cases to alter the intracellular potential and excitability (31). The opportunity is in hand with the fimbria because of slight differences in the conduction velocity among the fibers. The slower fibers may be influenced by the passage of action potentials along the faster fibers. An alternate explanation, and perhaps the most salient one, is that the prolonged recovery is somehow due to cells making up the "bed nucleus" of the fimbria-fornix system (9).

Most of the septally directed axons of the CA 3-4 pyramidal cells course in the fimbria. Fig. 2F-H is concerned with the direction of axons of CA1 pyramidal cells in the septal and intermediate parts of the hippocampus. A very small fiber response was seen when stimulating the CA1 alveus and recording in the fimbria opposite the lamella (Fig. 2F). On the other hand, a large fiber response was seen when recording in the fornix (Fig. 2G) or along the alveus (Fig. 2I, J), more or less, perpendicular to the lamella. Moreover, when stimulating perpendicular to the lamella (i.e., as in Fig. 2 point G), an antidromic pyramidal cell field response was recorded in CA1 of the hippocampus (Fig. 2H). Hence, at least some of the axons of the CA1 cells course septalward in a direction almost perpendicular to the long axis of the lamella. The conduction velocity of CA1 axons ranged from 1.6 - 2.0 M/sec. The responses shown in Fig. 2I, J gave a calculated conduction velocity of 1.8 msec. Typically, the conduction velocity of dorsal CA1 axons were slightly slower than the axons making up much of the fimbria. The conduction velocity of both CA1 and CA 3-4 axons becomes slower within the septum. The intraseptal conduction velocity was estimated to be from 0.8 - 1.0 M/sec.

The hippocampal origin of the fibers making up different parts of the IFim were grossly studied with macrostimulating electrodes at various points in both the dorsal and ventral hippocampus (See Fig. 3). In complete agreement with the findings of Siegel and Tassoni (49), it was found that fibers from the temporal portions of the hippocampus travel most laterally in the IFim. By contrast, fibers from dorsal hippocampal regions occupy the more medial aspects of the IFim. To illustrate this point, Fig. 3 shows the results of an experiment where macrostimulating electrodes were placed in the dorsal hippocampus (HPPS [d]) and ventral hippocampus (HPPS [v]). Figs. 3A, C and D show responses from the IFim following supramaximal HPPS (d) stimulation while Figs. 3B, D, F show responses following supramaximal HPPS (v) stimulation. The fiber responses to HPPS (v) stimulation were greatest in the IFim (L). On the other hand, HPPS (d) stimulation evoked the largest response in the IFim (M). These results, thus, support the anatomical studies of Siegel and Tassoni (49), and the electrophysiological studies of Andersen et al. (1,3) indicating that the fibers making up the more lateral parts of the fiber originate from the HPPS (v). Therefore, the hippocampal output appears to be ordered such that fibers arising from the more septal portions of the HPPS project via the more medial parts of the IFim, while fibers arising from the more temporal portions of the HPPS project via the more lateral parts of the IFim.

While the hippocampsal septal output was found to be grossly organized along a septal - temporal axis, a more refined organization became apparent. Raisman (41) has argued that the hippocampal-septal system was organized on the basis of cytoarchitectural fields. This organization has been denied by Siegel and his coworkers (21,48-50). While the date shows in Fig. 3 are in general accord with the findings of Siegel et al. (48-50), the data making up Figs. 4 and 5 suggest a compromise: A specific field of origin difference appears to be imposed upon the more general dorsal-ventral distinction.

To look for possible hippocampal field differences in the septal distribution of output, microstimulation electrodes were placed, in tandem, within the 'same' lamella of either the dorsal or intermediate section of the hippocampus. Figs. 4 and 5 show the result of the typical experiment using a tandem electrode arrangement. With supramaximal stimulation of CA3, the maximum evoked negative response was recorded 1250μ from the midline (Fig. 4). On the other, supramaximal CA1 stimulation evoked a response that was maximal 750μ from the midline. The generalization which follows from this, and similar experiments, is that for any given lamella, the CA1 projections tend to terminate medial to the CA3 distribution. But, as the tandem of stimulation electrodes moved from the septal to the temporal portions of the hippocampus, the array of ordered responses march from medial to lateral across the septum. That is to say, the more septalward a lamella, the more medial will be the projections from CA1 and CA3. But, for that same lamella, the CA1 projection will be medial to the CA3 projection. Hence, a compromise: There appears to be a cytoarchitectural field distinction, but there is a more general (and perhaps more important) dorsal-ventral distinction.

This ordersing of CA3 and CA1 septal input is consistent with earlier findings (16,17) showing the convergence upon single cells in septum for the two subfields, but marked convergence between the ipsilateral and contralateral CA3 subfields. This may provide the basis for an important functional difference between the two hippocampal output systems.

Organization of the Lateral Septal Region

Previous studies (14-18) in the rat and in the cat have found that hippocampal-septal projections to be monosynaptically excitatory. In those studies, however, the dorsal septal regions were the focus of attention. In the present study, the entire septal region dorsal to the plane of the anterior commissure was investigated. To facilitate regional comparisons, most of the field potential data to be presented Figs. 6-7, were obtained from a single preparation.

Fig. 6 shows field profiles of responses in the lateral septal region (1200 from the midline) evoked by IFim (L) and IFim (M) stimulation. Fig. 6A is a schema drawn from the histological section locating the lateral probe. Comparison of Fig. 6B and Fig. 6D shows that, within the lateral septal region, the IFim (L) stimulation evoked the largest response. However, responses to both IFim (L) and IFim (M) were maximal at a depth of 1500μ below the dorsal surface. Also, the N_1 component did not appear until a depth of 2000μ. With the appearance of N_1, the N_{2-3} components are maximal in the dorsal-lateral regions where there is no N_1 component in the field response. Another thing to note is that with IFim (L) stimulation the N_{2-3} components can be followed down to a depth of 4500μ, while with IFim (M) stimulation distinguishable N_{2-3} negativities cease at approximately 3000μ. Although not well illustrated here, the N_1, component tends to be maximal between 3000μ - 3500μ, thereafter the component disappears. The average conduction velocity of N_1 fibers was calculated to be 4.0 msec.

For the probe of Fig. 6, the DSN was sampled between 0μ and 2000μ and the LSN between 2000μ and 5000μ. The dorsal-ventral extent of the septal region showing the above properties was a function of the distance of the probe from the midline, as is the ventral boundary of the DSN.

Despite the fact that the fimbria-fornix fibers maintain the integrity of their position within the tract, IFim (M) stimulation evoked responses in the lateral septal region and vice versa (see Fig. 6). The problem might resolve itself if the extent of the dendritic field for a single neuron is taken into consideration. Septal neurons may show a spread of dendritic processes up to 2/3 of the septal hemisphere (10). It is not inconceivable then, that input arriving via IFim (M) should influence, if not activate, laterally placed neurons in the septum. Therefore, with a fimbria fiber distribution outlined by Siegel and Tassoni (49), it still should be possible to influence neurons residing outside of the focus of the input.

In the DSN, a total of 140 instances of PST with field potentials were examined using the IFim (L) - IFim (L) combination. This procedure was used to test the amount of available inhition. In 134, or 96% of the cases the N_3 component commenced to recover at a 100 msec ISI or greater. The most popular interval was 200 msec; accounting for 50% of the cases. In the LSN, by contrast, the test field responses started to show N_3 recovery at intervals of 50 msec, or less, in 63 of the 88 (i.e., 72%) testings. The interval which most often showed beginning N_3 recovery was 20 msec (35%), followed by 10 msec (28%). Fig. 7 illustrates a typical example of the regional differences observed with the PST procedure.

The testing procedure was carried out with the field potentials shown in Fig. 6B. Examples are taken from series at depths of 1000μ, 2000μ, and 3500μ, respectively. The time-course of N_3 recovery for each series is plotted in Fig. 7D.

The main feature shown in Fig. 7 is that the amount of N_3 suppression varies inversely with depth. At a depth of 1000μ (closed circles), the action currents of the principal cells did not commence to recovery until a 100 msec ISI. The test response did not show complete recovery until a 500 msec interval was reached. At an intermediate depth (2000μ, closed squares), the N_3 component of the test response showed some reocvery in a 50 msec interval. Still deeper, at 3500μ (closed triangles), the recovery of the N_3 component of the test response commenced in a 10 msec interval, and became fully recovered in a 100 msec interval.

Figs. 6 and 7 illustrated the distribution and behavior of field responses in the lateral septal region to IFim stimulation. Figs. 8-10 are concerned with unitary activity and the correlation with the regional differences suggested by the field analyses.

Fig. 8A is a histogram showing the number of units recorded in the DSN recovering in different categories of ISIs. Of 44 units tested, 39 (i.e., 90%) showed recovery of unitary firing at intervals of 100 msec or greater. This distribution of recovery intervals for unitary discharges corresponds nicely with the typical duration of suppression of field responses in the DSN.

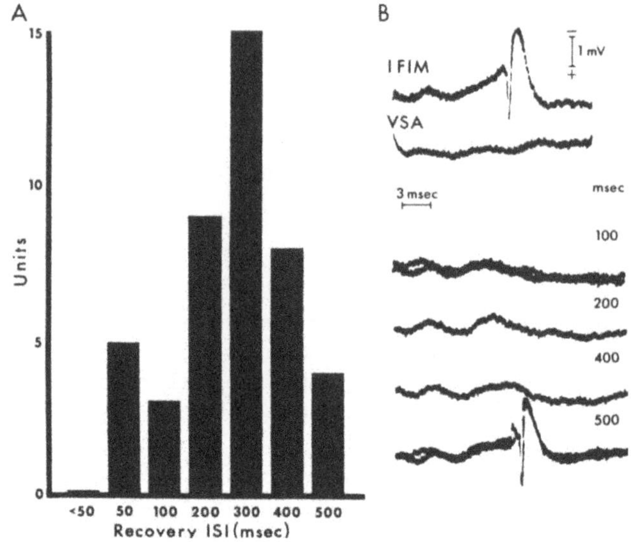

Figure 8. Analysis of unitary principal cell responses in the DSN. A. Histogram showing number of extracellular unitary discharges recovering in each category of ISI. B. Extracellular unitary discharge responding to IFim (L) stimulation but not VSA stimulation and paired-stimulus testing with the IFim-IFim combination. Test responses are shown for intervals ranging from 100 msec - 500 msec.

An example of a PST series for a unit isolated in the DSN is shown in Fig. 4B. With combined IFim-IFim stimulation, the unitary discharge was suppressed in intervals up to 500 msec.

To be noticed, also, is that the cell did not respond synaptically to VSA stimulation. A total of 28 units responding to IFim stimulation were tested for convergence of IFim and VSA input, but none was found.

One additional important aspect of the dorsolateral septal region is the absence of antidromically activated responses to fimbria-fornix stimulation. The relationship between the DSN and the hippocampus appears to be uni-directional.

Using a IFim-IFim stimulus combination, test field responses in the LSN showed only a short-term inhibition of the N_3 component. Likewise, extracellularly recorded unitary responses showed suppression only in the brief ISIs. A histogram of recovery intervals for units isolated in the LSN is shown in Fig. 9A. A total of 54 units were tested with IFim-IFim stimulation and of those, 48 (i.e., 88%) recovered in intervals of 50 msec or less.

Fig. 9B shows an extracellular unit which has a reocvery ISI of 20 msec. This is typical because of the 54 extracellular units tested, 30 (i.e., 56%) recovered at an ISI of 20 msec.

In further contrast to the dorsolateral septum, certain cells of the ventrolateral region (i.e., 3000μ - 5000μ) can be antidromically activated following IFim (L) stimulation. Recall that field responses to IFim (L) stimulation in this part of the lateral septal region included the N_1 component (see Fig. 6D). Fig. 9C shows an extracellularly recorded unitary response which arose from the N_1 field. The unitary responses of this variety are positive - negative. The apparent triphasic nature of the unitary response is due to the negativity of the N_1 field component. The cell depicted in Fig. 9C responded antidromically, but not synaptically. The refractory period of these cells ranged from 1.5 - 2.8 msec. On the other hand, cells which responded synaptically did not respond antidromically (Fig. 9B). No cases were encountered where a cell responded both antidromically, and synaptically, to fimbria-fornix stimulation. A total of 12 cells were antidromically identified following IFim stimulation as compared to 54 synaptically activated units. Thus, the ventrolateral septal cells which project onto the hippocampus are not under direct hippocampal influence.

Intracellular recordings from principal cells in the ventrolateral region also demonstrated that test responses are capable of issuing action potentials in a 50 msec interval or less. For example, in Fig. 9D, the action potential of the test response was present when tested in a 20 msec interval; however, it was blocked when the interval was adjusted to 10 msec (Fig. 9E). Fig. 9F is the extracellular control for Fig. 9D in which the N_3 component of the test response (refer to lower trace) shows little suppression of the N_3 component.

The typical behavior of the EPSPs within the lateral septal region was studied with PST (Fig. 10A, B). Fig. 10A shows the control IFim response and Fig. 10B shows the result with a 10 msec interval. Two things are to be noted with respect to the test response (Fig. 10B): (1) The slope of the rise-time of the EPSP remained essentially unchanged. (2) There was virtually no increase in the EPSP amplitude in spite of being superimposed upon a hyperpolarizing potential. Both of these phenomena would be predicted if the EPSPs are monosynaptic and are generated in the dendrites with the IPSPs generated at the soma (43). PST also showed the lack of summation of the hyperpolarizing potential. The

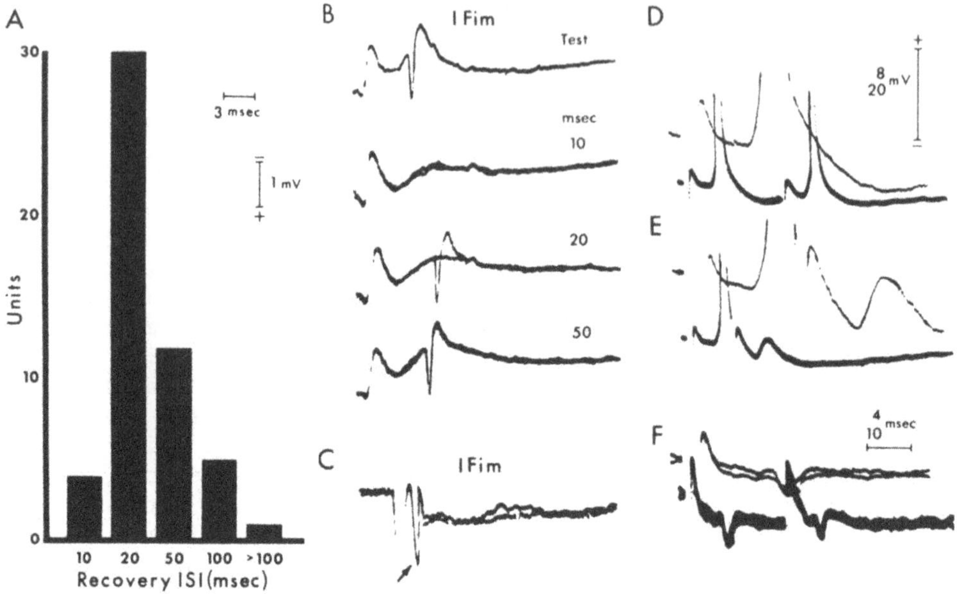

Figure 9. Unitary analysis of principal cells in the LSN. A. Histogram showing number of units recovering in each category of ISI. B. Orthodromic extracellular unitary discharge to IFim (L) stimulation in a PST paradigm. The test control (test) and test responses are shown for 10 msec, 20 msec, and 50 msec ISI. C. Antidromic extracellular unitary discharge (arrow) to IFim stimulation. D., E. Intracellular recordings from a principal cell, with PST for a 20 msec and 10 msec intervals, respectively. F. Extracellular control for D.

trailing hyperpolarization appears to be a continuation of the IPSPs associated with the conditioning response. The lack of summation is not attributable to the approach of the IPSP reversal potential since summation is still absent in intervals where the membrane potential is substantially recovered. Perhaps there is an inhibition of the inhibitory interneurons. An analogous situation might be the Renshaw cell inhibition of Renshaw cells in the spinal cord (47).

From the evidence thus far presented regarding extrinsic inputs, the lateral septal region consist of two physiologically different regions: the DSN and the LSN. Exploration of the intrinsic

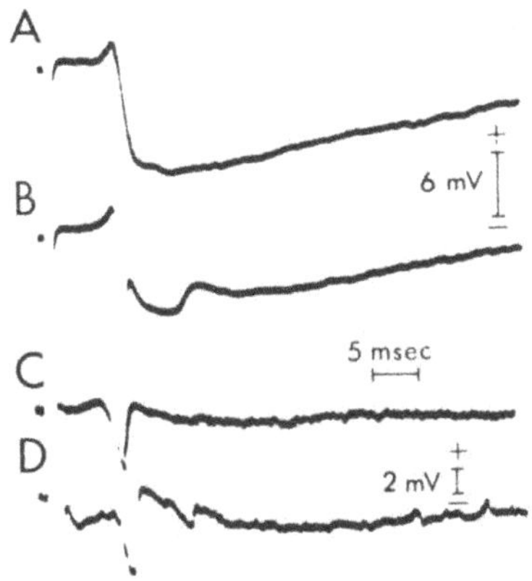

Figure 10. Example of the behavior of intracellularly recorded synaptic potentials in lateral septal region neurons. A. Intracellularly recorded EPSP-IPSP sequence to IFim (L) stimulation. B. PST of the response in A at a 10 msec ISI. C., D. Extracellular controls for A, B, respectively.

organization of these two regions was then undertaken. Fig. 11 deals with the pathways involved in the inhibition of the principal cells in the DSN. In the rat (14,17), evidence was adduced for the existence of a recurrent collateral pathway mediating the inhibition of the N_3 component. Similar results were obtained in the present study on the cat. Fig. 11A illustrates the effects of conditioning the N_{2-3}-P_1 response to IFim stimulation with an antidromic response elicited by LOC microstimulation. The LOC stimulus activated the descending axons of the principal cells, thereby antidromically activating the cell soma. The antidromic field evoked by LOC stimulation typically presented a large negative potential change followed upon by a smaller but more prolonged positivity (Fig. 11A, LOC, downward oblique arrow). The antidromic invasion of the cell soma itself did not always occur since the negativity would occasionally change in a step-like manner, indicating failure at invasion. But, the fact that the LOC induced negativity was indeed a collection of principal cell soma discharges was determined by occlusion with the N_3 component to IFim stimulation (now shown).

For the LOC-IFim combination, the IFim test response was not fully recovered until a 400 msec ISI was reached. For a 20 msec ISI, the N_3 component was almost completely suppressed. Therefore, by activating the axons of the principal cells a powerful inhibitory circuit can be activated. The duration of this suppression

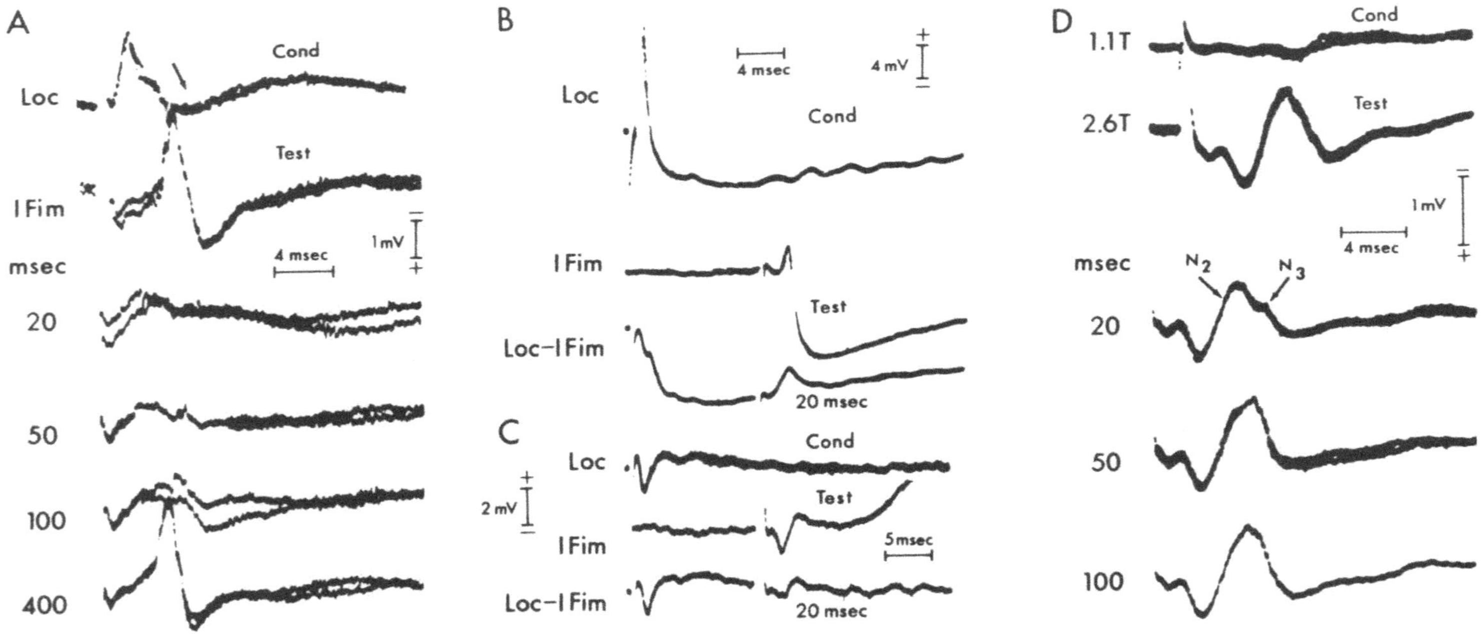

Figure 11. Study of the inhibitory circuits in the dorsal part of the lateral septal region. A. PST conditioning the antidromic (IFim) evoked field response with the antidromic (LOC) in evoked field response. Arrow points to the positivity associated with the antidromic negativity. Test responses are shown for 20 msec, 50 msec, 100 msec, and 400 msec interstimulus intervals. B. Intracellular correlates to A; LOC is an antidromic spike and hyperpolarization to LOC stimulation. IFim is the control test response, and PST with a LOC-IFim combination is shown for a 20 msec intervals. C. Extracellular controls for B. D. PST with a IFim (1.1T) - IFim (2.6%) combination. Test responses at 2.6T are shown for 20 msec, 50 msec, and 100 msec intervals.

approximated the duration of N_3 suppression for IFim-IFim combinations. In the shorter ISIs, however, the suppression was seldom complete.

Intracellular correlations to the field analysis presented in Fig. 11A are shown in Fig. 11B. LOC stimulation evoked an antidromic spike and subsequent hyperpolarization. After decay of the spike mechanism, IFim stimulation evoked the typical EPSP-IPSP sequence. The IFim response was then conditioned by the LOC response at a 20 msec ISI. In the LOC conditioning response of the LOC - IFim combination, there was, of course, no EPSP but the hyperpolarizing potential is seen. Conditioning with the LOC response 20 msec prior to the IFim response shows the correspondence between this hyperpolarization and the IFim evoked IPSP. This is demonstrated by the lack of summation of the hyperpolarizations. This indicates that the IFim induced polysynaptic hyperpolarization, which is an IPSP is of a similar nature as the LOC induced hyperpolarization. The extracellular controls are shown in Fig. 11C.

Fig. 11 A-C offered evidence that a recurrent collateral pathway is involved in the production of principal cell inhibition. Not withstanding the recurrent collateral pathway, there is evidence for a second pathway for the activation of the inhibitory interneurons in the dorsolateral septal region. The latter may be referred to as afferent collateral pathway. There are two principal reasons for suggesting that the recurrent collateral system is not the only way of activating the inhibitory interneurons. First, when using a conditioing stimulus reduced to where there is not recordable N_3 component of the conditioing response, a suppression of the N_3 component of the test response still results. In Fig. 11D, the conditioing stimulus was 1.1T and was paired with a test stimulus of 2.6T. In a 20 msec ISI, for the PST paradigm, there was clear but not total suppression of N_3. Complete recovery of the test response was accomplished in a 100 msec ISI. Since there was no apparent principal cell discharging, due to an insufficient stimulus, there should have been no opportunity for recurrent collateral activation of interneurons. Since the inhibition is polysynaptic as determined by intracellular studies (16,17), interneurons might be activated by a relatively low threshold afferent collateral system. A second line of evidence resulted from pharmacological studies in this septal region (19). There is pharmacological evidence to suggest that there are two separate inhibitory pathways in the DSN to bring about the inhibition of principal cell firing.

Experimental observations suggest different functional responsibilities for the two inhibitory pathways. When pairing field responses in LOC-IFim combinations, suppression of the test response did not ensue unless there was an antidromic negativity of marked amplitude in the conditioning response. That is, when recording outside the region of an invading antidromic volley, there was no inhibitory pathway activated for those cells. Therefore, the inhibition established for the recurrent collateral system appears to be spatially restricted to the vicinity of focus of excitation. The main responsibility of the recurrent collateral system may be to pace the output of the principal cells (4). On the other hand, the main responsibility of the afferent collateral pathway may be to establish "surrounding" regions of inhibition (33). It was possible to see in intracellular recordings, for instance, lone but still polysynaptic IPSPs with very low stimulus intensities. This should be the intracellular correlate to the field studies of Fig. 11D. Therefore, with orthodromic activation as the conditioing response, it appears possible to set up relatively wide spread regions of inhibition. Since it is not likely that the recurrent collateral system is involved here, it might be postulated that it is the afferent collateral pathway which establishes extended domines of inhibition.

The intrinsic circuitry involved in principal cell inhibition was also studied for the LSN (Fig. 12). Recall that in the LSN the amount of N_3 suppression with PST was markedly less than in the dorsolateral region (see Fig. 7). A microstimulation electrode was positioned ventral to the recording site to activate the descending axons of the cells receiving hippocampal input.

As in the DSN, microstimulation (LOC) can be used effectively to antidromically activate the descending axons of LSN cells. Fig. 12A-D shows occlusion between the LOC evoked negative response (downward arrow in Fig. 12A) and the N_{2-3} component of the IFim evoked response (Fig. 12B). In the LOC-IFim PST combinations where the peak negativities coincide (Fig. 14C), the lack of summation implies that the LOC evoked negativity was generated by somatic action currents and not, say by passing fibers. Using the LOC response of Fig. 12A to condition the orthodromic IFim response of Fig. 12B found that the N_{2-3} component of the test response was somewhat suppressed in brief ISIs. Fig. 12E shows the test responses to IFim stimulation for different ISIs in a PST. In Fig. 12E, the test control (downward arrow) appears as the light trace and the conditioned test response (upward arrow) as the dark trace. There was clear suppression of the N_{2-3} component in both 10 msec and 15 msec ISIs. This suggests the possibility of a recurrent collateral pathway being involved in the inhibition of principal cells in the LSN. Therefore, the whole lateral septal region might utilize a recurrent collateral mechanism for cell inhibition, with the dorsal part of the lateral septal region having an afferent collateral mechanism as a complement.

Based upon variations in the amount of response suppression

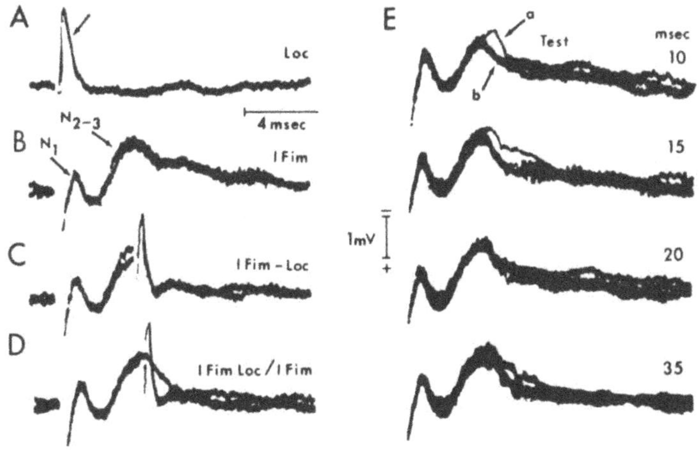

Figure 12. Study of the inhibitory circuit in the ventral part of the lateral septal region. A. Response to LOC stimulation. Arrow indicates the antidromic negativity. B. Antidromic IFim evoked field response, typical of this region. C. PST with IFim-LOC combinationa. D. PST with IFim-LOC combination with the IFim conditioning control superimposed. E. Test responses for the LOC-IFim combination showing 10 msec, 15 msec, 20 msec, and 35 msec interstimulus intervals. The test controls are superimposed. "a" (light trace) represents the test control whereas "b" (dark trace) is the actual test response. The test control is superimposed for 20 msec and 35 msec intervals but is obscured because the test response has fully recovered. The LOC microstimulating electrode was situated ventral to the recording point.

and input-output relations, the lateral septal region appears to consist of two functionally distinct regions. These functional regions have a marked degree of correspondence to the anatomical regions outlined by Andy and Stephan (5-8); a dorsal region (the DSN) and a ventral region (the LSN). While some investigators (e.g., Fox [24]) have not discriminated between dorsal and ventral portions of the lateral septal region, orhters (e.g., Lauer [34], Loo [35]) have noted morphological differences. Fig. 13 summarizes the findings of this portion of the present study.

The DSN appears to receive the heaviest hippocampal projection of all septal regions (41). The result that the N_{2-3} component may, instead, become small in the LSN because of the temporal dispersion of EPSPs. There are no myelinated fibers in the lateral septal region (42) making the intraseptal conduction slow. This reduction in conduction velocity from the fibers of the IFim to their intraseptal terminals would tend to exaggerate the dispersion of synaptic potentials. This might account for the amplitude discrepancy, but not the functional differences noted between the DSN and LSN.

The DSN appears to receive no VSA input that is monosynaptically excitatory. The DSN does not project back through the fimbria-fornix system nor does it seem to project out of the septum proper. Based upon the ability to antidromically activate DSN cells, the principal cells of this region appear to project primarily to the nucleus of the diagonal band of Broca (NDBB) and to the LSN.

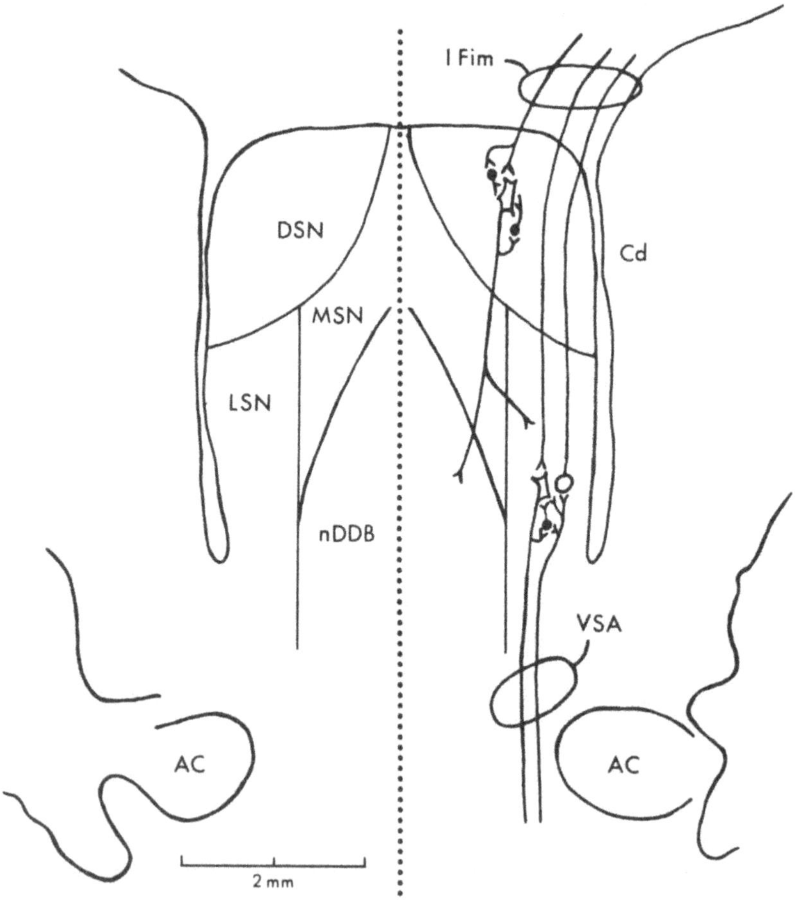

Figure 13. Summary diagram of major findings pertaining to the lateral septal region.

Using PST paradigms, the DSN showed a much greater capacity for these response suppression than the LSN, perhaps suggesting a higher integrative character.

The ventral part of the lateral septal region, the LSN, is distinct from the DSN. The LSN appears to receive significantly less hippocampal input than the DSN. But, the LSN also appears to send a projection back through IFim. Consistent with the latter observation, Siegel and Tassoni (50) have reported a projection from the lateral septal region to the hippocampus.

An important feature of the hippocampal-LSN relationship is demonstrated by finding that those cells in the LSN which are the recipient of hippocampal input project ventrally out of the region, rather than dorsally back to the hippocampus. Conversely, those cells in the LSN which project via the IFim do not receive direct hippocampal input. In the LSN, the ascending and descending pathways appear to be independent.

The Caudal Septal Region

It has become custom to consider the caudal septal region as an extension of the lateral septal region (24, 35). Andy and Stephan (5-8), however, consider this region to be distinct from both the lateral septal and medial septal regions. Physiologically, also, the caudal septal group, principally the nucleus septal fimbrialis, (nSF) appears distinct. As evidence, consider the data of Figs. 14 and 15.

Fig. 14A is a schema locating the probe where the data for this figure and that for Fig. 15 was taken. To both IFim (L) and IFim (M) stimulation, a conspicuous synaptic responses does not occur until depths of 1000μ and 1500μ. Above those levels there was only a short latency, N_0 (fiber), response. The synaptic field response then gradually developed with depth.

A distinguishing feature of the caudal septal region was uncovered by the PST paradigm (Fig. 15A, B). Fig. 15A shows IFim (M) evoked field responses at a depth of 2500μ from the dorsal surface. Fig. 15A shows record specimens for test responses at 20 msec, 50 msec, and 100 msec. The test response of the IFim (M) - IFim (M) combination, as shown in Fig. 15B, recovered in 50 msec. Of 15 PST series examined, in 11 instances the test field response recovered in 50 msec or less.

This short duration of test response inhibition likely reflects relatively low level and short duration IPSPs (Fig. 15G). Fimbria stimulation leads to a EPSP-IPSP sequence, but the IPSP

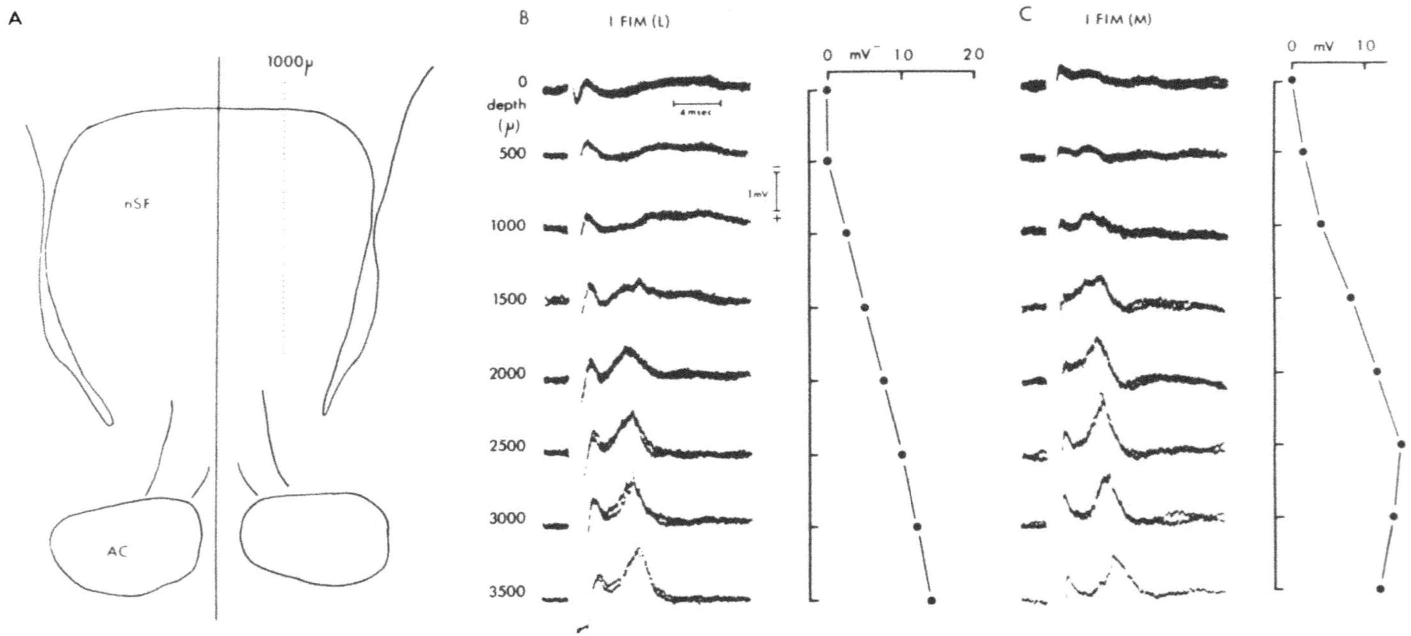

Figure 14. Field profiles through the nucleus septalis fimbrialis (nSF) following IFim (L) and IFim (M) stimulation. A. Sketch drawn from the histological section locating the recording probe. B. Field responses following IFim (L) stimulation at the depths indicated. C. Field responses following IFim (M) stimulation at the indicated depths.

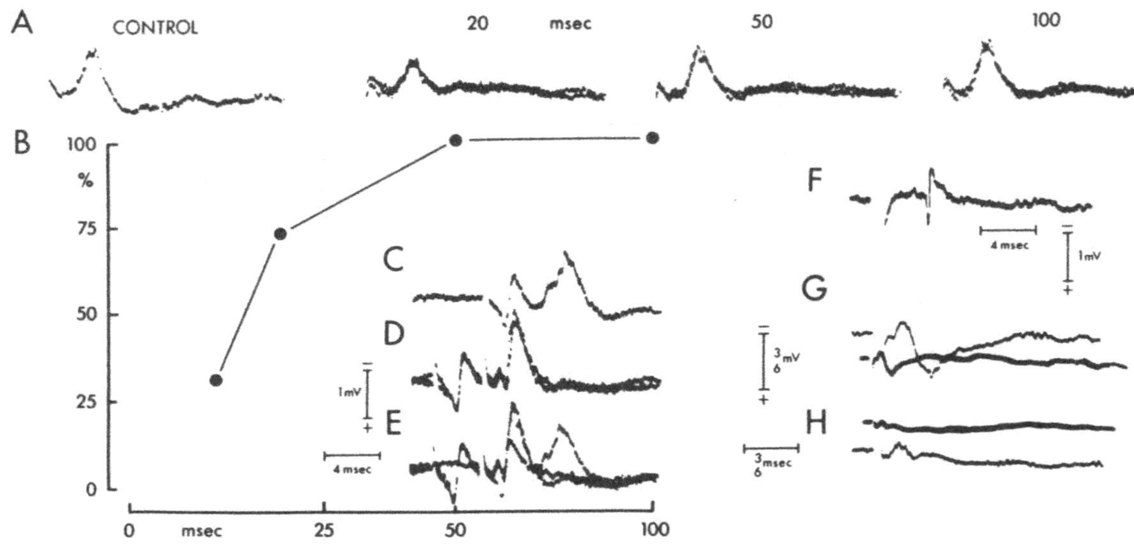

Figure 15. PST of field responses in nSF and unitary analysis. A. PST of field responses recorded at a depth of 3000μ subsequent to IFim (M) stimulation. Record specimens are the test control and test responses at 20 msec, 50 msec, and 100 msec ISIs. B. Plot of the peak amplitude as a function of ISI of the tests responses in the series of A. C–E. PST to field responses following IFim (M) stimulation. C. Test control. D. PST where the first negative component is superimposed upon the synaptic negativity. E. PST with the test control superimposed. F. Extracellular unitary discharge following IFim (M) stimulation. G. Intracellularly recorded EPSP-IPSP sequence following IFim (M) stimulation. H. Extracellular control for G.

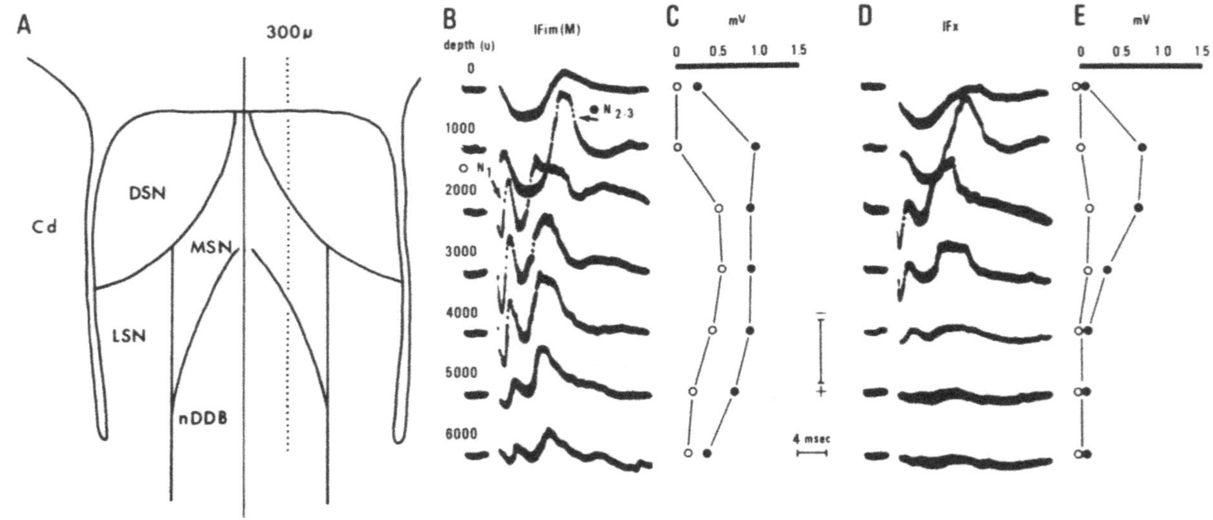

Figure 16. Field profiles through the medial septal region. A. Sketch drawn from the histological section locating the probe at 300μ from the midline. B., C. Field profile of responses to IFim (M) stimulation. The depths are indicated. D., E. Field profile of responses to IFx stimulation. The responses are the average of 8.

component is small compared to the highly developed IPSPs of the lateral septal region proper.

The caudal septal region might be the point of origin of the septal component of the stria medullis (SM) (37). This, and the short duration of test response suppression suggests that the caudal septal group is distinct from the lateral septal region.

Organization of the Medial Septal Region

To get an overview of the response characteristics of the medial septal region to IFim (M) and IFx input, field profiles were recorded and analyzed. A typical instance is shown in Fig. 16, 300μ from the septal midline and 2000μ rostral to the reference zero point. Because the responses in the medial septal region were smaller and tended to be more variable, those presented in Figs. 16, 17 were averaged 8 times. Fig. 16 B, C shows a profile of responses following IFim (M) stimulation. From the surface (0μ), to a depth of 1000μ, the response was dominated by a large negativity (N_{2-3}) with a peak latency of 9.3 msec. At a depth of 2000μ however, a short latency negativity (N_1) appeared. As in the lateral septal region, the N_1 component is the antidromic activation and the N_{2-3} corresponds to the synaptic activation of cells (18). The amplitude of the N_1 - N_{2-3} complex remained stable until a depth of 5000 was reached. Both components then diminished in amplitude.

There is a noticable latency shift of the N_{2-3} component between the depths of 1000μ and 2000μ. This is the depth where, with a probe this distance from the midline, the boundary between the dorsal septal nucleus (DSN) and the medial septal nucleus (MSN) should occur. The responses dorsally in the medial septal region (i.e., DSN) appeared and behaved like those recorded in the study of the lateral septal region. Therefore, the response characteristics of this dorsal zone will not be described further.

It might also be mentioned here that stimulation of the IFim (L) evoked responses in the medial septal region (not shown) which are similar in waveform and behavior to those following IFim (M) stimulation. However, the responses were of smaller amplitude with respect to both N_1 and N_{2-3} components. But the responses to IFim (L) stimulation were usually traced to greater depths.

IFx stimulation typically evoked synaptic responses restricted to the upper 3000μ. Figure 16 C, D shows the companion profile with IFx stimulation. As with IFim (M) stimulation, IFx stimulation led to maximal responses at a depth of 1000μ. From there the N_{2-3} component became smaller and is almost totally absent at 4000μ. In addition, there was usually little or no N_1 component associated with the activation of the IFx. These data indicate that IFx input is restricted to what should correspond to the MSN of Andy and Stephan (5-8). Furthermore, it does not appear as if the MSN sends a projection via the IFx, but only via the fimbria.

Since cells in the MSN and nDBB which are the targets for IFx and IFim (M) input show EPSP-IPSP sequences (18), the suppressive effects of the IPSPs were tested with PST paradigm. Fig. 17A, B shows the effects upon test field response from a PST procedure using an IFim (M) - IFim (M) combination. The data for Fig. 17 A, B was obtained from the same probe. Fig. 17A shows the typical result in the MSN and Fig. 17B shows the result in the nDBB.

In the MSN test field responses showed full recovery at relatively brief ISIs. In Fig. 17A, the test response is fully recovered in a 50 msec interval and is 88% recovered in a 20 msec interval. In contrast, these responses in the nDBB did not show full recovery until longer ISIs were reached. In Fig. 17B, the test response was not quite fully recovered in a 300 msec interval with a marked suppression of the test response at intervals less than 300 msec.

Pairing stimuli with respect to extracellular unitary discharges tell a similar story (Fig. 18, 19). The recovery intervals for test unitary discharges with a IFim (M) - IFim (M) combination were compiled for 71 MSN neurons. According to the breakdown of Fig. 18A, 79% of the unitary discharges recovered in intervals of 20 msec or less. Fig. 18B shows a typical result where the unitary discharge was suppressed in a 10 msec interval but was present in the 20 msec interval. There was usually a slight increase in the discharge latency at the recovery ISI (see Fig. 18B, 20 msec). This probably reflects an incomplete dissipation of IPSPs.

A total of 155 units were likewise examined in the nDBB (Fig. 19). As predicted by the field analysis, many nDBB unitary discharges showed prolonged recovery intervals. For example, 65% recovered in a 100 msec ISI or greater. However, the modal recovery interval was, again, 20 msec. Fig. 19B shows an example of a nDBB cell showing a prolonged recovery interval, in this case, 200 msec. The results from the analysis of the field and extracellular unit data indicate that the MSN and nDBB have potentially different functional properties.

It might well be pointed out that for both cells of the MSN and nDBB the recovery ISIs were stable for a given cell. That is, if a cell demonstrated recovery in a 20 msec interval, it would continue to do so with repeated testing.

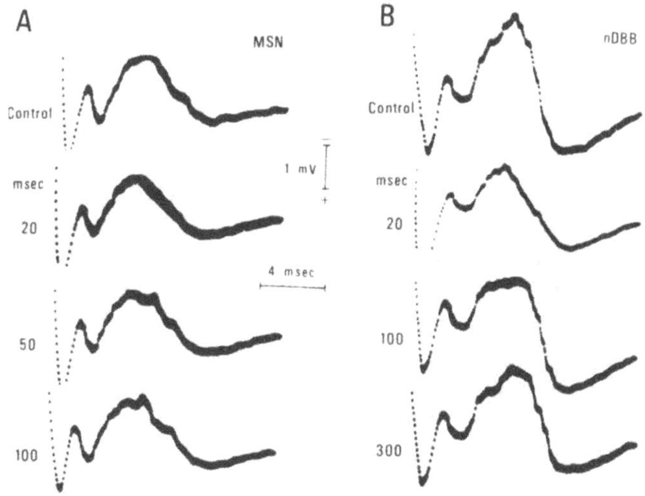

Figure 17. PST of field responses in the MSN and nDBB. A. PST of field responses recorded in the MSN following IFim (M) stimulation. The record specimens show the test control and the test responses at 20 msec, 50 msec, and 100 msec ISIs. B. PST of field responses recorded in the nDBB. IFim stimulation. The record specimens shown are the test control and test responses at 20 msec, 100 msec, and 300 msec ISIs. Averaged 8 times. The baseline for responses of A and B correspond to the midpoint of the numbers indicating the ISI.

Some MSN cells receive monosynaptically excitatory input via the fimbria-fornix system. A separate population of cells project back via the IFim (M) and are thus capable of being antidromically activated. An example of such a cell is shown in Fig. 20A. Here, the cell was antidromically activated as evidenced by a refractory period of 4.0 msec. At the same time, the cell is not synaptically activated. On the other hand, as seen earlier, cells which were aynaptically activated were not antidromically activated. The latency for antidromic discharges for 36 units test ranged from 1.9 msec to 4.2 msec, showing refractory periods from 2.7 msec to 4.2 msec.

Other cells of the MSN could be antidromically activated by VSA stimulation (Fig. 20B). The discharge latency for 17 of these units ranged from 1.5 msec to 6.8 msec. The refractory periods ranged from 2.5 msec to 3.9 msec. In most instances, the cells antidromically activated by VSA stimulation could also be synaptically activated by fimbria-fornix stimulation. Therefore, the

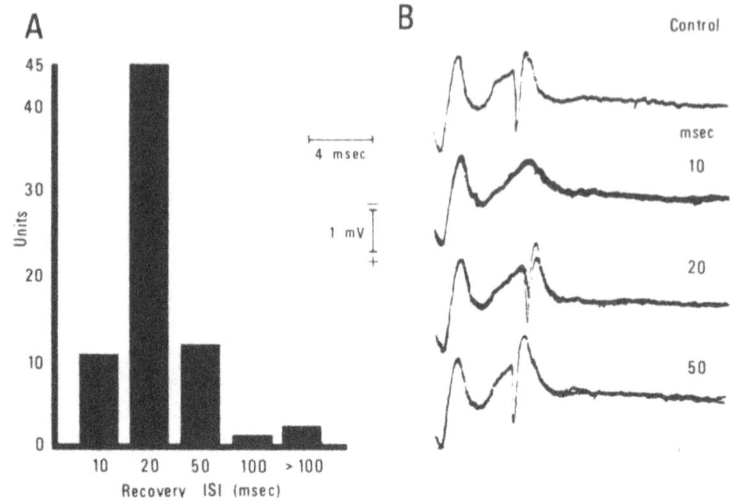

Figure 18. Unitary analysis of principal cells in the MSN. A. Histogram showing number of units recovering in each category of ISI. B. Orthodromic extracellular unitary discharge to IFim (M) stimulation. Record specimens for the test control and test responses at 10 msec, 20 msec, and 50 msec are shown.

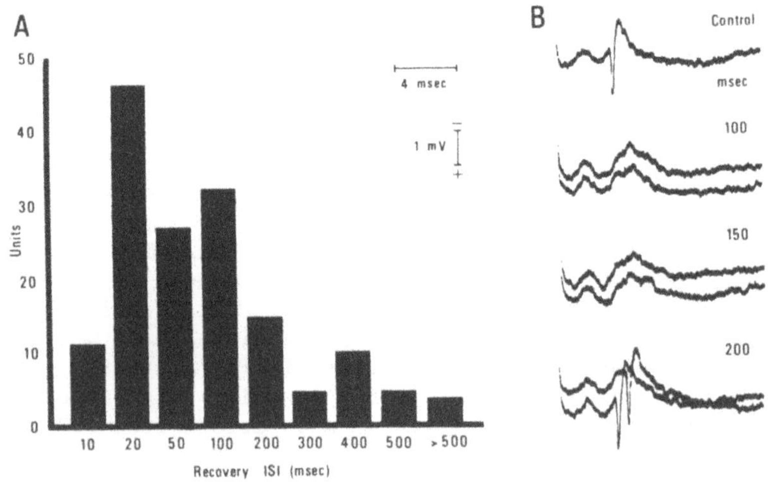

Figure 19. Unitary analysis of principal cells in the nDBB. A. Histogram showing number of units recovering in each category of ISI. B. Orthodromic extracellular unitary discharge to IFim (M) stimulation. Record specimens for the test control and test responses at 100 msec, 150 msec, and 200 msec are shown.

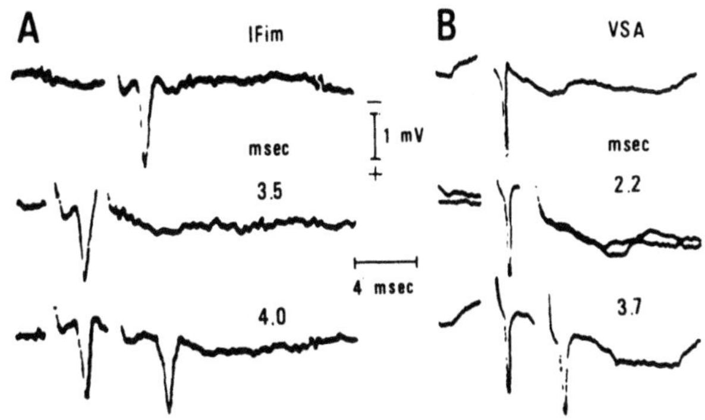

Figure 20. Antidromic activation of cells in the MSN. A. Extracellular unitary discharge following IFim (M) stimulation showing a recovery ISI of 4.0 msec. B. Extracellular unitary discharge following VSA stimulation showing a recovery ISI of 3.7 msec.

MSN appears to be a synaptic station for ascending and descending input.

Just as cells in the MSN were found to have projections into the IFim (M) and VSA systems, so were cells in the nDBB. Fig. 21A (upper trace) shows a short latency extracellular unitary discharge to a VSA stimulus. PST of the unit showed it to have a refractory period of 4.0 msec. Cells which were antidromically activated to VSA stimulation usually did not show an IS-SD break. At recovery ISIs, a reflection on the early phase of the spike was often exaggerated (Fig. 21A, arrow). It was also common to be able to both antidromically and snyaptically activate cells by VSA stimulation (Fig. 21B).

Another population of cells was antidromically activated by IFim (M). A total of 47 were examined and judged to be antidromic. Fig. 21C show an example of one such cell presenting a recovery ISI of 3.0 msec. The cells projecting into the IFim (M) typically did not show an IS-SD break. Intracellular records of antidromically activated cells also showed no evidence of IS-SD breaks. At threshold for an antidromic spike is no apparent inflection on the rising phase (Fig. 21D). For the extracellular units, the latency of antidromic unitary discharges to IFim (M) stimulation ranged from 1.9 msec to 5.2 msec. The average value was 3.5 msec. The average refractory period of the cells was 3.2 msec.

The majority of the cells in the nDBB were either antidromically or synaptically activated by IFim (M) stimulation but not both. But, there was a number of cells that showed both varieties of activation. One such nDBB cell, of 28 encountered, is shown in Fig. 21 E-F. The isolated cell issues two discharges at a stimulus intensity of 3.2T. As the intensity is reduced to 2.4T, the synaptically activated disappears, leaving the antidromically activated discharge. After penetration of the cell at 2.4T, the antidromic spike was seen to be followed by an EPSP (arrow). With further reduction in the stimulus intensity, to 1.8T, both the antidromic spike and EPSP was eliminated, leaving an IPSP. Therefore, in this instance, the antidromic spike had a lower threshold than the orthodromic spike, but at sufficient stimulus strength an antidromic and synaptic spike may appear.

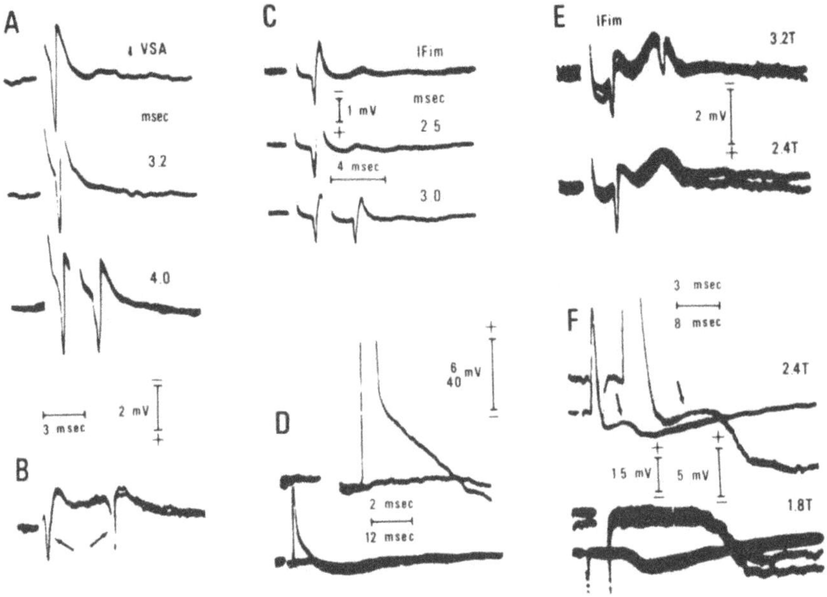

Figure 21. Antidromic responses in the nDBB. A. Extracellular unitary discharge to VSA stimulation. The recovery ISI was 4.0 msec. B. Cell responding both antidromically and synaptically to VSA stimulation (arrows). C. Extracellular unitary discharge to IFim stimulation. The recovery ISI of this unit was 3.0 msec. D. Intracellularly recorded spike following IFim (M) stimulation. At threshold for the spike, no EPSPs are apparent. The response is shown at the different amplifications and sweep speeds. E. Cell showing antidromic and synaptic extracellular discharges to IFim stimulation at 3.2 times threshold (3.2T) stimulus intensity. At 2.4T, only the antidromic discharge remains. F. Intracellular recordings from cell recorded from in E. At 2.4T, there is the antidromic spike and later EPSPs (arrow). At 1.8T, only the IPSP remains. Responses are shown at two different amplifications and sweep speeds.

The results of this portion of the study indicate that the medial septal region is comprised of two physiologically distinct nuclei--the medial septal nucleus (MSN) and the nucleus of the diagonal band of Broca (nDBB). But, as Fox (24) recognized, the medial septal region, as a whole, is a synaptic station for ascending and descending input. Fig. 22 summarizes the findings.

The MSN receives hippocampal input via the IFim (L), IFim (M), and IFx. The analysis of field potentials suggest that the IFx projection is restricted to the medial part of the DSN and to the MSN. Of course, it was not determined in this study which portion of the hippocampal formation (i.e., CA1, CA3 or subiculum) contributes to the IFx projection.

Cells in the MSN appear to project out of the septal region in both dorsal and ventral directions. The population of projection neurons appeared to be mutually exclusive.

The typical duration of test response suppression is relatively short (i.e., less than 50 msec) in the MSN, indicating perhaps that this region performs a 'relay' function.

The nDBB, like the MSN, receives hippocampal input primarily via the IFim (M) and IFim (L). It does not appear to be the target for much of the IFx projection. In further contrast to the MSN, test responses recorded in the nDBB show a wide range of recovery ISIs.

The complexity of the nDBB is re-inforced by the observations that the projection systems are themselves complex. For example, there is a population of cells projecting to the hippocampal formation, but without direct hippocampal input. There is a second population projecting into the hippocampus formation, but which receives hippocampal input. There is yet another major projection system--the ventral septal efferent (VSE) system. With respect to the VSE system, if the efficacy of antidormically activating cells is a valid measure, then it appears that the majority of nDBB ventral output is through the diagonal band to the anterior amygdaloid region.

To add to the complexity to the nDBB output system, in a small number of instances (i.e., 9), nDBB neurons were antidromically activated by both IFim (M) and VSA stimulation. This would seem to suggest that a small population of nDBB cells send axons dorsally through the fimbria and ventrally through the diagonal band system.

The strong output systems of the medial septal region has been well documented with anatomical techniques (12,13,50). What

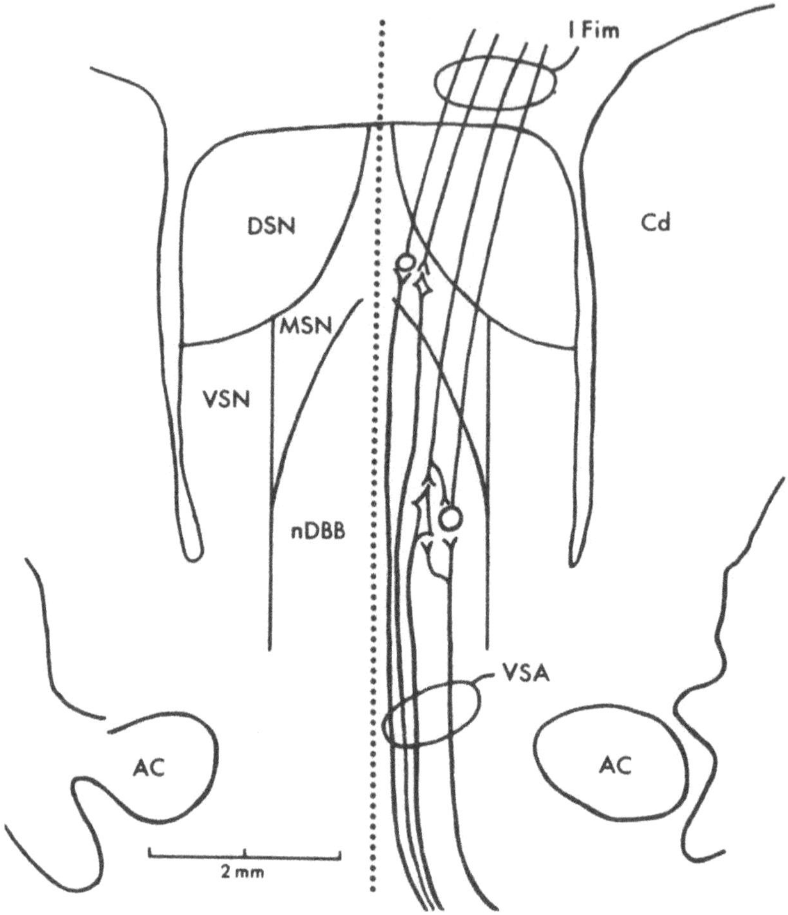

Figure 22. Summary diagram of the major findings pertaining to the medial septal region.

these physiological studies indicate is that there are several input-output pathways in the medial septal region. The degree of their synaptic intimacy is yet to be determined. Also yet to be determined is the basic knowledge concerning the intrinsic organization of the MSN and nDBB. The nDBB, especially, gives the impression that it is going to be difficult to accurately describe local interrelationships. Thus far, studies paralleling those of the lateral septal region have not been fruitful.

Final Comment

Cajal (44) was one of the first to guess at the intimate functional relationship between the hippocampus and septum. Raisman (41) later emphasized the feedback circuit existing between the two limbic structures in which the lateral septal region played an important part. But what is the nature of the responsibility of the septum? Ranck (45) has made an interesting observation in that the cells in the septal region within the context of his experimental conditions do not show the same behavioral specificity as do hippocampal cells; instead, firing in a more general way. It is likely that it is the hippocampal CA1-subiculum complex which shows the greatest development in phylogeny (11). The septal nuclei, having less specificity, may function to influence diencephalic activity in a general way while, at the same time, organizing the feedback control of hippocampal activity also in a general way. That is, septal cells and their activity may not be uniquely involved with any particular behavior or autonomic activity. Instead, reacting to a multifarious collection of hippocampal and VSA inputs, septal cells respond and set into action a chain of events which modifies both the general activity of the hippocampus and the diencephalon in behaviorally consistent ways.

SUMMARY

Electrophysiological studies of the septum were performed on acutely prepared cats. Functional subdivisions were sought using as criteria: (1) differences in input-output relation, and (2) differences in the amount of available inhibition. The septal nuclei showed to have at least 5 major, physiologically distinct, regions--corresponding to the major anatomical groupings of Andy and Stephan (5-8).

The lateral septal region appears to be comprised of two functionally distinct zones: a dorsal zone (i.e., n. septalis dorsalis, DSN) and a ventral zone (8.e., n. septalis lateralis, LSN).

The DSN appears to receive the heaviest ipsilateral fimbria (IFim) projection, but receives no projection via the ventral septal afferent (VSA) system. Cells in the DSN, in turn, project to both the nucleus of the diagonal band of Broca and to the LSN. There is no return projection through the fimbria.

Test responses recorded from the DSN show prolonged period of suppression. The suppression seems to result from the action of inhibitory interneurons which may be activated by two distinct pathways: a recurrent collateral pathway and an afferent collateral pathway.

The LSN receives a lesser share of the IFim input. Cells in the LSN project out of the lateral septal region in both dorsal and ventral directions. The LSN appears to depend upon a recurrent collateral pathway for the activation of inhibitory interneurons.

The medial septal region, likewise, was found to have two distinct zones: n. septalis medialis (MSN) and n. of the diagonal band of Broca (nDBB).

The MSN project out of the septal region in both dorsal and ventral directions. This, coupled with short durations of response suppression (similar to the LSN) gives this region the appearance of a relay center.

The nDBB also projects out in both dorsal and ventral directions. But, in contrast to the MSN, the durations of test response suppression were highly variable. Therefore, in addition to performing relay-like functions, the nDBB gives the impression of being capable of high orders of integration.

The caudal septal region, referred to here as the nucleus septalis fimbrialis (nSF), receives a heavy fimbria-fornix input. The short duration of response inhibition gives this area, like the MSN and LSN, the character of the relay center.

ABBREVIATIONS

ME	microelectrode
LOC	microstimulating electrode
VSA	ventral septal afferents
AC	anterior commissure
Cd/pt	caudate/putamen
ref. zero	reference zero
DSN	nucleus septalis dorsalis
LSN	nucleus septalis lateralis
MSN	nucleus septalis medialis
nSF	nucleus septalis fimbrialis
nDBB	nucleus of the diagonal band of Broca
IFim (L)	ipsilateral fimbria, lateral portion
IFim (M)	ipsilateral fimbria, medial portion
IFx	ipsilateral fornix longus
HPPS(d)	dorsal hippocampus
HPPS(v)	ventral hipposampus
PST	paried-stimulus testing
ISI	interstimulus interval
IFim	ipsilateral fimbria (no medial - lateral distinction made)

REFERENCES

1. Andersen, P., Bland, B. H., and Dudar, J. D. Organization of the hippocampal output. Exptl. Brain Res., 17 (1973) 152-168.

2. Andersen P., Bliss, T. V. P., and Skrede, K. K. Unit analysis of hippocampal population spikes. Exptl. Brain Res., 13 (1971) 208-221.

3. Andersen, P., Bliss, T. V. P., and Skrede, K. K. Lamellar organization of hippocampal excitatory pathways. Exptl. Brain Res., 13 (1971) 222-238.

4. Andersen, P., and Eccles, J. D. Inhibitory phasing of neuronal discharge. Nature, 195 (1962) 645-647.

5. Andy, O. J. and Stephan, H. The nuclear configuration of the septum of galago demidovii. J. Comp. Neurol., 111 (1959) 503-545.

6. Andy, O. J. and Stephan, H. Septal nuclei in the Soricidae (Insectivores): Cytoarchitectonic study. J. Comp. Neurol., 117 (1961) 251-273.

7. Andy, O. J. and Stephan, H. The septum of the cat. Charles C. Thomas, Springfield, Ill. (1964).

8. Andy, O. J. and Stephan, H. The septum in the human brain. J. Comp. Neurol., 133 (1968) 383-410.

9. Chronister, R. B., Sikes, R. W., Farnell, K., White, L. E., and DeFrance, J. F. The bed nucleus of the fornix longus: Septal or Hippocampal? (Submitted for publication).

10. Chronister, R. B., Sikes, R. W., White, L. E. and DeFrance, J. F. The septal region of the telencephalon: a Golgi analysis (in preparation).

11. Chronister, R. B., and White, L. E. Comparative analysis of allocortex. Paper presented at the Richard Lende Memorial Symposium. Albany, (1974).

12. Crosby, E. C. The forebrain of Alligator mississippienis. J. Comp. Neurol., 27 (1917) 325-402.

13. Daitz, H. M. and Powell, T. P. S. Studies of the connexions of the fornix system J. Neur. Neurosurg. and Psychiat., 17 (1954) 75-82.

14. DeFrance, J. F., Shimono, T., and Kitai, S. T. Hippocampal inputs to the lateral septal nucleus: Patterns of facilitation and inhibition. Brain Res., 37 (1972) 333-339.

15. DeFrance, J. F., Kitai, S. T., Hatada, K., and Christensen, C. Contrasting effects of fimbria stimulation in the septum of the cat. Brain Res., 58 (1973) 240-244.

16. DeFrance, J. F., Kitai, S. T., and Shimono, T. Electrophysiological analysis of the hippocampal-septal projections: I. Response and topographical characteristics. Exptl. Brain Res., 17 (1973) 447-462.

17. DeFrance, J. F., Kitai, S. T., and Shimono, T. Electrophysiological analysis of the hippocampal-septal projections: II. Functional characteristics. Exptl. Brain Res., 17 (1973) 463-476.

18. DeFrance, J. F., Yoshihara, H., and Chronister, R. B. Electrophysiological studies of the septal nuclei: II. The Medial Septal Region (in preparation).

19. DeFrance, J. F., Yoshihara, H., McCrea, R. A., and Kitai, S. T. Pharmacology of the inhibition in the lateral septal region. Exptl. Neurol., 17 (1975) 502-523.

20. Eccles, J. C. and Krnjevic, K. Presynaptic changes associated with post-tetanic potentiation in the spinal cord. J. Physiol. (Lond.), 149 (1959) 274-287.

21. Edinger, H., Siegel, A., and Trioana, R. Single unit analysis of the hippocampal projections to the septum in the cat. Exptl. Neurol., 41 (1973) 569-583.

22. Elliot Smith, G. The fornix superior. J. Anat. (Lond.), 31 (1897) 80-94.

23. Elliot Smith, G. The relation of the fornix to the margin of the cerebral cortex. J. Anat. (Lond.), 32 (1898) 23-58.

24. Fox, C. A. Certain basal telencephalic centers in the cat. J. Comp. Neurol., 72 (1940) 1-62.

25. Fox, C. A. The stria terminalis, longitudinal association bundle and precommissure fornix in the cat. J. Comp. Neurol., 79 (1943) 277-295.

26. Freygang, W. H., and Frank, K. Extracellular potentials from single spinal motorneurons. J. Gen. Physiol., 42 (1959) 749-760.

27. Green, J. D., and Arduini, A. Hippocampal electrical activity in arousal. J. Neurophysiol., 17 (1954) 533-557.

28. Herrick, C. J. The morphology of the forebrain in amphibia and reptilia. J. Comp. Neurol., 20 (1910) 413-547.

29. Herrick, C. J. The amphibian forebrain. Vi. Necturus. J. Comp. Neurol., 58 (1933) 1-288.

30. Hille, B. The selective inhibition of delayed potassium currents in nerve by tetraethyl-ammonium ion. J. Gen. Physiol., 50 (1967) 1287-1302.

31. Katz, B. A note on itneraction between nerve fibers. J. Physiol., 100 (1942) 369-371.

32. Kölliker, A. Über den Fornix longus (Forel) und die Beziehungen desselben zum Marke des Gyrus fornicatus. Anat. Anz., 9 (1894) 516

33. Kuffler, S. W. Discharge patterns and functional organization of mammalian retina. J. Neurophysiol., 16 (1953) 37-68.

34. Lauer, E. W. The nuclear pattern and fiber connections of certain basal telencephalio centers in the macaque. J. Comp. Neurol., 82 (1945) 215-254.

35. Loo, Y. T. The forebrain of the opossum, Didelphis virginiana. Part II. Histology J. Comp. Neurol., 52 (1931) 1-148.

36. McLennan, H., and Miller, J. J. The hippocampal control of neuronal discharges in the septum of the rat. J. Physiol., 237 (1974) 607-624.

37. Nauta, W. J. H. An experimental study of the fornix system in the rat. J. Comp. Neurol., 104 (1956) 247-271.

38. Petsche, H., Gogolak, G., and Van Zwieten, P. A. Rhythmicity of septal cell discharges at various levels of reticular excitation. EGG Clin. Neurophysiol., 19 (1965) 25-33.

39. Petsche, H., Stumpf, Ch., and Gogolak, G. The significance of the rabbit's septum as a relay station between the midbrain and the hippocampus. I. control of hippocampal arousal activity by the septal cells. EGG Clin. Neurophysiol., 14 (1962) 202-211.

40. Poletti, C. E., Kinnard, M. A., and MacLean, P. D. Hippocampal influence on unit activity of hypothalamus, pre-optic region, and basal forebrain in awake, sitting squirrel monkeys. J. Neurophysiol., 36 (1973) 308-324.

41. Raisman, G. The connexions of the septum. Brain, 89 (1966). 317-348.

42. Raisman, G. A comparison of the mode of termination of the hippocampal and hypothealamic afferents to the septal nuclei as revealed by electron microscopy of degeneration. Exptl. Brain Res., 7 (1969) 317-343.

43. Rall, W. Distinguishing theoretical synaptic potentials computed for different soma-dendritic distribution of synaptic input. J. Neurophysiol., 30 (1967) 1138-1168.

44. Ramon y Cajal, S. Studies on the cerebral cortex. Translated by Kraft. The Year Book Publishers, Chicago (1955) 135-163.

45. Ranck, J. B., Jr. Studies on single neurons in dorsal hippocampal formation and septum in unrestrained rats. Exptl. Neurol., 41 (1973) 461-555.

46. Rose, J. E., and Woolsey, C. N. A study of thalamus-cortical relations in the rabbit. Bull. John Hopkins Hosp., 73 (1943) 65-128.

47. Ryall, R. W. Renshaw cell mediated inhibition of Renshaw cells: patterns of excitation and inhibition from impulses in motor axon collaterals. J. Neurophysiol., 33 (1970) 257-270.

48. Siegel, A., Edinger, H., and Ohgami, S. The topographical organization of the hippocampal projections to the septal area: a comparative neuroanatomical analysis in the gerbil, rat, rabbit, and cat. J. Comp. Neurol., 157 (1974) 359-378.

49. Siegel, A., and Tassoni, J. R. Differential efferent projections from the ventra- and dorsal hippocampus of the cat. Brain Behav. Evol., 4 (1971) 185-200.

50. Siegel, A., and Tassoni, J. P. Differential efferent projections of the lateral and medial septal nuclei to the hippocampus in the cat. Brain Behav. Evol., 4 (1971) 201-219.

51. Terzuolo, C. A., and Araki, T. An analysis of intra- versus extracellular potentials changes associated with activity of single spinal motoneurons. Ann. N. Y. Acad., 94 (1961) 547-558.

52. Thomas, R. C., and Wilson, V. J. Precise localization of Renshaw cells with a new marking technique. Nature, 206 (1965) 211-213.

53. Young, M. W. The nuclear pattern and fiber connections of the non-cortical centers of the telencephalon of the rabbit (Lepus cuniculus). J. Comp. Neurol., 65 (1936) 295-401.

54. Zuckerkandl, E. Das Riechbündel des Ammonshornes. Anat. Anz., 3 (1888) 425-434.

UNIT ACTIVITY IN THE SEPTAL NUCLEI DURING WATER DEPRIVATION, DRINKING, AND REHYDRATION[1]

John G. Bridge

Vassar College

INTRODUCTION

Extensive evidence implicates the septal nuclei in consummatory behavior. Lesions in the septal area cause hyperdipsia in rats (3,9,13). Electrical stimulation of the septal area reduces the water intake both of rats with water available *ad libitum* and of those on 23 hour water deprivation schedules (15). Carbachol stimulation of the septal area increases drinking (7), and atropine blockage of the medial septal nucleus reduces drinking (8). According to Bridge and Hatton (5), septal unit firing in rats anesthetized with urethane is usually faster during and after stimuli which are associated with or which induce dehydration (water deprivation; subcutaneous and carotid injections of hypertonic saline) than during and after stimuli which are associated with or which induce hydration (the 0.5 hour drink period of a 23.5 hour water deprivation schedule; carotid injections of hypotonic saline; stomach loads of tap water).

The septal area also appears to be involved in eating and/or in controlling food intake, though not to the extent that it is involved in drinking. When the lateral septal area is stimulated with carbachol, rats eat more than before stimulation. Food deprivation (or water deprivation) increases septal self-stimulation

---

[1]This research was supported by Grant No. NS 09140 from the National Institute of Neurological Diseases and Stroke to Dr. Glenn I. Hatton. Portions of these data were reported at the May, 1974 Midwestern Psychological Association Meeting in Chicago, Illinois.

rates in cats and rats (4). According to Johnson and Thatcher (12), as food deprivation time increases, septal lesioned rats increase lever-pressing rates for food reward significantly faster than do unlesioned control rats. Although hyperdipsia has been reported as a consequence of septal lesions more often than has hyperphagia, one study (14) found that septal lesioned Holtzman rats eat more but do not drink more than sham lesioned or unlesioned controls.

There are at least two inconsistencies in the literature about the septal area and drinking behavior. First, while studies showing that lesions often produce hyperdipsia (9), that stimulation can reduce water or food intake (15) and that unit activity is greater during dehydration than during hydration (5) imply that the septum is hyperactive during deprivation and relatively inactive during satiation, other studies indicate the septal area is a potent site of self-stimulation (12). If electrical stimulation activates septal cells and if active septal cells are associated with deprivation, why then would animals eagerly self-stimulate? Another discrepancy is that while septal lesions produce hyperdipsia and electrical septal stimulation reduces water intake, carbachol stimulation of the septal area also results in hyperdipsia (7). At present it is impossible to infer how the septal nuclei function during water regulation and other consummatory behaviors. That is, are septal cells relatively active during hydration, as the self-stimulation and carbachol stimulation studies suggest; or are the septal units relatively active during dehydration, as the electrical lesion and stimulation studies indicate? A description of septal unit activity during naturally occurring consummatory behavior may help us to understand these apparent discrepancies and related phenomena. In the present experiment, septal units were recorded in rats: 1) During ad lib food and water conditions; 2) During the sequence of 23.5 hour water deprivation--0.5 hour drink period--rehydration; and, 3) During the sequence of 23.0 hour food deprivation--1.0 hour eat period--food satiation.

METHOD

Thirty-two adult male Holtzman rats were each implanted with an assembly of six recording electrodes of either 62.5 μm diameter nickel-chromium conductor insulated with enamel except at the tip or 25 μm diameter platinum-iridium wire insulated, except at the tip, with teflon. One 125 μm diameter nickel-chromium ground electrode insulated with enamel except for approximately 0.5 mm at the tip was lowered to a position just above the septal area. Rats were housed singly in cylindrical plexiglas cages of approximately 35 cm diameter. The light-dark cycle in the colony room was 14 hours light: 10 hours dark. No female rats were housed in this colony room.

Rats scheduled to be tested in water-deprived or food-deprived conditions were adapted to a 23.5 hour water deprivation schedule or a 23.0 hour food deprivation schedule for 10 days before the implantation surgery. Maintained on water or food deprivation, rats were allowed a minimum of 10 days of post-operative recovery before the first recording day. Rats scheduled for _ad lib_ recording had food and water available for at least 10 days before surgery and were allowed 10 days of post-operative recovery with food and water available continuously.

A recording session began when the rats were two hours into the light portion of their light-dark cycle. Home cages served as recording chambers, and the procedure involved carrying the rat in his home cage to the recording room, removing the rat from his cage, plugging the leads from the recording apparatus into the implanted socket, and returning the rat to the cage. Electrical activity at each electrode tip was observed on a dual-beam oscilloscope, and the activity of electrodes with good signal-to-noise ratios were recorded on one and/or channel two of a stereo tape recorder. Most units were recorded bipolarly, using one of the remaining five electrodes as a reference electrode.

Impulses from the brain that were sensed by the recording electrode went to an off-lesion-record junction box via three-strand phono wire. In the lesion position, the junction box could supply current to the electrode tip for a marking lesion; in the recording position, impulses from the electrode went to channel one and channel two, respectively, of a stereo tape recorder via a low level preamplifier set to amplify times 1000. Impulses were monitored with a loudspeaker, and an on-line counter was available to record the rates of impulses whose amplitudes exceeded the threshold of the oscilloscope trigger.

Recording sessions which focused on drinking and on the consequences of water intake lasted approximately one hour: 15 minutes of 23.5 hour water-derprived base rate were recorded, then the 30 minute drink period, and, finally, a 15 minute rehydration period. During the half hour drink period, approximately half the rats were allowed to eat; the food was removed from the cages of the remaining rats during this period. Recording sessions under _ad lib_ conditions lasted one hour. To observe the effects of food deprivation and eating on septal unit activity required 90 - 120 minutes of recording: 15 minutes of food-derprived base rate activity, a one hour eat period, and a 15 minute satiation period. Recording sessions began at 9:00 A.M. (two hours into the light period of the light-dark cycle).

Recording sites were marked with direct current lesions ranging from 8 μA for 6 seconds to 30 μA for 15 seconds. Animals were perfused transcardially with 0.9% saline solution followed by a mixture of one part 37% formaldehyde and nine parts saline. To confirm the location of the marking lesion, brains were frozen and sectioned at 50 μm intervals and the cells stained with cresyl violet.

RESULTS

When unit activity was measured during dehydration, drinking, and rehydration, the cells characteristically fired at a faster than average rate during the dehydrated period (Fig. 1). The mean firing rates did not change significantly as the animals drank, but the variance of this mean increased markedly. That is, some cells fired much more rapidly and others slowed considerably, while the mean did not change significantly; approximately half the cells increased and half decreased firing rates during the initial drink period. When the rat terminated its initial drinking bout, discharge rates returned during the non-consummatory behavior that followed to near deprived levels. Discharge rates during the remainder of the drink period then decreased until they reached approximately 57% of the water-deprived base rate. When the rats were allowed to eat during the drink period, the non-consummatory discharge rates decreased only to 67% of the water-

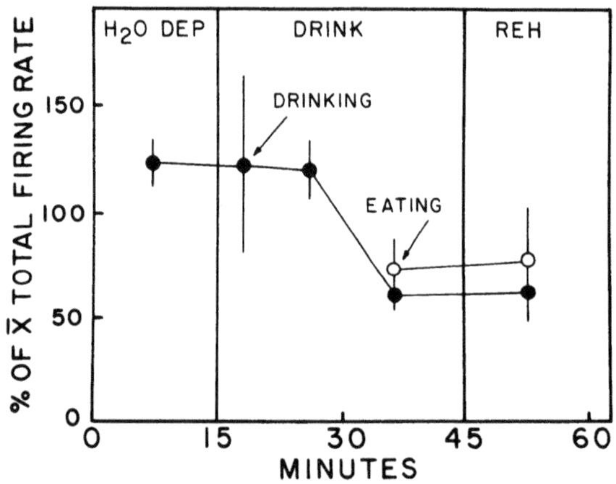

Figure 1. Indicated are mean discharge rates per second expressed as a percentage of the mean firing rate for all data points (within animals). On the abscissa is the recording time in minutes. Panel one (H_2O DEP) shows 15 minutes recorded when rats were water-deprived for 23.5 hours. Panel two (DRINK) shows 30 minutes recorded when rats were allowed free access to water. Panel three (REH) shows a 15 minute rehydrated period recorded after the 0.5 hour drink period.

deprived base rate. Mean firing rates did not change when the animals were on ad lib food and water throughout the recording session, nor did they change when food-deprived rats were allowed to eat during the recording session (see below).

The mean absolute change of discharge rates (without regard to direction) from water-deprived base to the first prolonged drinking bout of the 0.5 hour drink period was the largest (45%) measured in any sequential comparison between non-consummatory and consummatory behavior (see Fig. 2). Unit activity was much more likely to change if an animal initiated drinking than if non-consummatory behavior continued. For example, Fig. 2, panel A shows that in 10 of 27 cases (37%) firing rates changed 30% or more and in 7 of 27 (26%), rates changed 50% or more. To determine how often changes of this magnitude occurred during non-

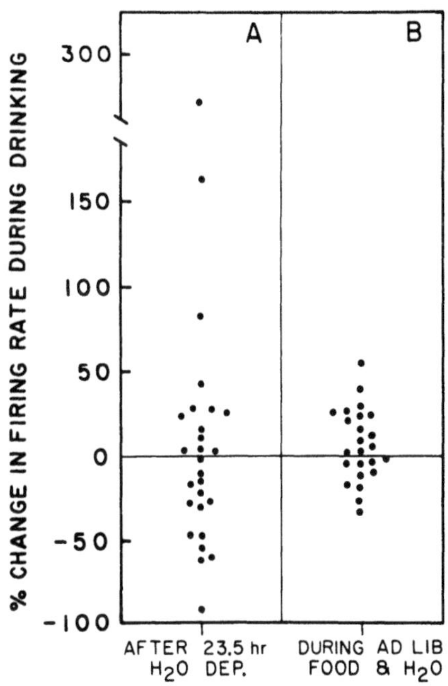

Figure 2. Panel A indicates the distribution of changes in the firing rates of cells when rats deprived of water for 23.5 hours began to drink. Unit activity was measured in animals that sustained drinking for a minimum of 5 minutes and was compared to the 5 minutes of unit activity during the non-consummatory behavior that immediately preceded the drinking. Panel B shows the distribution of changes in firing rates at the onset of drinking in rats with free access to food and water. Drinking occurred for a minimum of 30 seconds and was compared to the 1 minute of non-consummatory behavior that immediately preceded the drinking.

consummatory behavior, firing rates in 200 sequential pairs of 5 minute recording sessions were compared: in only 5% of these pairs did the firing rates change by 30% or more, andonly 2% of them differed by as much as 50%.

Under all conditions, the mean changes in firing rates while rats were eating were much smaller than those during drinking, even when animals had been deprived of food. During the eating period, it was possible to measure the discharge rates of components of the eating behavior: rooting about in the food pile, picking up and carrying food, chewing, and swallowing. Cells often fired faster during a specific component of this behavioral sequence, usually swallowing (see Fig. 3). Cells also fired faster during swallowing in the _ad lib_ periods and after 23.0 hours of food deprivation. The lick-swallow sequence of drinking was too rapid to measure each component separately.

For food-deprived rats, the apparent trend (Fig. 4) of septal unit activity during eating and during the post-eat period was not statistically significant. On some occasions rats were still eating at the end of the one hour eat period and appeared to be not yet satiated. Discharge rates of cells in these rats did not differ from the rates of those that had finished eating. And discharge rates during eating did not consistently differ from those during non-consummatory behavior. Food-deprived rats drank in short bouts which occurred at such different times throughout the recording session that representing them on single data points was impossible. Generally, the firing rates of these food-deprived animals changed slightly to moderately; the direction of the change was consistent within but not among animals.

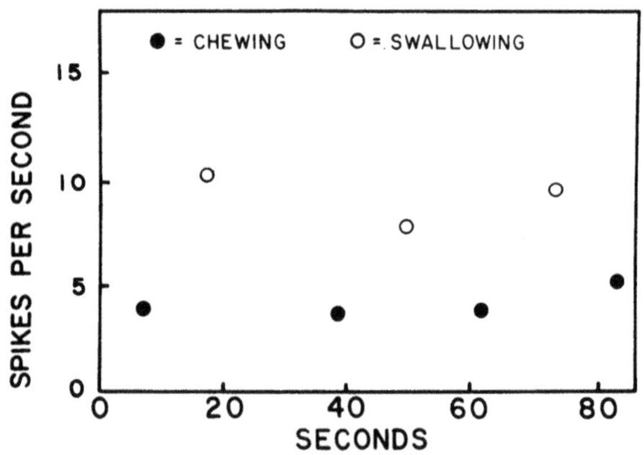

Figure 3. Mean discharge rates of the chewing and swallowing components of eating by one rat. One swallow lasts approximately 700 ms.

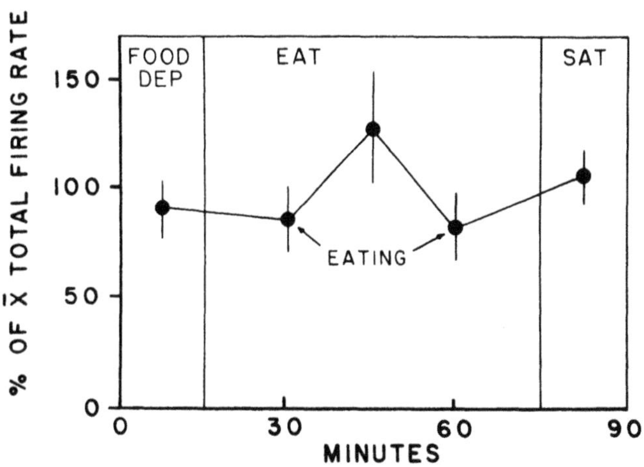

Figure 4. Mean discharge rates per second expressed as a percentage of the mean discharge rate for all data points (within animals). On the abscissa is the recording time in minutes. Panel one (FOOD DEP) shows a 15 minute period recorded when animals had been deprived of food for 23.0 hours. Panel two (EAT) shows a one hour period recorded when food was available _ad lib_. Panel three (SAT) shows a 15 minute post-eat period. N = 11 during non-consummatory behavior; during eating, N = 6.

DISCUSSION

The two strongest findings in this experiment were the changes in septal unit activity in water regulation and in drinking behavior. As water-deprived rats became rehydrated during the 0.5 hour drink period, septal unit activity decreased significantly. And during the initial drinking bout by water-deprived rats, individual septal units changed markedly: many increased and many decreased firing rates, although the mean remained relatively stable.

The 23.5 hour water deprivation schedule was chosen for this experiment because it is a natural sequence of dehydration and rehydration, because many other researchers use this schedule, and because other studies have reported the details of its osmotic and volemic correlates. Hatton and Bennett (10) have shown that as rats adapt to a 23.5 hour water deprivation schedule over a 10 day period: 1) the animals drink their fill progressively earlier in the half hour session until, by the 10th day, they drink in the first 15 minutes nearly all the water they will drink, and 2) the difference between pre-drink and post-drink osmotic pressure increases as a function of days of deprivation. The difference between pre-drink and post-drink plasma protein did not change over days of adaptation. Hatton and Bennett also reported that by the time rats stopped drinking, their osmotic pressure had dropped to ad lib levels or below. (After 6 minutes of access to water, rats are still hyperosmotic; from 8 - 10 minutes, osmotic pressure is near ad lib levels, and by 13 minutes the osmotic pressure has dropped below ad lib levels.) This is a crucial observation: if osmotic pressure decreases before drinking ends, there is no need for an "anticipatory" mechanism in water regulation. Almli and Gardina (1) have shown that smaller (but significant) increases precede initiation of drinking by ad lib rats.

Rats in the present experiment that were adapted to 23.5 hour water deprivation and received no food during the drink period were under conditions similar to those analyzed by Hatton and Bennett. By the end of the drink period, discharge rates in this group had decreased significantly, though they dropped more slowly than osmotic pressure decreased or volume increased in the Hatton and Bennet study. Unit activity in the group for which food was available during the drink period decreased less markedly (Fig. 1), and firing rates of cells in the ad lib group did not change. These results indicate that, in general, septal cells are hyperactive during hyperosmotic (dehydrated) conditions and hypoactive during hypo-osmotic (rehydrated) stimuli; they corroborate an earlier report (5) of this effect in urethane-anesthetized rats.

Firing rates changed considerably during the first drinking bout by rats adapted to 23.5 hours of water deprivation. These changes were smaller later on in the 0.5 hour drinking period and when rats were on an ad lib, rather than a water deprivation schedule. Furthermore, when rats are deprived of water for only 2 - 3 hours, and not adapted to this schedule, the changes observed during drinking are also small (Ranck, J. B., personal communication). Therefore, these results strongly suggest an interaction between drive state (water deprivation) and the effect drinking has on septal unit activity.

Hayward and Smith (11) found that septal stimulation decreases urine flow by, they suggested, activating supraoptic neurons and increasing the release of antidiuretic hormone. The rapid discharge rates of septal cells during dehydration and their slower firing rates during rehydration found in the present study also strongly support this hypothesis.

But the septal area is not involved only in water saving; it is also involved in the behavior of drinking. Wishart and Mogenson (15) found that septal stimulation can inhibit ongoing drinking; Miller and Mogenson (1971) reported that septal stimulation can alter discharge rates of lateral hypothalamic neurons, but the effect depends on the rate of the lateral hypothalamic unit. If the unit is in a fast phase, septal stimulation usually decreases its discharge rate; and septal stimulation activates the firing of a lateral hypothalamic cell in a slow phase. That is, septal stimulation moderates the discharge rates of lateral hypothalamic neurons. This dual effect makes the septal function in drinking behavior difficult to understand. It is interesting that the septal area also moderates the intake of aversive solutions (2,16). Unfortunately, the bi-directional change of septal discharge rates complicates matters more, and no simple explanation is available.

The present experiment indicates there are two distinct functions of septal units in water regulation. First, septal units fire significantly faster during water deprivation than during rehydration. Second, at the onset of drinking septal cells are more likely to change firing rates if the animal has been water deprived than if the animal is on *ad lib* food and water or on food deprivation. Because the septum has two distinct roles in water regulation, the apparently contradictory results of electrical stimulation and lesion experiments versus chemical stimulation and blockage studies are not necessarily incompatible. The septum's excitation of the supraoptic nucleus most parsimoniously explains why electrical septal stimulation reduces water intake while electrical septal lesions increase drinking. In one role, then, the septum governs water retention. But changes of septal firing rates during drinking may also influence lateral hypothalamic cells and may be mediated by cholinergic synapses. In this role the septum influences water ingestion.

SUMMARY

To investigate changes in the firing rates of septal cells during and after drinking, single and multiple unit activity in the septal nuclei of unanesthetized and unrestrained rats was monitored with multiple electrodes. In rats adapted to a 23.5 hour water deprivation schedule, cells discharged nearly twice as fast before the 0.5 hour drink period as they did afterward. Electrical activity recorded for one hour in a control group of rats on *ad libitum* food and water did not change significantly. Similarly, in a group adapted to a 23 hour food deprivation schedule, one hour of eating, and 15 minutes of food satiation showed no significant changes. Septal unit activity changed markedly during drinking if the animal had undergone 23.5 hours of water deprivation, but the change was not as great if the animal had drunk water on an *ad lib* schedule; unit activity in rats deprived of food did not change significantly during eating.

The finding that septal cells are much more active during dehydration than rehydration supports the hypothesis generated by studies in which electrical lesion or stimulation techniques were used: that one septal role in water regulation is to stimulate during dehydration supraoptic cells which then release more antidiuretic hormone, causing the animal to conserve water.

REFERENCES

1. Almli, C. R. and Gardina, J. Ad libitum drinking of rats and vascular osmolality changes. Physiology and Behavior, 12 (1974) 231-238.

2. Beatty, W. W. and Schwartzbaum, J. S. Enhanced reactivity to quinine and saccharine solutions following septal lesions in the rat. Psychonomic Science, 8 (1967) 483-484.

3. Blass, E. M., Hanson, D. G. Primary hyperdipsia in the rat following septal lesions. Journal of Comparative and Physiological Psychology, 70 (1970) 87-93.

4. Brady, J. V., Boren, J. J., Conrad, D., Sidman, M. The effect of food and water deprivation upon intracranial self-stimulation. Journal of Comparative and Physiological Psychology, 50 (1957) 134-137.

5. Bridge, J. G., Hatton, G. I. Septal unit activity in response to alterations in blood volume and osmotic pressure. Physiology and Behavior, 10 (1973) 769-774.

6. Donovick, P. J., Burright, R. G., Lustbader, S. Isotonic and hypertonic saline ingestion following septal lesions. Communications in Behavioral Biology, 4 (1969) 17-22.

7. Fisher, R. E., Coury, J. N. Cholinergic tracing of a central neural circuit underlying the thirst drive. Science, 138 (1962) 691-693.

8. Grossman, S. P. Effects of chemical stimulation of the septal area on motivation. Journal of Comparative and Physiological Psychology, 58 (1964) 194-200.

9. Harvey, J. A., Hunt, H. F. Effect of septal lesions on thirst in the rat as indicated by water consumption and operant responding for water reward. Journal of Comparative and Physiological Psychology, 59 (1965) 49-56.

10. Hatton, G. I., Bennett, C. T. Satiation of thirst and termination of drinking: roles of plasma osmolality and absorption. Physiology and Behavior, 5 (1970) 479-487.

11. Hayward, J. N., Smith, W. K. Influence of limbic system on neurohypophysis. Archives of Neurology, 9 (1963) 171-177.

12. Johnson, D. A., Thatcher, K. Differential effects of food deprivation on the fixed ratio behavior of normal rats and rats with septal lesions. Psychonomic Science, 26 (1972) 45-46.

13. Lubar, J. F., Schaefer, C. F., Wells, D. G. The role of the septal area in the refulation of water intake and associated motivational behavior. Annals of the New York Academy of Science, 157 (1969) 875-893.

14. Stein, L., Ray, O. S. Self-regulation of brain-stimulating current intensity in the rat. Science, 130 (1959) 570-571.

15. Stoller, W. L. Effects of septal and amygdaloid lesions on discrimination, eating, and drinking. Physiology and Behavior, 8 (1971) 823-828.

16. Wishart, T. B., Mogenson, G. J. Reduction of water intake by electrical stimulation of the septal region of the rat brain. Physiology and Behavior, 5 (1970) 1399-1404.

FUNCTIONAL ASPECTS OF THE HIPPOCAMPAL-SEPTAL AXIS

H. Edinger and A. Siegel[1]

New Jersey Medical School, Newark, New Jersey 07103

INTRODUCTION

In recent years several investigators have studied the effects of electrical stimulation of the hippocampus (7) or of the fiber pathways of the hippocampus (3,4,10) on single neuron activity in the septum. It has been shown that activation of the hippocampal efferents produces excitation as well as inhibition of long duration in septal cells. Our main concern has been with the electrophysiological consequences of the topographical organization of the hippocampal efferents.

Investigations in our laboratory of the neuroanatomical organization of the efferent pathways from the hippocampus to the septum in the primate as well as in other mammalian species has been reviewed (Siegel and Edinger, this symposium). These findings can be summarized as follows: In non-primate mammals efferent fibers from the dorsal hippocampus pass in the medial portion of the fornix to terminate in the medial portion of the septum, while efferents originating in the ventral hippocampus pass through the fimbria to terminate in the lateral septal area. In the primate, the anterior hippocampus projects to the lateral septum while the posterior hippocampus projects to the medial septum. In the experiments to be described below unit responses were recorded in the medial and lateral septum to electrical stimulation of the dorsal and ventral hippocampus.

---

[1]Supported by NIH Grant NS 07941-04 and by the Benevolent Foundation of Scottish Rite Freemasonary, Northern Jurisdiction, U.S.A. The authors acknowledge the contributions of Dr. Raymond Troiano and the technical assistance of Miss Ann Yoanides.

Our results indicated that short-latency excitation of medial septal neurons was evoked by stimulation of the dorsal hippocampus while short latency excitation of lateral septal neurons was produced by stimulation of the ventral hippocampus.

METHODS

Cats anesthetized with Methoxyflurane gas anesthesia were used in these experiments. Single unit activity was recorded with tungsten microelectrodes and conventionally amplified. Correct placement of hippocampal stimulating electrodes was achieved by the use of an array of stainless steel electrodes, insulated with varnish except for the regional 0.5 mm near the tip. The array was mounted in a stereotaxic carrier and lowered through the cortex in 1 mm steps. One of the electrodes in the array was used to record EEG and evoked potentials at each step. Upon penetration of the hippocampus, seizure patterns could be recorded the characteristic large potentials evoked by stimulation through another electrodes in the array could be detected.

RESULTS

Focal Potentials

Focal potentials with superimposed multiple unit activity were recorded in various portions of the septum following hippocampal stimulation. In the medial septum maximum amplitude was seen following the stimulation of the ventral hippocampus. The earliest unit potentials seen upon stimulation of the hippocampus were of small amplitude and double pulse stimulation revealed that they had short refractory periods. This suggests that these early potentials represent fornix fiber activity. The latency of these fiber potentials following dorsal hippocampus stimulation was 10 msec, while following ventral hippocampal stimulation the earliest latency was 12 msec. The calculated conduction velocity of the fornix fibers is therefore about 1-2 m/sec.

Unit Responses

Trans-synaptic excitation of septal units following hippocampal stimulation was found mainly at short latency, but long latency excitation was also encountered. Short latency excitation occurred at 10-30 msec and the unit responses were often superimposed on the septal focal potential. The short latency responses were not antidromic since they did not follow high frequency hippocampal stimulation in a one-to-one manner and the latency of excitation was not fixed. Thus, it may be assumed that these units are activated monosynaptically (or possibly oligo-synaptically) by the hippocampal projection fibers. This excitation was invariably followed by a prolonged phase of suppressed neuronal activity or inhibition. The duration of this inhibition ranged from 50 to 250 msec.

Stimulation of the dorsal hippocampus produced short latency unit driving in the medial septum. On the other hand, stimulation of the ventral hippocampus produced short latency driving in the lateral septum (see Fig. 1).

Medial septal units exhibiting short latency driving followed by inhibition in response to stimulation of the dorsal hippocampus show only inhibition in response to stimulation of the ventral hippocampus. An example of a medial septal unit response to hippocampal stimulation is seen in Fig. 2. Conversely, units located in the lateral septum driven at short latency (followed by suppression of activity) by ventral hippocampal stimulation exhibited only inhibition of long duration upon stimulation of the dorsal hippocampus (Fig. 3). This suggests that a reciprocal inhibitory network is present in the septum. (See Discussion.)

Many units in the septum show suppression of activity following dorsal or ventral hippocampal stimulation without a prior phase of excitation. Since extracellular microelectrodes were used in these experiments, the possibility that there is subthreshold depolarization following stimulation of the hippocampus in some of

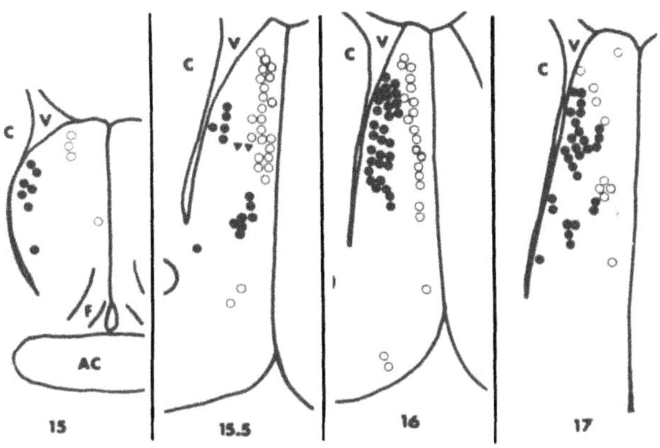

Figure 1. Location of septal units driven by stimulation of the hippocampus. Open circles indicate units driven at short latencies by stimulation of the dorsal hippocampus; closed circles indicate units driven at short latencies by ventral hippocampal stimulation; closed triangles indicate convergence-cells driven at short latencies by stimulation of both dorsal and ventral hippocampus. Numbers at bottom of figure indicate the level of the frontal plane for each of the sections according to the Atlas of Jasper and Ajmone Marsan (1954).

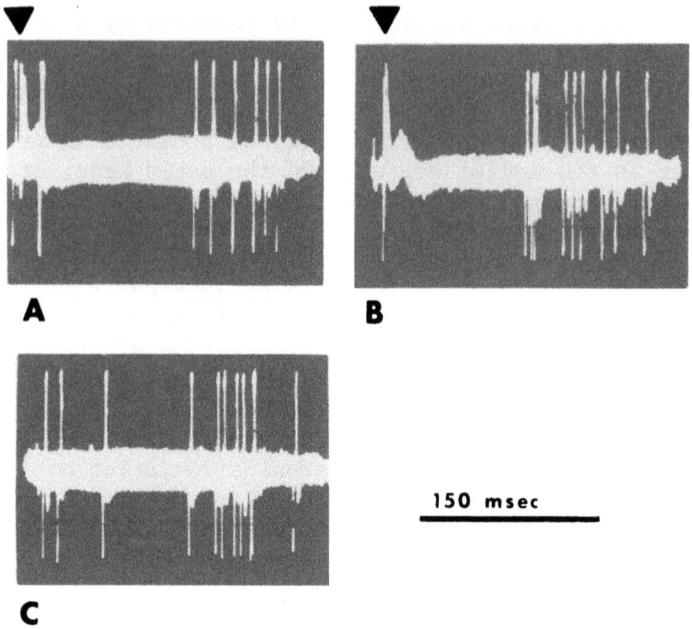

Figure 2. Response of unit in medial septum to dorsal and ventral hippocampal stimulation. Stimulus onset is indicated by symbol. Three superimposed traces. (A) 0.2 mamp stimulation of dorsal hippocampus produces excitation-inhibition sequence; (b) 0.2 mamp stimulation of ventral hippocampus produces only inhibition; (c) spontaneous activity.

these cases can not be excluded. It should be noted, however, that increasing stimulus intensity produced a prolongation of the inhibitory phase in these units. Nevertheless, units showing only inhibition to either dorsal or ventral hippocampal stimulation showed no topographic restriction as was so clearly evident for units driven by hippocampal stimulation (See Fig. 4).

Units in the septum that were driven at longer latencies by hippocampal stimulation (i.e. > 30 msec) likewise showed no preferential topographic distribution. It is suggested that these units are activated by multisynaptic pathways in the septum.

The possibility that the differential topographical distribution of evoked unit responses in the septum is due to stimulation of different CA fields in the hippocampus is obviated by two facts. Firstly, the tips of the stimulating electrodes were, in the course of these experiments, distributed throughout all CA fields (see Fig. 5). Secondly, the rather large stimulating currents (0.2-0.8 ma) often used in these experiments could be

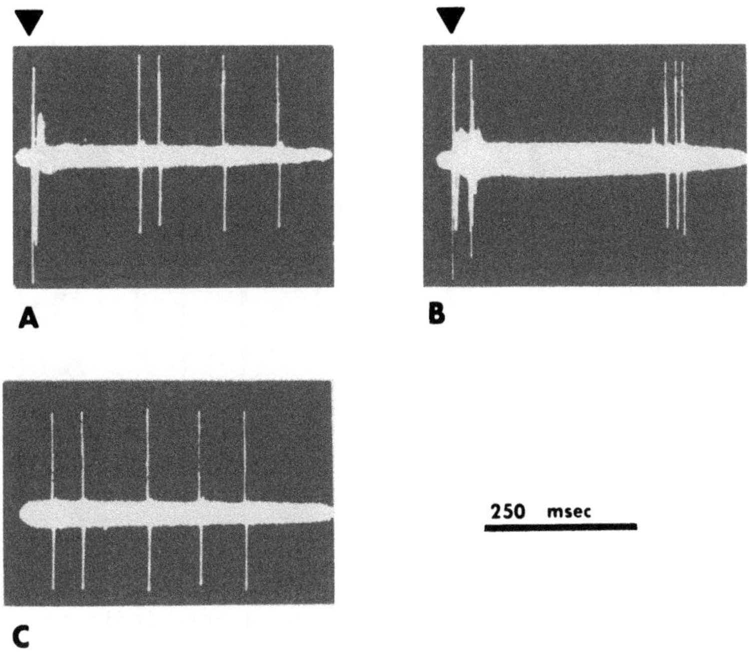

Figure 3. Response of a unit in the lateral septum to dorsal and ventral hippocampal stimulation. Stimulus onset indicated by symbol. Three superimposed traces. (A) dorsal hippocampus produces inhibition of unit response; (B) ventral hippocampus produces excitation-inhibition sequence; unit response is superimposed on focal potential; (C) spontaneous activity.

expected to spread through all cytoarchitectonic regions of the hippocampus especially since stimuli were usually delivered by monopolar electrodes.

DISCUSSION

The application of neuroanatomical and neurophysiological techniques to the hippocampal-septal axis now permits us to present a coherent picture of its organization. The most striking feature of this complex system is the manner in which topographical orientation is maintained through a multisynaptic pathway. Lomo (9) has demonstrated electrophysiologically that the fibers of the perforant pathway from the entorhinal cortex enter the dentate gyrus in a topographically organized manner. The perforant fibers depolarize the dendrites of the granule cells of a segment of the dentate gyrus that lies perpendicular to the long

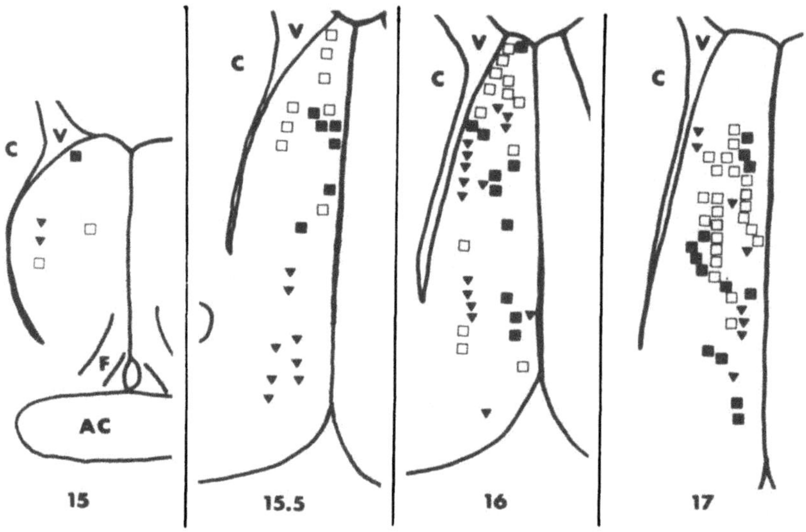

Figure 4. Location of units displaying suppression of activity following hippocampal stimulation. Open squares indicate units suppressed by dorsal hippocampal stimulation; closed squares indicate units suppressed by ventral hippocampal stimulation; closed triangles indicate units suppressed by both dorsal and ventral hippocampal stimulation. Numbers at bottom of figures indicate the level of the frontal plane for each of the sections according to the Atlas of Jasper and Ajmone Marsan (1954).

axis of the dentate gyrus. The mossy fibers of the granule cells, in turn, pass to the dendrites of the pyramidal cells of CA4 and CA3 in the same lamella (2). The Schaeffer collaterals from CA3 also are oriented in this plane and project to CA1. Andersen, et al. (1) showed that the axons of the hippocampal pyramidal cells enter the fimbria in a manner that preserves the topographical relationships of the various hippocampal lamellae. The findings obtained in our laboratory extends the concept of hippocampal lamination to the septum. (Siegel and Edinger, this volume). At this point, the output of the dorsal lamellae terminate in the medial septum while the output of the ventral lamellae terminate in the lateral septum. Thus, the lamellar organization of the system extends from the preforant pathway fibers to the septum.

The data presented in this report show that the hippocampal efferent fibers produce excitation of the septal receiving cells. The widespread inhibition seen in the septum following stimulation implies, that in addition to the focal excitatory response, an extensive inhibitory system is activated as well. We visualize

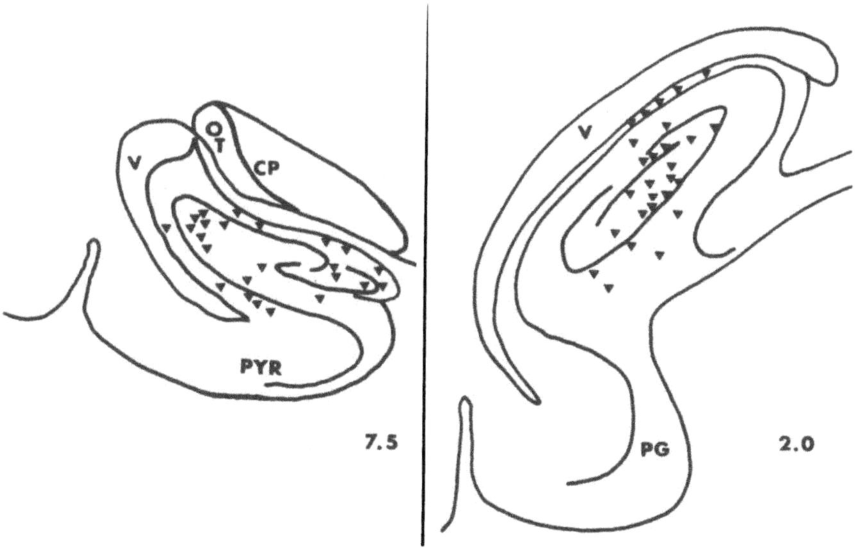

Figure 5. Locations of hippocampal stimulating electrodes. Triangles indicate locations of the tips of the ventral hippocampal (frontal plane 7.5) and dorsal hippocampal (frontal plane 2.0) stimulating electrodes. Abbreviations: CP, cerebral peduncle; OT, optic tract; PG, parahippocampal gyrus; PYR, pyriform cortex; V, ventricle.

this inhibition as playing a role similar to surround inhibition seen commonly in the CNS--sharpening a focus of excitation with an inhibitory surround (6). Thus, excitation of medial septal neurons will inhibit surrounding neurons particularly those in the lateral septum. A reciprocal relationship holds for the excitation of lateral septal cells by the ventral hippocampus, in which case medial septal neurons are inhibited.

In addition, septal unit driving is invariably followed by inhibition of long duration suggesting a recurrent inhibitory circuit. DeFrance, et al. (5) have recorded unit responses consistent with the expected properties of inhibitory interneurons. Although we have been able to record units responding in bursts to stimulation of the hippocampus we have not been able to positively identify these cells as inhibitory interneurons. Tombol and Petsche (12), using the Golgi technique, have described small ovoid neurons in the medial septal area whose axons are probably confined within the septum. In Fig. 6 we propose a model of the neuronal circuitry of the septum. In this model the small neuron represents the hypothesized inhibitory interneuron which produces recurrent, as well as surround, inhibition. This, in our opinion is

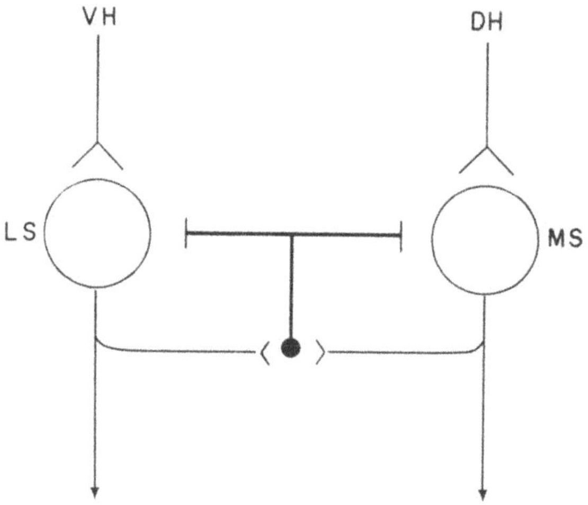

Figure 6. Hypothetical model of neuronal circuitry of the septum. Efferents from the dorsal hippocampus (DH) terminate on medial septal neurons (MS) while efferents from ventral hippocampus (VH) terminate on lateral septal neurons (LS). These synapses are excitatory as are the synapses made by collaterals of target neurons on the interneuron (dark). Discharge of the interneuron produces inhibition of the target neurons.

the most parsimonious hypothesis available to explain our data. More elaborate schemes consistent with the data are, of course, possible. In particular, the interneuron represented in the diagram may in fact be a complex network but the concept of inhibitory interneurons receiving multiple inputs from numerous septal neurons and producing widespread inhibition should be tested.

Although no experimental data on the extent of convergence of hippocampal efferents on septal neurons exists, it is evident from geometric considerations that convergence must be considerable. Thus, we visualize the septum to function in the following manner. Activity in adjacent groups of hippocampal efferent pyramidal cells could result in discharge of a septal neuron. The activation of this neuron will excite the inhibitory interneuronal network to block activity in other portions of the septum. The septum therefore acts as a filtering mechanism to permit transmission of information only from that portion of the hippocampus whose pyramidal neurons are most active at that point in time. Information transmitted from less active areas of the hippocampus will be suppressed by the inhibitory network of the septum. In terms of behavior this hypothesis obviously places critical stress on the differential effects of excitation of neurons in different portions of the septum. Differential projections from medial and lateral septum to the preoptic area and hypothalamus are not evident (11), but this does not preclude the possibility that the efferent projections from these two areas have different physiological effects on downstream neurons. Differential projections from medial and lateral areas have been found in the reciprocal connections back to the hippocampus. The medial septum projects to the dorsal hippocampus while the lateral septum projects to the ventral hippocampus. The functional significance of these two feedback pathways awaits further investigation.

REFERENCES

1. Andersen, P., Bland, B. H., and Dudar, J. D.: Organization of the hippocampal output. Exp. Brain Res. 17:152-168, (1973).

2. Blackstad, T. W., Brink, D., Hem, J. and Jeune, B.: Distribution of hippocampal mossy fibers in the rat. An experimental study with silver impregnation methods. J. Comp. Neurol. 138:433-450 (1970).

3. DeFrance, J. P., Kitai, S. T. and Shimono, T.: Electrophysiological analysis of the hippocampal-septal projections: I. Response and topographical characteristics. Exp. Brain Res. 17:447-462 (1973a).

4. DeFrance, J. F., Kitai, S. T. and Shimono, T. Electrophysiological analysis of the hippocampal-septal projections: II. Functional characteristics. Exp. Brain Res. 17: 463-476 (1973b).

5. DeFrance, S. F., Shimono, T. and Kitai, S. T. Hippocampal inputs to the lateral septal neucleus: Patterns of facilitation and inhibition. Brain Res. 37:333-339 (1972).

6. Eccles, J. C. The Inhibitory Pathways of the Central Nervous System. Liverpool Univ. Press (1969).

7. Edinger, H., Siegel, A. and Troiano, R. Single unit analysis of the hippocampal projections to the septum in the cat. Exptl. Neurol. 41:569-583 (1973).

8. Jasper, H. H. and Ajmone Marsan, C. A. Stereotaxic Atlas of the Diencephalon of the Cat. Nat. Res. Council Canada, Ottawa (1954).

9. Lomo, T. Patterns of activation in a monosynaptic cortical pathway: the perforant path input to the dentate area of the hippocampal formation. Exp. Brain Res. 12:18-45 (1971).

10. McLennan, H. and Miller, J. J. The hippocampal control of neuronal discharges in the septum of the rat. J. Physiol. 237:607-624 (1974).

11. Siegel, A. and Tassoni, J. P. Differential efferent projections of the lateral and medial septal nuclei to the hippocampus in the cat. Brain Behav. Evol. 4:201-219 (1971).

12. Tombol, T. and Petsche, H. The histological organization of the pacemaker for the hippocampal theta rhythm in the rabbit. Brain Res. 12:414-426 (1969).

SEPTAL UNIT RESPONSES TO HIPPOCAMPAL AND HYPOTHALAMIC STIMULATION IN THE CHLORALOSE-ANAESTHETIZED OR *CERVEAU ISOLÉ* CAT

J. S. McKenzie, J. McEwen and D. K. Rogers

McKenzie and Rogers, Department of Physiology, University of Melbourne, Parkville, 3052, Victoria, Australia

INTRODUCTION

There is substantial anatomic evidence for projections to septal nuclei from the hippocampus (2,7) via the fimbria-fornix system, and from the tegmentum and hypothalamus via the medial forebrain bundle (2,14,15). On the basis of these inputs, the septal area has been described as a region of confluence, convergence and correlation between neural activities ascending in the medial forebrain bundle (MFB) and descending from the limbic system (9). By implication, neurones of the septal nuclei should give evidence of interactions between hypothalamic mechanisms and modulating influences from the limbic cortex.

Accordingly, one would be led to predict significant convergence on to individual septal neurones of inputs from the hippocampus (HC) and MFB region of the lateral hypothalamus (HYP). Such convergence need not be monosynaptic but ought to be demonstrable neurophysiologically, either as convergent activation, or as inhibitory interactions between the effects of each input on septal units.

Until recently, published evidence for effects on septal neurones of synchronous input activation were lacking. A number of recent investigations, particularly in the rat by De France, Kitai and Shimono (5) and by McLennan and Miller (12) have demonstrated the direct excitatory effect of fimbria and fornix axons on septal neurones, and a similar finding was made in the cat by De France *et al.* (4) and by Edinger, Siegel and Troiano (6). Evidence for septal unit responses to HYP stimulation was published in a preliminary abstract (10) and is now being extended by further investigations.

In order to test the prediction of HC-HYP convergence in the septum, we surveyed the septal nuclei of the cat for unit responses to single stimuli in the dorsal HC and MFB. With a limited number of responsive units, we looked for inhibitory interactions between the two anatomical inputs. Further, as a contribution to assessing the influence of anaesthesia, we compared the response distributions obtained under chloralose anaesthesia with those in unanaesthetized cerveau isolé brains.

METHODS

Results were obtained from 14 chloralose and 6 cerveau isolé experiments. After ether or halothane induction, chloralose was administered (50 mg/Kg i.v.), or else the brain stem was divided by pretentorial radiofrequency coagulation or by post-tentorial, pretrigeminal transection with a spatula.

Extracellular unit potentials were recorded with tungsten microelectrodes, using an oblique approach in coronal planes to avoid electrode deflection by intact sagittal sinuses, falx, or medial hemisphere surfaces. Ablation exposure of the septum was not adopted, so avoiding possible damage to it or to the dorsal fornix.

Ipsilateral HC and HYP were stimulated through stainless steel electrodes, either blunt concentric with tip protruding 1 mm, or 4-pole multiple-electrode sticks with 1 mm vertical spacing between pole centres. Stimulation polarity was cathodal at the tips of concentric pairs and at the deeper of successive pairs along 4-pole sticks. Peak currents were less than 2 mA.

Recording and stimulating loci were identified in frozen sections. Microelectrode positions were reconstructed from cathodal lesions at deepest points. Stimulation points were identified by the Prussian blue reaction for anodally-deposited Fe.

Locations of septal units were designated by reference to the nomenclature and atlas of Andy and Stephan (1), plotted on enlarged tracings of the sections, and transferred to composite diagrams at selected atlas planes.

RESULTS

Anatomical Distribution of Responsive Units

For purposes of analysis, dorsal and lateral nuclear complexes were grouped together, but are distinguished in the composite diagrams drawn from Andy and Stephan (1). All units responding to stimulation are shown in Fig. 1.

Chloralose. Under chloralose anaesthesia, units responding to HC were distributed among the three major nuclei: dorsal-lateral (DL), medial (M), and diagonal band (nDB). The only discernible concentration of units in the rostro-caudal axis was in the caudal pole of medial anterior (MA) nucleus (Fig. 1, chloralose: 15.6). At other levels there were more responding units in nDB than in other nuclei. Some were observed in a few tracks through medial areas of n.accumbens.

Units responding to HYP were also most frequent in nDB, but in contrast to the above were predominant at rostral levels.

A minority of units, scattered sparsely through the septal regions, responded to both HC and HYP (Fig. 1).

Unanaesthetized cerveau isolé. In these preparations few units responded to HC but many to HYP (Fig. 1, C.I.). For HYP stimulation the responding units were, in contrast with chloralose preparations, concentrated rostrally for DL and M nuclei. Only one unit responded to both HC and HYP. (Fig. 1, C.I.; 15.1)

Topographic Properties of Input

Evidence of spatial segregation within input channels from either HC or HYP was obtained with some units. In experiments where concentric stimulating electrodes were placed in both lateral and ventromedial HYP, at 6 recording loci units were evoked from both sites, in each case at shorter latency (2-7 msec) to lateral than to medial stimulation; in only one case was the same unit responding to each input. Experiments using vertical multipolar stimulating electrodes provided further evidence for topographic discreteness in projections to septal neurones. In one microelectrode track traversing DL, M and nDB (Fig. 2) under chloralose, units in DL responded respectively to: HYP1 and HYP2; HYP1 only; HYP3 only. In M, units responded respectively to: HYP1, HC1, HC2; HYP1 and HYP3; HYP1, HYP3 (4-pulse stimulation) and HC1 (Fig. 2, 1-3; Fig. 3); HC1 and HC2; HYP1, 2 and 3 (Fig. 2, 1-6; Fig. 4). Equivalent unit-to-unit variations were observed in the continuation through nDB.

Response Properties

Units commonly responded with single spikes or several spikes at variable intervals. Rapid burst responses typical of

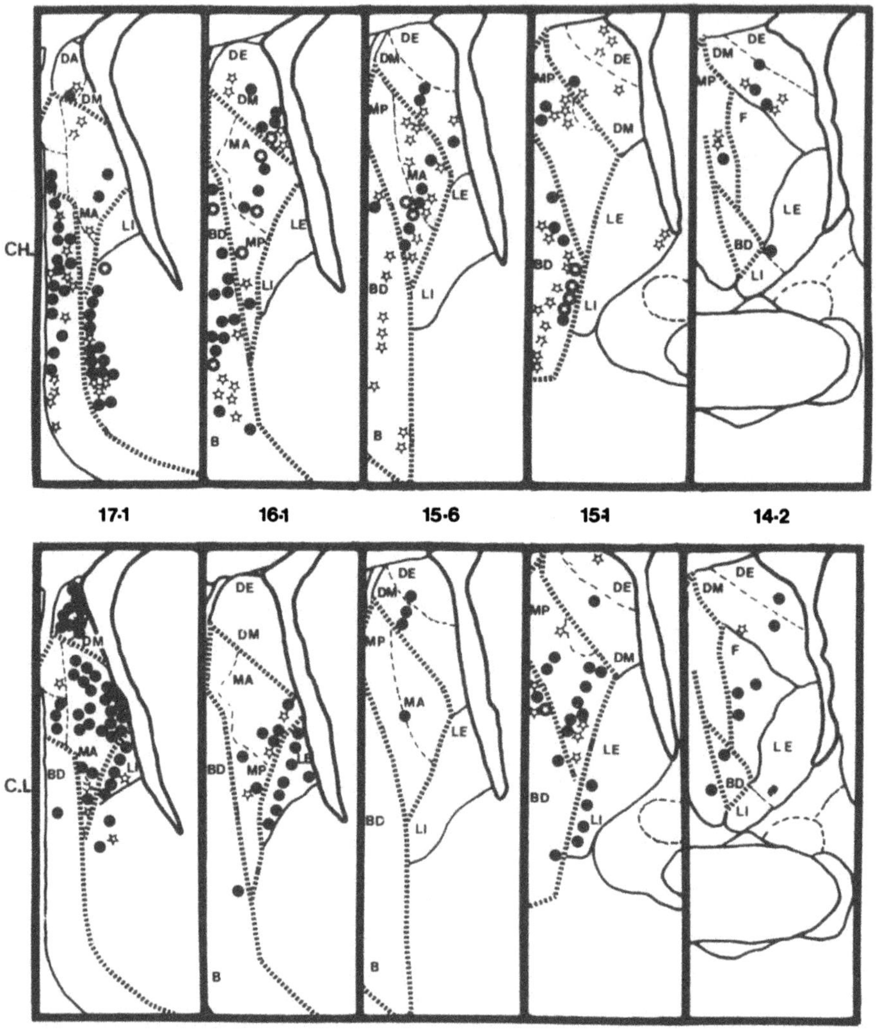

Figure 1. Anatomical distribution of unit responses to HYP (filled circles) or DHC stimulation (stars) or to both (stars within circles). The reconstructed positions of units have been transferred to nearest equivalent points relative to nuclear borders as designated in transverse atlas planes (17.1, 16.1, 15.6, 15.1, 14.2) of Andy and Stephan (1964). Data from chloralose-anaesthetized preparations (CH) and cerveau isolé preparations (C.I.) are represented in upper and lower rows respectively. Nuclear groups are delineated by the thick hatched lines into dorsal, medial and lateral groups and diagonal band (BD and B). Other abbreviations follow Andy and Stephan (1964).

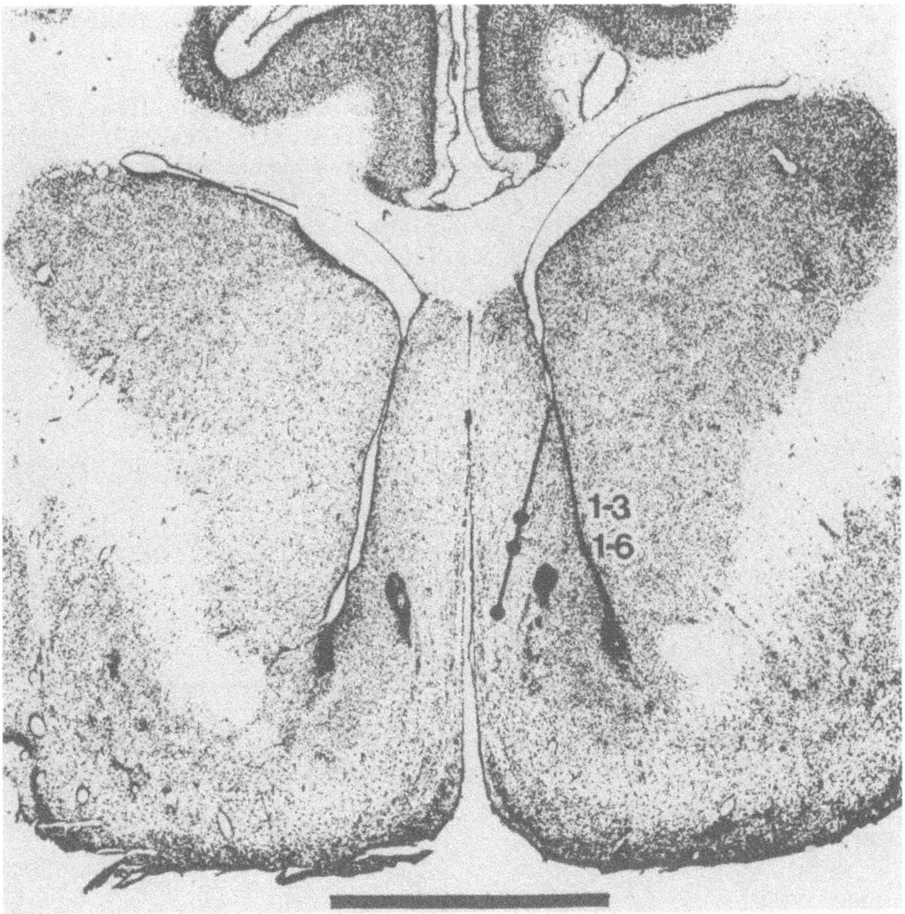

Figure 2. Transverse thionine-stained section through the septal area, with microelectrode track marked by oblique black line. A marking lesion at the deepest point is located by the lowest black dot. The upper two dots represent calculated positions of units, labeled 1-3 and 1-6, in nucleus M, and depicted in Figs. 3 and 4 respectively. Calibration: 5 mm.

intralaminar thalamic neurones (11) were rare. Often several different units, easily distinguishable, responded at one septal locus.

Response latencies varied over a wide range (Fig. 7), with differences between nuclei, inputs and preparations. Short-latency responses (up to 5 msec) to HC or HYP were common in M and nDB under chloralose, but infrequent in DL. On the other hand, in cerveau isolé cats, M units responded to HYP typically at long latency, while DL responses to HYP were common at short latency (Fig. 7).

Some short-latency responses were tested by following-rate and by the collision technique (3) for antidromicity (Fig. 4). Ability to follow stimulation rates above 50 Hz was found to be no infallible index of antidromic invasion, nor was a short and invariant responses latency (Fig. 4). It appears hazardous to infer this property from simple indices, particularly in view of the finding that the earlier, antidromic member of a unit's double discharge was suppressed by a conditioning stimulus at long intervals, but not the immediately subsequent orthodromic response (Fig. 4).

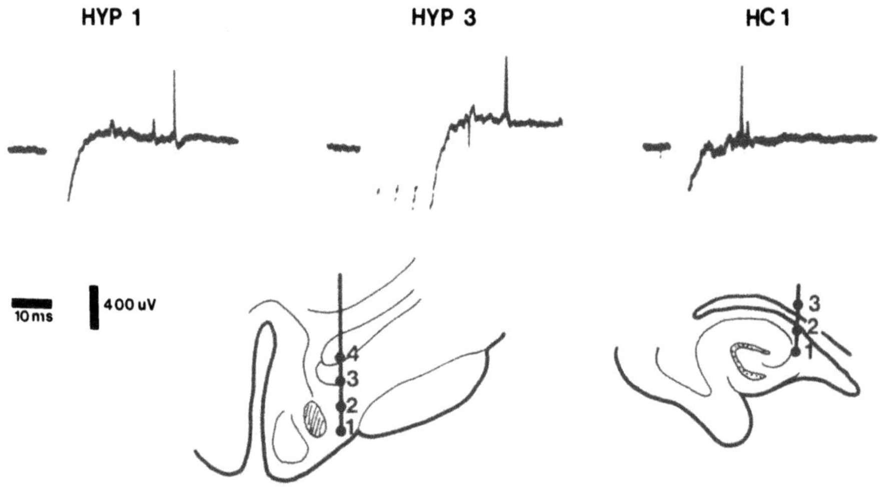

Figure 3. Units evoked in the one septal locus (Fig. 2: 1-3) by stimulation at different positions in HYP and DHC. In the left-hand trace (HYP1), a unit responds to single pulse HYP stimulation (marked by large baseline deflection). The position of the stimulating cathode is labeled 1 on the left hand vignette. In the middle trace (HYP3), 4-pulse cathodal stimulation at position 3 evokes the same unit as well as a smaller, shorter-latency unit. Cathodal stimulation at position 2 did not evoke a unit response. In the right-hand trace (HC 1), single pulse stimulation of DHC at cathode position 1 of right-hand vignette also evokes the same larger unit. Stimulation at cathode position 2 did not. In this and following Figs. negativity is upwards.

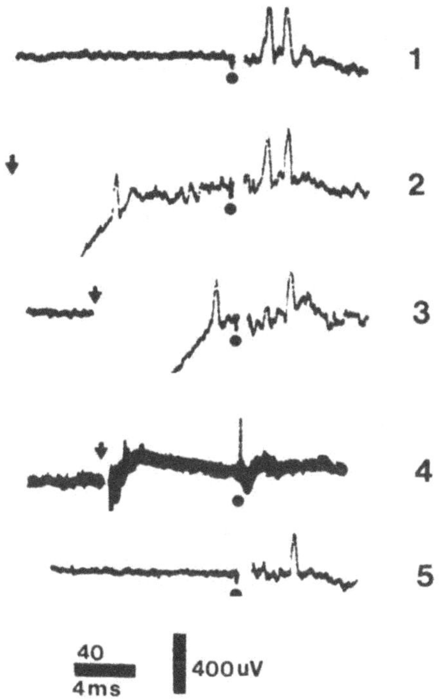

Figure 4. Responses of unit in medial septal nucleus to electrical stimulation of HYP and DHC. (Chloralose anaesthesia). In line 1, the unit responds with 2 spikes, at 2 msec and 3 msec latencies, to HYP stimulation (at dot). The unit also responds to HC (arrow) at 7 msec latency (line 2). In line 3, the 2 msec latency response to HYP suffers collision extinction when the unit's response to DHC precedes it within a period equalling twice its initial latency to HYP, and is thus antiodromic. The unit's response to HYP at 3 msec latency does not show collision extinction, and is orthodromic. In line 4, DHC conditioning stimulation precedes HYP test stimulation by 90 msec, and suppresses the antidromic, but not orthodromic spike, as seen in faster display (line 5.) Time calibrations: 40 msec for line 4, 4 msec for other lines.

Interactions Between Inputs

A minority of responding units were tested for conditioning interactions between HC and HYP inputs (Figs. 4,5,6). Of these, a minority showed conditioned suppression by one input of response to the other. Under chloralose, 5 of 28 HC-evoked units were suppressed by conditioning HYP stimuli, and 9 of 26 HYP-evoked responses by HC stimulation. In *cerveau isolé* experiments, none of 13 HC-evoked, but 6 of 58 HYP-evoked units were suppressed by HYP and HC stimulation, respectively. The time-course of suppression was not studied systematically, but durations of 300 msec were observed following single conditioning stimuli to HC or HYP.

Spontaneous Units

In the major septal nuclei spontaneous units were encountered most commonly in nDB and MP. Under chloralose, 18 of 39 tested were suppressed by HC and 3 by HYP. In a few instances suppression lasted several seconds following single stimuli. Rebound excitory phases as described by Gergen (8) were not observed.

In *cerveau isolé* preparations, none of 15 spontaneous units was suppressed.

Effects of Chloralose Anaesthesia

Differences in unit latency distribution between chloralose and *cerveau isolé* preparations, mentioned above, were overshadowed by differences in unit responsiveness between prepara tions and nuclei (Fig. 7). The total numbers of units responding over M, DL and nDB, independently to each input, were: HC:72, HYP:66 under chloralose, and HC:15, HYP:80 in *cerveau isolé* ($X^2 = 30.3$, $p < 0.001$). Thus, over all nuclear groups combined, and in each of M and DL nuclei, unit responses to HC were more numerous under chloralose anaesthesia than in cerveau isolé preparations. However, responses to HYP in each of M and DL groups were more numerous in the unanesthetized *cerveaux isolés* . These differences between input effectiveness for the respective types of preparation were highly significant. (M: $X^2 = 10.0$, $P < 0.01$; DL: $X^2 = 15.4$, $P < 0.001$). In contrast to these nuclei, excitability of nDB units, for each input, was much greater under chloralose than in *cerveau isolé* preparations (Fig. 7).

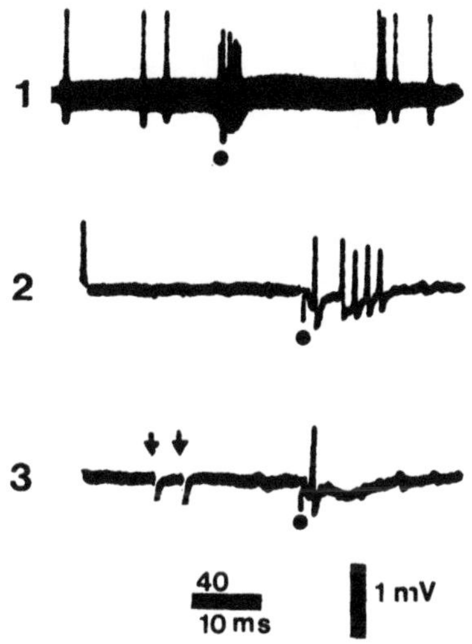

Figure 5. Effect of HYP and DHC electrical stimulation on a unit in nDB (cerveau isolé preparation). Trace 1 at slow sweep speed. The unit fires spontaneously. HYP stimulation (dot) consistently excites the unit. Traces 2 and 3 at faster sweep speeds. Trace 2 shows the response to HYP--an early single spike, followed by a later burst. The early spike followed stimulation rates of 200/sec and was probably antidromic. Trace 3, double pulse DHC stimulation (arrows) preceding HYP stimulation suppresses later burst component of the response to HYP, but not the antidromic spike. Time calibration: 40 msec line 1, 10 msec lines 2,3.

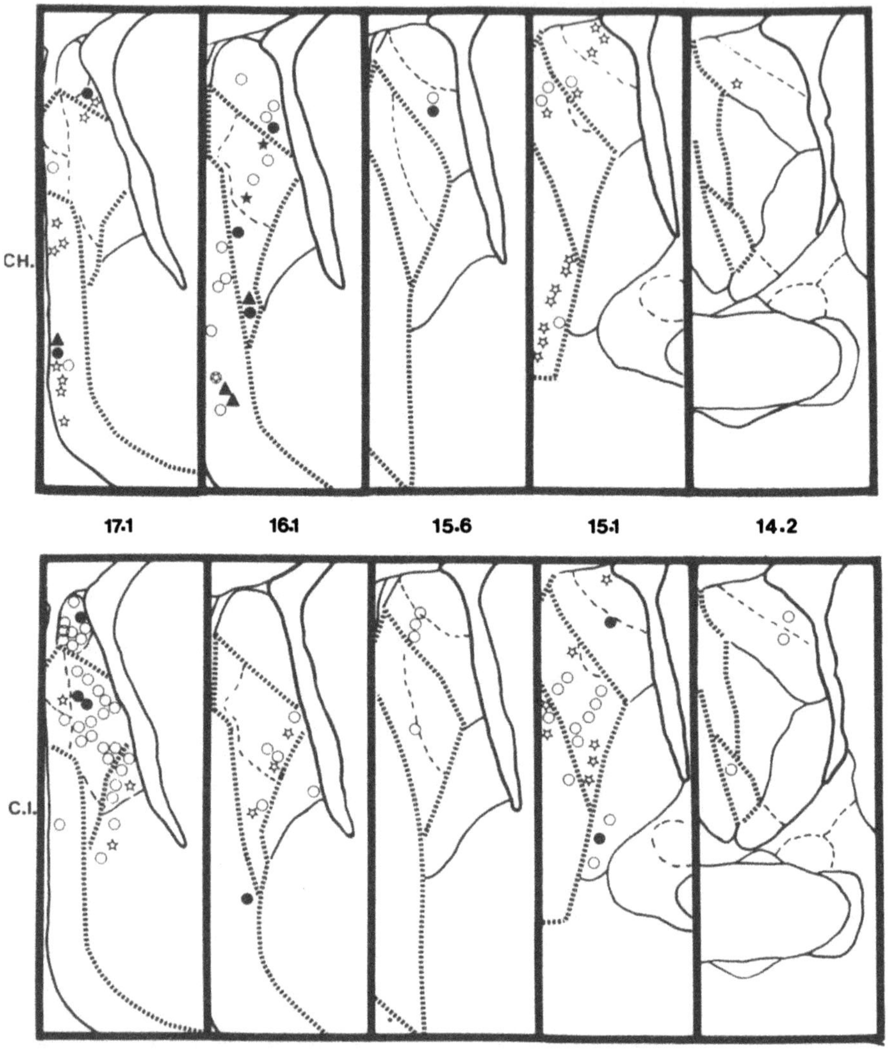

Figure 6. Anatomical distribution of evoked units tested for conditioning interactions. Open circles: units evoked by HYP stimulation and unaffected by conditioning HC stimulation. Filled circles: units evoked by HYP stimulation and suppressed by conditioning HC stimulation. Open stars: units evoked by HC stimulation and unaffected by conditioning HYP stimulation. Triangles: units evoked by HC stimulation and suppressed by conditioning HYP stimulation. Closed stars: units evoked by both HC and HYP stimulation, and suppressed by conditioing HYP and HC stimulation respectively. Nuclear subdivisions as in Fig. 1. Upper row: chloralose anaesthesia (CH.). Lower row: cerveau isolé preparations (C.I.).

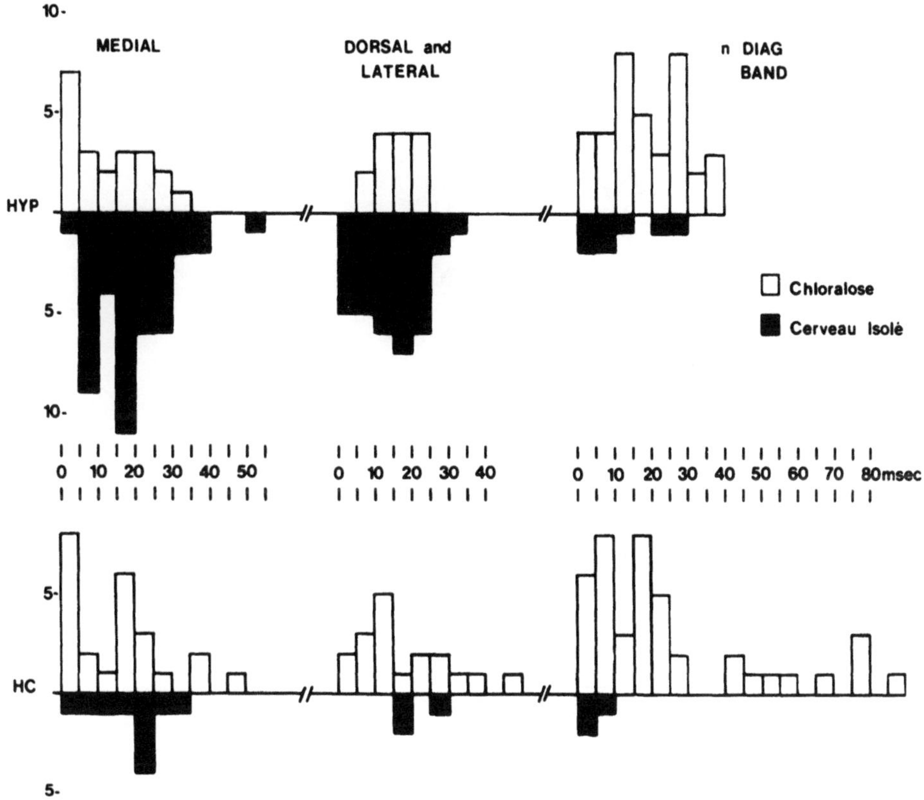

Figure 7. Latency distributions of units evoked by HYP or HC stimulation in chloralose-anaesthetized (white columns) or cerveau isolé (black columns) preparations, at class intervals of 5 msec (abscissae). Number of units in each interval shown by ordinates in two directions (scales at left). Results plotted separately for M, DL and nDB nuclear groups.

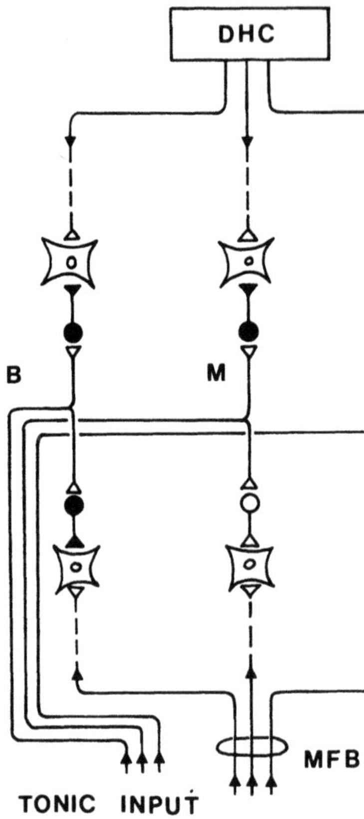

Figure 8. Schematic model to interpret differential effects of chloralose anaesthesia compared to cerveau isolé conditions, on unit numbers in diagonal band (B), medial (M) and dorsal-lateral (DL) septal nuclear groups, responding to dorsal hippocampus (DHC) and lateral hypothalamus (MFB) electrical stimulation. Inhibitory interneurones depicted as full black circles with short axon and filled triangle representing inhibitory synaptic terminals. Excitatory interneurones and synapses represented as open circles and triangles. Further explanation in text.

DISCUSSION

By adopting the anatomical classification of Andy and Stephan (1), the main difference from nomenclatures used in other recent investigations probably consists in distinguishing the dorsal and caudal extension of nDB from the medial nuclei. Here we have been guided by the distinctive histological picture of nDB neurones at all levels, observed in our Nissl sections.

A considerable problem for analyzing nuclear variations in responsiveness is posed by variations in effectiveness of stimulation positions. Evidence has been presented that in some situations there may be very narrow localization of projections to individual spetal neurones.

Given this limitation to sampling generality, and perhaps reinforced by it, the evidence obtained did not support the concept that the septal area constitutes a mixing centre, or zone of confluence, for limbic and hypothalamic circuits; for if so, its neurones would be expected to show widespread signs of convergent functional inputs from the two major fibre streams described. It is noteworthy, however, that the anatomic basis for the concept is derived mainly from studies on the rat, leaving open the possibility that species differences are responsible for the discrepancy. The paucity of convergent units is not attributable to chloralose anaesthesia, for they were even less abundant in *cerveau isolé* preparations. However, in the above experiments we did not stimulate the ventral HC. Siegel and Tassoni (16) described ventral HC projections to lateral septal nuclei and to nDB in the cat, and it is quite possible that the predicted convergence with hypothalamic inputs would be observed on activating those projections.

The comparison of results obtained under chloralose and in *cerveau isolé* cats raises problems of practical significance to studies of the organization of septal inputs. The general effect of chloralose appeared to be a facilitation of HC excitatory inputs, but a relative depression of HYP inputs (if *cerveau isolé* be accepted tentatively as "more normal") except those to nDB. Explanation of these effects requires further investigation to distinguish between several possibilities. The HC and HYP inputs largely go to independent neurones, and these could possibly constitute pharmacologically distinct classes with respect to the effects of chloralose on them or on their associated neuronal microcircuits. Alternatively, the effects of chloralose could be exerted on the input sources. Thus, since HC output axons are excitatory (5), a facilitation of HC stimulation effects by chloralose might be due to depression of inhibitory interneurones either in the HC itself (13), or in the neighborhood of the septal neurones to which it projects. The latter seems more likely with activation of septal afferent pathways by electrical stimulation.

In that case a schema is able to explain these findings. Pathways from DHC supply excitatory input to neurones recorded in the 3 septal nuclear groups (Fig. 8, upper). In addition, some other source of input, for example a ventral diencephalic extension of ascending reticular pathways, is assumed to drive tonically inhibitory interneurones that antagonise DHC input. The same tonic input drives interneurones associated with units responding to HYP input, again inhibitory in nDB, but excitatory in M and DL nuclei. In the presence of chloralose, all interneurone firing is assumed to be depressed, so the main inputs to recorded neurones operate without modification. In the absence of anaesthesia (*cerveau isolé*), inhibitory or excitatory interneuronal synaptic drives produce the observed difference in response density between HC and HYP inputs.

Since the model postulates hitherto - unresolved interneurones, it can be tested only by improved methods for sampling small-neurone activities. It would then be possible to determine whether an association of tonic interneurones with target neurones of DHC and HYP occurs in the septum, whether their tonic activities have the required correlations with postulated excitatory and inhibitory actions, and whether the tonic firing is suppressed by chloralose, as proposed in the model.

Whatever the mechanisms involved, the present results demonstrate that results obtained on anaesthetized preparations are of limited generality in studies of septal area organization.

SUMMARY

1. A survey was made of the septal nuclei in the cat for unit responses to dorsal hippocampal (DHC) and lateral hypothalamic (HYP) stimulation, in chloralose-anesthetized and unanaesthetized *cerveau isolé* preparations.

2. Units responding to one or other of HC and HYP occurred throughout the septal region, but few displayed convergent responses to both inputs, in either type of preparation.

3. For some recording loci, there was evidence of a localized spatial organization of HC and HYP excitatory inputs to septal units. A few short-latency unit responses were shown to be antidromic, while some were orthodromic, but this question was not studied systematically.

4. In a minority of tested units, conditioing HC stimulation suppressed responses to HYP, and *vice versa*, under chloralose. In *cerveau isolé* preparations, HC but not HYP inhibited a minority of tested unit responses.

5. The use of chloralose or cerveau isolé preparations was associated with differences in unit latencies to HC compared with HYP stimulation, but more significantly with differences in numbers of responding units. Relative to cerveau isolé conditions, chloralose appeared to facilitate unit responsiveness to HC in medial, dorsal-lateral and diagonal band nuclear groups, and to HYP in the last-named, but to produce a suppression of responses to HYP in medial and dorsal-lateral nuclei. These differences are explained in terms of a postulated differential organization of septal interneurones.

REFERENCES

1. Andy, O. J. and Stephan, H. The Septum of the Cat. C. Thomas, Springfield, 1964.

2. Cajal, S. Ramon Y. Studies on the Cerebral Cortex (Limbic Structures), transl. by L. M. Kraft. Lloyd-Luke, London, 1955.

3. Darian-Smith, I., Phillips, G. and Ryan, R. D. Functional Organization in the trigeminal main sensory and rostral spinal nuclei of the cat. J. Physiol. (Lond.) 168 (1963) 129-146.

4. De France, J. F., Kitai, S. T., Hatada, K., and Christensen, C. Contrasting effects of fimbria stimulation in the septum of the cat. Brain Research, 58 (1973) 240-244.

5. De France, J. F., Kital, S. T. and Shimono, T. Electrophysiological analysis of the hippocampal - septal projections: I. Response and topographical characteristics. Exp. Brain Res., 17 (1973) 447-462.

6. Edinger, H., Siegel, A. and Troiano, R. Single unit analysis of the hippocampal projections to the septum in the cat. Exp. Neurol., 41 (1973) 447-462.

7. Fox, C. A. Telencephalic centres in the cat. J. Comp. Neurol. 72 (1940) 1-62.

8. Gergen, J. A. Functional properties of the hippocampus in the subhuman primate, in W. R. Adey and T. Tokizane (Ed.s) Structure and Function of the Limbic System, Progress in Brain Research, 27 (1967) 442-461.

9. MacLean, P. D. Contrasting functions of limbic and neocortical systems of the brain and their relevance to psychophysiological aspects of medicine. Amer. J. Med., 25 (1958) 611-626.

10. McEwen, J. and McKenzie, J. S. Responses of single neurones in the septal nuclei to different inputs. Aust. J. exp. Biol. med. Sci., 45 (1967) 5P

11. McKenzie, J. S. and Rogers, D. K. Hippocampal suppression of intralaminar thalamic unit responses in cats and comparison with suppression evoked from the caudate nucleus. Brain Research, 64 (1973) 1-15.

12. McLennan, H. and Miller, J. J. The hippocampal control of neuronal discharges in the septum of the rat. J. Physiol, 237 (1974) 607-624.

13. Molnar, P. The hippocampus and the neural organization of mesodiencephalic motivational function. Recent Developments in Neurobiology in Hungary, 4 (1973) 93-173.

14. Nauta, W. J. H. Hippocampal projections and related pathways to the mid-brain in the cat. Brain, 81 (1958) 319-340.

15. Raisman, G. The connexions of the septum. Brain, 89 (1966) 317-348.

16. Siegel, A. and Tassoni, J. P. Differential efferent projections from the ventral and dorsal hippocampus of the cat. Brain Behav. Evol., 4 (1971) 185-200.

CHOLINERGIC ENZYME ACTIVITY IN THE SEPTUM AND RELATED BRAIN AREAS IN DECEASED MENTALLY NORMAL, CHRONIC SCHIZOPHRENIC AND ORGANIC BRAIN SYNDROME PATIENTS[1]

Edward F. Domino

Department of Pharmacology, The Lafayette Clinic
Detroit, MI 48207 and The University of Michigan
Ann Arbor, MI 48109

INTRODUCTION

The relationship between the activity of the cholinergic enzymes and mental illness is far from clear. However, we do know a lot concerning the neurochemistry of the cholinergic system. It is now well known that there are two major types of cholinesterases able to hydrolyze acetylcholine (ACh). One type "true" or acetylcholinesterase (AChE) is found primarily in "cholinergic" neurons throughout the body, but also in non-neuronal tissues including red blood cells and the placenta. The other type "pseudo" or butyrylcholinesterase (ChE) is found throughout the body, especially in the brain and in the plasma.

We have reviewed the literature to 1971 on the activity of both AChE and ChE in the blood, cerebrospinal fluid (CSF) and brain of mentally ill patients (3). From reports in the literature CSF and brain AChE activity may be enhanced in some schizophrenic patients.

Our own studies on red cell AChE and plasma ChE activity have shown that the former is normal and the latter decreased in one-third of adult male chronic drug-free schizophrenics (4) but not in male childhood schizophrenics (14). In an independent replication we have confirmed our original findings in adult chronic schizophrenic patients (7). The decrease in plasma ChE activity

[1]Supported in part by a grant from the State of Michigan to The Lafayette Clinic for schizophrenia research.

seemed unrelated to nutritional status or genetic factors (4), in schizophrenic patients with normal liver function as judged by the usual clinical and laboratory criteria.

In view of the fact that many schizophrenic patients are on neuroleptics such as chlorpromazine, it is important to know how neuroleptics affect cholinesterase activity. Chlorpromazine has been reported to be an AChE and ChE inhibitor *in vitro*. In our study guinea pig red cell and brain AChE activity was only slightly decreased by this drug while plasma ChE showed a much greater decrease. Chlorpromazine therapy of schizophrenic patients for 28 days showed only slight but not statistically significant changes in red cell AChE and plasma ChE (11,12).

In this manuscript I would like to report on our data on brain choline acetyltransferase (ChAc), total cholinesterase (TChE), AChE and ChE activity in the frontal cortex, head of the caudate nucleus, septal area, medial amygdala, hippocampus and posterior hypothalamus of three different patient groups. We have had an opportunity to study the gross distribution of various enzymes in the brains of patients who, prior to death, were ascertained to be mentally normal, had chronic schizophrenia, or an organic brain syndrome. Brain areas were chosen because of their theoretical interest.

METHODS

Tissue Sources

Seventeen brains were obtained from autopsies performed on patients who died at Northville State Hospital, Detroit General Hospital, or University of Michigan Hospital. After death, the bodies were kept at 4° C. Autopsies were performed within 36 hours and usually within 24 hours. At autopsy the brain was removed and frozen. One to four days after death the brain was transferred to a dry ice freezer (-50°C) at the Lafayette Clinic and kept there until 24 hours before dissection when it was gradually thawed in a cold room. After dissection, the tissues were either homogenized the same day and the homogenates frozen for assay at a later date or the tissues were packed in plastic bags and frozen for later homogenization and assay.

Patient Diagnosis

Three patients were diagnosed as having an organic brain syndrome. Seven patients were diagnosed as chronic schizophrenics given recent psychotropic drugs and one a chronic schizophrenic who was drug free. The rest (six) were mentally normal prior to death. All diagnoses were made by the attending psychiatrist and verified by review of the records.

Drug History

Four subjects were drug free for several months prior to death. Five patients received psychotropic drugs (usually antipsychotics) within one week prior to death. All others received various non-psychoactive drugs (e.g., digitoxin, antibiotics) prior to death. Most of the drugs were discontinued from one week to hours before death.

Cause of Death

Ten patients died of bronchial pneumonia and/or pulmonary embolism. One died of acute peritonitis and perforated gastric ulcer. One died of a myocardial infarction and terminal pneumonia. One suffered congestive heart failure and one died of an acute myocardial infarction. Another suffocated, one died of renal failure, and one died of lung cancer.

Physical Characteristics

Seven patients were males and ten were females. Their ages ranged from 14 to 90 years. Four were black, the rest white.

There were unavoidable delays in obtaining the brains or performing the enzymatic studies. Delays of from four days up to one year occurred between date of death and date of enzyme assay. All tissue was frozen during the interim.

Chemical Assays

Protein standards were bovine serum albumin and Dade protein standard. Rat brain was the standard enzyme source of ChAc, TChE, AChE, and ChE.

Protein Determination

Proteins were determined according to the method of Lowry _et al_. (13) using the Folin Cu-EDTA method. Details can be found in Domino _et al_. (6).

Choline Acetyltransferase

ChAc activity was assayed using a combination of the methods of McCaman and Hunt (15) and Schreier and Shuster (16). Enzyme activity was expressed as micromoles ACh/gm protein/min. See Domino *et al.* (6), for further details.

Acetylcholinesterase and Butyrocholinesterase

TChE was measured using the method of Siakotos *et al.* (19). This is a radiochemical method in which acetyl-1-^{14}C-choline was used as a substrate. The principle of the assay is that ACh is hydrolyzed and after ten minutes the reaction stopped by the addition of a strongly acidic cationic exchange resin which binds the unhydrolyzed substrate. TChE activity was measured without any inhibitor. ChE activity was measured using the inhibitor, BW 284c51 (1:5-bis-[4-allyl methylammoniumphenyl] pentane-3-one dibromide), at 1×10^{-6} M final concentration, dissolved in 0.1 M phosphate buffer. AChE was obtained by subtracting the residual ChE from TChE. All enzyme values were expressed in micromol ACh hydrolyzed/gm of brain protein/min. For further details see Domino *et al.* (6).

RESULTS

Age Differences

The mean ± S.E. age of the mentally normal group (N = 6) was 53.3 ± 9.9, that of the chronic schizophrenic patients (N = 8) was 62.2 ± 4.9, and that of the organic brain syndrome group (N = 3) 83.3 ± 5.2 years. Using a group comparison *t*-test the age difference between the normals and the chronic schizophrenics was not statistically significant. As might be expected, the organic brain syndrome group was significantly older than both the normals ($P < .03$) and the chronic schizophrenic patients ($P < .02$).

Protein

The concentration of protein/gm of wet brain tissue among the various brain areas was approximately 10%. There were no significant differences between the three diagnostic categories of mentally normal, chronic schizophrenic, or organic brain syndrome patients in the areas studied in this report.

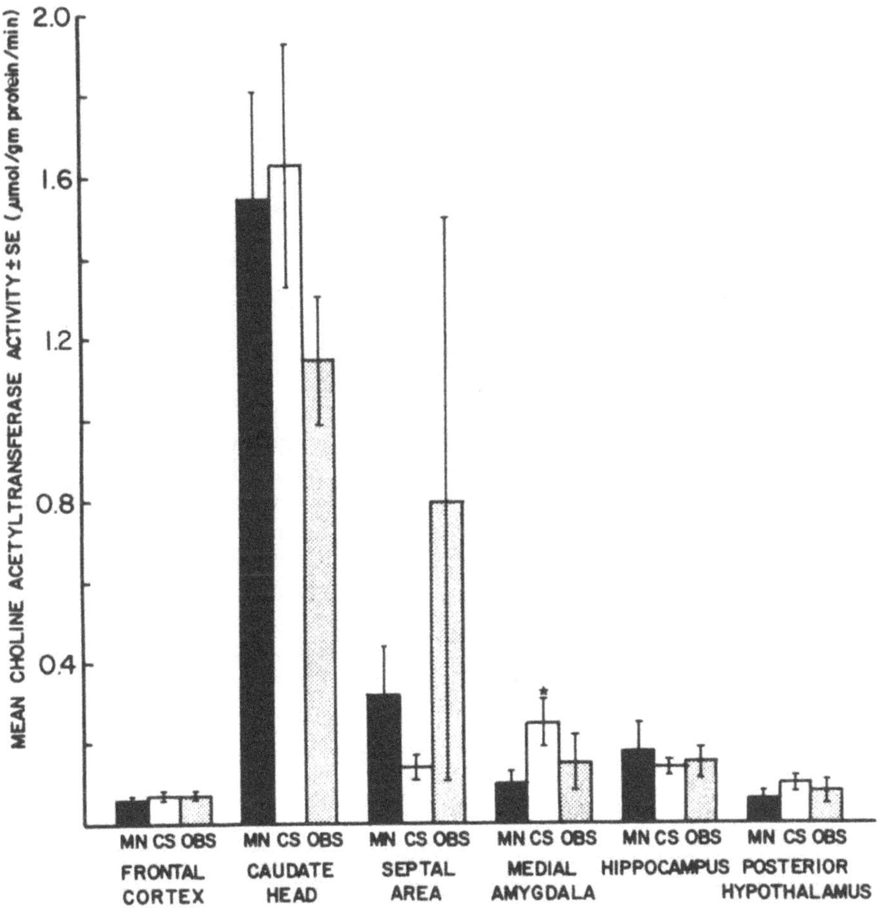

Figure 1. Regional brain choline acetyltransferase activity of deceased mentally normal, chronic schizophrenic, and organic brain syndrome patients. Mean ChAc activity ± S.E. in μmol/gm protein/min are given on the Y axis. The height of the bar represents the mean and the thin line ± the S.E. Six different brain areas were compared as noted. The symbols are as follows in this and subsequent figures: MN-mentally normal, CS-chronic schizophrenic and OBS-organic brain syndrome patients. Three to 12 subjects are given per group. Group comparison <u>t</u> test* $P < .05$. **$P < .01$. ***$P < .001$.

Choline Acetyltransferase

The distribution of ChAc activity varied widely among the brain areas studied (Fig. 1). In general, the lowest enzyme activity was in the cerebral cortex and the highest in the caudate nucleus. A similar pattern of distribution was seen among the mentally normal, chronic schizophrenic, and organic brain syndrome patients. There was no significant difference between any of these subgroups where three or more were tested except that the medial amygdala of the chronic schizophrenics showed twice the activity when compared to the normals ($P < .05$). This difference was still present when the data were normalized. Even when normalized there was no difference in septal ChAc between these two groups.

Total Cholinesterase

Using ^{14}C-ACh as substrate, marked regional differences in TChE were found (Fig. 2). The lowest activity (mean ± S.E.) was present in the frontal cortex (12.0 ± 1.2 μmol/gm protein/min) and the highest activity in the body of the caudate nucleus (478.4 ± 39.3 μmol/gm protein/min) for the mentally normal. There were no significant regional brain differences between the mentally normal, chronic schizophrenic, and organic brain syndrome patients except that the septal area of chronic schizophrenics showed significantly less TChE ($P < .05$). This finding was also significant when the TChE was normalized by comparing each value to the rat brain standard run simultaneously.

Acetylcholinesterase

Using the specific AChE inhibitor, BW 284c51, and subtracting the residual ChE activity, an estimate of AChE activity was obtained (Fig. 3). The qualitative distribution of AChE in the brain was similar to that of TChE with the frontal cortex showing the lowest and the caudate the highest activity in both the mentally normal and chronic schizophrenic groups. In addition to a significant decrease ($P < .05$) in AChE in the septal area, the caudate of chronic schizophrenics also showed significantly lower activity ($P < .05$). When the data were normalized by comparing each value to the rat brain standard run simultaneously only the difference in the septum remained significant ($P < .05$). In view of the small number for the organic brain syndrome group, no statistical comparisons were possible.

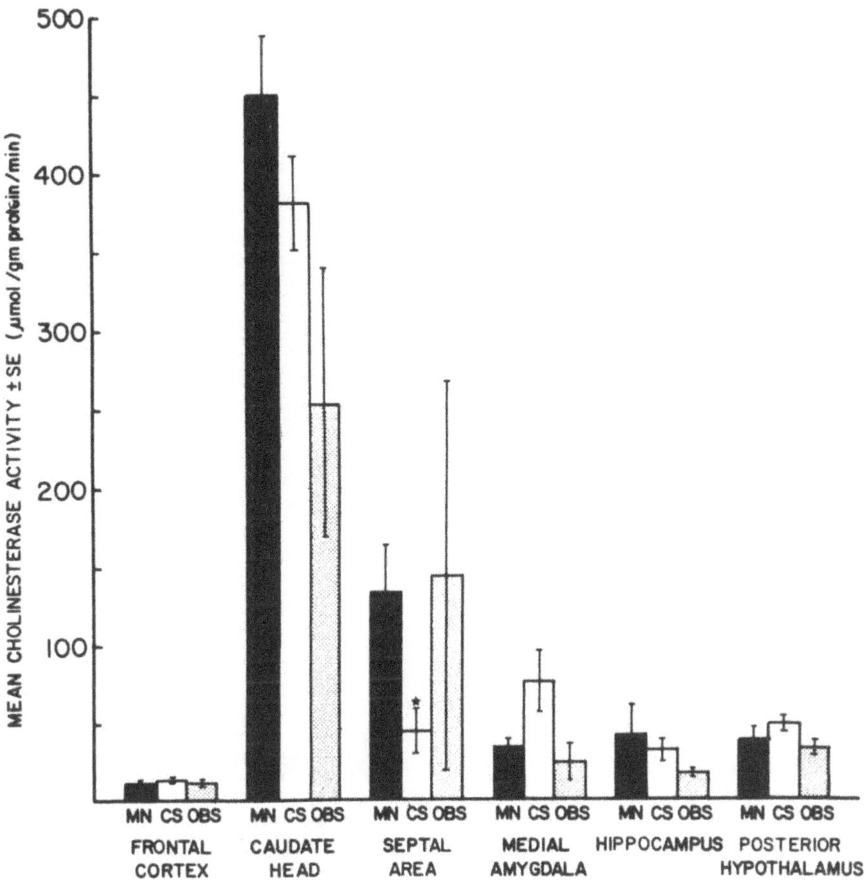

Figure 2. Regional brain total cholinesterase activity of mentally normal, chronic schizophrenic and organic brain syndrome patients. Note that there was a barely significant ($P < .05$) decrease in septal area TChE activity in the chronic schizophrenic patients.

Butyrocholinesterase

ChE activity was much less than AChE activity (Fig. 4). The relative activity of ChE with frontal cortex being lowest and caudate nucleus highest remained as with AChE. In view of the small number for the organic brain syndrome group, no statistical comparisons were possible. Surprisingly, ChE was enhanced compared to the mentally normal group in the posterior hypothalamus,

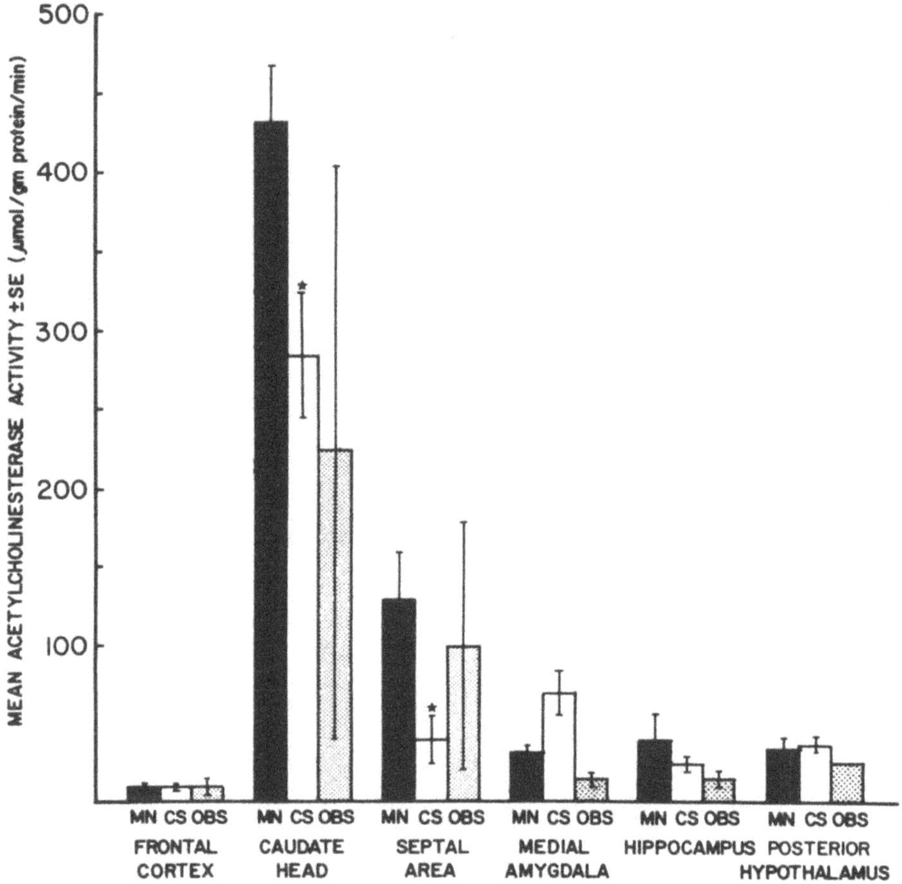

Figure 3. Regional brain acetylcholinesterase activity of mentally normal, chronic schizophrenic and organic brain syndrome patients. Note that there was a barely significant ($P < .05$) decrease in septal area AChE in the chronic schizophrenic patients.

hippocampus, medial amygdala, and caudate nucleus but not in the septum and frontal cortex of the chronic schizophrenic patients. These differences were significant ($P < .05$ to $< .01$). When the data were normalized, that is, expressed as a fraction of the ChE standard for each assay, the differences were still statistically significant ($P < .05$ to $< .01$).

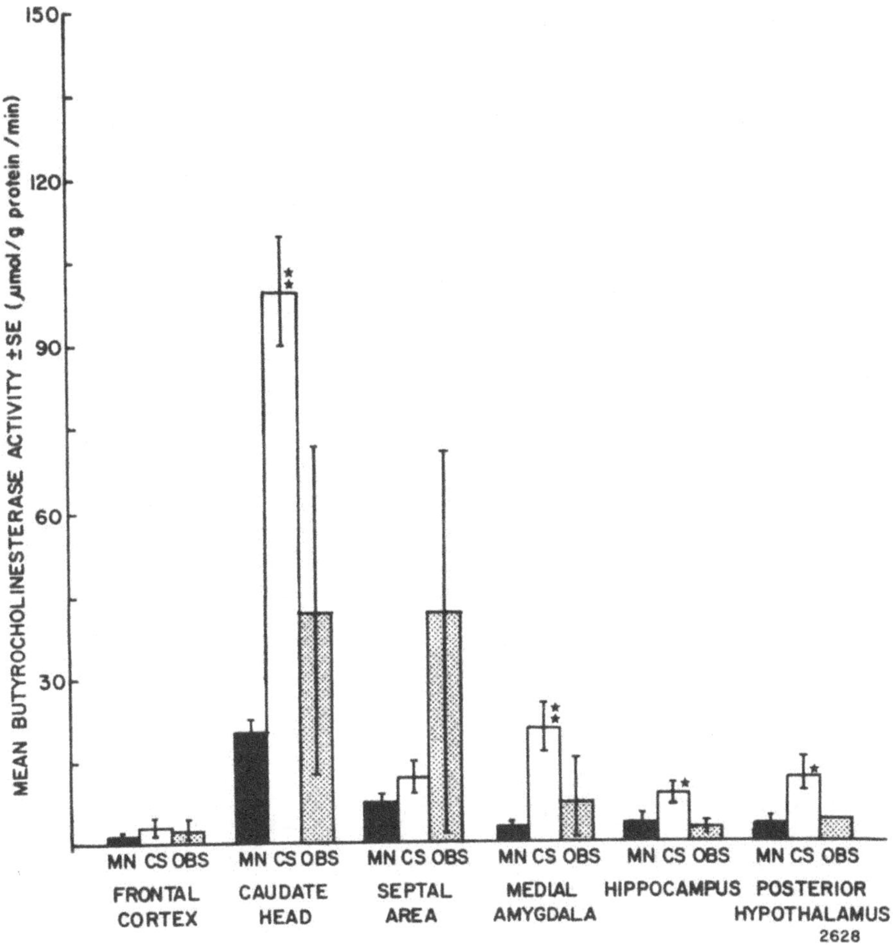

Figure 4. Regional brain butyrocholinesterase of mentally normal, chronic schizophrenic and organic brain syndrome patients. Note that ChE was generally much lower than the levels of AChE in similar brain areas. Surprisingly, a number of brain areas showed greater ChE activity in the chronic schizophrenic patients.

DISCUSSION

The major finding of this preliminary study is that most brain areas of the chronic schizophrenic patients show normal protein, ChAc, TChE, and AChE activity. Only the septal area of the schizophrenics showed decreased TChE and AChE. Protein, ChE and ChAc activity in this region were normal. The selective decrease in AChE is, of course, surprising. While not highly significant ($P < .05$), it raises the question of possible artifacts due to dissection error. This is especially important in view of the vague status of the septal area of the human brain. Without a doubt this brain area is least well defined. The area which was dissected in the present study was at the base of the septum pellucidum, in front of the anterior commissure and consisted of midline frontal basal cortex. Heath and Gallant (9) and Sem-Jacobsen et al. (17,18) have reported abnormal electrical discharge in this general area and other portions of the brain in chronic schizophrenics. Heath and Krupp (10) have proposed an autoimmunologic theory of schizophrenia involving septal protein.

Another unexpected finding was that ChE in many of the brain areas of the chronic schizophrenic patients was significantly elevated. It must be remembered that all of the schizophrenics except one were on neuroleptic mediation prior to death. It is well known that neuroleptic agents tend to activate ChE activity in small concentrations, and inhibit in large concentrations (2). The one drug free patient had relatively lower ChE activity than the other schizophrenics, suggesting that psychotropic drugs may interfere with this enzyme. In general, all of the chronic schizophrenic patients had various neuroleptic agents given until shortly prior to the terminal illness. All patient groups, as a result of their terminal illness, were given various medications including antibiotics, cardiovascular drugs, etc., until just before death. In view of these many variables that could not be controlled, the data obtained are perhaps surprising because of their consistency with known literature, especially in the distribution of these enzymes in the nervous system of animals where such variables can be controlled.

There are many other objections to studies of this type. One is faced with the practical considerations of patients dying from different causes. The time from death to autopsy is a variable that cannot be controlled under most circumstances because of the need to obtain family permission, autopsy scheduling, etc. Furthermore, once an autopsy was performed all brains were frozen until subsequent thawing for dissection and regional assay. Thus, the question of enzyme stability with freezing and thawing becomes paramount.

Foldes et al. (8) have observed that up to six months of freezing at -20° C does not destroy TChE, AChE, and ChE activity of human brain. However, Bull et al. (1) have reported very low brain ChAc activity of cadavers. Another criticism of this research is the small number of patients, especially for the organic brain syndrome group. This prevented a statistical analysis of most of the findings compared to the mentally normal subjects.

It should be noted that the organic brain syndrome patients were significantly older than the mentally normal subjects and chronic schizophrenics. In comparing ChAc, TChE, AChE, and ChE with age no significant trend was observed. However, there was a tendency for ChE to increase with age in the chronic schizophrenic patients ($r = .8114$, $P < .05$). Thus, while our small series of patients did not show striking enzyme differences with age, this factor confounds our results.

Because of the many uncontrolled variables, this study must be regarded as extremely preliminary. Studies of this type, especially incorporating controls for patient diagnosis, age, cause of death, medication, role of freezing and thawing, dissection, and assay errors must be continued. It is hoped that this study will encourage continued and better controlled research to confirm or deny the observed results.

SUMMARY

The activity of choline acetyltransferase (ChAc), total cholinesterase (TChE), acetylcholinesterase (AChE), and butyrocholinesterase (ChE) were determined in various brain areas of patients disagnosed as chronic schizophrenics (CS), organic brain syndrome (OBS), or mentally normal (N) prior to death. The septal area of CS patients ($P < .05$) showed significantly less TChE activity than the mentally normal group. The septal area of the CS also had less AChE activity ($P < .05$). ChE activity in most areas was greater than normal in the CS patients previously on neuroleptic drugs. ChAc activity was significantly increased ($P < .05$) in the medial amygdala of CS patients. The majority of brain areas sampled showed no significant differences in enzyme activity among the three groups. The differences that were observed must be considered preliminary and subject to further validation under better controlled conditions.

REFERENCES

1. Bull, G., Hebb, C. and Rathovic, D. Choline acetyltransferase activity of human brain tissue during development and at maturity. J. Neurochem. 17 (1970) 1505-1516.

2. Domino, E. F., Hudson, R. D. and Zografi, G. Substituted phenothiazines: Pharmacology and chemical structure. In Drugs Affecting the Central Nervous System, Berger, A., Ed., pp. 327-397, Marcel Dekker, Inc., New York, 1968.

3. Domino, E. F. and Krause, R. Cholinesterase activity and mental disease: A literature review. Michigan Mental Health Bulletin. 5 (1971) 3-18.

4. Domino, E. F. and Krause, R. Reexamination of red cell and plasma cholinesterase activity in drug-free chronic schizophrenic patients and normals. Biol. Psych. 4 (1972) 17-31.

5. Domino, E. F., Krause, R. R. and Bowers, J. Regional distribution of some enzymes involved with putative neurotransmitters in the human visual system. Brain Research, 58 (1973a) 179-189.

6. Domino, E. F., Krause, R. R. and Bowers, J. Various enzymes involved with putative neurotransmitters - regional distribution in the brain of deceased mentally normal, chronic schizophrenics or organic brain syndrome patients. Arch. Gen. Psychiat. 29 (1973b) 195-201.

7. Domino, E. F., Krause, R. R., Thiessen, M. and Batsakis, J. G. Blood protein fraction comparisons of normal and chronic schizophrenic patients. Arch. Gen. Psychiat., 32 (1975). 717-721.

8. Foldes, F. F., Zsigmond, E. K., Foldes, V. M. and Erdos, E. G. The distribution of acetylcholinesterase and butyrylcholinesterase in the human brain. J. Neurochem., 9 (1962) 559-572.

9. Heath, R. G. and Gallant, D. M. Activity of the human brain during emotional thought. In Role of Pleasure in Behavior, Heath, R. G., Eds., pp. 83-109, Hoeber Medical Division, Harper and Row Publishers, Inc., New York, 1964.

10. Heath, R. G. and Krupp, I. M. Schizophrenia as a immunologic disorder. Arch. Gen. Psychiat., 16 (1967) 1-9.

11. Krause, R. R. and Domino, E. F. Effects of acute and chronic chlorpromazine in vivo on red cell, plasma and brain cholinesterase. Res. Comm. Chem. Path. and Pharmacol, 1 (1970) 653-666.

12. Krause, R. A. and Domino, E. F. Addendum to "Effects of acute and chronic chlorpromazine in vivo on red cell, plasma and brain cholinesterase." Res. Comm. Chem. Path. and Pharmacol., 2 (1971) 339-342.

13. Lowry, O. H., Rosebrough, N. J., Farr, A. L. and Randall, R. J. Protein management with the Folin phenol reagent. J. Biol. Chem., 193 (1951) 265-275.

14. Lucas, A. R., Krause, R. and Domino, E. F. Biological studies in childhood schizophrenia: Plasma and RBC cholinesterase activity. J. Autism and Childhood Schizophrenia, 1 (1971) 72-81.

15. McCaman, R. E. and Hunt, J. M. Microdetermination of choline acetylase in nervous tissue. J. Neurochem., 12 (1965) 253-259.

16. Schrier, B. K. and Shuster, L. A simplified radiochemical assay for choline acetyltransferase. J. Neurochem., 14 (1967) 977-985.

17. Sem-Jacobsen, C. W., Petersen, M. C., Lazarte, J. A., Dodge, H. W., Jr. and Holman, C. B. Electroencephalic rhythms from the depths of the frontal lobe in 60 psychotic patients. EEG Clin. Neurophysiol., 7 (1955) 193-210.

18. Sem-Jacobsen, C. W., Petersen, M. C., Dodge, H. W., Jr., J. A. and Holman, C. B. Electroencephalographic rhythms from the depths of the parietal, occipital and temporal lobes in man. EEG Clin. Neurophysiol. 8 (1956) 263-278.

19. Siakotos, A. N., Filbert, M. and Hester, R. A specific radioisotopic assay for acetylcholinesterase and pseudocholinesterase in brain and plasma. Biochem. Med., 3 (1969) 1-2.

Part III
Endocrinology

A NEUROENDOCRINOLOGICAL APPROACH TO THE INVESTIGATION OF SEPTUM[1]

Nobuyoshi Hagino, M.D. and Sadao Yamaoka, M.D.

Department of Anatomy, The University of Texas Health Science Center at San Antonio, Texas and Department of Physiology Saitama Medical University, Moro, Saitama, Japan

INTRODUCTION

Induction of ovulation in female rats could be driven by stimulation of the septum (13) and blocked by local anesthesia of the septum (23). In infertile rats, an interruption of the dorsal connection of the preoptic area (efferents and afferents paths of the septum) caused a delay in onset of puberty. Bilateral lesions of the medial forebrain bundle or both corticomedial and basomedial regions of the amygdala also resulted in delayed onset of puberty. Once sexual maturity is reached in the corticomedial amygdaloid lesioned rats, the ability to run completely normal estrus cycles, to respond to constant illumination, and to mate, become pregnant and rear litters is retained (5). Lesions severing the dorsal connection of the preoptic area (68), or complete bilaterial section

---

[1]This study was supported by AID Contract csd/2821, NSF Grant GB-28817X, NIH Contract 69-2133, NIH Grant HD-06950, National Academy of Science USA and Hungarian Academy of Science. The authors wish to express their gratitude to Miss T. Monkovsky and Mr. H. W. Nentwitch and L. C. Pleasant for their excellent technical assistance, and to Mrs. E. M. Menchaca for determination of plasma LH. Determination of acetylcholine contents in the brain was performed with collaboration of Drs. A. Modak and W. Stavinoha, Department of Pharmacology, The University of Texas Health Science Center at San Antonio, Texas. The electron microscopic pictures were taken by Dr. B. Kosaras, Department of Anatomy (Director, Prof. B. Flerkó) University Medical School, Pécas, Hungary.

of the stria terminalis and the fimbria in the dorsal part of their course (54) would interrupt the major direct fiber tracts from the forebrain to the medial hypothalamus. After both types of lesion regular cyclic ovulation continued (following an initial diestrus period). In addition, complete removal of neocortex could not block regular estrus cycles and normal pregnancies (10).

It, therefore, leaves open the question of what role the forebrain signal may play in modulating gonadotrophin secretion and regular cyclic ovulation in the female rat. An attempt has been made to describe the relevant functional evidence of the septum, hippocampus and amygdala, which have been important in guiding the course of neuroendocrinological investigation of the regulation of gonadotrophin secretion and ovulation in the female rat.

RESULTS AND DISCUSSION

Spontaneous Activity of the Septal Neurons

The septum is divided into three parts: medial, lateral and basal areas. The dorsal area is included in the lateral area. The ventral, trigonum septi and bed nucleus of the anterior commissure are included in the basal area. A tungsten Microelectrode (insulated with epoxylite or insulex) was inserted into the septum (frontal area to the bed nucleus of anterior commissure) twelve hours after urethane anesthesia and extracellular recording of the spontaneous activity of the septal neurons was begun. Criterion of change of spontaneous discharge unit firing was counted every 10 seconds during 1 minute intervals; an average mean value with standard error was calculated. A change of more than 50% in mean firing rate during 1 minute intervals was taken as an indicator of significant change. The majority of the septal neurons in the medial, lateral and basal areas discharge between 5 and 10 pulses per second during EEG sleep (Fig. 1). Two characteristic types of neurons are revealed in the septum: one is characterized by bursts of discharge firing and the other is a simple and tonic discharge firing neuron (Fig. 2). A total of 95.9% of the medial septal neurons are correlated with EEG sleep and arousal while other septal neurons (lateral 65%, basal 75%) show less correlation (Fig. 3, Table 1). The appearance of rhythmic bursts of septal neuron discharge regulates the appearance of hippocampal theta wave activity in the rabbit (40) and rat (46). Lesioning of the septum or fornix abolished the appearance of the hippocampal theta wave activity (19). In our study, it was difficult to find any significant correspondence between spontaneous bursts of septal neuron activity and the appearance of hippocampal theta wave activity. However, as the tip of the microelectrode contacted the

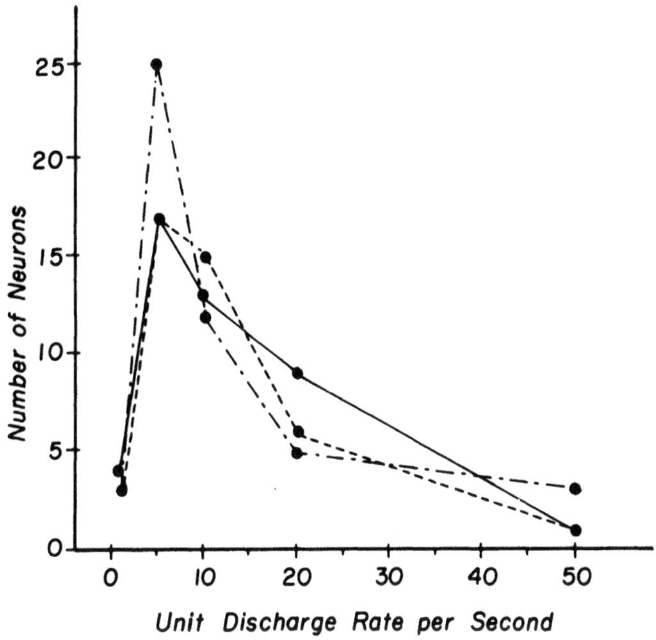

Figure 1. The distribution of firing rate per second of the medial (•———•), lateral (•-----•), and basal (•-.--•) septal neurons. Vertical axis represents the number of neurons and horizontal axis represents the number of neurons and horizontal axis represents the unit discharge rate per second. The majority of the septal neurons discharge between 5 and 10 pulses per second.

fornix fibers (determined by Weil's stain), the bursts of activity were correlated with the hippocampal theta wave activity. This indicates that the appearance of hippocampal theta wave activity may be controlled by the ascending afferent fibers of the hippocampus through the medial and lateral hypothalamus (1) instead of the role of septum as a pacemaker (19,49,65).

It was found that the mean firing rate of septal neurons declined at 7 p.m. and remained low and stable until just prior to resumption of light under a light-dark schedule (light on at 5 a.m. and off at 7 p.m.) (Figs. 2,4). Diurnal change of the mean firing rate of septal neurons corresponds to a cirdacian distribution of EEG activity (frontal cortical and hippocampal EEG activity). When rats are maintained under continuous illumination (24 hours illumination) a circadian distribution of EEG arousal and sleep does not continue (27) and dirunal changes in mean firing rate of septal neurons is also not observed (76) and (Table 2). Under this

TABLE 1: CORRELATION OF NEURON ACTIVITY TO PERIOD OF EEG AROUSAL AND SLEEP IN NORMAL CYCLING FEMALE RATS

	Neurons in					
Changes in unit activity from EEG arousal to sleep	Basal Septum Number	%	Lateral Septum Number	%	Medial Septum Number	%
Increased firing rate	14	28.6	10	20.4	21	42.9
Decreased firing rate	23	47.0	22	44.9	26	53.0
Changed (Subtotal)	37	75.6	32	65.3	47	95.9
Unchanged	12	24.4	17	34.7	2	4.1
Total	49	100.0	49	100.0	49	100.0

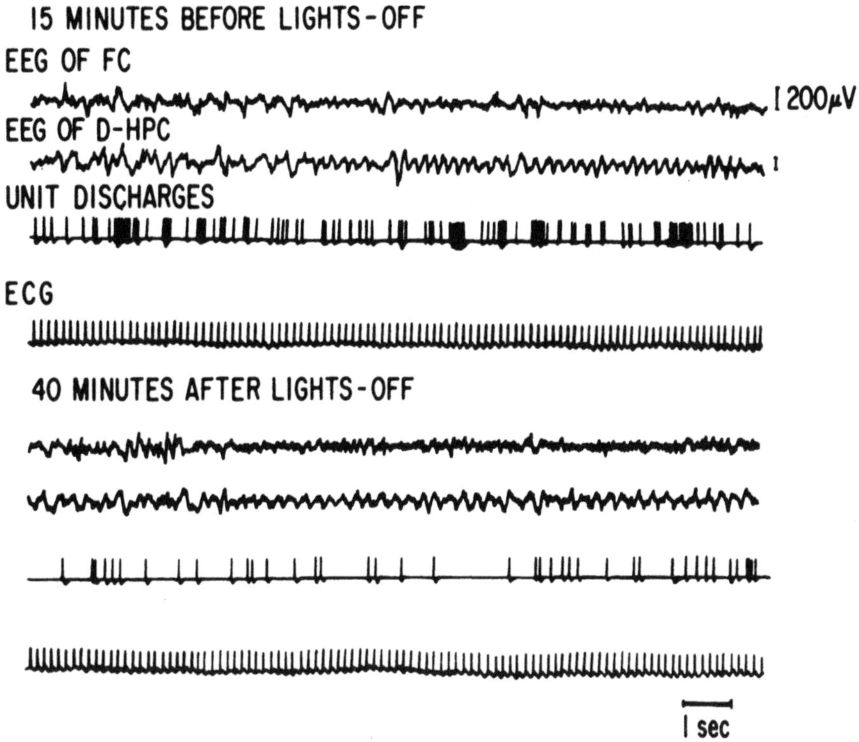

Figure 2. The pattern of unit discharges: EEG recording from frontal cortex (FC), dorsal hippocampus (D-HPC), and ECG, during a 15 second period 15 minutes before lights-off or 40 minutes after lights-off in the diestrus rats. A neuron shows two characteristic types of discharge firing: one is characterized by bursts of discharge firing (lower column). The bursts of discharge firing disappears after lights off, in addition, the dorsal hippocampal stimulation produces the bursts of discharge firing (Fig. 9).

circumstance, percent of mean firing of septal units during EEG arousal or sleep is similar in the rat maintained under light-dark schedule (Table 3).

Spontaneous activity of the septal neurons corresponds with cerebral cortical as well as hippocampal activity. Furthermore, exteroceptive factors, such as light, maneuver the activity of the reticular-thalamco-cortical activation system and further, regulate the spontaneous activity of the septal neurons.

TABLE 2: INFLUENCE OF ILLUMINATION ON THE DIRUNAL VARIATION OF THE ACTIVITY OF SEPTAL NEURONS

Changes in unit activity around 7 p.m.	Controlled Illumination No. of rats	Controlled Illumination %	Constant Illumination No. of rats	Constant Illumination %
Increased firing rate	3	21	0	0
Decreased firing rate	10	79	0	0
Changed (Subtotal)	13	100	0	0
Unchanged	0	0	8	100
Total	13	100	8	100

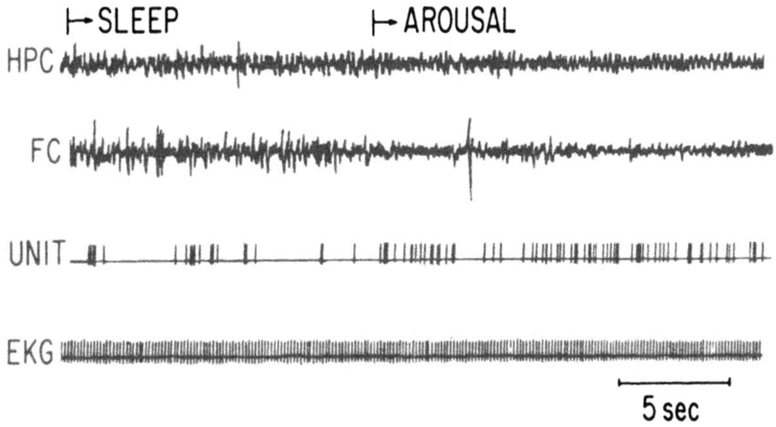

Figure 3. Changes in unit activity of septal neuron from EEG sleep to arousal. The unit firing (unit) increases during EEG arousal stage.

Ascending Afferent Paths

Thalamic septal connection. The mean firing rate of the septal unit corresponds with the cortical EEG activity of arousal and sleep. Furthermore, it was found that hippocampal stimulation was less effective on the mean firing rate of septal unit during cortical EEG arousal stage (Table 4).

It suggests that the reticular ascending pathway exert its effect on the septum to alter the threshold of spontaneous activity of septal neurons. To pursue this problem, the repetitive stimulus was applied for the anterior thalamus where it regulates emotional behavior through the cingulate gyrus and hippocampus (47,48). Repetitive stimulation of the anterior ventral thalamic nucleus (AV) reduced the mean firing rate of the septal unit. An application of 80μA with 100 Hz to the AV caused EEG arousal activity on the cerebral cortex and reduced the septal unit. However, an application of 80μA with 60 Hz to the AV produced a reduction of septal unit without evoking cortical EEG arousal activity (Fig. 5, ref. 21). Single stimulation of the AV produced an inhibitory effect on spontaneous activity of the septal neurons without evoking cortical EEG arousal.

The connection of anterior thalamic nucleus to the hippocampus through the cingulate gyrus is designated as a circuit of emotional behavior (48) and the stimulation of the septum causes pleasure

TABLE 3: CORRELATION OF NEURON ACTIVITY TO PERIOD OF EEG AROUSAL AND SLEEP IN PERSISTENT ESTRUS RATS PRODUCED BY CONSTANT ILLUMINATION

Changes in unit activity from EEG arousal to sleep	Controlled Illumination Neurons in Septum		Constant Illumination Neurons in Septum	
	Number	%	Number	%
Increased firing rate	45	30.6	27	30.3
Decreased firing rate	71	48.3	41	46.1
Changed (Subtotal)	116	78.9	68	76.4
Unchanged	31	21.1	21	23.6
Total	147	100.0	89	100.0

TABLE 4: EFFECT OF HIPPOCAMPAL OR AMYGDALA STIMULATION (0.5 msec 300 A 1 msec INTERVAL 3 PULSES) ON SPONTANEOUS ACTIVITY OF SEPTAL NEURONS DURING EEG AROUSAL AND SLEEP

Status of firing rate	Hippocampal St. Stage of EEG		Amygdala St. Stage of EEG	
	Sleep	Arousal	Sleep	Arousal
Excitation	2	0	2	0
Inhibition	17	2	4	0
Combined Response	9	5	5	1
Subtotal	28	7	11	1
No response	11	23	1	10
Total	39	30	12	11

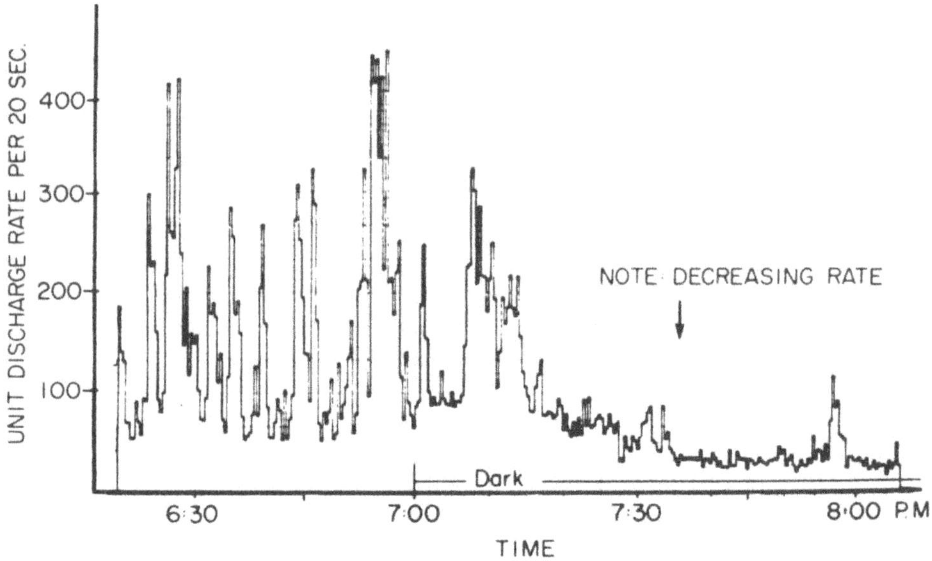

Figure 4. The frequency histogram of unit discharge rate per 20 seconds in a septal neuron. Vertical axis represents unit discharge rate, and horizontal axis represents the time of day and the time of lights-off. The mean firing rate of septal neurons declined at 7 p.m.

behavior (47). It was found that the direct connection exist between the anterior thalamus and the septum; and stimulation of the anterior thalamus exerts its inhibitory influence on the septal neurons.

Therefore, the ascending reticular activation system may exert their effect on the spontaneous activity of septal neurons through the anterior thalamus.

Midbrain limbic septal connection. The existence of ascending connection of the anterior thalamus to the septum does not necessarily indicate that no other connection is available for activating the spontaneous activity of septal neurons. It was found that an anatomical reciprocal neural circuit exist betwee- the central gray matter and the septum through the medial forebrain bundle (16,20,22).

There are two types of neural interconnections between the caudal central gray matter (CCG) and the septum: one is a monosynaptic excitatory connection interspersed with inhibitory neurons and the other is a polysynaptic connection of inhibitory neurons.

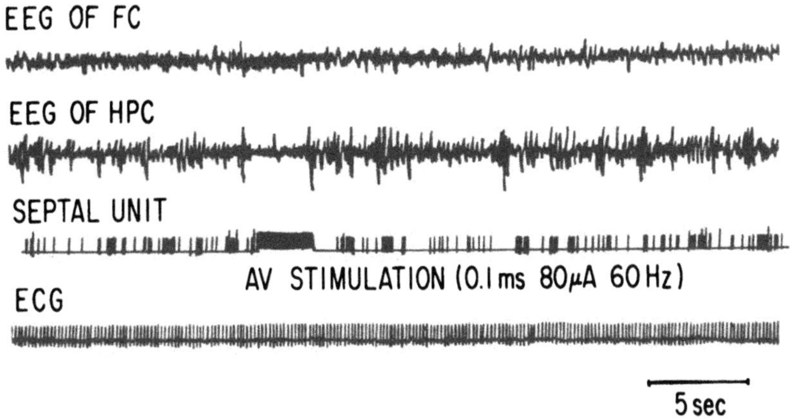

Figure 5. The picture shows the effect of electrical stimulation (0.1 msec 80μA 60 Hz 3 sec) of the anterior ventral thalamic nucleus on the spontaneous activity of septal neuron (septal unit). EEG recording was obtained from frontal cortex (FC) and dorsal hippocampus (D-HPC). ECG: electrocardiogram. Stimulation of the anterior thalamic nucleus reduced the unit firing without evoking cortical activity.

Stimulation of CCG with 0.5 msec 300 Aμl msec interval 3 pulses produces either long latency of inhibition of spontaneous activity of septal neuron following the monosynaptic excitation or direct inhibition (Figs. 6,8). Repetitive stimulation (80μA, 0.1 msec 100 Hz for 3 sec) of CCG also produces a predominantly inhibitory effect on spontaneous activity of the septal neurons (Fig. 7). Figure 7 shows that stimulation of CCG decreases neural activity of the septal neuron within 40 seconds after stimulation (primary response), and further, spontaneous activity of the septal neuron tends to reduce its activity again within 3 minutes after stimulation (secondary response).

Stimulation of the reciprocating neural closed circuit of CCG and septal interconnection (16,20,22) may play a role in primary and secondary responses of septal neuron activity. A medial portion of the medial forebrain bundle projects to the medial nucleus of the septum and a lateral portion sends fibers to the lateral nucleus of the septum (20). The monoamine-containing neurons in the pontine and mesencephalalic regions of the brain (16) have a possible correlation of their aggregate distributions with slow wave sleep and paradoxical sleep (30). Monoamines have an inhibitory effect on synaptic transmission in the cortex (44,45) and on spontaneous activity of cortical neurons (56). Impulses from the midbrain limbic area travel along with the medial forebrain bundle

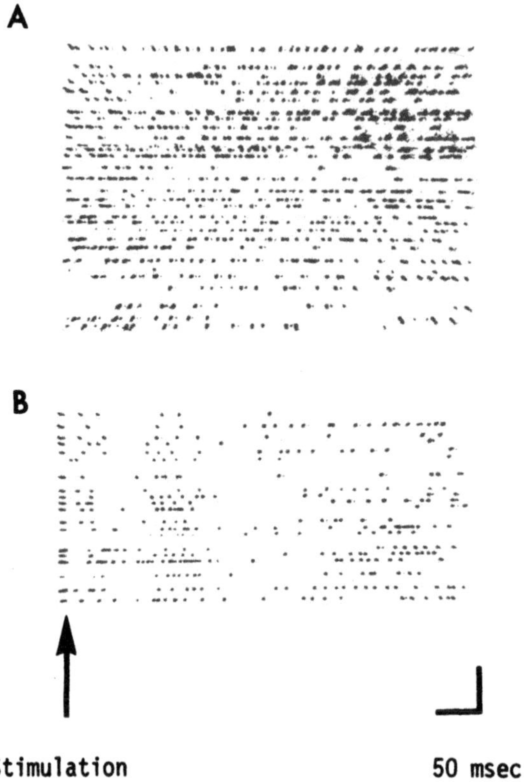

Figure 6. Spontaneous unit firing of septal neuron shows on upper picture (A) and effect of electrical stimulation (0.5 msec 300μA 1 msec interval 3 pulses) of caudal portion of central gray mater (behind the trochlear nucleus) on the unit activity in the septum shows on lower picture (B). The stimulation produces combined responses characterized by facilitation and inhibition of neuron activity following the direct inhibition.

and reach the septum, and have an inhibitory effect on spontaneous activity of the septal neurons. Monoaminegic neurons may be more closely related to above region in which stimulation causes a predominantly inhibitory effect on the septum. The existence of monoaminergic neurons of the midbrain limbic septal interconnection does not necessarily indicate that no other mechanism is available for activating this neural interconnection. Among the nervous

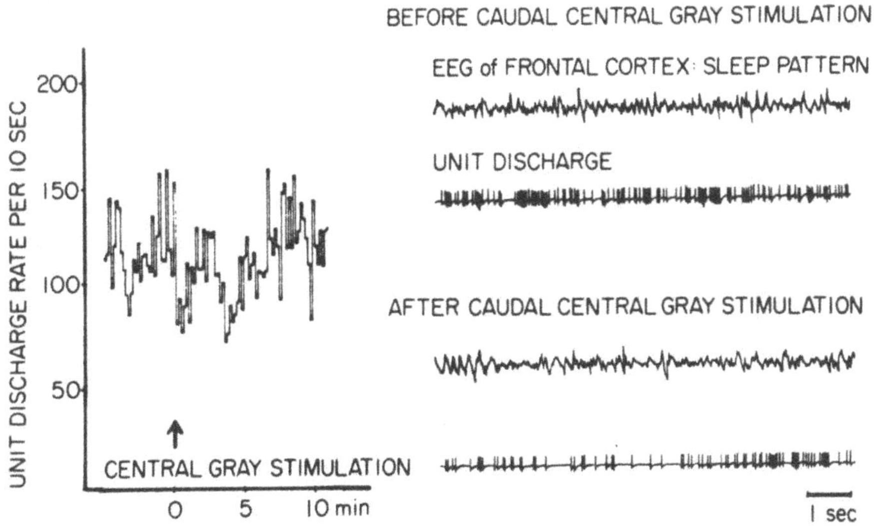

Figure 7. Repetitive stimulation (80µA 0.1 msec 100 Hz 3 sec) of caudal portion of central gray produces primary and secondary reduction of spontaneous activity of the septal neuron. The stimulation of caudal central gray mater reduces the appearance of bursts of discharge firing.

pathways which may exert their effect on the septal function, the widespread system of ascending cholinergic reticular system may be included (39). However, the precise distribution of the cholinergic reticular system and monoamine containing fibers is not identical (63). In the normal brain, there are many acethylcholine containing tracts which cannot be traced back to their nuclei of origin, so that it is often difficult to decide in which direction the constituent fibers are traveling. It was demonstrated the polarity of the acetylcholine containing fibers: when acetylcholine estrase (AChE) containing tracts were involved in surgical lesions, the interrupted ends reacted differently on opposide sides of the cut. On one side the terminal portion of certain axons became distended and varicose, and stained much more heavily for AChE than did normal fibers. On the opposite side of the cut, it was found, by contrast, that interrupted axons of the central nervous system lost their normal staining for AChE (60,61,62). AChE-containing fibers are plentiful in the fornix between the hippocampus and the medial septal nucleus and nucleus of diagonal band which give rise of the hippocampal afferents, but not on the descending columns of the fornix beyond the septum. Therefore, an interruption of these fibers leads to accumulation of AChE on the septal side as well as a loss of AChE on the hippocampal side. With the above information in mind, an attempt was made to rationalize the correlation of polarity of AChE and acetylcholine content. This study was performed in association with Dr. Stavinoha, Department of Pharmacology, using technique of pyrolysicgas chromatography after periodide precipitation (37). Lesion of the medial and lateral nuclei of the septum causes a marked reduction of acetylcholine content in the dorsal hippocampus, less reduction is found in the ventral hippocampus (Table 6). Furthermore, when the septal lesion is extended into the nucleus of diagonal band the acetylcholine content of the hippocampus is extensively reduced (64). It was, therefore, found that polarity of the AChE corresponds with polarity of aceylcholine content. This type of reduction of aceylcholine is not found after abolition of the olfactory bulb, and further there is no projection fiber from the olfactory bulb to the septum (51,59). So cholinergic fibers in the hippocampus must be considered either to rise in the septum or to pass through the septum. The ventral pathway of the ascending cholinergic reticular system distributes its fibers into the hippocampus through the lateral preoptic area (39); in addition, lesion of the septum increases acetylcholine content in the hypothalamus (Table 6). However, this septal lesion does not significantly change the content of acetylcholine in the thalamus and midbrain (Table 6).

The primitive conscious propioception which generate the mechanism of emotion (2) travels along with the spinoperiaquoductal tract (K. Otani, personnel communication) and reach the midbrain limbic area. Ascending impulses from the midbrain limbic area travel along with either the medial forebrain bundle or the dorsal longitudinal fasciculus and influence on the septal function. Given this situation, the monoaminergic and the cholinergic pathways of the ascending midbrain limbic system are involved.

Descending Afferent Paths

In the early stage of embryonic development the septum forms an island together with the hippocampus and amygdala. In the later stage, both hippocampus and amngdala migrate into the telencephalon.

Hippocampal-Septal Interconnections. In recent years, attention has been drawn to the possibility of the closed circuit of hippocampal-septal interconnection (52,53). CA_1 of the anterior hippocampal field project their fibers to the medial septal nucleus. The fibers from the medial septal nucleus project to the field of CA_2, CA_3 and CA_4, and then, the fibers from these fields project to the field of CA_1 through the Schaffer's collateral system. In addition, from the field of CA_3 and CA_4 project to the

TABLE 5: EFFECT OF ANTERIOR THALAMIC (AV) STIMULATION ON THE NEURAL ACTIVITY OF THE SEPTUM

(80μ A, 0.1 msec, 3 sec)

cps (Hz)	Neural activity of septum (No. of units)				
	Increased	Decreased	Subtotal	Unchanged	Total
30	2	9	11	5	16
60	2	7	9	1	10
100	2	8	10	3	13
Total:	6	24	30	9	39

TABLE 6: EFFECT OF SEPTAL LESION ON ACETYCHOLINE CONTENTS IN THE BRAIN

Region	Side	Acetylcholine contents (n mol/mg) tissue	
		Sham operation	Septal lesion
Hippocampus	dorsal-left	23.86±4.93	15.68±2.30
	dorsal-right	27.37±6.89	18.84±5.18
	ventral-left	27.49±4.94	22.04±4.02
	ventral-right	27.88±3.99	21.83±4.73
Amygdaloid		39.98±0.55	52.72±5.27
Hypothalamus	medial	39.81±4.15	53.09±11.49
	lateral	38.49±5.32	38.72±6.88
Thalamus		37.20±5.18	35.19±5.39
Midbrain		32.82±9.23	37.25±7.44
Striatum		66.40±8.06	55.83±10.23
Cerebral Cortex		16.95±4.28	22.26±2.03

Sham operation - 6 rats
Septal lesion - 6 rats
The animals were sacrificed one week after lesion for determination of acetylcholine.

lateral septal nucleus. The fibers from the lateral nucleus project to the medial septal nucleus, from where fibers are sent to the field of CA_3 and CA_4 (8).

Stimulation (0.5 msec 300μA/msec interval 3 impulses) of dorsal hippocampus (anterior part of field CA. and CA_2) produces either an excitation of spontaneous neuron activity in the septum following the evoked response or combined responses characterized by inhibition and facilitation of neuron activity following the evoked response (Fig. 8). In addition, hippocampal stimulation causes purely prolonged inhibition. Stimulation of the fornix exerts a purely excitatory effect on the medial septal nucleus, and stimulation of this input does not produce any effect on the lateral septal nucleus (43). Furthermore, fimbrial stimulation causes an excitation in the lateral septal nucleus which is followed by a period of inhibition mediated by the recurrent axons collateral impinging on septal interneurons (8,43). However, a marked latency difference exists between the onset of the EPSPs and IPSPs in the septal neurons upon hippocampal stimulation (8). In addition,

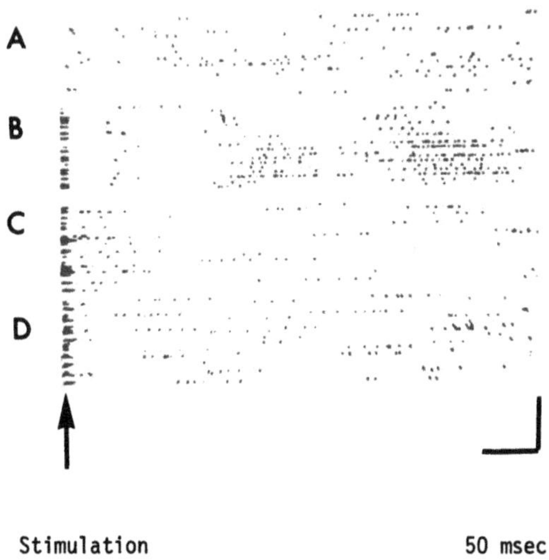

Figure 8. Spontaneous unit firing of septal neuron shows the upper column (A). The stimulation of caudal portion of central gray matter (B) produces combined responses characterized by inhibition and facilitation of neuron activity following the direct inhibition. This evidence indicates that the stimulation causes an inhibition in the septal neuron which is followed by a period of excitation mediated by the recurrent axons collateral impinging on septal interneurons (inhibition). The stimulation of the anterior amygdala (D) has an antagonistic effect between the septal neuron response and the dorsal hippocampal stimulation (C).

IPSPs are recorded during all hippocampal stimulation. Gating mechanism by the recurrent axons collateral (inhibitory interneuron) may make it possible to produce combined responses characteristic by inhibition and facilitation of septal neuron activity upon hippocampal stimulation.

Repetitive stimulation (0.1 msec 80μA 100 Hz 3 sec) of dorsal hippocampus increase the firing rate in 27 of 70 septal neurons and decrease it in 23 neurons (Table 7). The majority of neurons exhibit an excitation upon repetitive dorsal hippocampal stimulation in the period of primary neural response, however, these neurons exhibit the opposite response in the period of secondary response even during EEG sleep (Fig. 9). However, dorsal hippocampal stimulation has less effect upon the spontaneous activity of septal neurons when the stimulus is applied during cortical EEG arousal stage (Table 4).

Dorsal hippocampus (filed of CA_1 and CA_2) has an excitatory or inhibitory influence on the septal function. An existence of gating mechanism (8,43) by the recurrant axon collateral (inhibitory interneuron) make it possible to produce combined response characteristic by inhibition and facilitation of septal neuron activity upon hippocampal stimulation. An appearance of secondary response of septal neuron activity can be driven by a reciprocating circuit of the hippocampal and septal interneuronal connection. Given this situation, the reticulo-thalamo-cortical activating system has an influence on the function of hippocampal-septal system.

<u>Amygdaloid-septal interconnections</u>. Occurrence of the secondary response in the neuron activity of the septum immediately following the primary response upon hippocampal stimulation raises the following question; whether a reciprocating circuit of hippocampal-septal interconnection (52,53) is solely responsible for activation of the secondary response or if other reciprocating circuit is involved, since the septum has a reciprocal connection with the anterior amygdala via the diagonal band of Broca (8,14,42). To pursue this problem, electrical stimulation (0.5 msec 30μA 1 msec interval 3 pulses) was applied for the anterior amygala and spontaneous activity of the septal neurons was recorded.

There are three types of neuron connections between the anterior amygdala neurons and septal neurons: 1) direct inhibitory interneuron connection prior to reaching the septal neuron, 2) pure direct excitatory connection, and 3) recurrent axon collateral connection (Fig. 8).

Repetitive stimulation (8μA, 0.1 msec, 100 Hz 3 sec - 300 impulses) of the anterior amygdala resulted in an increase in firing rate in 15 of 53 neurons and a decrease in 23 neurons (Fig. 10,

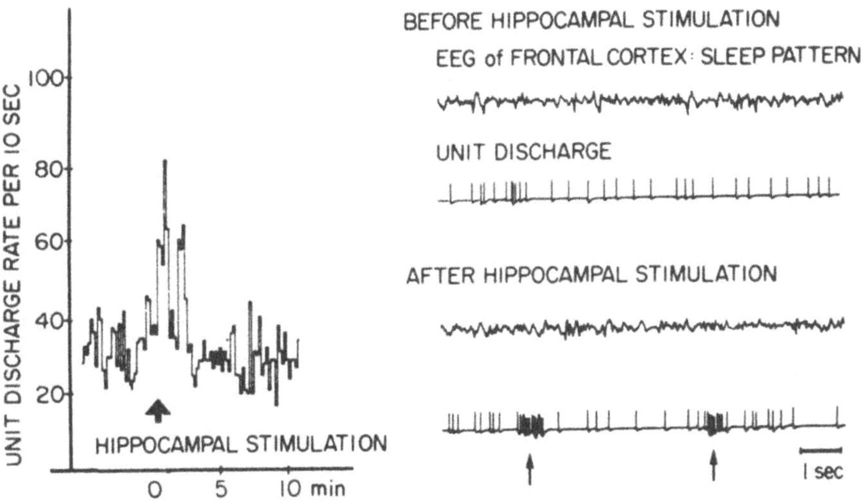

Figure 9. Repetitive stimulation (80 A 0.1 msec 100 Hz 3 sec) of dorsal hippocampal stimulation (CA_1 and CA_2) produces primary and secondary response on the spontaneous activity of the septal neuron. The stimulation of dorsal hippocampus produces the bursts of discharge firing (arrows).

Table 7). The anterior amygdala stimulation predominantly produced an inhibitory effect on spontaneous septal neuron activity, however, Figure 10 shows that the stimulation of the anterior amygdala resulted in decreased neuron activity in the septum within 10 seconds after stimulation. Further, during the period of 8 to 25 seconds after stimulation an increased secondary response appears. By all means, this anterior amygdala stimulus has an antagonistic effect between the septal neuron response and the dorsal hippocampal stimulation. When the hippocampal stimulus causes an excitation of neuron activity in the septum, stimulation of the anterior amygdala stimulus induces an inhibitory effect on the same neuron.

Application of 3 pulses stimulation to the anterior amygdala also shows the antagonistic effect between the septum response and hippocampal stimulation (Fig. 8). An antagonistic effect on the septal neuron activity by the dorsal hippocampus and anterior amygdala suggests the existance of intermingled reciprocating neural circuit between the hippocampus and amygdala through the septum.

<u>Olfactory septal connections</u>. The fiber connection between the olfactory bulb and the septum has not been found (2,51). The

TABLE 7: EFFECT OF LIMBIC STIMULATION ON THE NEURAL ACTIVITY OF THE SEPTUM

(80μA, 100 Hz, 0.1 msec, 3 sec)

Area of stimulation	Neural activity of septum (no. of units)				
	Increased	Decreased	Subtotal	Unchanged	Total
D-Hippocampus CA_1-CA_2	27	23	50	20	70
Ant. Amygdala	15	23	38	15	53
Midbrain limbic area (CG)	4	10	14	3	17

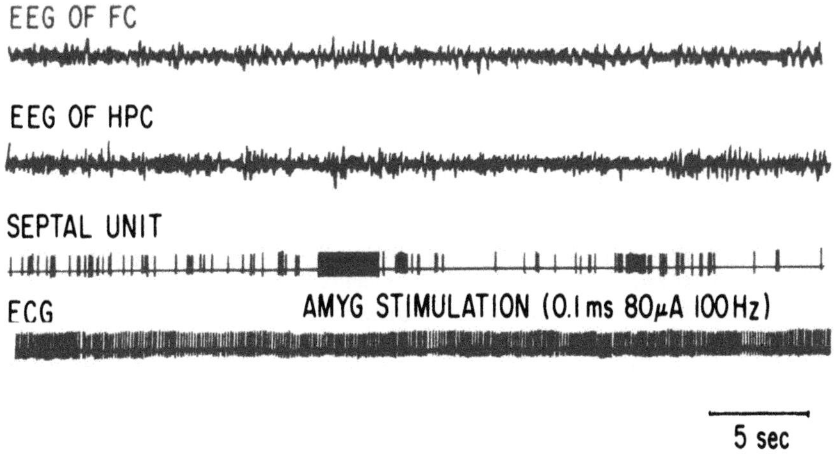

Figure 10. Repetitive stimulation (0.1 msec 80μA 100 Hz 3 sec) of the anterior amygdala produces primary reduction of spontaneous unit firing of the septum during first 10 seconds. However, spontaneous unit firing increases following 10 seconds. EEG recording is obtained from frontal cortex (FC) and dorsal hippocampus (HPC).

olfactory bulb projects its fibers to the anterior lateral part of the medial septum (51). Abolition of the olfactory bulb does not produce any significant change of acetylcholine content in the hippocampus (Table 8). Furthermore, stimulation of olfactory bulb induces only excitatory postsynaptic potentials (EPSPs) without cellular discharge on the hippocampal neurons; however, stimulation of the septum is highly effective in eliciting EPSPs and discharge of the cell (15).

The olfactory bulb may not have direct influence on the septal function; however, stimulation of olfactory bulb by odors may reach the septum through the olfactory tubercle or amygdala. The details need to be investigated further.

Descending Efferents of the Septum Concerning Regulation of LH Release

Septo-preoptico-tuberal fibers. In the standardized procedure performed on the day of proestrus, the spontaneous activation of the anterior pituitary which leads to release of LH from the anterior pituitary is blocked by pentobarbital anesthesia during the critical period of gonadotrophin secretion that begins about 2 p.m. (12) This procedure was employed by Critchlow (4) in demonstrating that electrical stimulation of the tuberal hypothalamus just above the median eminence would frequently induce ovulation in the cyclic rat. Similar results can easily be obtained with electrodes placed in the medial preoptic area; however, ovulation has not been elicited by placing a large irritative focus in the lateral preoptic area. In addition, in a more caudal region above the caudal margin of the optic chiasma the frequency of positive results was greatly reduced and in the tuberal region itself, negative results were the rule (13). Blockade of ovulation by administration of pentobarbital would be due to an elevated threshold in the reticular formation of the mid-brain (57); however, local anesthesia of the septum, medial preoptic area, and basomedial hypothalamus, also revealed a blockade of ovulation (23).

In the experiments of Hillarp (29), the state of constant estrus could be induced in rats by either the bilateral destruction of large areas in the anterior hypothalamic area or very restricted basal area in the anterior tubercinereum. These results and above facts indicate that fibers controlling ovulation originate as a diffuse system in the medial preoptic area and in a restricted manner in the basomedial hypothalamus which lies above the anterior pituitary. Recently, the fiber distribution of the forebrain pathway has been studied extensively by Tajasen and Everett (69) in rats using combination technique of deafferentation and stimulation. Fibers of the afferent system involved in ovulation are more widely dispersed in the forebrain, including preoptic area, than in the area of the basomedial hypothalamus. Therefore, the frontal deafferentation of the preopticotuberal pathway blocked spontaneous ovulation and produced polyfollicular ovaries with constant estrus (28), in addition, when the frontal cut was made anterior to the preoptic area in order to interrupt the anterior, lateral and superior connections of the preoptic area, ovulation with irregular vaginal cycles was observed (35). An interesting development was the finding that rostral and dorsal to the preoptic region, positive results for induction of ovulation could be obtained by means of the large foci placed in the medial or lateral septal nuclei, in the nucleus accumbens or in the medial parolfactory area (13); however, small foci similarly placed were not effective. Ovulation was not elecited by placing large or small irritative foci in the caudate-putamen, the anterior olfactory area, the olfactory tubercle, the far lateral preoptic area, or dorsal thalamus adjacent to the foramen of Monro. The same is true for the lateral hypothalamus posterior to the optic chiasma. It may be supposed that, if there are any connections from or through these areas into the septal complex and medial preoptic area which normally play a role in LH release, they are blocked out by the barbiturate.

The bursts of high amplitude, fast activity in EEG records of

TABLE 8: EFFECT OF OLFACTORY BULBECTOMY ON ACETYLCHOLINE CONTENTS IN THE BRAIN

Region	Acetylcholine contents (n mol/mg tissue)	
	Sham operation	Olfactory bulbectomy
Hippocampus	22.11±4.29	24.70±1.20
Amygdala	33.48±5.10	35.63±2.85
Hypothalamus	31.29±4.35	29.99±5.78
Thalamus	32.57±2.42	35.07±1.87
Striatum	60.38±9.57	74.25±6.83

Sham operation = 5 rats
Olfactory bulbectomy = 4 rats
The animals were sacrificed one week after lesion for determination of acetylcholine.

TABLE 9: EFFECT OF mPOA STIMULUS (30μA-60 sec. center-cathode) ON MULTINUNIT ACTIVITY IN ARCUATE NUCLEUS (ARC) UNDER URETHANE ANESTHESIA

Area of Stimulation	Area of Recording	Neural Activity	
		Increased	Decreased
L-mPOA	L-ARC	8	4
R-MPOA	L-ARC	3	1
L-mPOA	L-R-ARC	3	1

R-mPOA: Right side of Medial Preoptic Area

L-mPOA: Left side of Medial Preoptic Area

L-R-ARC: Left and Right sides of Arcuate Nucleus

unanesthetized rats were observed in the lateral preoptic area and anterior hypothalamic area lasting for about 20 minutes on the afternoon of proestrus (6), in addition, similar activity was found in the multiunit recording of the medial preoptic area and anterior hypothalamus (34). The findings, however, have yet to be verified.

Electrochemical stimulation of the septum facilitates the multiunit activity (MUA) of the medial preoptic area and arcuate nucleus, in addition, the threshold for facilitation of neural activity of the arcuate nucleus in the septum is higher than that in the medial preoptic area (34). This result is comparable to the results of Everett (13) in which a larger stimulation (irritation) is required in the septum for the induction of ovulation as compared with the medial preoptic area. Electrochemical stimulation (60μA - 60 sec DC) of the medial septal area caused an elevation in plasma level of LH, but it began within 60 minutes, and the slope of elevation increased gradually (Fig. 11). A delayed elevation of plasma LH in septal stimulated rats has been reported.

On the line of forebrain afferent system in the hypothalamus, stimulation of the medial preoptic area facilitated the multiunit activity of the arcuate nucleus in nembutalized rats (70) and urethane anesthetized rats (Fig. 12), and furthermore, electrochemical stimulation of the medial preoptic area in nembutalized proestrus rats facilitated the release of LH within 30 minutes (Fig. 13, and ref. 11,73).

There are two hypotheses concerning the hypothalamic regulation of the anterior pituitary LH release; 1) dural fiber system (11) and 2) a point-to-point fiber system for LH release in the hypothalamus (43). In our experiment, unilateral stimulation of the medial preoptic area facilitated the multiunit activity of the arcuate nucleus bilaterally (Table 9). This result support more likely dual fiber system for LH release in the hypothalamus rather than a point-to-point fiber system.

<u>Septotuberal tract</u>. Projections from the lateral septum reach the dee- islands of cells of the olfactory tubercle, and join the medial forebrain bundle, where they are continuous with degeneration entering this bundle from the diagonal band of Brocae at more caudal levels. This type of degeneration is never found after hippocampal lesion, so it must be considered either to arise in the septum or to pass through the septum from a non-hippocampal source. The ventral part of the septum projects its fibers to the medial forebrain bundle and they pass through the lateral preoptic area and reach the level of midbrain in the rat (52). In the rabbit, the medial forebrain bundle projects its fibers into the caudal part of the central gray matter, and the fibers from this area project back to the septum through the medial forebrain bundle and diagonal band of Broca (22).

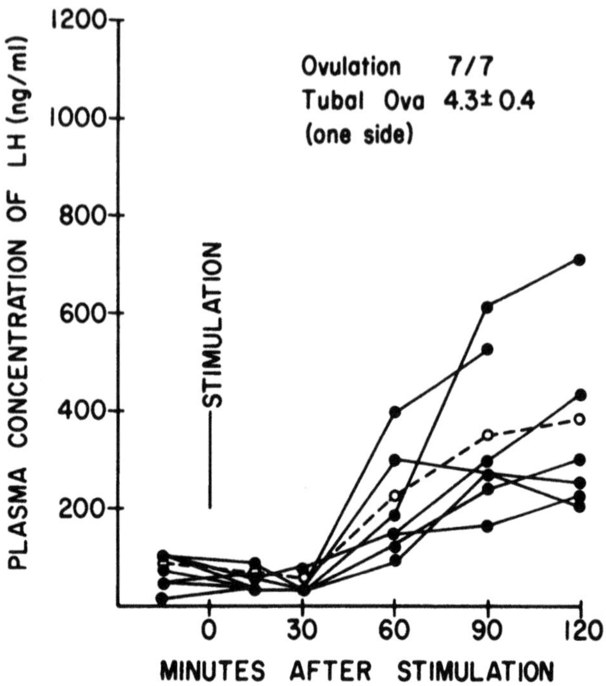

Figure 11. Effect of electrochemical stimulation (60μA - 60 sec) of medial septum on plasma level of LH in nembutalized proestrus rats. Vertical line represents plasma concentration of LH (ng per ml plasma) and horizontal line represents minutes after stimulation. Ovulation occurred in all 7 rats and an average number of tubal ova was 4.3± 0.4 per rat.

Unilateral lesion of dorsolateral, and/or ventrolateral parts of the septum produced degenerating axon terminals in the anterior and medial area of the arcuate nucleus bilaterally (24). Unilateral lesion of the medial forebrain bundle at the level of lateral preoptic area produced digenerating axon terminals in the entire area of the arcuate nucleus bilaterally. It should be noted that the process of astrocytes enfolded the degenerating axon terminals in the arcuate nucleus of the hypothalamus two days after lesion of the septotuberal tract, and axons from adjacent regions tended to invade the vacated terminal space and compete with other axons for exclusive occupancy of this space (Fig. 14)

Fiber connection of the septum to the arcuate nucleus through the medial forebrain bundle make a cross at the retrochiasmatic area and reach bilateral sides of the arcuate nucleus. Considering its termination in the arcuate nucleus of the hypothalamus, it would be better to name it septo-tuberal tract.

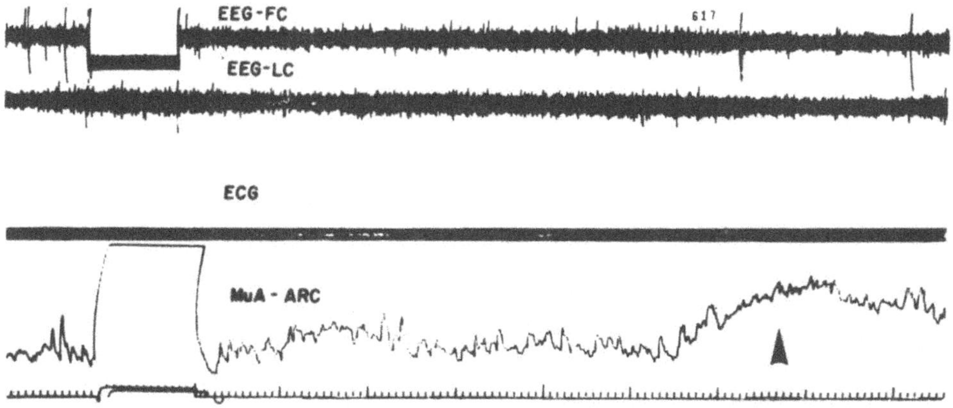

Figure 12. Effect of medial preoptic stimulation (30μA - 60 sec) on multiunit activity in the arrcuate nucleus under urethane anesthesia in the rat. EEG-FC: electroencephalograph from frontal cortex. EEG-LC: EEG from limbic cortex. ECG: electrocardiogram. MUA-ARC: multiunit activity in arcuate nucleus. The stimulation of the medial preoptic area facilitates the neural activity in the arcuate nucleus (arrow).

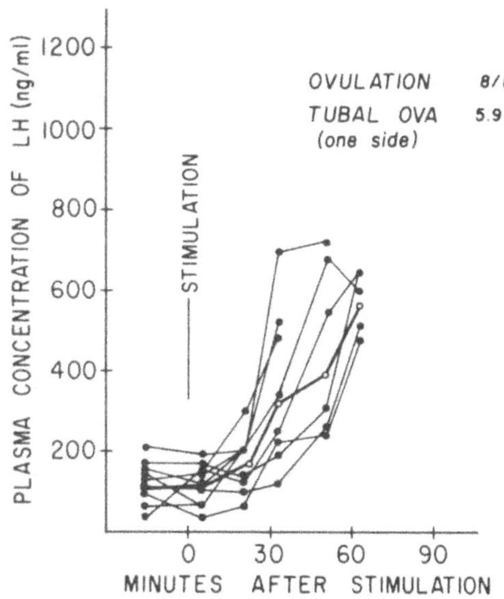

Figure 13. Effect of electrochemical stimulation (60μA-60 sec) of medial preoptic area on plasma level of LH in nembutalized proestrus rats. Vertical line represents plasma concentration of LH 9ng per ml plasma) and horizontal line represents minutes after stimulation. Ovulation occurred in 8 of 10 rats and an average number of tubal ova was 5.9 ± 0.75.

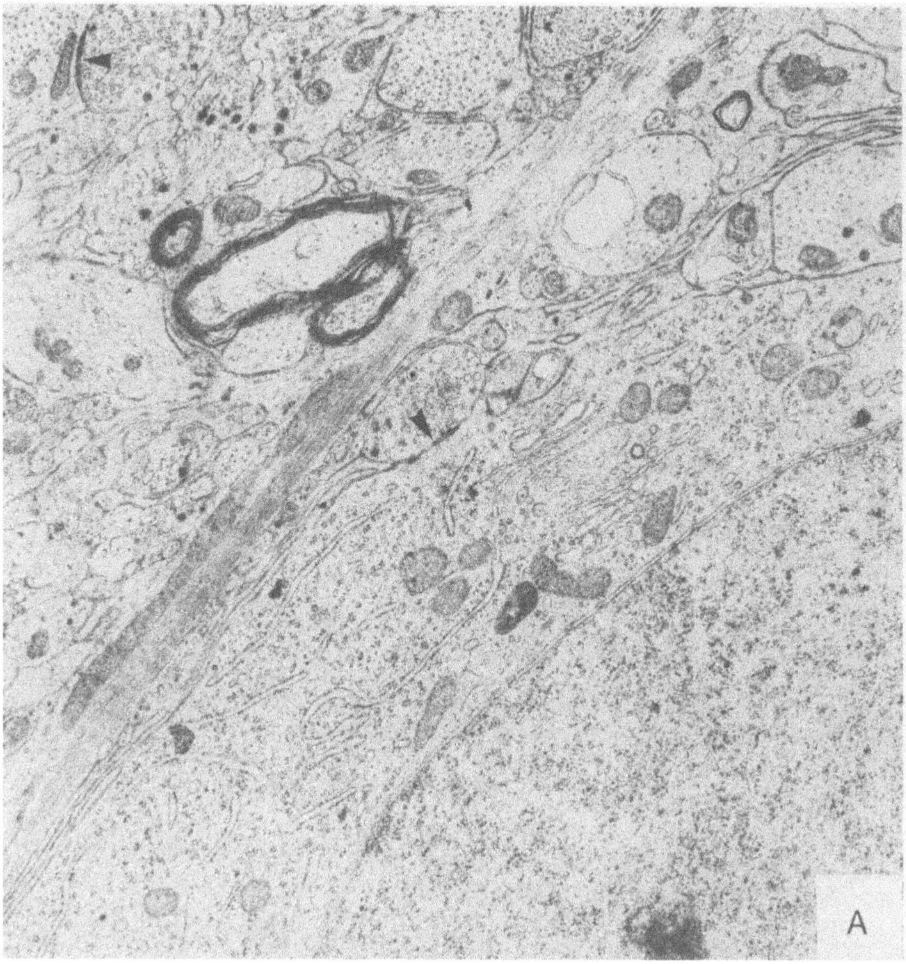

Figure 14. The degenerating axon terminals were examined in 2 or 4 days after septal lesion in electron microscope preparation. The semi-thick sections were stained with toludine blue for histological examination of the arcuate nucleus. The ultra-thick sections were contrasted with uranyl acetate and lead citrate for examination of ultrastructure of the arcuate nucleus.

A. Two major distinguished synapses (arrows) can be seen in the arcuate nucleus of the intact control rat. One is the axon-somatic synapse and the other is the axon-dendritic synapse.

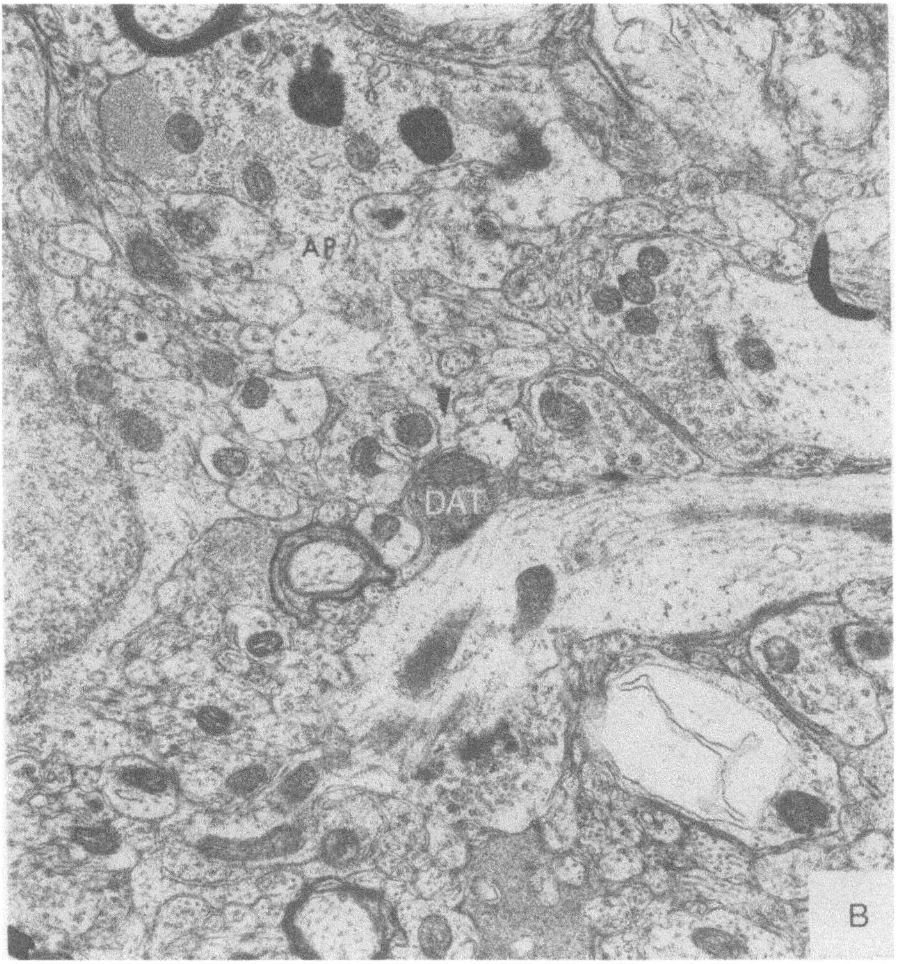

Figure 14 B. The earlier degeneration of axon-dendritic synapse are seen in 2 days survival rats after septal lesion. The astroglial process (AP) attached to the degenerating axon terminal (DAT).

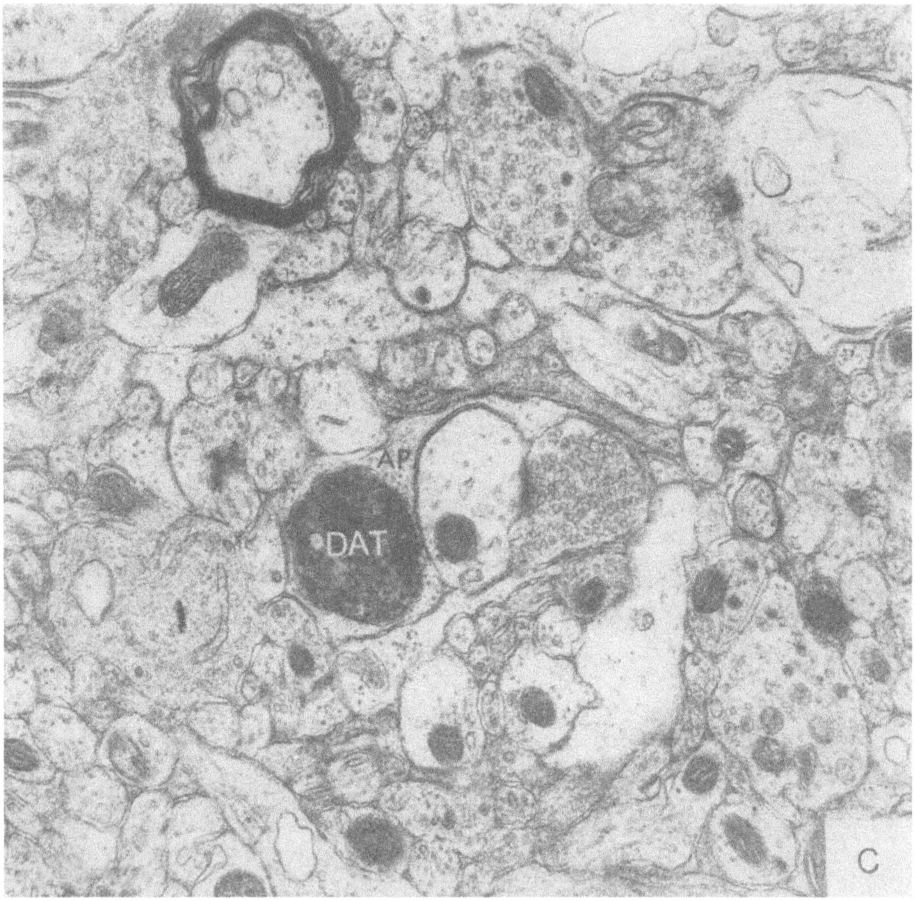

Figure 14 C. The complete degenerating axon terminal are also seen in 2 days survival rats. In addition, the glial process (AP) conceivably become interposed between the terminal bouton (DAT) and the post-synaptic element during degeneration.

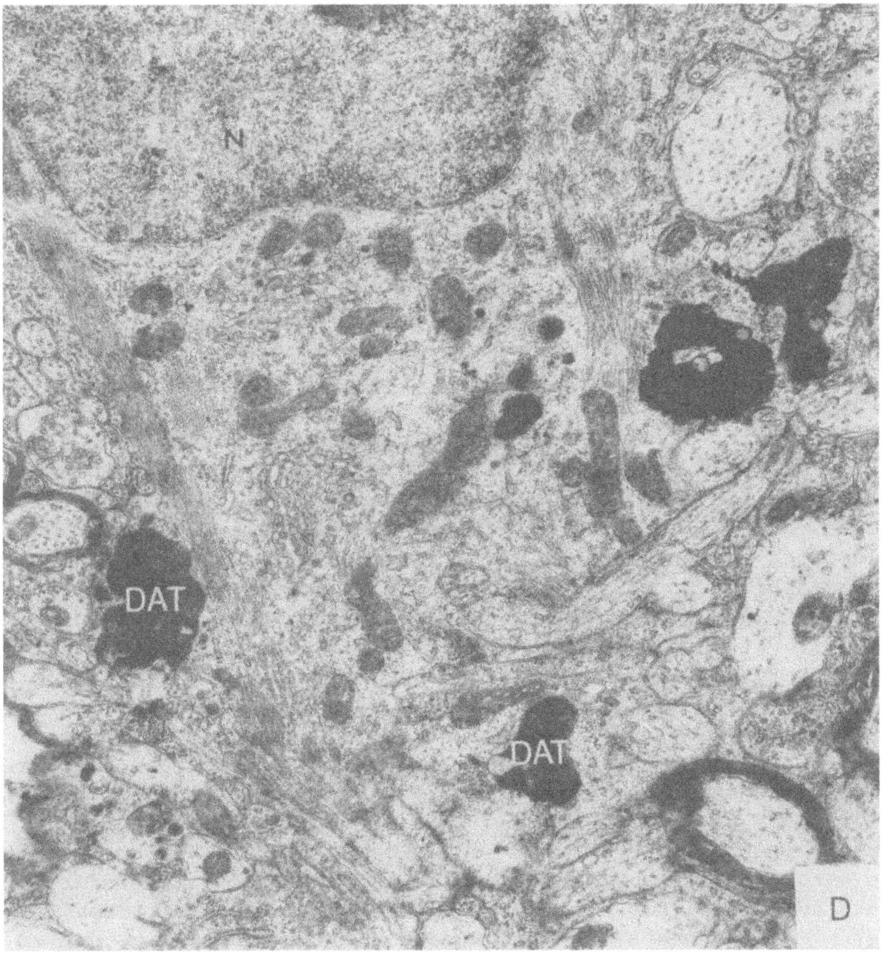

Figure 14 D. In 4 days survival rats, several electron dense masses (degenerating axon terminals, DAT) are seen completely enclosed in astroglial process and perinuclear cytoplasm.

Dual fiber systems of both preopticotuberal and septo-tuberal tracts are intermingled at the basomedial hypothalamus. Stimulation of the medial forebrain bundle resulted predominantly in an inhibitory effect on the neural activity of the arcuate nucleus (Fig. 15); however, electrochemical stimulation of the medial forebrain bundle at the lateral preoptic level had no effect on the release of LH in nembutalized proestrus rats (Fig. 16). The technique of radioimmunoassay for plasma LH used in this study does not detect a reduction in plasma LH from basal level, therefore, an inhibitory effect of the septo-tuberal tract on LH release can not be substantiated. To pursue this problem, electrochemical stimulation (60 A for 60 sec) was performed in the lateral preoptic area 2 minutes prior to stimulation of the medial preoptic area in nembutalized proesrus rats and determined plasma level of LM. Stimulation of the lateral preoptic area caused a delay of LH release induced by the medial preoptic stimulation (LH was released within 60 minutes after medial preoptic stimulation) and plasma level of LH gradually elevated (Fig. 17). The results are comparable to that in the septal stimulation. Impulses which were generated by electrical stimulation of medial preoptic area were retained in the medial preoptic area and passed through the retrochiasmatic area within 5 minutes after stimulation (73). Therefore, stimulation of the lateral preoptic area may interfere with neural activity of the medial preoptic area directly rather than direct inhibition of the arcuate nucleus. To test this possibility, similar stimulation was applied for the lateral preoptic area at 2, 15 or 30 minutes after medial preoptic stimulation. The results obtained in the experiments are comparable to that in the previous experiment (Fig 18). It should be noted that even lateral preoptic stimulation performed 30 minutes after the medial preoptic stimulation produced more prominent suppression than that at 2 or 15 minutes after the medial preoptic stimulation. The evidence from these experiments indicate that stimulation of the lateral preoptic area (medial forebrain bundle) exerts its inhibitory effect on the neural activity of the arcuate nucleus to suppress the release of LH. Therefore, stimulation of the septum makes it possible to facilitate both excitatory efferents, such as septo-preoptico-tuberal fibers and inhibitory efferents, such as septo-tuberal tract. A delay of LH release by stimulation of the septum could be due to superinduction of the inhibitory influence on the excitatory efferents of LH release.

Septal Modulating System for Exteroceptive and Interoceptive Pathways

Does the septum act as a modulator for the ascending and descending fiber systems, regarding regulation of LH release? The cholinergic system has long been known to play a significant role

A

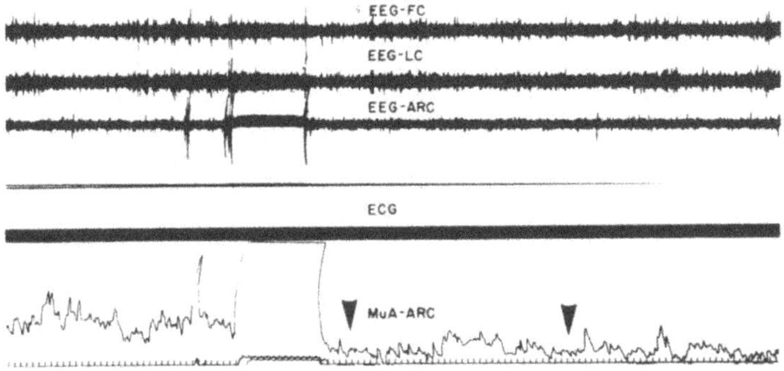

B

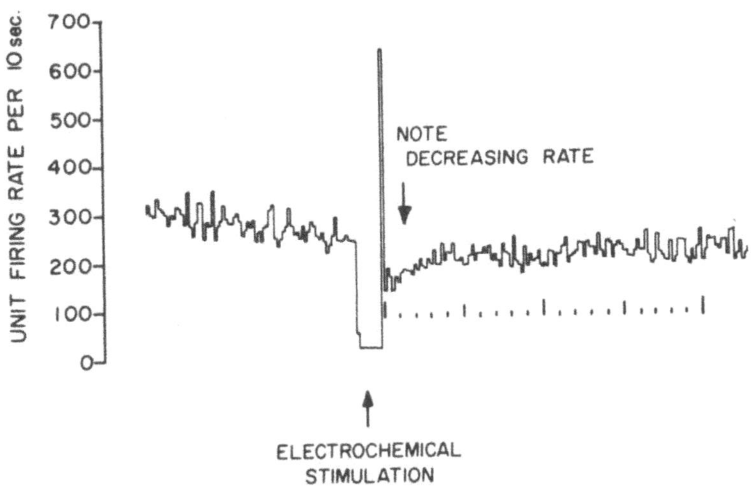

Figure 15 A. Effect of lateral preoptic stimulation (30μA-60 sec) on multiunit activity in arcuate nucleus under urethane anesthesia in the rat. EEG-FC: electroencephalograph from frontal cortex. EEG-LC: EEG from limbic cortex. EEG-ARC: EEG from arcuate nucleus. ECG: electrocardiogram. MUA-ARC: multiunit activity in arcuate nucleus.

Figure 15 B. Effect of lateral preoptic stimulation (30μA-60 sec) on neural activity in arcuate nucleus under urethane anesthesis in the rat. Vertical axis represents unit firing rate per 10 second and horizontal division represents minutes after stimulation. The stimulation of the medial forebrain bundle suppresses the neural activity in the arcuate nucleus (arrow).

in both synaptic function and membrane phenomena in the central nervous system. The application of acetylcholine to the brain or to the pituitary causes modification in the regulation of gonadotrophin secretion (10. Cyclic fluctuation of cholinestrase activity in the hypothalamus has been correlated with the strous cycle in rats (38). Anticholinergic agents are found to be effective in preventing release of gonadotrophin (12), as well as ovulation, induced by estrogen treatment in the rat (58).

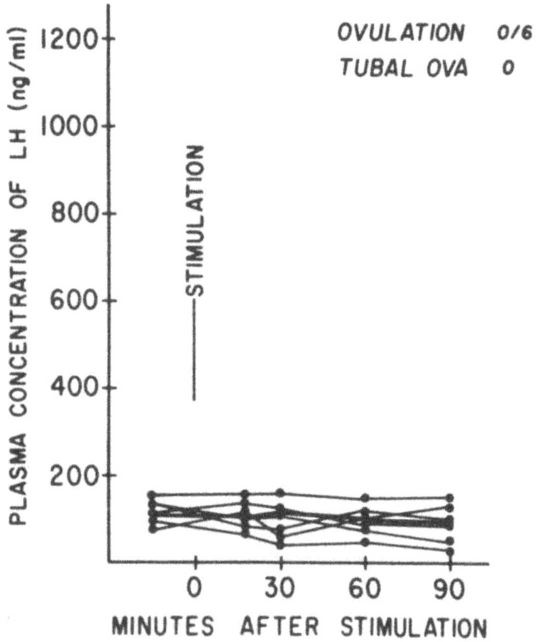

Figure 16. Effect of electrochemical stimulation (60μA-60 sec) of lateral preoptic area on plasma level of LH in nembutalized proestrus rats. Vertical axis represents plasma concentration of LH (ng per ml plasma) and horizontal line represents minutes after stimulation. None of 6 rats was induced ovulation by stimulation.

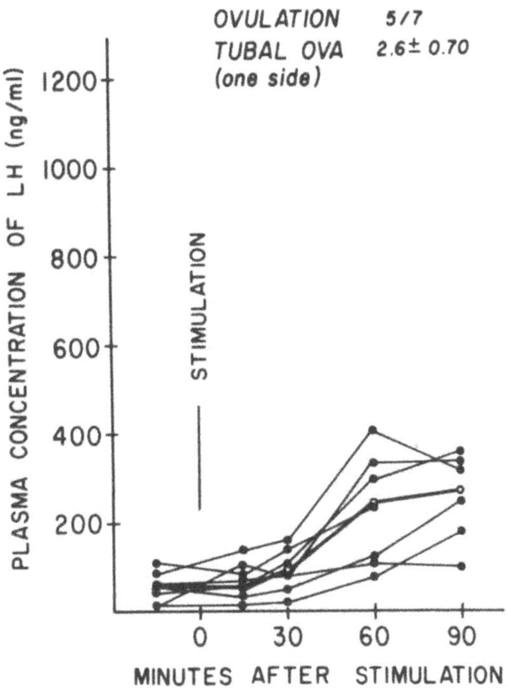

Figure 17. Effect of electrochemical stimulation (60μA-60 sec) of lateral preoptic area on plasma levels of LH in medial preoptic stimulated (60μA-60 sec) nembutalized proestrus rats. Vertical axis represents plasma concentration of LH (ng per ml plasma) and horizontal axis represents minutes after stimulation. Ovulation occurred in 5 of 7 rats and an average number of tubal ova was 2.6 ± 0.7 per rat.

It is known that estrogen administered to infantile female rats plays an important role in the onset of puberty and that the hypothalamus is very sensitive to this estrogen (7,25,55). Administration of small quantitites of estrogen (0.05μg/rat daily for 4 days) causes a marked reduction of acetylcholine content in both the hypothalamus and amygdala, including priform cortex and brings precocious puberty. Treatment with this regimen and even larger quantitites of estrogen (2μg/rat) does not produce any significant change in acetylcholine content in both the hippocampus and corpus striatum (Table 10).

Either the hypothalamus or amygdala contains cholinergic neurons, while neither the hippocampus nor the corpus striatum contains

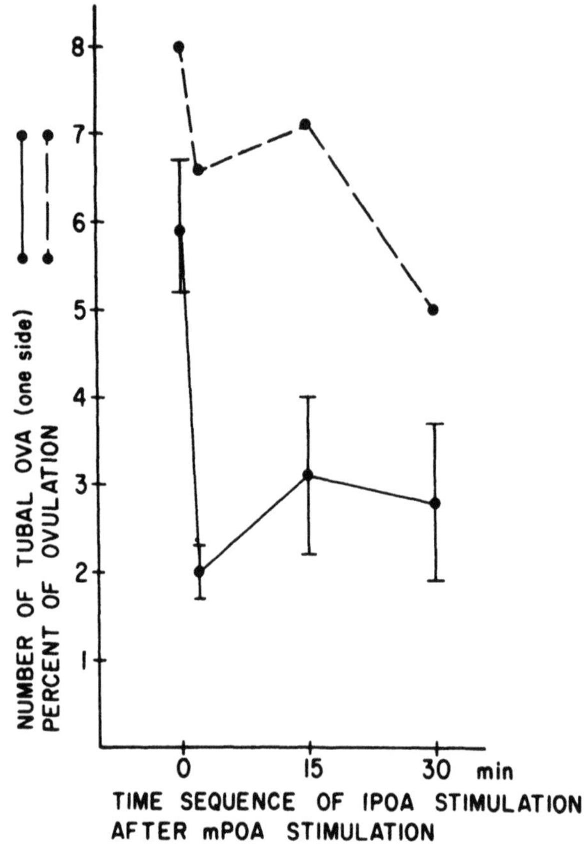

Figure 18. Effect of electrochemical stimulation (60μA-60 sec) of lateral preoptic area on ovulation in medial preoptic area stimulated (60μA-60 sec) nembutalized proestrus rats. A solid line (o———o) represents number of tubal ova and a broken line (o-----o) represents percent of ovulation. Vertical axis represents number of tubal ova and percent of ovulation, and horizontal line represents time sequence of lateral preoptic stimulation after medial preoptic stimulation.

cholinergic neurons. There are only cholinoceptive noncholinergic neurons in the regions of the hippocampus and corpus striatum; suggesting that cholinoceptive noncholinergic neurons in the hippocampus may not respond to estrogen treatment. A more likely possibility may be considered; that activation of the cholinergic system in the hypothalamus and amygdala by estrogen treatment may be associated with the presence of a specific protein which binds estrogen (9,32,33,50,66,77), since nuclear uptake of estradiol in the hippocampus is 100 fold lower than in the hypothalamus. On the other hand, the hippocampus contains a specific protein which binds corticosterone (41). To test this possibility, injections of corticosterone were given subcutaneously daily for 4 days, in the same manner as estrogen treatment, at 24 days of age and acetylcholine content in the hippocampus was determined on days 30. Administration of corticosterone (2μ or 20μg/rat daily for 4 days) in infantile female rats caused markedly reduction of the acetylcholine in the hippocampus, with less reduction in the amygdala (Table 11).

This evidence indicates that steroid hormones (estrogen or corticosterone) activate both cholinergic neurons and cholinoceptive noncholinergic neurons. Given this situation, the presence of a specific protein which binds either estrogen or corticosterone has an important role in the activation of cholinergic system in both hypothalamus and limbic areas. EEG arousal is associated with increased cortical activity, and it has been shown that acetylcholine can have a facilitatory effect on cortical neurons (45) and high frequency stimulation of the midbrain reticular formation in cats causes a five to six fold increase in the output of acetylcholine from the cerebral cortex (31). It was found that a decrease in the seizure threshold of the amygdala (71) and a marked increase incorporation of S^{35} methionine into protein in the amygdala (72) were observed prior to vaginal opening in rats. These phenomena could be due to the activation of cholinergic fibers in the amygdala induced by circulating level of estrogen prior to onset of puberty.

Influence of the amygdala on the hypothalamus regarding LH release and ovulation travels with the stria terminalis instead of the ventral amygdalofugal fibers (75). The postcommissureal component of the stria terminalis, which projects to the bed nucleus and to the anterior hypothalamus, originates in a widespread zone within the amygdaloid complex, comprising the basal, medial and central nuclei. In contrast with respect to the precommissural component of the stria terminalis which projects to the ventromedial hypothalamus, only a lesion in the cortical nucleus of the amygdala produced degeneration of this pathway (36). These efferent fibers of the amygdala contain estrogen sensitive neurons (67). Cholinergic neurons are located in regions of the anterior amygdala and lateral preoptic area. In addition, ascending cholinergic fibers through the lateral preoptic area project to the regions of medial and cortical nuclei of the amygdala (63). The efferent fibers from the anterior amygdaloid nucleus distribute mainly into the pars ventralis of the lateral amygdaloid nucleus with less distribution into the central amygdaloid nucleus (63); these efferent axons also distribute into the septum (Table 3). The lateral striae bundle of cholinergic efferents originate or pass through the lateral preoptic area projects to the regions of the medial and cortical nuclei of the amygdala and the efferents from

TABLE 10: EFFECT OF ESTROGEN ON ACETYLCHOLINE CONTENT OF IMMATURE FEMALE RAT

Treatment regimen	No. of rats	Acetylcholine content (nmol/gm tissue ± S.D.)			
		Hippocampus	Amygdala	Hypothalamus	Corpus Striatum
Control	4	25.6 ± 2.1	40.9 ± 4.4	33.0 ± 6.6	
Estrogen* 0.05μg	4	23.8 ± 3.9	28.4 ± 7.7†	22.7 ± 4.9	
Estrogen 2μg	4	24.5 ± 5.1	28.0 ± 11.8	20.0 ± 5.6	
Control	5	25.69± 1.89			65.27±9.66
Estrogen 0.05μg	5	27.64±2.97			64.16±3.51

† P <.025

* Injections of estradiol benzoate were given subcuatneously daily for 4 days beginning at 24 days of age. The animals were sacrificed after 30 days.

TABLE 11: EFFECT OF CORTICOSTERONE ON ACETYLCHOLINE CONTENT OF IMMATURE FEMALE RATS

Treatment	No. of rats	Acetylcholine content (um/gm tissue ± S.D.)		
		Hippocampus	Amygdala	Hypothalamus
Control	4	26.36±0.49	48.13±4.41	39.43±4.20
Cmpd B-2μg	4	21.99±1.47 †	35.12±2.01 †	32.96±3.23
Cmpd B-20μg	4	20.25±3.18 †	38.72±4.56 †	33.51±4.03

† $p < 0.025$ n less

cmpd-2 g 2 μg treatment with corticosterone for 4 days
cmpd-20 g 20 μg treatment with corticosterone for 4 days

TABLE 12: EFFECT OF ELECTROLYTIC LESIONS OF LATERAL PREOPTIC AREA, AMYGDALOID BODY, OR MEDIAL PREOPTIC AREA ON THE ONSET OF PUBERTY

Lesioned area †	No. of rats	Days of Vaginal Opening (Days of age±S.D.)	Ovulation
Intact control	12	34.2±1.1	12 rats
Lateral preoptic area	12	41.8±3.5	12 rats
Cortico-medial and basal amygdala	8	44.5±5.2 No opening on days 50	3 rats 5 rats
Medial preoptic area	7	29.0±1.0	No ovulation ††

† Electrolytic lesion was performed at 24 days of age

†† All of 7 rats showed persistent vaginal cornification after vaginal opening

the lateral preoptic area also distribute into the septum. With the above information, an attempt was made to rationalize the correlation of the neural activity of ascending cholinergic reticular system and the onset of puberty.

Both estrogen sensitive neurons and cholinergic neurons in the amygdala exert their effect on the hypothalamus to participate the onset of puberty. Therefore, either an interruption of ascending cholinergic input to the amygdala by lesions of the lateral preoptic area or disturbance of interaction of estrogen sensitive neurons and cholinergic neurons by lesions of amydgala may cause a delay of onset of puberty.

Lesions of the lateral preoptic area bilaterally, including medial forebrain bundle, in infantile female rats caused a delay of onset of puberty. Furthermore, bilateral lesions of the corticomedial and basomedial nuclei of the amygdala brought about similar results. However, similar bilateral electrolytic lesions in the medial preoptic area causes precocious vaginal opening and cornification, but ovulation does not occur. In addition, these rats show persistent vaginal cornification following vaginal opening (Table 12). In adult female rats, bilateral electrolytic lesions facilitates LH release (26).

The existence of amygdalo-hypothalamus interconnection concerning the regulation of gonadotrophin secretion does not necessarily indicate that no other limbic system, such as hippocampus, is available for the regulation of gonadotropin secretion.

In immature rats, a lowering of local seizure thresholds in the medial amygdala, but not the hippocampus is associated with both natural puberty and precocious ovulation (31). In addition, the timing of onset of puberty was not affected by electrical stimulation of the hippocampus (74). Whereas, electrochemical stimulation of the hippocampus during proestrus suppresses multiunit activity in the arcuate nucleus and prevents ovulation (17); in addition, stimulation of hippocampus is able to block the ovulation expected upon stimulation of the amygdala or preoptic area (75).

Horizontal deafferentation of forebrain afferents at the level of the anterior commissure suppresses gonadotrophin secretion in adult rats (34). In addition, deafferentation of the anterior efferents to the hypothalamus through the septum (horizontal knife was placed above the anterior commissure) produces more prominent effect than deafferentation of the fornix on the delay of onset of puberty (Table 13). So, it must be considered this efferent fibers either to raise in the septum or to pass the septum from non-hippocampal source for regulation of gonadotrophin secretion.

TABLE 13: EFFECT OF HORIZONTAL DEAFFERENTATION OF SEPTAL EFFERENTS AND AFFERENTS ON THE ONSET OF PUBERTY

Region	No. of rats	Recognition of the day of			
		Vaginal opening	Vaginal cornification	Ovulation	No. of Tubal ovc.
Intact control	11	34.6± 0.6	34.7± 0.6	35.1± 0.3	5.1
Operated control	9	36.1± 0.5	36.6± 0.6	37.1± 0.9	5.2
Frontal cut*	9	36.1± 0.5	38.8± 0.7	40.1± 0.7	4.5
Back cut**	10	37.1± 0.8	38.3± 0.9	38.8± 0.9	4.0

*Frontal cut: Horizontal deafferentation was placed in front of the anterior commessure.

**Back cut: Horizontal deafferentation was placed in back of the anterior commessure.

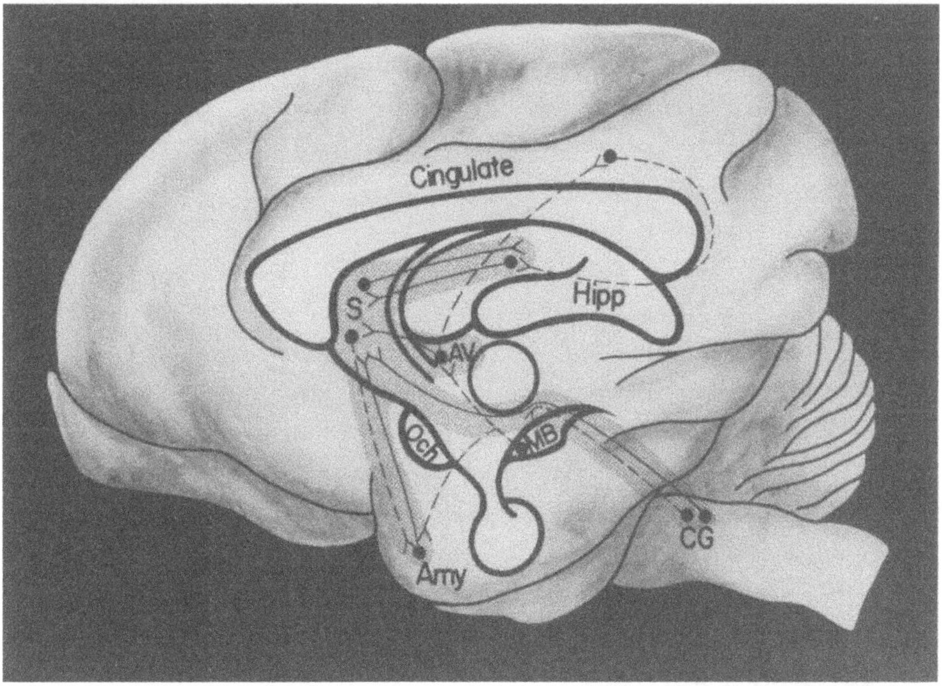

PROPOSED PATHWAY

PROSPECTIVE PATHWAY

Figure 19. Lateral view of rat's brain, showing the ascending septal modulating system in the limbic area, receiving the reciprocating neural interconnection from the midbrain limbic area, the anterior amygdal, the hippocampus.

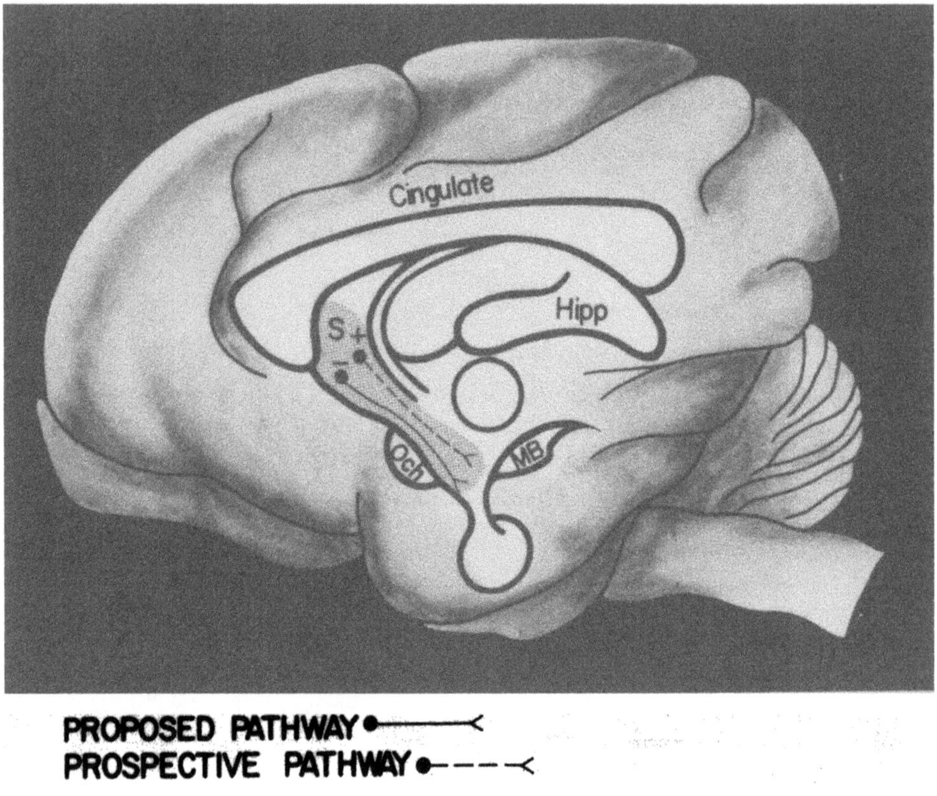

Figure 20. Lateral view of rat's brain, showing the facilitatory pathway (+) (septo-preoptico-tuberal fibers) and the inhibitory pathway (-) of LHRH release (septo-tuberal tract) in the septum.

Exteroceptive (ascending) sensory input from midbrain limbic area travels along with the ventral ascending cholinergic fibers to reach the amygdala and hippocampus. At this level, exteroceptive sensory input is combined with interoceptive factors to form the forebrain signal. And further, the forebrain signal reaches the hypothalamus to regulate gonadotrophin secretion. The activation of ascending sensory input and the forebrain signal appears to be modulated by the activation of the septal function. In addition, the circulating level of plasma estrogen activates the ascending sensory input and the forebrain signal to initiate the onset of puberty.

CONCLUSION

The application of the technique of electrophysiology and neurochemistry to the analysis of neuroendocrine mechanisms of the septum is still in its incipient stage.

Exteroceptive sensory input from the midbrain limbic area travels along with the medial forebrain bundle and/or the dorsal longitudinal fasciculus of ascending reticular projection to reach the septum. At this level, this input signal is combined with information from the dorsal hippocampus and/or anterior amygdala, (Fig. 19), and further, the forebrain signal reaches the basomedial hypothalamus through the septo-preoptico-tuberal fibers (facilitatory pathway) and septo-tuberal tract (inhibitory pathway) (Fig. 20). Subsequently, the forebrain signal makes it possible to regulate LHRH secretion in the hypothalamus. Given this situation, the hippocampal input has an antagonistic effect between the septal neurons and the anterior amygdala. This gives us thought for a neural control mechanism for the cyclicity of ovulation in the female rats.

It, therefore, is reasonable to hypothesize that exteroceptive sensory input which regulates the cyclicity of ovulation appears to be modulated by the activity of the septum.

REFERENCES

1. Anchel, H., and Lindsley, D. B. "Differentiation of Two Reticulo-Hypothalamic Systems Regulating Hippocampal Activity", Electroencephalography and Clinical Neurophysiology, 32 (1972) 209-226.

2. Clark, W. E., LeGros, L. C., and Meyer, M. "The Terminal Connextions of the Olfactory Tract of the Rabbit", Brain, 70 (1947) 304-328.

3. Clemens, J. A., et. al. "Areas of the Brain Stimulatory to LH and FSH Secretion", Endocrinology, 88, 180-184.

4. Critchlow, V. "Ovulation Induced by Hypothalamic Stimulation in the Anesthetized Rat", Amer. J. Physiol., 195 (1958) 171-174.

5. Critchlow, V. and Bar-Sela, M. E. "Control of the Onset of Puberty." In Neuroendrocrinology, (Eds.) L. Martini and W. F. Ganong., Vol. 2, pp. 101-162, New York Academic Press.

6. Critchlow, V., and Sawyer, C. H. "Electrical Activity of the Rat Brain Correlated with Neurogenic Stimulation of the Adenohypophysis", Federation Proceedings, 14 (1955) 32-33.

7. Davidson, J. M. "Feedback Control of Gonadotropin Secretion", In Frontiers in Neuroendocrinology, (Eds.) W. F. Ganong and L. Martini, pp. 343-388, Oxford Press, 1969.

8. De France, J. F., Kitai, S. T., and Shimono, T. "Electrophysiological Analysis of the Hippocampal-Septal Projections: II Functional Characteristics", Exp. Brain Research, 17 (1973) 463-476.

9. Eisenfeld, A. J. "H^3-estradiol: In Vitro Binding to Macromolecules from the Rat Hypothalamus, Anterior Pituitary and Uterus", Endocrinology, 86 (1970) 1313-1318.

10. Everett, J. W. "Central Neural Control of Reproductive Functions of the Adenohypophysis", Physiological Rev., 44, 373-431.

11. Everett, J. W., Krey, L. C., and Tyrey, L. "The Quantitative Relationship Between Electrochemical Preoptic Stimulation and LH Release in Proestrous Versus Late-Diestrous Rats", Endocrinology, 93 (1973) 947-953.

12. Everett, J. W., Sawyer, C. H., and Markee, J. E. "A Neurogenic Timing Factor in Control of the Ovulatory Discharge of Luteinizing Hormone in the Cyclic Rat", Endocrinology, 44 (1949) 234-250.

13. Everett, J. W., et.al. "Electrolytic Irritative Lessons in the Hypothalamus and Other Forebrain Areas: Effect of Luteinizing Hormone Release and the Ovarian Cycle", In Hormonal Steroids, (Eds) L. Martini and A. Pecile, Vol. I, pp. 235-250, Academic Press, 1964.

14. Fox, C. A. "Certain Basal Telencephalic Centers in the Cat", J. Comp. Neurol., 72 (1940) 1-62.

15. Fox, C. A. "The Stria Terminalis: Longitudinal Association Bundle and Precommissural Fornix in the Cat", J. Comp. Neurol., 79 (1943) 277-295.

16. Fuxe, K. "The Distribution of Monoamine Terminals in the Central Nervous System", Acta. Physiol. Scand., 64, suppl. 247, p. 37, 1965.

17. Gallo, R. V., et. al. "Effects of Electrochemical Stimulation of the Ventral Hippocampus on Hypothalamic Electrical Activity and Pituitary Gonadotropin Secretion in Female Rats", Endocrinology, 89, 704-713.

18. Gellhorn, E. "Recent Contributions to the Physiology of the Emotion", Psychiat. Res. Rep. Amer. Psychiat. Asso. 12: 209-223.

19. Green, J. D. and Arduini, A. A. "Hippocampal Electrical Activity in Arousal", J. Neurophysology, 17 (1954) 533-557.

20. Guillery, R. W. "Degeneration in the Hypothalamic Connexions of the Albino Rat", J. Anat., (Lond.), 91 (1957) 533-557.

21. Hagino, N. "Forebrain Regulation of ACTH Secretion in the Rat", In Brain-Pituitary-Interrelationship, (Eds) A. Brodish and E. S. Redgate, pp. 334-336, Karger-Basel, 1973.

22. Hagino, N. "Medial Bundle and Stria Medullous Thalamous", Brains and Nerves, 13, 8 (1961) 587-597.

23. Hagino, N. "Site of Positive Feedback Action Upon the Onset of Puberty", In Brain Function and Reproduction, (Ed). M. Kawakami, Kyodo Igaku Shoin Co. Tokyo, Japan, 1969.

24. Hagino, N. Kosaras, B., and Flerko, B. "Neuroendocrinological Approach to Investigation of the Septo-tuberal Tract in the Rat", Anat. Record, 181, 369.

25. Hagino, N., Ramaley, J. A., and Gorski, R. A., "Inhibition of Estrogen-induced Precocious Ovulation by Phenobarbital in the Rat", Endocrinology, 79, 451-454.

26. Hagino, N., Wagner, J. W., and Gorski, R. A. "Ovulation Induced by Hypothalamic Lesioning in the Rat", Anat. Record, 148 (1964) 288.

27. Hagino, N. and Yawaoka, S. "Role of Biorhythm of Sleep and Wakefulness in the Periodicity of Gonadotrophin Secretion in the Baboon (un-human primate)", In Biological Rhythm in Neuroendocrine Activity, (Ed.) M. Kawakami, pp. 326-337, Igaku-Shoin Ltd., Tokyo, Japan, 1974.

28. Halasz, B. and Pupp. L. "Hormone Secretion of the Anterior Pituitary Gland After Physical Interruption of All Nervous Pathways to the Hypophysiotrophic Area", Endocrinology, 77 533-562.

29. Hillarp, N. A. "Studies on Relocations of Hypothalamic Centers Controlling the Gonadotrophic Function of the Hypophysis", Acta Endocrin. Kgh. 2 (1949) 11-23.

30. Jouvet, M. "A Possible Role of the Catedh Containing Neurons of the Brain-stem of the Cat in the Sleep-waking Cycle", Acta Physiol. pol, 24 (1972) 5-19, Vol. 23.

31. Kanai, T. and Szerb, J. C. "Mesencephalic Reticular Activating System and Cortical Acetylcholine Output", Nature, London, 205 (1965) 80-82.

32. Kato, J. and Villee, C. A. "Preferential Uptake of Estradiol by the Anterior Hypothalamus of the Rat", Endocrinology, 80 (1967) 567-575.

33. Kato, J., et.al, "Nuclear Estradiol Receptor in Rat Anterior Hypophysis", J. Biochem. 67 (1970) 871-872.

34. Kawakami, M., et. al. "Role of Limbic Forebrain Structures on Reproductive Cycles", In Biological Rhythms in Neuroendocrine Activity (ed.) M. Kawakami, Igeku Shion Ltd., pp. 197-219, 1974.

35. Koves, K. and Halasz, B. "Localization of the Neural Structures Triggering Ovulation in the Rat", Neuroendocrinology, 6 (1970) 180-193.

36. Leonard, C. M. and Scott, J. W. "Origin and Distribution of the Amygdalofugal Pathways in the Rat: An Experimental Neuroanatomical Study", J. Comp. Neuro., 141 (1971) 313-330.

37. Lewis, P. R., Shute, C. C. D., and Silver, A. "Confirmation From Choline Acetylase Analyses of a Massive Cholinergie Innervation to the Hippocampus", J. of Physiol., 172 (1964) 9P-10P.

38. Libertum, C., Timiras, P. S., and Kragt, C. L. "Sexual Differences in the Hipothalamic Cholinergic System Before and After Puberty: Inductory Effect of Testosterone", Neuroendocrinology, 12 (1973) 73-85.

39. Macadar, A. W., et.al. "The Functional Relationship Between Septal and Hippocampal Theta Rhythm", Physiol. and Behavior, 5 (1970) 1443-1449.

40. Macadar, O., et.al. "The Functional Relationship Between Septal and Hippocampal Theta Rhythm", Physiol. and Behavior, 5 (1970) 1443-1449.

41. McEwen, B. S., Zigmond, R. E., and Geilach, J. L. "Site of Steroid Binding and Action in the Brain", In Structure and Functions of Nervous Tissue, Vol. 5, (Ed.) G. H. Bouine, pp. 205-291. Academic Press, N.Y.

42. McLean, P. D. "Implication of Micro-electrode Findings on Exteroceptive Input to the Limbic Cortex", In Limbic System Mechanisms and Autonomic Function, (Ed.) C. H. Hockman, pp. 115-136, Charles Thomas Publishers, 1972.

43. McLennan, H., and Miller, J. J. "The Hippocampal Control of Neuronal Discharges in the Septum of the Rat", J. Physiol., 237 (1974) pp. 607-624.

44. Marrazzi, A. S. "Some Indications of Cerebral Humoral Mechanisms", Science, 118 (1953) 367-370.

45. Marrazzi, A. S. and Hart, E. R. "Relationship of Hallucinogens to Adrenergic Cerebral Neurohumors", Science, 121 (1955) 365-367.

46. Morales, F. R., et.al. "Septal Unit Activity and Hippocampal EEG During the Sleep-wakefulness Cycle of the Rat", Physiol. and Behavior, 6, pp. 563-567.

47. Olds, J. and Milner, P. "Positive Reinforcement Produced by Electrical Stimulation of Septal Area and Other Regions of Rat Brain", J. of Comp. and Physiol, Psychol., 47 (1954) 419-427.

48. Papez, J. W. "A Proposed Mechanism of Emotions", Arch. Neurol. Psychiatry., 38 (1937) 725-743.

49. Petsche, H., Stumpf, C., and Gogolak, G. "The Significance of the Rabbit's Septum as a Relay Station Between the Midbrain and the Hippocampus: I. The Control of Hippocampus Arousal Activity by the Septum Cells", Electroenceph. Clin. Neurophysol., 14 (1962) 202-211.

50. Pfaff, D. and Keiner, M. "Atlas of Estradiol-concentrating Cells in the Central Nervous System of the Female Rat", J. Comp. Neurol., 151 (1973) 121-157.

51. Powell, T. P. S., Cowan, W. M., and Raisman, G. "The Central Olfactory Connexions", J. Anat., London, 99 (1965) 791-813.

52. Raisman, G. "The Connexions of the Septum", Brain, 89 (1966) 317-348.

53. Raisman, G., Cowan, W. M., and Powell, T. P. S. "An Experimental Analysis of the Efferent Projection of the Hippocampus", Brain, 89 (1966) 83-108.

54. Raisman, G. and Field, P. M. "Functional Implications of a Quantative Ultrastructural Analysis in the Preoptic Area and Ventromedial Nucleus" In The Neurobiology of the Amygdala. (Ed) B. E. Eleftheuon, pp. 83-94, Plenum Press, N. Y., 1972.

55. Ramirez, V. D. and Sawyer, C. H. "Advancement of Puberty in the Female Ray by Estrogen", Endocrinology, 76 (1965) 1158-1168.

56. Salmoiraghi, G. C., and Stefanis, C. N. "Patterns of Central Neurons' Responses to Suspected Transmitters", Arch. Ital. Biol., 103 (1965) 705-724.

57. Sawyer, C. H., Critchlow, V., and Barracxlough, C. A. Mechanism of Blockade of Pituitary Activation in the Rat by Morphine, Atropine, or Barbituates", Endocrinology, 57 (1955) 345-354.

58. Sawyer, C. H., Markee, J. E., and Townsend, B. F. "Cholinergic and Adrenergic Components of the Neurohumoral Control of the Release of LH in the Rabbit", Endocrinology, 44 (1949) 18-37.

59. Sethy, V. H., Roth, R. H., Kuhar, M. J., and Van Woert, M. H. "Choline and Acetylcholine: Regional Distribution and Effect of Degeneration in Cholinergic Nerve Terminals in the Rat Hippocampus", Neuropharmacol., 12 (1973) 819-823.

60. Shute, C. C. D. and Lewis, P. R. "Cholinergic and Monoaminergic Systems of the Brain", Nature (London), 212 (1966) 710-711.

61. Shute, C. C. D., and Lewis, P. R. "Cholinergic Nervous Pathways in the Forebrain", Nature (London), 189 (1961) 332-333.

62. Shute, C. C. D. and Lewis, P. R. "Cholinesterase-containing Systems of the Brain of the Rat", Nature, (London), 199 (1963) 1160-1164.

63. Shute, C. C. D. and Lewis, P. R. "The Ascending Cholinergic Reticular System: Neocortical, Olfactory, and Subcortical Projections", Brain, 90 (1967) 497-520.

64. Stavinoha, W. B., and Weintraub, S. T. "Estimation of Choline and Acetylcholine in Tissue by Pyrolysis Gas Chromatography" Anal. Chem., 46 (1974) 757-760.

65. Stumpf, C., Petsche, H., and Gogolalc, G. "The Significance of the Rabbit's Septum as a Relay Station Between the Midbrain and the Hippocampus: II. The Differential Influence of Drugs Upon Both the Septal Cell Firing Pattern and the Hippocampus Theta Activity", Electroenceph. Clin. Neurophysiol., 14 (1962) 212-219.

66. Stumpf, W. E. "Estradiol-concentrating Neurons: Topography in the Hypothalamus by Dry-mount Autoradiography", Science, 162 (1970) 1001-1003.

67. Stumpf, W. E. "Estrogen-neurons and Estrogen-neuron Systems in the Periventricular", Am. J. Anatomy, 129 (1970) 207-217.

68. Taleisnik, S., Velasco, M. E., and Astrada, J. J. "Effect of Hypothalamic Deafferentation on the Control of Luteinizing Hormone Secretion", J. Endocrinology, 46 (1970) 1-7.

69. Tejatat, T. and Everett, J. W. "Surgical Analysis of the Preoptictuberal Pathway Controlling Ovulatory Release of Gonadotropins in the Rat", Endocrinology, 81 (1967) 1387-1398.

70. Terasawa, E. and Sawyer, C. H., "Electrical and Electrochemical Stimulation of the Hypothalamo-adenohypophysical System With Stainless Steel Electrodes", Endocrinology, 84 (1969) 918-925.

71. Terasawa, E. and Timiras, P. S. "Electrophysiological Study of the Limbic System in the Rat at Onset of Puberty", Am. J. Physiol., 215 (1968) 1462-1467.

72. Ter Haar, M. B. and MacKinnon, P. C. B. "Changes in [^{35}S] Methionine Incorporation Into Protein of the Rat Brain During Puberty", Brain Research, 85 (1975) 535-538.

73. Turgeon, J. and Barraclough, C. A. "Temporal Patterns of LH Release Following Graded Preoptic Electrochemical Stimulation in Proestrous Rats", Endocrinology, 92 (1973) 755-761.

74. Velasco, M. E. and Taleisnik, S. "Effect of Hippocampal Stimulation on the Release of Gonadropin", Endocrinology, 85 (1969) 1154-1159.

75. Velasco, M. E. and Taleisnik, S. "Release of Gonadotropins Induced by Amygdaloid Stimulation in the Rat", Endocrinology, 84 (1969) 132-139.

76. Yamaoka, S. and Hagino, N. "Spontaneous Septal Neuron Activity in the Rat", Brain Research, 67 (1974) 147-152.

77. Zigmond, R. E. and McEwen, B. S. "Selective Retention of Oestradiol by Cell Nuclei in Specific Brain Regions of the Ovariectomized Rat", J. Neurochem., 17 (1970) 889-899.

THE EFFECTS OF ABLATION OF THE SEPTAL NUCLEI IN THE RAT ON CIRCADIAN VARIATION AND STRESS RESPONSE PATTERN OF CORTICOSTERONE, GROWTH HORMONE AND PROLACTIN[1]

Jo Seggie and Gregory M. Brown

Neuroendocrine Research Section, Clarke Institute of Psychiatry, Toronto, Ontario, Canada

It has long been known that destruction of the septal nuclei in rats is followed by a very dramatic change in affective behavior (3). This change has frequently been dubbed the septal syndrome and is characterized by behavioral overreaction to most environmental stimuli. Once present, however, the septal syndrome is a self-limiting condition observed to persist for postoperative periods of 7-75 days (12,17). Normal attenuation of the syndrome is facilitated by environment stimulation (11,18). One way environmental factors could influence the behavior of septal rats is through the brain-pituitary-endocrine axis as hormones have direct influences on brain activity and behavior (10,13,29).

Coupled with the fact that many behavioral disturbances are associated with disorders of the endocrine system, it seemed reasonable to suppose that the septal syndrome might have an endocrine component. Indeed, stimulation of the septal nuclei inhibits adrenal activity in the rat (26) and increased secretion of ACTH (2) and adrenal hypertrophy has been found after septal lesions (18,19, 28). However, this adrenal response does not occur if the animals are not disturbed postoperatively (16). Accordingly, we investigated the effects of some endocrine manipulation on behavioral manifestations following septal lesions. We found first that

---

[1]This paper is not intended as a comprehensive review of the role of the septal nuclei in endocrine function. Rather, it provides an overview of a series of investigations we have done with the objective of determining the role of the septum in stress hormone responses.

neither adrenalectomy nor gonadectomy altered the septal syndrome (20). Nor was the syndrome due to increased levels of ACTH since blockade of ACTH release with dexamethasone also had no effect on the septal syndrome (22). Melanocyte stimulating hormone was also ruled out as having a role in the septal syndrome (6). Thus the possibility that behavioral manifestations in the septal rat were dependent on circulating corticosteroid levels was ruled out. It appeared that endocrine changes observed following septal lesion must be secondary phenomena related either to the lesion itself or to the behavioral syndrome. Due to the crudeness and insensitivity of behavioral measures, we focused our primary attention on direct measurement of hormones in the blood with the objective of defining the endocrine manifestations more precisely.

In the normal rat, adrenal steroid levels exhibit a 24 hour rhythm (7). Responses to stimulation are more liable at the trough values of corticosterone (21). Normal, sham-operated and septal animals showed equivalent levels at crest. However, at trough septal animals had a five-fold increase over non-lesioned rats. A similar observation was found by Endroczi and Myakas (9). The question naturally arose as to whether this finding was due either to an altered adrenal rhythm or alternatively to an exaggeration of the trough sensitivity. In order to clarify this issue, we observed resting corticosterone levels every 3 hours for 24 hours under housing conditions in which extraneous disturbances from environmental conditions were rigidly controlled. Normal, sham-operated and septal lesioned rats exhibited identical 24 hour rhythms (23). So, it remained to be explained why we originally found elevated trough levels in septal rats and in addition why septal rats do not consistently have adrenal hypertrophy. Perhaps septal rats do in fact have increased sensitivity to stress. Preliminary data suggested this might be so (22). Septal rats had a greater corticosterone response following behavior rating than non-lesioned controls.

We next undertook a detailed study of stress responses. Corticosterone was measured at five different times during the first hour after either 5 seconds of handling or 3 minutes exposure to a novel environment at crest and trough of the adrenal rhythm (25). Following stress, adrenal response patterns vary in a characteristic fashion which is determined by both the time in the cycle at which stress is applied and also by the type of stimulus used (4, 24). Normal and sham-operated rats responded in the expected manner under all four conditions. In contrast, septal rats showed specific peculiarities in their response pattern which confirmed the hypothesis that their sensitivity to stress is altered.

Fig. 1 shows corticosterone response to 3 minutes of novel environment at crest of the adrenal cycle and is representative of the response profiles seen under the other three experimental conditions. Septal rats responded to stress in a significantly shorter time interval and with significantly higher corticosterone levels than non-lesioned rats during the first 15 minutes after the end of stimulation. In addition, considering all four conditions (trough vs crest, 5 sec stimulus vs 3 min), the degree of overresponsiveness was far greater at trough than crest. Thus, it was demonstrated that although septal rats had normal baseline adrenal function, they were overresponsive to stress as measured by corticosterone levels. This overresponsiveness is due in part to a lowered threshold of activation since septal rats show adrenal activation to levels of electric shock at which normal and sham-operated rats show no response (27). Here we had an endocrine measure which agreed well with previous reported finding of increased sensitivity to shock in septal rats (14).

So it was established that, with respect to adrenal function, undisturbed septal rats are normal. However, environmental disturbances will cause adrenal activation: levels of disturbance above the normal threshold will cause exaggerated responses while disturbance levels which are below the threshold for the intact animal will nontheless produce a response in a septal animal. The continued occurrence of these adrenal responses is the probable cause of the adrenal hypertrophy. This interpretation is given additional support by the observation that hypertrophy can be prevented by eliminating postoperative environmental disturbance (16).

Having clarified the nature of the adrenal effects of septal lesions, it was of considerable interest to examine other stress responsive hormones. Next we investigated the effect of septal lesions on growth hormone levels. To the authors knowledge, there is little information available on the role of septum in growth hormone regulation.

In the rat, growth hormone is released in a pulsatile manner (15) and falls dramatically in response to stress (5). Thus resting growth hormone levels show high variation and frequently need to be transformed before analysis can be done. We first investigated the resting diurnal variation of growth hormone by sampling every 3 hours in the day. Data was subjected to a long transformation (Fig. 2). The data from normal and sham-operated rats were equivalent and were, therefore, combined. This control group evidenced a 24 hour rhythm with crest and trough values occurring around the middle of the day and the middle of the night respectively under a 12 hour light 12 hour dark cycle. A septal lesion did not significantly influence this 24 hour cycle. In this respect, growth hormone and adrenal findings are similar.

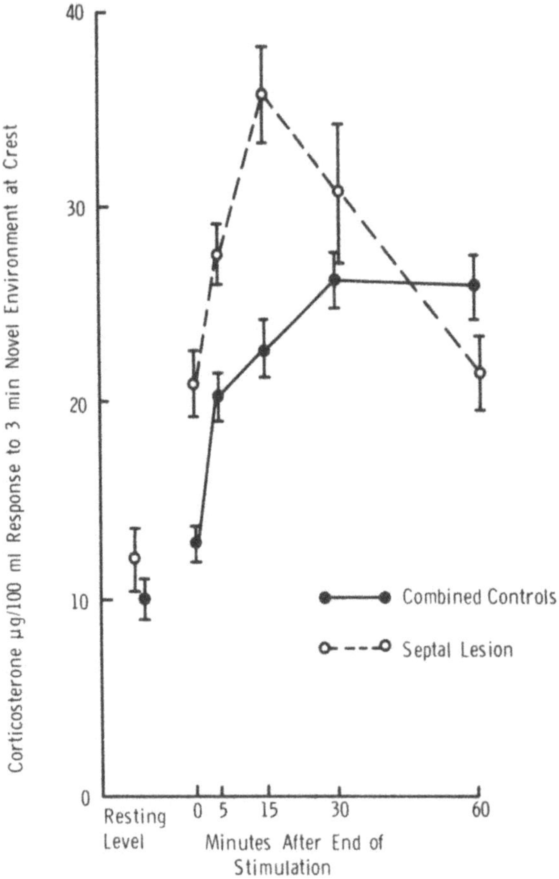

Figure 1. Adrenal stress response of septally-lesioned and non-lesioned control rats at crest of adrenal cycle from Seggie, Uhlir and Brown, Neuroendocrinology, in press. Vertical bars represent standard error.

Growth hormone responses to stress were also studied. In order to facilitate comparison of responses with the well documented adrenal system, crest and trough of the adrenal cycle were used as times of observation. And, like in the corticosterone study, 5 seconds of handling or 3 minutes of novel environment were the stimuli used. As it turned out, neither time of day nor type of stimulation had any effect on the normal pattern of growth hormone stress responses in this study and so the data from these conditions were combined (Fig. 3). As expected in intact and sham animals, growth hormone levels responded to stress with a marked

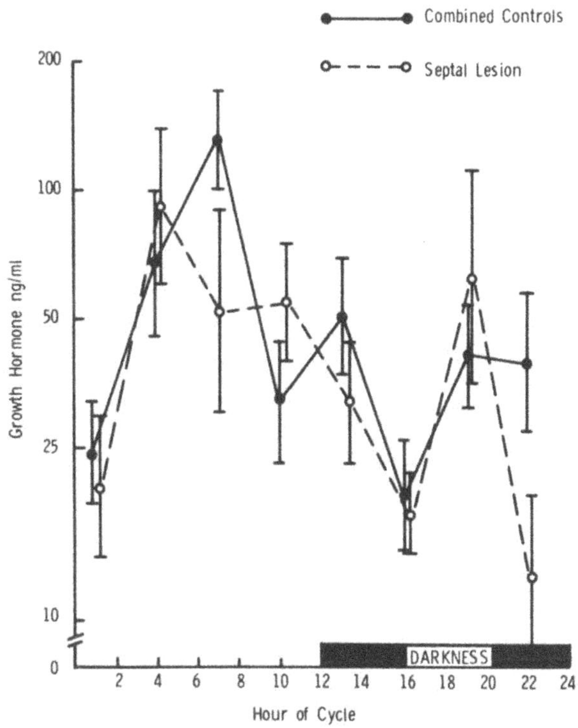

Figure 2. Growth hormone diurnal variation in non-lesioned controls and rats with a septal lesion. Vertical bars represent standard error.

fall and began to return to resting levels by one hour after stimulation. A septal lesion significantly affected this response. Septal rats showed no evidence of recovery of resting levels of growth hormone in the hour of observation. Thus, a septal lesion appears to potentiate the growth hormone as well as the adrenal response to stress. Potentiation occurs with both hormones, even though the direction of response for the two hormones is opposite. This observation suggests that the effect of the lesion is specifically to exaggerate activity in stress responding components of some endocrine control system rather than simply to modify other regulatory mechanisms such as feedback control.

Lastly, we have some preliminary information on the effect of a septal lesion on prolactin responses in the rat (6,27). Fig. 4 shows that a septal lesion had no effect on resting prolactin levels nor was the stress response altered by a septal lesion.

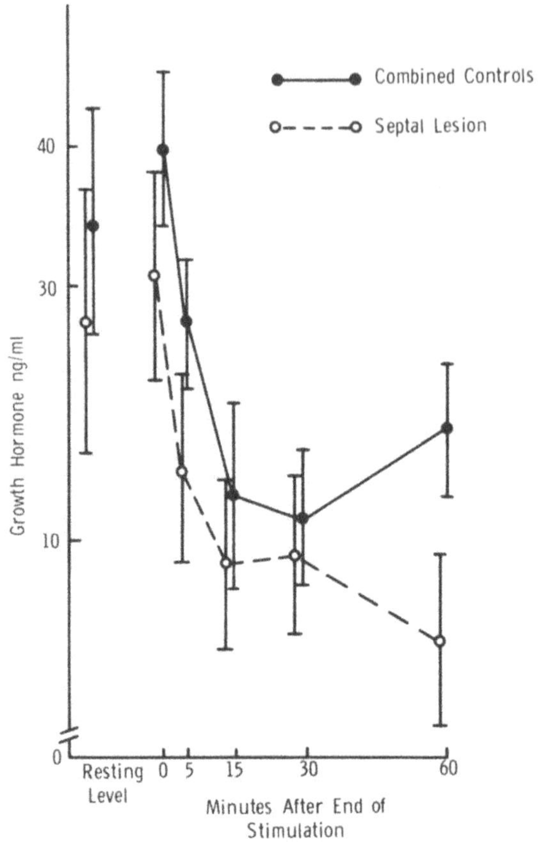

Figure 3. Growth hormone stress response in non-lesioned controls and rats with a septal lesion. Vertical bars represent standard error. N per point is 65-70.

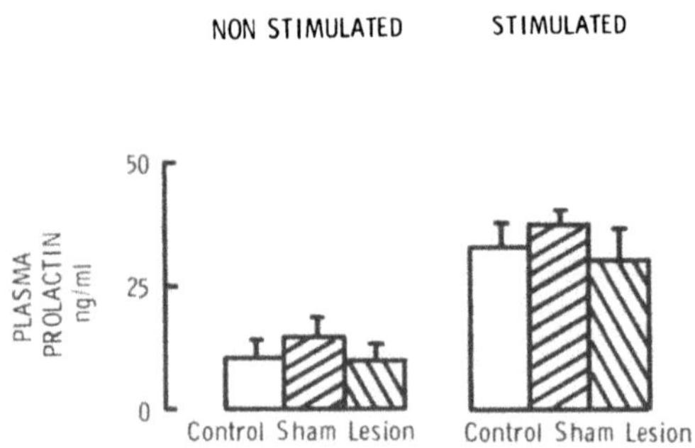

Figure 4. Prolactin levels in normal control, sham operated and septal lesion rats at rest (non-stimulated) and 15 minutes after 3 minutes exposure to a novel environment (stimulated) at peak of the adrenal cycle. Vertical bars represent standard error.

A detailed study of prolactin stress response patterns using the same design as in the corticosterone and growth hormone study was also done. Fig. 5 shows a sample of these results. A septal lesion had no significant effect on the pattern of prolactin stress responses. This finding would suggest that the controlling mechanisms for prolactin differ from that of corticosterone and growth hormone.

Investigation of the effect of septal lesions on circadian variation of prolactin has also been undertaken but is not yet complete.

In conclusion, we have investigated 24 hour resting levels and the stress response patterns of corticosterone and growth hormone and have preliminary information on prolactin in the rat. Resting levels of these three hormones are unaffected by damage to the septal area. However, septal lesions potentiated the stress response of corticosterone and growth hormone without affecting the stimulus and cycle dependent characteristics of the normal response. Based on our findings, it is hypothesized that in the intact organism the septal nuclei may perform an evaluative or modulative function in stress situations which then determines such things as threshold, latency and magnitude of stress responses.

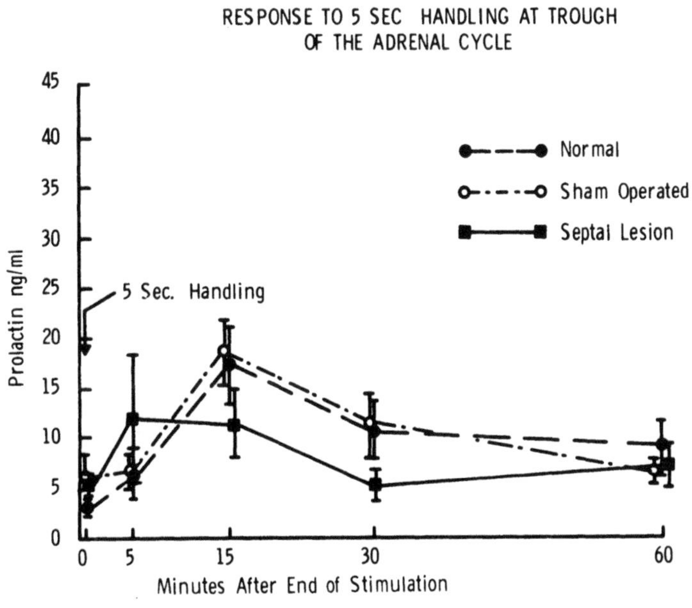

Figure 5. Prolactin stress response in normal, sham operated and rats with a septal lesion. Vertical bars represent standard error.

Whether the septum functions independently in its endocrine effects or in cooperation with other limbic structures as would be suggested by many behavioral studies remains to be seen.

REFERENCES

1. Ader, R. and Friedman, S. B. Plasma corticosterone response to environmental stimulation and the 24-hour adrenocortical rhythm. Neuroendocrinology, 3 (1968) 378-386.

2. Bohus, B. The effect of central nervous lesions on pituitary adrenocortical function in the rat. Acta Physiological Academy of Science in Hungary, 20 (1961) 373-377.

3. Brady, J. V. and Nauta, W. J. H. Subcortical mechanisms in emotional behavior: affective changes following septal forebrain lesions in albino rats. Journal of Comparative and Physiological Psychology, 46 (1953) 339-346.

4. Brown, G. M. and Martin, J. Corticosterone, prolactin and growth hormone responses to handling and new environment in the rat. Psychosomatic Medicine, 36 (1974) 241-247.

5. Brown, G. M. and Reichlin, S. Psychologica and neural regulation of growth hormone secretion. Psychosomatic Medicine, 34 (1972) 45-61.

6. Brown, G. M., Uhlir, I. V., Seggie, J., Schally, A. V. and Kastin, A. J. Effect of septal lesions on plasma levels of MSH, corticosterone, GH and prolactin before and after exposure to novel environment: role of MSH in the septal syndrome. Endocrinology, 94 (1974) 583-587.

7. Critchlow, V., Liebelt, R. A., Bar-Sela, M., Mountcastle, W. and Lipscomb, H. S. Sex differences in resting pituitary-adrenal function in the rat. American Journal of Physiology, 205 (1963) 807-815.

8. Dunn, J. and Scheving, L. Plasma corticosterone levels in rats killed sequentially at the trough or peak of the adrenocortical cycle. Journal of Endocrinology, 49 (1971) 347-348.

9. Endroczi, E. and Nyakas, C. Effect of septal lesions on exploratory activity, passive avoidance learning and pituitary-adrenal function in the rat. Acta Physiology of Hungary, 39 (1971) 351-360.

10. Ford, D. H. (ed). Influence of Hormones on the Nervous System, S. Karger, Basel, 1971.

11. Gotsick, J. and Marshall, R. Time course of the septal rage syndrome. Physiology and Behavior, 9 (1972) 685-687.

12. Krieckhaus, K., Simmons, H., Thomas, G. and Kenyon, J. Septal lesions enhance shock avoidance behavior in rats. Experimental Neurology, 9 (1964) 107-113.

13. Lissak, K. Hormones and Brain Function, New York: Plenum Press, 1973.

14. Lorens, S. A., Sorensen, J. P. and Harvey, J. A. Lesions in the nuclei accumbens septi of the rat: behavioral and neurochemical effect. Journal of Comparative and Physiological Psychology, 73 (1970) 284-290.

15. Martin, J. B. Pulsatile GH secretion in the unanesthetized rat: effects of stress, morphine and somatostatin (SRIF) on the response. Endocrine Society 56th Annual Meeting, Abs. #197, 1974.

16. Montgomery, R. L. and Berkut, M. K. Behavior and adrenal weight of non-stressed rats with selected limbic-lobe lesions. Physiology and Behavior, 4 (1969) 745-748.

17. Reynolds, R. Equivalence of radio-frequency and electrolytic lesions in producing septal rage. Psychonomic Sciences, 2 (1965) 35-36.

18. Seggie, J. Effect of somatosensory stimulation on affective behaviour of septal rats. Journal of Comparative and Physiological Psychology, 66 (1968) 820-822.

19. Seggie, J. Endocrine and circadian variables in manifestation of affective behavior following septal ablation in rats. Proceedings of the 78th Annual Convention of the American Psychological Association, 5 (1970) 199-200.

20. Seggie, J. Effect of adrenalectomy or gonadectomy on affective behavior changes following septal lesions in the rat. Journal of Comparative and Physiological Psychology, 74 (1971) 11-19.

21. Seggie, J. and Brown, G. M. Septal lesions and resting adrenal function: a possible explanation of conflicting findings. Neuroendocrinology, 8 (1971) 367-374.

22. Seggie, J. and Brown, G. M. Effect of dexamethasone on affective behavior and adrenal reactivity following septal lesions in the rat. Journal of Comparative and Physiological Psychology, 83 (1973) 60-65.

23. Seggie, J., Shaw, B., Uhlir, I. and Brown, G. M. Baseline 24-hour plasma corticosterone rhythm in normal, sham-operated and septally-lesioned rats. Neuroendocrinology, 15, (1974) 51-61.

24. Seggie, J., Uhlir, I. V. and Brown, G. M. Stress response patterns of plasma corticosterone, prolactin and growth hormone in the rat. Proceedings of the Canadian Federation of Biological Societies (Abstract) 17, #326.

25. Seggie, J., Uhlir, I. and Brown, G. M. Adrenal stress responses following septal lesions in the rat. Neuroendocrinology, in press.

26. Taylor, A. N., Branch, B. J., Casady, R. L. and Turner, B. B. Septal inhibition of pituitary-adrenal activity in freely behaving rats. Paper presented at the 55th Annual Meeting of the Endocrine Society, Chicago, 1973.

27. Uhlir, I., Seggie, J. and Brown, G. M. The effect of septal lesions on the threshold of adrenal stress response. Neuroendocrinology, 14 (1974) 351-355.

28. Usher, D. P., Kasper, P. and Birmingham, M. Comparison of pituitary-adrenal function in rats lesioned in different areas of the limbic system and hypothalamus. Neuroendocrinology, 2 (1967) 157-174.

29. Woodbury, D. M. and Vernadakis, A. Influence of hormones on brain activity. In: L. Martini and W. F. Ganong (Eds) Neuroendocrinology, Vol. 2, New York: Academic Press 1967.

PLASMA CORTICOSTERONE LEVELS DURING ACTIVE AVOIDANCE LEARNING IN RATS WITH SEPTAL LESIONS[1]

Marc DeRyck, Christer Køhler, Holger Ursin and Seymour Levine

Institutes of Physiology and Psychology, University of Bergen and Department of Psychiatry, Stanford University School of Medicine

INTRODUCTION

One of the characteristics of rats with lesions in the septal area is a marked enhancement of 2-way avoidance learning (6,8). Amongst the various hypotheses which have been advanced to account for this finding are: loss of inhibitory responses (9), reduced freezing (12) and decreased fear (8).

These explanations fall roughly into two categories. The first postulates changes in the septal lesioned animal which involve alterations in the problem-solving capacity in 2-way avoidance. The second proposes changes in motivation as a consequence of septal lesions such as decreased fear. One possible way of distinguishing between these concepts would be to use an independent measure of arousal. Thus greater or lesser arousal would suggest a motivational change. If there is no difference in the physiological response during the early phases of avoidance learning in spite of a difference in the acquisition of avoidance, an explanation involving problem-solving ability would be suggested. However, changes in the physiological responses should occur at later stages

---

[1]Sponsored in part by the Norwegian Research Council for Science and Humanities, European Brain and Behaviour Training Program, The Swedish Medical Research Council, National Institutes of Health, Bethesda, Maryland USA (NICH&HD 02281) and the National Institute of Mental Health, Chevy Chase, Maryland USA (K5-MH-19936 to SL).

in training, since performance is altered in spite of intact motivational mechanisms.

One such indicator of arousal is the activity of the pituitary-adrenal system as measured by concentrations of plasma corticosterone. Normal rats show a decrement in plasma corticosterone levels when a 2-way avoidance response has been well established (1). The poor acquisition of active avoidance in rats with amygdala lesions may be attributed to reduced arousal, since the initial corticosterone level following the first day of training is lower in rats with amygdala lesions than in normals (2). Using a similar paradigm, it should be possible to discriminate between the major hypotheses in animals with septal lesions. If the septal lesion produces enhanced 2-way active avoidance learning by changing fear or arousal, then the response of the pituitary-adrenal system during the early phase of conditioning would differ from controls. In contrast, if the septal lesioned animal is superior in the problem-solving aspects of 2-way avoidance acquisition it should show normal activation of the pituitary-adrenal system, but show more rapid decrements following superior performance.

The present experiment was undertaken to examine and distinguish those aspects of septal function which result in enhanced 2-way avoidance learning (Experiment 1). By a fortuitous procedural accident we also produced an unexpected impairment of 2-way active-avoidance learning in one group of septal animals (Experiment 2). The results are included since they demonstrate that the performance of the septal lesioned rat seems to be the decisive factor determining corticosterone levels during avoidance learning.

METHOD

Subjects

In <u>Experiment 1</u>, 16 male Møll-Wistar rats weighing 330-395 gr. were used. Bilateral electrolytic lesions were produced in the septum in 9 rats, 7 were sham operated. Following histology, all lesioned rats were accepted. In <u>Experiment 2</u>, the subjects were 19 Møll-Wistar rats weighing 280-320 gr. Electrolytic lesions were produced in the same area in 10 rats, 9 were sham-operated. One of the operated rats was excluded following histology, leaving a total of 9 lesioned rats in the material. All animals were run together, without knowledge of the surgical treatment of the individual rat. Rats were run in two waves; Experiment 1 took place several months after Experiment 2. The subjects were given three weeks to recover from surgery before behavioral testing. The rats were well handled before and after surgery before the testing began. Food and water were given ad lib.

All rats were housed individually in a sound-insulated room. Lights were controlled rigidly, off at 5 p.m., on at 8 a.m. Only the experimenters were allowed to enter this room. Feeding and cleaning procedures took place at the same time each afternoon. Testing took place at the same time each day, between 9 a.m. and 1 p.m.

Surgery

All surgery was performed under pentobarbital sodium anesthesia. The rats were placed in a stereotaxic instrument with a 5 mm tilt. The electrode was inserted through a trephine-hole 1.8 mm in front of the bregma, 0.5-1 mm lateral to the midline on both sides, and 5.5 and 4.5 mm down from the surface of the dura. The electrode was a Teflon-coated stainless steel electrode, with a .5 mm bare tip. The lesion was produced by anodal DC current, 2.5 ma., for 20 sec. The sham-operated animals received the same treatment up to and including drilling of the skull.

Apparatus

The shuttlebox (24 cm wide, 29 cm tall, and 100 cm long) was divided into two compartments by a barrier 8 cm high. In Experiment 2, the two compartments were separated above the barrier by a plexiglass door hinged in the roof of the apparatus. A while noise (10 db above background level of 60 db) serving as the CS was delivered through speakers at either end of the shuttlebox. A Grason Stadler scrambled shock source provided the US of 0.5 ma. electric shock.

Avoidance Learning and Plasma Corticosterone

Twenty-one acquisition sessions, each consisting of 10 trials at 60 sec intervals, were given with intersession intervals of 24 hr for five days a week; there was no testing on the weekend (72 hr interval). Blood samples were never taken immediately following the weekend. The CS-US interval was 5 sec. If the rat did not avoid, the CS plus US remained on until the rat escaped. Maximum shock duration was 25 sec. In Experiment 2, a shaping procedure was tried on Day 10. For rats still not performing any escape on the first two trials on Day 10, plexiglass plates were installed in each compartment, reducing the size approximately 50%. This increased the probability that the rat escaped over the hurdle. On Day 11, the door was removed for all animals. Except for the door used for the first 10 days in Experiment 2, and the shaping on Day 10 in Experiment 2, all parameters and procedures were identical in the two experiments.

Following the acquisition period, a single blocked-extinction sessions was subsequently given (the 22nd session). This involved 5 sec presentations (10 trials at 60-sec intervals) of the CS only and prevention of barrier crossing by a plexiglass wall placed over the shuttle barrier.

Blood samples for corticosterone assay (5) were drawn from the rats following five of the sessions: the 1st session, 6th or 7th session (balanced design), 13th session, 21st session, and blocked-extinction session. Three basal samples (subject sampled directly from home cage) were taken during the experiment: 1 wk prior to the 1st avoidance session, 24 hr after the 5th or 6th session (balanced design), and 4 days after the extinction session. Finally, two "ether-stress" samples were also taken on the days when the first and final basal samples were taken.

The basal samples were taken by quietly entering the animal room and removing the rat to be sampled. The rat was anesthetized by ether within 1 min from first disturbing it, and within another 1/4-1 min a blood sample had been drawn from the surgically exposed jugular vein. These manipulations cause pituitary-adrenal activation, but there is a 3-5 min delay in rise of plasma corticosterone concentration (3). If no blood was obtained in less than 3 min from first disturbing the rat, no sample was taken. To obtain the ether-stress sample, the animal was returned to his cage until 20 min had passed from the time it was first disturbed and anesthetized. At this time it was re-anesthetized and another blood sample drawn. Post-avoidance session sampling was carried out in the same manner, the subject being anesthetized beginning 20 min after introduction to the avoidance apparatus. The avoidance session lasted only 10-11 min and the rat was returned to its cage and the animal room until the required 20 min from stimulus onset had passed. Ether-anesthetized rats were never put back into the animal room until all rats had been sampled that day.

Behavioral measures observed during each avoidance session included number of avoidance responses, number of bolus droppings, and number of intertrial barrier crossings. Also, the intertrial behavior was recorded during the sessions following which the rat was blood sampled: the 1st, 6th or 7th, 13th, 20th, and blocked-extinction sessions. In each five sec of the intertrial time the following behavior categories were recorded for occurrence vs. nonoccurrence: immobile, moving head, changing position within box (but not sniffing), sniffing, grooming, or freezing (defined as piloerection).

Histology

The brains were fixed in formalin and imbedded in paraffin, sectioned frontally at 20μ, and every 20th section stained with thionin.

RESULTS

The lesions were large, and included the lateral, dorsal and the medial nuclei, and parts of the diagonal band. The lesions affected the anterior commissure, and also extended slightly into the tissue ventral to the commissure, particularly in rostral sections. The postcommissural fornix fibers were only partially affected (see Fig. 1).

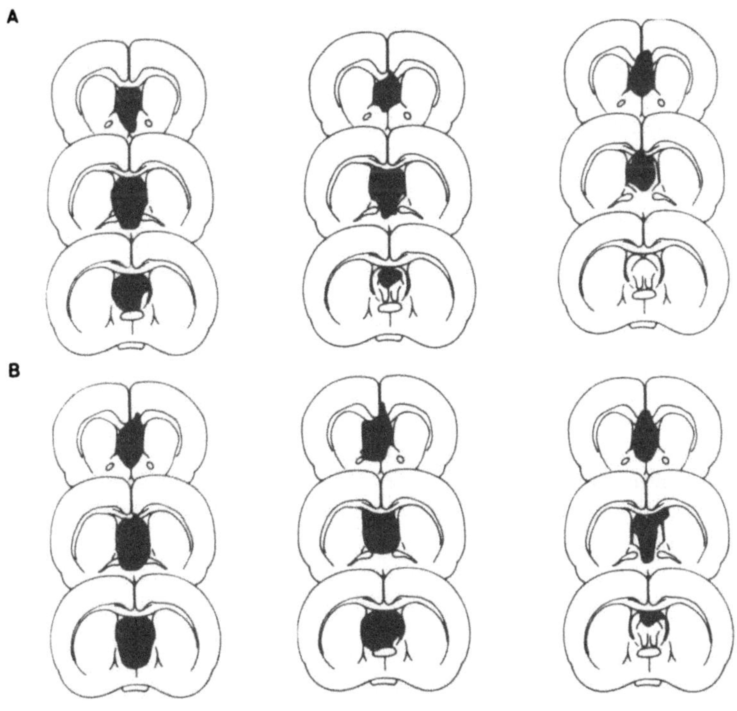

Figure 1. Frontal sections from Experiment 1 (upper row) and Experiment 2 (lower row). Top section: rostral level; lower section: caudal level; black area: lesioned area. The three brains selected represent the largest (left), medium sized and the smallest lesion (right).

Experiment 1

Avoidance learning. The septum-lesioned group acquired the avoidance response faster than their controls (Fig. 2). Analysis of variance for session 1, 6/7, 13 and 21 showed a significant group ($F = 10.91$, $df = 1/14$, $P < 0.01$) and session effect ($F = 42.99$, $df = 3/42$, $P < 0.001$). Group-session interaction was not significant. The same findings were evident from an analysis based on blocks of session. The difference between the learning curves was significant for all four sampled points at the .05 or .01 levels (*t*-test).

The septal lesioned rats had shorter latencies on the first three trials on Day 1, but this tendency was statistically significant only on trial 1 ($P < 0.02$).

Septal lesioned rats did not differ from normals in number of fecal boli dropped during the session. There was significantly less "freezing" among the septal lesioned rats than in their controls, except on Day 1. Analysis of variance showed significant group effect ($F = 13.27$, $df = 1/14$, $P < 0.005$) and session effect ($F = 10.9$, $df = 3/42$, $P < 0.001$), but no significant interaction effect. Probably parallel to this, sniffing increased, while there did not seem to be any change in grooming.

During the forced extinction session, both groups showed a significant increase ($P < 0.01$, Newman-Keuls test) in freezing from the preceding avoidance session (21). There was significantly less "freezing" in the septal lesioned rats during the intertrial intervals (*t*-test, $P < 0.01$).

Corticosteroids. There was a tendency of higher basal levels in rats with septal lesions,but this tendency was not significant for any of the basal levels sampled. For the ether stress sample, there was no such tendency before the experiment, and an insignificant tendency for the ether stress sampled following the avoidance procedure (Fig. 2).

The postsession samples were not different for the first day. For the other three sample points there were highly significant differences. An analysis of variance showed significant treatment effect ($F = 20.05$, $df = 1/13$, $P < 0.001$), sessions effect ($F = 18.88$, $df = 3/39$, $P < 0.001$) and interaction effect ($F = 5.06$, $df = 3/39$, $P < 0.005$).

Following the extinction session, there was a significant increase in steroids from the preceding avoidance session (21) for rats with septal lesions ($P < 0.01$). For controls, there was no such increase, but they had a fairly high level on the sessions preceding the extinction session.

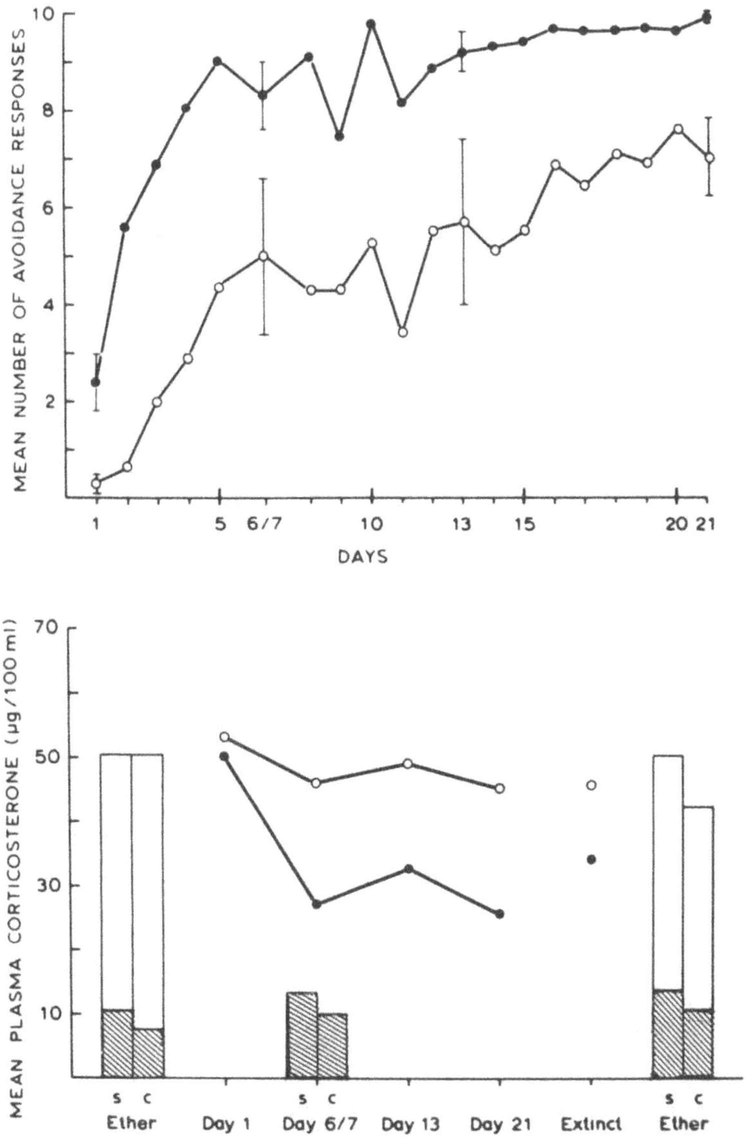

Figure 2. Experiment 1: Open circles: normal rats; filled circles: rats with septal lesions. Upper figure: Performance in the 2-way active avoidance box. Vertical lines indicate standard error. Lower figure: Plasma corticosterone levels before, during and after avoidance learning. Shaded bar indicates basal levels; open bars: ether exposure samples. The filled circles indicate post-session samples for rats with septal lesions; the open circles indicate post-session samples for normal rats.

Experiment 2

Avoidance learning. An analysis of variance showed no treatment effect when calculations were based on avoidance score from Day 1, 6, 13 and 21 (sample days). On Day 6 the normal group avoided more than the septal lesioned rats ($P < 0.05$, *t*-test). The sessions effect was highly significant ($F = 63.33$, $df = 3/48$, $P < 0.001$), but no significant treatment-session interaction was found. As is evident from Fig. 3, removal of the door dramatically improved performance of the septals ($P < 0.05$, *t*-test). The change from Day 10 to Day 11 was not significant for the controls.

The learning curve was split up and an analysis of variance was made for sessions 1, 6, 10, and sessions 11, 15 and 21 separately. Only the sessions effect proved to be significant ($F = 17.61$, $df = 2/32$, $P < 0.001$ and $F = 6.89$, $df = 2/32$, $P < 0.005$, respectively). An analysis of the latencies (Fig. 3) demonstrated that the rats with septal lesions had lower latencies on the very first trial on session 1, ($P < 0.05$, *t*-test). They were also faster on trial 3 on session 11 ($P < 0.05$), just after the door had been removed.

Septal lesioned rats dropped less fecal boli only during the first session ($P < 0.05$). There was no significant difference between the groups on time spent freezing in intertrial interval when measured on sample days. Both septal lesioned rats and controls showed a significant increase in the time spent freezing during the blocked extinction session ($P < 0.01$, $P < 0.001$, respectively).

Corticosteroids. The basal levels were higher for the rats with septal lesions on the first measure taken ($P < 0.01$), but there was no significant difference for any of the other basal samples. Both normals and septals showed a tendency of a slightly higher value on this first sample day. There was also a tendency of a higher ether stress value on this same day, but this did not reach statistical significance at the 0.05 level. The ether stress following the whole procedure did not show any statistical significant difference between groups (Fig. 3).

There was no difference between the groups during avoidance training. An analysis of variance showed only a significant session effect ($F = 12.40$, $df = 3/48$, $P < 0.001$). During blocked extinction there was no statistically significant difference from session 21 for rats with septal lesions, while controls proved to increase their plasma corticosterone level during this session ($P < 0.02$).

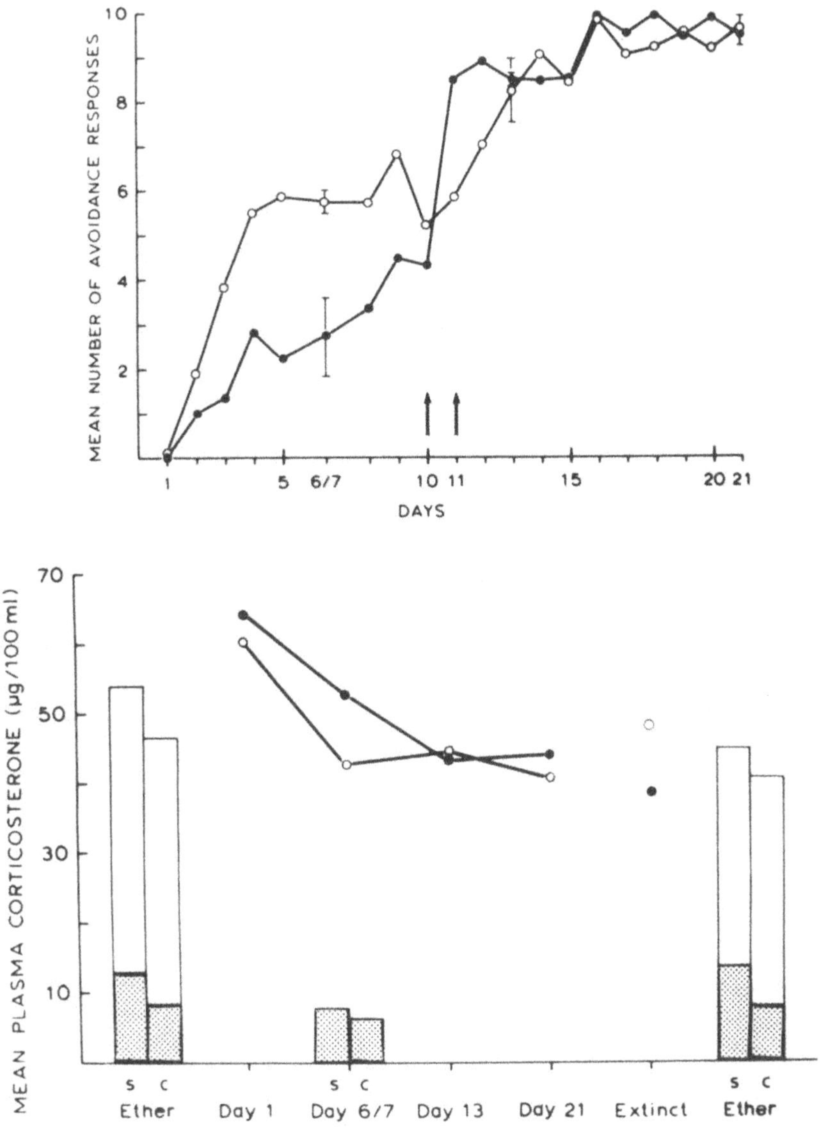

Figure 3. Experiment 1: Open circles: normal animals; filled circles: septal lesioned rats. Upper figure: Performance in the 2-way active avoidance apparatus. On Day 10 a plexiglass plate was installed in each compartment. On Day 11, the door between the two compartments was removed. Vertical lines indicate standard error. Lower figure: Plasma corticosterone levels before, during and after avoidance learning. Shaded bars indicate basal levels; open bars: ether exposure samples. The filled circles indicate post-session samples for rats with septal lesions; the open circles post-session samples fornormal rats.

DISCUSSION

Previous reports on the septal nuclei (4,13) have claimed that the septal nuclei influence pituitary-adrenal activity. The data presented in this paper do not reveal any indications of a direct influence of the septal nuclei on the ultimate release of adrenal corticosterone. Lesions in the septal nuclei of the rat do not in our hands change the basal level except for one measurement. In experiments reported elsewhere no changes in basal levels were found when lesions were restricted to the dorsolateral nuclei (14), and no changes in basal levels or corticosteroid levels following ether exposure were apparent when both the dorsolateral and medial nuclei were damaged (11). The lack of effect on basal levels of the septal lesions is confirmed by Seggie (see her chapter in this book). It should be noted, however, that Seggie does report that there are significant differences in the response of the pituitary-adrenal system in septal animals under certain conditions; under moderate stress a greater adrenal response in septal lesions was observed in the immediate postoperative period. Their interpretation is that the increases in plasma corticosterone following exposure to moderate stress are due to the well-documented hyper-reactivity observed in these animals. The tendency for septal animals to show generally higher steroid values under basal conditions in our experiments which proved to be significant in one measurement, could be due to less rigid control of extraneous laboratory noise. Seggie et al. (1974) have demonstrated that only under rigid control of laboratory conditions do septal and normal rats show similar basal levels.

The discrepancy between our failure to show differences in response to stress and the data reported by Seggie are probably due to the fact that the stress that we have used--ether or electric shock in the avoidance situation--was severe enough in intensity to obliterate the differences in response to a more moderate stress observed in septal animals. It is also possible that Seggie's phenomena only exist in the immediate postoperative period (Day 3-5). Where we do observe a difference in plasma corticosterone in septal animals we conclude that this difference is primarily due to a marked change in the behavior during the course of avoidance conditioning rather than any alteration in the regulation of the pituitary-adrenal system. In the conventional 2-way active avoidance situation (Experiment 1), where rats lesioned in the septal area where significantly superior in the acquisition of the problem, they also showed a correspondingly striking decrement in plasma corticosterone levels that occurs much earlier than observed in normals. Whereas there were no differences on Day 1, there are differences on Days 6 and 7, at which time there are significant differences in the number of avoidance responses made by septal animals. It appears that when septal animals attain the capacity to cope with the environment in a 2-way active avoidance situation, they show a marked decrement of arousal or "fear" during these latter stages of avoidance training. There are also behavioral indices of a reduced fear during the later stages of avoidance training, as observed by less freezing and more sniffing in the intertrial intervals. This reduction in "fear" as observed by both behavioral and physiological indices appears to be a direct result of their capacity to acquire the active avoidance performance much earlier.

There is further evidence to support the conclusion that the steroid change follows behavior. When a door was located between the two compartments the septal lesioned rats were no longer behaviorally superior to the controls; on the 6th day they even performed more poorly. In this situation, the steroids do not drop faster than normal. When the door was removed, the situation again favored the way the septal lesioned rats solved this problem. The lesioned rats then learned rapidly, and showed the expected fall in corticosteroids. Endroczi and Nyakas (4) reported augmented levels of plasma corticosterone following passive avoidance. Since the septal animals are inferior in learning the passive avoidance task, the increased number of shocks could account for these differences.

Lubar and Numan (8) have suggested that the most parsimonious explanation of all of the avoidance changes observed in the septal lesioned rats is that there is decreased fear. The retarded ability of septal animals to form a conditional emotional response (CER) might explain the passive avoidance deficit and the retarded 1-way active avoidance learning. It has been observed that animals learn 2-way active avoidance under moderate shock more rapidly than under intense shock and the relationship between fear and active avoidance does seem to follow the now traditional curvilinear relationship between emotionality and performance (7). Our data, however, do not support a decreased fear interpretation and are more consistent with an interpretation which postulates that the absence of the septal nuclei changes the manner in which animals cope and solve a 2-way active avoidance problem. We have not observed any indices of differences in arousal and fear on the first day of avoidance training. What changes are observed do appear to follow the alterations in behavior. Thus, the reduction of fear, as observed by both behavioral and physiological indices are obliterated when the behavioral differences are no longer present. We believe, therefore, that the endocrine changes and the emotional changes during avoidance learning are secondary to the alterations in the tactics that septal animals use to master this particular problem.

SUMMARY

Rats with septal lesions showed the same basal plasma corticosterone levels as normal rats and also the same response as normals following exposure to ether, or the first day, in the avoidance apparatus. The corticosterone levels during acquisition of a 2-way active avoidance depended upon the performance in the shuttlebox. In the usual 2-way avoidance, rats with septal lesions showed significantly superior acquisition. These animals also showed a marked decrement on Day 6 or 7 of training which was not present in normal rats that had not reached the same level of performance. This decrement in plasma corticosterone in the septal animals appears to be a direct result of the more rapid acquisition of the avoidance response. When a door was located between the two compartments the septal lesioned rats were no longer behaviorally superior to the controls; on the sixth day even performed more poorly. In this situation, the steroids do not drop faster than normal. Thus, the reduction of fear, as observed by both behavioral and physiological indices, are obliterated when the behavioral differences are no longer present. The differences observed in endocrine and emotional responses during 2-way active avoidance learning seem secondary to the differences in the way septal lesioned and normal rats solve this type of problem.

REFERENCES

1. Coover, G. D., Ursin, H. and Levine, S. Plasma-corticosterone levels during active-avoidance learning in rats. Journal of Comparative and Physiological Psychology, 82 (1973) 170-174.

2. Coover, G., Ursin, H. and Levine, S. Corticosterone and avoidance in rats with basolateral amygdala lesions. Journal of Comparative and Physiological Psychology, 85 (1973) 111-112.

3. Davidson, J. M., Jones, L. E. and Levine, S. Feedback regulation of adrenocorticotropin secretion in "basal" and "stress" conditions: Acute and chronic effects of intrahypothalamic corticoid implantation. Endocrinology, 82 (1968) 655-663.

4. Endroczi, E. and Nyakas, C. Effect of septal lesion on exploratory activity, passive avoidance learning and pituitary-adrenal function in the rat. Acta Academiae scientiarum Hungaricae, 39 (1971) 351-360.

5. Glick, D., von Redlich, D. and Levine, S. Fluorometric determination of corticosterone and cortisol in 0.02-0.05 milliliters of plasma or submilligram samples of adrenal tissue. Endocrinology, 74 (1964) 653-655.

6. King, F. A. Effects of septal and amygdaloid lesions on emotional behavior and conditioned avoidance responses in the rat. Journal of Nervous and Mental Disease, 126 (1958) 57-63.

7. Levine, S. UCS intensity and avoidance learning. Journal of experimental Psychology, 71 (1966) 163-164.

8. Lubar, J. F. and Numan, R. Behavioral and physiological studies of septal function and related medial cortical structures. Behavioral Biology, 8 (1973) 1-25.

9. McCleary, R. A. Response-modulating functions of the limbic system: Initiation and suppression. In: Progress in physiological psychology, Stellar, E. and Sprague, J. M. (eds.), pp. 209-272. New York: Academic, 1966.

10. Seggie, J., Shaw, B., Uhlir, I. and Brown, G. M. Baseline 24-hour plasma corticosterone rhythm in normal, sham-operated and septally-lesioned rats. Neuroendocrinology, 15 (1974) 51-61.

11. Srebro, B., Ellertsen, B. and Ursin, H. Deficits in avoidance learning following septal lesions in the albino rat. In preparation, 1975.

12. Thomas, G. J., Hostetter, G. and Barker, D. J. Behavioral functions of the limbic system. In: Progress in physiological psychology, Stellar, E. and Sprague, J. M. (eds.), Vol. 2. New York: Academic, 1968.

13. Usher, D. R., Kasper, P. and Birmingham, M. K. Comparison of pituitary-adrenal function in rats lesioned in different areas of the limbic system and hypothalamus. Neuroendocrinology, 2 (1967) 157-174.

14. Ursin, H., Coover, G. D., Köhler, C., DeRyck, M., Sagvolden, T. and Levine, S. Limbic structures and behavior: Endocrine correlates. In: Neuropsychopharmacology Symposium in Utrecht, 1974, Gispen, W. H. (ed.).

Part IV
Behavior

BEHAVIORAL FUNCTIONS OF THE SEPTUM: A RE-ANALYSIS

Sebastian P. Grossman

The University of Chicago

THE MANY BEHAVIORAL EFFECTS OF SEPTAL LESIONS

Introduction

Behavioral investigations of the functions of the septum have a very brief but stormy history. Prior to the early 1960's, there was only a gentle interest in the septum as a possible relay station along the circuit which Papez (107) had postulated as a substrate for "emotions" or "affect." The few experimental investigations which were reported in the 1950's generally agreed with such an interpretation in showing "rage," increased startle reactions (9, 10), and faster acquisition of conditioned avoidance responses (CAR) (75) in rats with septal lesions. It was only after McCleary (91) followed up some of the implications of earlier electrophysiological (64) and behavioral (111) observations which suggested that the septal area might exercise important general inhibitory influences on autonomic functions as well as behavior, and demonstrated apparent inhibitory effects on passive avoidance behavior which were not readily compatible with the classic picture, that psychologists became seriously concerned about the septum. In the few years which have elapsed since then, a staggering number of reports of experimental work on the septum as well as a flood of theoretical interpretations of these data have been published. Just about every behavior and/or psychological function which has been investigated to date has been shown to be affected in some way by septal lesions, and attempts to account for these diverse effects in terms of meaningful neural dysfunctions have become increasingly convoluted and improbable. It is not possible to review this literature in any depth in the context of the present discussion, but I shall briefly summarize some of the major empirical observations and proposed interpretations before discussing some of our own attempts to understand this obviously very complex problem.

Aggressive Behaviors

To begin with the classic symptom of "rage" reactions to normal handling--they disappear, often within a few days, when the animals are handled, (9,10) when their cage is repeatedly shaken (125), when they are housed in pairs (1) or even when nothing at all is done to speed recovery (76,144). Moreover, there are marked species and strain differences with respect to this effect of septal damage. Septal rage rarely occurs in the cat (100), rhesus monkey (140), hamster (192), or cotton rat (11) which all show other components of the septal lesion syndrome. Indeed, we have consistently found that the incidence of septal "rage" is so low in the albino rats of the Sprague-Dawley strain which are used in our own laboratory that we no longer include the phenomenon in our systematic investigations.

It is, moreover, not clear that the septal "rage," when it occurs, represents an increase in aggressive behavior, as is commonly assumed. Rats with septal lesions do indeed "overrespond" to painful electric shock although their detection threshold is not affected (78,79) as well as to other potentially noxious stimuli such as loud noises (9) or bright lights (105), but the nature of their response is not at all clear. Our own extensive observations indicate that rats which display the "rage" syndrome will indeed bite (i.e., "aggress") when cornered in their cage but go to great lengths to avoid contact when the situation permits it. We (96) as well as others (1,7,31) have found that rats with septal lesions engage more readily in shock-induced "fighting" with members of the same species. However, we share the opinion of many investigators in this field that the threatening postures which are counted as aggressive interactions in these tests may, more properly, be considered defensive rather than aggressive reactions. We (96) find, in fact, that clearly aggressive interactions which occur in a food competition situation are reduced or totally abolished by septal lesions, apparently because the previously dominant member of a pair of male rats refuses to engage in fights. Our observations are in good agreement with reports of the fighting behavior of mice (94) and quail (110) with septal lesions, but it is only fair to point out that an increase in aggressive interactions has been reported in hooded rats under some conditions which did not involve painful shock (12).

Interspecies aggressive interactions (e.g., muricide) have been reported to be unaffected by septal lesions (66,67) in test paradigms where the potential antagonists are permitted to live together for some time prior to surgery. We (96) have found that septal lesions do dramatically increase the frequency of lethal attacks between rats and mice when the animals are first introduced to each other shortly after surgery. The different results of these two sets of studies may further support our notion that rats with septal lesions are not generally "hyperemotional" but respond irritably towards novel stimuli. It may be interesting to note, in this context, that all changes in aggressive behavior have a very clear time course--they are most pronounced immediately after surgery and disappear 10-15 days later (96). This transience is, of course, reminiscent of the "rage" effect itself, but the relationship between the two phenomena is not clear since the changes in intraspecific aggressive interactions as well as the increased muricide are often seen in animals which display little or no overt evidence of septal "rage."

A similar time course has been reported for exaggerated reactions to intense lights (105) or foot shock (78), and it is possible that the increased reactivity to various environmental stimuli, although perhaps not causally related, may, at least in part, reflect a depletion of forebrain serotonin which has been shown to follow lesions in the medial forebrain bundle and its target organs, including the septum (60,61). The time course of this depletion is similar to that seen in many of the behavioral measures which we have just discussed, and a direct relationship between pain sensitivity and brain serotonin has been established in several studies (59,126).

Avoidance Behaviors

King's (75) classic observation that rats with septal lesions acquire a shuttle box CAR faster than controls has been replicated in numerous laboratories in the rat (72,97), mouse (18), hamster (90), and cat (145), although there is at least one report of retarded CAR acquisition in a shuttle box, using intermittent shock applications (23). The effect appears to be independent of the septal "rage" syndrome since it has been seen in animals which do not overreact to normal handling (72,76,90). It also does not appear to be related to the effects of septal lesions on pain sensitivity or brain serotonin since its time course is markedly different (97).

However, a major question has arisen concerning the proposed interpretation of this effect of septal damage since it appears to be peculiar to the shuttle box apparatus. Septal lesions retard CAR acquisition in multichambered alleys (95), T-mazes (105), shelf-jumping or lever-pressing (52), and even in double-grill boxes when the problem is arranged so that the rat or cat always avoids by leaving one distinct compartment and enters another and clearly different "safe" compartment (72,116,138). In the cat, CAR acquisition has been reported to be normal or even better than normal in an unusual "one-way" shuttle box situation which avoids handling between trials (85,145). Hamilton (52), working in my laboratory, replicated this observation but found only impairments in CAR acquisition in this species in several other CAR paradigms, and we have come to the conclusion that the shuttle box may give "atypical" results for some as yet poorly understood reasons.

The distinction between shuttle box and other avoidance paradigms appears to hold even when the reinforcement contingencies are significantly altered. We (70,116) consistently find that rats with septal lesions emit many more spontaneous avoidance responses than controls in a shuttle box programmed so that an animal can postpone CS as well as UCS presentations by spontaneously crossing from one compartment to the other at least once in each predetermined interval (see Fig. 1). This agrees well with many other observations which indicate that rats with septal lesions find it difficult to withhold potentially reinforced responses (see below), but there are at least two reports of slower and hence more efficient responding under similar "Sidman" free-operant conditions not involving a shuttle box (101, 128).

To complicate matters a bit further, there are reports that septal lesions in rats (114) as well as cats (100) impair the performance of preoperatively acquired CAR's in a shuttle box (100) and other avoidance paradigms (114). Since there are also reports of poor postoperative performance in simple (e.g., T-maze) appetitive paradigms when the behavior was trained prior to the lesion (16), it is not clear that the deficits seen in CAR situations have any specific relationship to avoidance behavior per se.

Conditioned Emotional Responses and Reactions to Punishment

In their historic investigation of the behavioral consequences of septal lesions, Brady and Nauta (9) included a test for conditioned emotional response (CER) which consisted essentially of a measure of freezing and defecation in response to a CS which had been associated with unavoidable painful shock. Brady and Nauta (9) reported that rats with septal lesions showed less CER than the controls. In subsequent investigations (1) the procedure was refined to make a disruption of ongoing lever pressing behavior the measure of the CER, but the result remained the same--septal lesions reduced or abolished the CER. Other

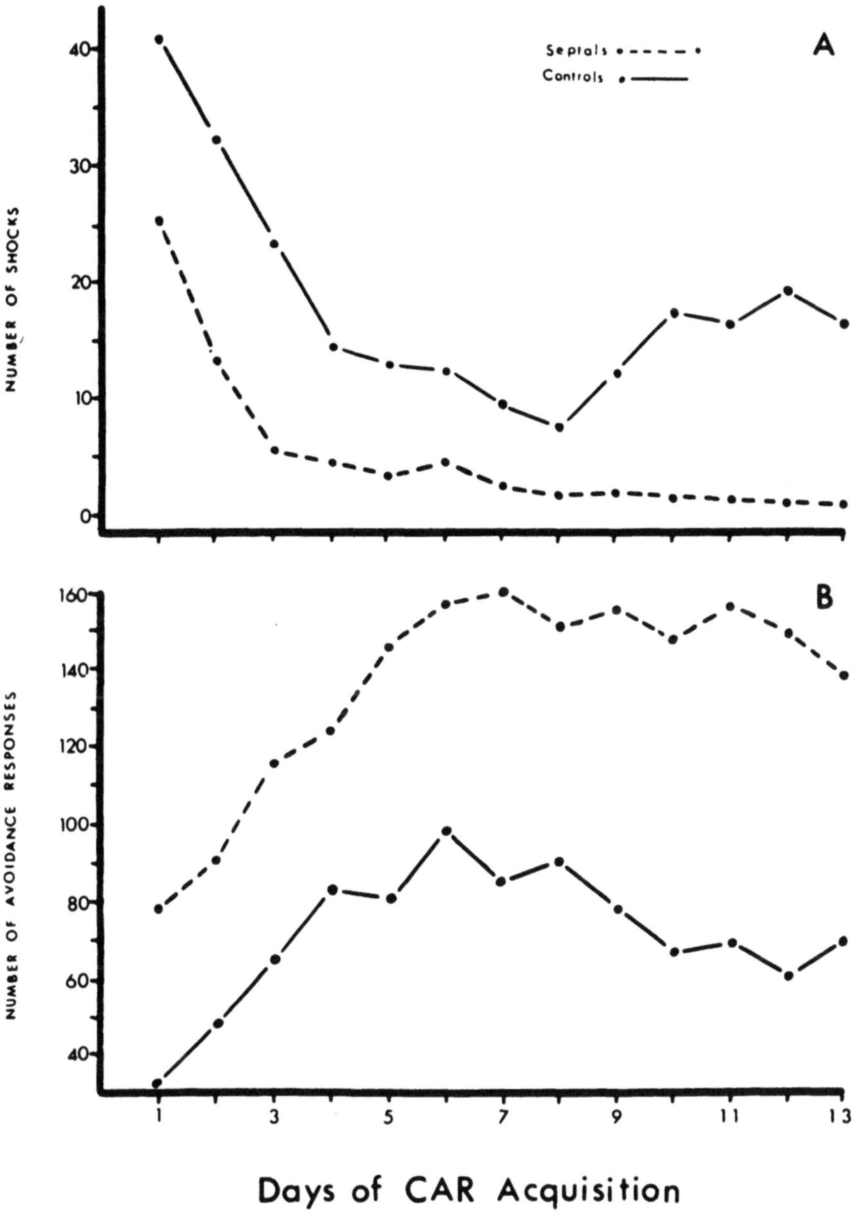

Figure 1. The effects of lesions in the septal area of rats on the acquisition of a discriminated free-operant avoidance response in a shuttle box. (A--The total number of shocks received each day. B--The total number of avoidance responses made each day.) From Kelsey & Grossman (70). Reprinted by permission of the American Psychological Association, Washington, D. C.

investigators (60,135) have shown that rats with septal lesions also do not acquire a CER as well as controls while working for food as well as water rewards. There is some indication (60) that this impairment is long-lasting and that punishing responses which occur during the presence of the CS does not abolish the deficit.

This brings us to a component of the septal lesion syndrome which many investigators have found baffling and even paradoxical. Since McCleary's first report (91) of a "passive avoidance" deficit in cats with septal lesions, the phenomenon has been replicated in cats (37,52,85) as well as rats (65,97,120). Although it is descriptively accurate, the term "passive avoidance" is unfortunate because it implies a kinship with other forms of avoidance behavior which may not, in fact, exist. I shall therefore discuss the phenomenon in terms of reactions to punishment in the following discussion.

Impaired reactions to punishment have been observed in a variety of paradigms after septal lesions, including food or water rewarded lever pressing (120) where an increase in appetitive motivation (see below) might play a role (60), but also in other situations where the punished response is not appetitively reinforced (80,95,142). There are some reports (99,120) that punishment may be normally effective unless unpunished appetitive responding is permitted within each test session, and a number of others suggest that the greater the amount of postoperative approach training, the smaller the effect of subsequent punishment on the behavior of rats with septal lesions (38,39,40,143). Matters become confounded when one considers punishments other than the ubiquitous footshock which is used in all of the experiments which we have so far discussed. There is some evidence that rats with septal lesions are as competent or better than their controls in suppressing the intake of a preferred sweet solution (although they normally ingest more of it than controls) when this becomes a CS for debilitating X-ray treatments (93).

Appetitively Reinforced Operant Responding

The impaired response to punishment which is seen in most experimental paradigms indicates that rats and cats with septal lesions find it difficult to withhold previously rewarded behaviors. McCleary (91,92) suggested some years ago that the behavior of such animals could best be described as "perseverative," and he related this to a general loss of inhibitory control over potentially reinforced behaviors. Numerous investigators have observed overresponding in various paradigms where such behavior is either inefficient (i.e., not rewarded) or subject to penalty. For example, rats trained to lever press for rewards which are delivered according to a fixed interval (FI) schedule (i.e., only those lever presses are reinforced which occur a predetermined time after the last reward) rapidly learn not to respond for some time after a reward has been obtained (the post-reinforcement pause), gradually begin to respond towards the middle of the delay interval, and work faster and faster as the time approaches when the next reward is scheduled. A number of investigators have found that rats with septal lesions over-respond in such a paradigm, and it has been suggested (32,112,119) that the principal effect of the lesion might be an increase in "preservative" responding during the normally quiet post-reinforcement pause. In more recent experiments (3), a change in both post-reinforcement responding and in terminal rate have been observed. When one takes the overall increase in response rate into account in these studies, it is, in fact, not clear that a specific increase in preserverative responding occurs. We (116) have recently re-examined this matter and find a very marked effect on responding during the first (i.e., post-reinforcement) half of the FI shortly after surgery but also observe that this effect disappears within a week after surgery. Increases in terminal rate, on the other hand, appeared to be permanent, suggesting that the rat with septal lesions may not perseverate but, rather, find it difficult to withhold the next potentially rewarded behavior. We shall return to this intriguing hypothesis in a moment.

The operant paradigm which has been used most extensively to examine the effects of septal lesions on "response inhibition" or "timing" behavior has been the DRL (differential reinforcement of low rate) schedule which programs reinforcements at regular intervals as in the FI but penalizes the animals for responses which occur before the required delay has elapsed by resetting the delay timer whenever such a response is made. To optimize rewards on such a schedule, a rat must learn to space its responses so that they occur just after each delay interval has elapsed. Most animals learn to perform this very difficult timing task astonishingly well. In view of the effects of septal lesions on FI performance, it is perhaps not surprising that a large number of investigators (14,15,16,17,33,88) have found that rats with large septal lesions consistently overrespond and thus earn few reinforcements when confronted with DRL contingencies. There is some indication that the effect may be ameliorated when the shift from the initial training conditions (reinforcement of every response or CRF) to the DRL contingency is made gradually (13,58). However, we (70) have found that rats with septal lesions perform inefficiently in a modified DRL paradigm which provides no operant training prior to the introduction of the final DRL contingencies and thus avoids any possibility of "perseveration" in previously reinforced response patterns. Since our modified DRL paradigm included the additional requirement that successive responses be made on different levers, we were able to separately analyze "perseverative" responses to the most recently correct lever and "anticipatory" responses to the lever which provided the next reinforcement. We found, in very good agreement with our recent observations of the distribution of responses in the FI, that rats with septal lesions did not emit more "perseverative" responses than controls (in fact, reliably fewer), but appeared to be impaired mainly in their ability to withhold the next, potentially rewarded response. They thus earned far fewer rewards than controls although the total number of responses emitted per session did not differ after the first five days of training, during which the rats displayed marked generalized overresponding (see Fig. 2). Ross and I (in press) have recently replicated these observations in some detail (see Fig. 22). It is interesting to note that our rats were capable of waiting the required time between responses when the end of the delay period was indicated by an auditory signal (days 8-10, Fig. 2).

Rats and cats with septal lesions also often overrespond during extinction (121) and in a variety of other cats (146) as well as rats (41,55) with septal lesions learn spatial or kinesthetic discriminations as well as controls but perform very poorly when the reinforcement conditions are reversed. Cats (145) and rats (121) learn "go/no go" discriminations poorly, apparently because they continue to respond on "no go" trials. In line with these observations, it is typically found that rats with septal lesions tend to display less spontaneous alternation behavior and are slower than controls in learning tasks which require alternation (24,30,118,134).

Septal lesions also increase responding on schedules such as continuous reinforcement (CRF) (83) or fixed ratio (FR) (17,62) which do not differentially reward "timing" or the inhibition of potentially correct responses and increase running speed in straight alleys which lead to food or water rewards (21,104,116). There is some evidence (17) that this may be independent of the motivational changes which are often seen after septal lesions (see below).

Changes in Food and Water Intake

Some septal lesions produce hyperdipsia (58,87) which does not appear to be related to prandial requirements (8) or be secondary to polyuria (86), although possible connections to pituitary dysfunctions have been investigated (5,87). The hyperdipsia appears to be related to an exaggerated response to hypovolemic conditions (8), but this interpretation has not gone unchallenged (6).

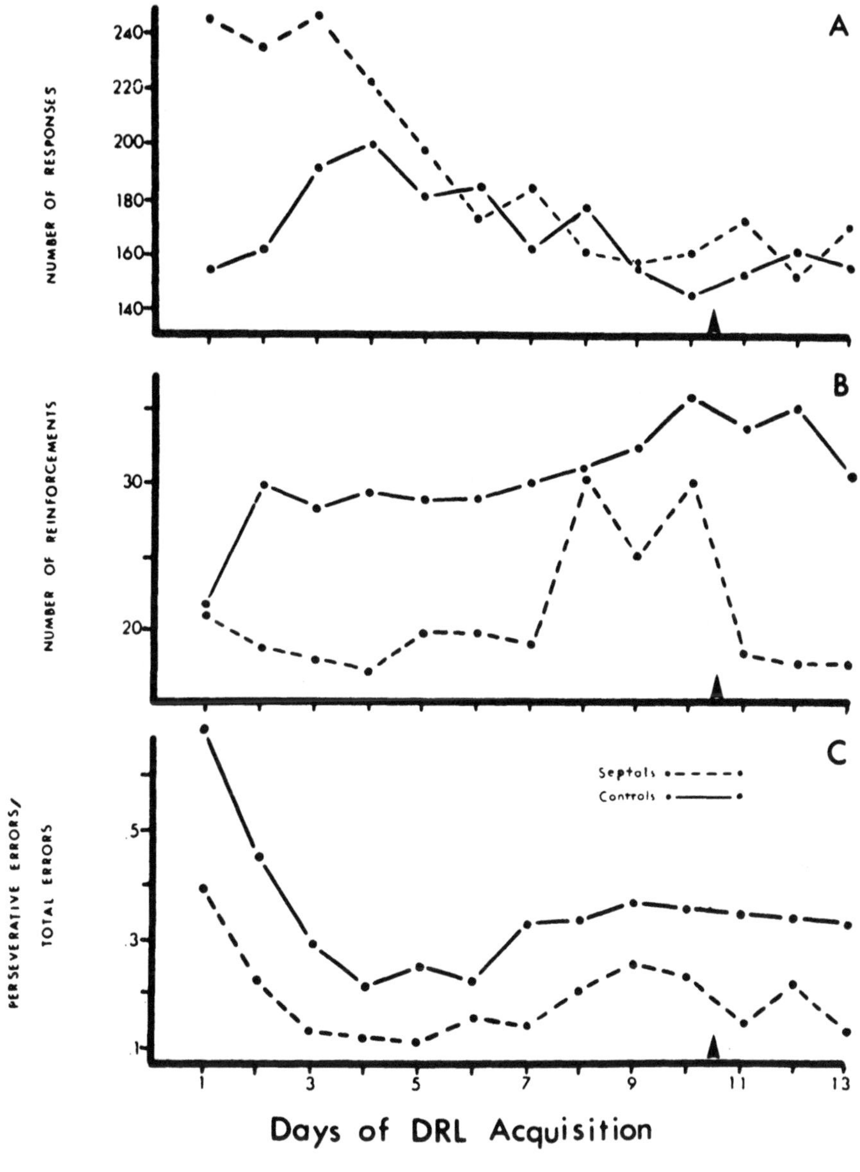

Figure 2. The effects of lesions in the septal area of rats on the acquisition of responding in a modified DRL paradigm where food reinforcements were contingent on alternation between two levers in addition to the withholding of successive responses for a delay period of 30 sec. (A--The total number of responses made each day. B--The total number of reinforcements obtained each day. C--The percentage of incorrect responses that were perseverative.) From Kelsey & Grossman (70). Reprinted by permission of the American Psychological Association, Washington, D. C.

Food intake, on the other hand, is generally normal after septal lesions in rats (69,83) as well as cats (145). A transient increase in food intake shortly after surgery has been reported by some investigators (28,127), and there is some indication that it may be correlated with hyperdipsia (143). However, the effects are very small and, in our experience, not greater than those seen after many other CNS lesions which have not been related specifically to the regulation of food intake. Other investigators (83), concur with our interpretation. Indeed, we (117) as well as others (3) have observed a very marked drop in body weight immediately after septal lesions which is not recovered in spite of the slight hyperphagia which some investigators have reported (28).

What does appear to be significantly affected by septal lesions is food, or more precisely, taste preference. Beatty and Schwartzbaum (2,3) have reported that rats with septal lesions drink more of a saccharin or sucrose solution than normal rats. The observation has been extensively replicated (28,35), and we (117) have recently observed similar effects with respect to sweetened milk. This exaggerated preference for sweet tasting solutions appears to be independent of the hyperdipsia which is seen in some rats with septal lesions (3), but there is some indication that the effect may disappear with prolonged exposure (2, 71).

A similarly exaggerated preference for saline solutions has been reported in rats with septal lesions (43, 103), even in animals which were not hyperdipsic as a result of the lesion (19). These effects do not seem to be the result of a general dulling of taste sensations since rats with septal lesions drink significantly less of a quinine adulterated solution than normal controls (14,15,29).

Locomotor Activity

Rats with septal lesions are less active than controls in wheels (21,30), tilt boxes (135), and home cages (133) but hyperactive in situations wuch as mazes (105,133), which encourage exploratory behavior. In novel open fields, rats with septal lesions are typically more active than controls (26) whereas the opposite effect has been reported in familiar test situations (22, 119). Rats with septal lesions also appear to habituate faster than controls (30,42) and this may contribute to their behavior in open fields. Another factor (26) may be photophobia (122), although we (69) have found this to be a very minor influence in our animals.

FUNCTIONS OF THE SEPTUM

Throughout the history of scientific interest in the functions of the septum, valiant attempts have been made to account for the diverse behavioral consequences of septal lesions as manifestations of a disturbance in a single global function such as "emotionality," "behavioral inhibition," "incentive motivation," etc.

Although parsimonious, such explanations are not very likely, a priori, because the septum is not structurally homogeneous (36) and because septal lesions interrupt a number of major fiber systems, some of which do not synapse in the area (113). It is thus not surprising that it has become increasingly difficult to account for the rapidly growing list of behavioral dysfunctions without taking resource to theoretical constructs which are so general as to become heuristically meaningless.

We as well as others have therefore attempted to "dissect" the septal lesion syndrome into independent components which might be more readily susceptible to interpretation. There are several approaches to this problem, including comparative time course studies, attempts to pharmacologically manipulate selected components of the septum, analyses of the effects of small lesions within the septum, and selective transection of specific afferent or efferent connections of the area. I will briefly review some examples of the work which we and others have done in each of these areas.

Time Course Studies

The "rage" response to normal handling which is sometimes seen in rats with septal lesions disappears, sometimes within a few days (see above). A number of investigators (76) have demonstrated that the facilitated CAR acquisition which was originally believed to be another manifestation of increased "affect" or "emotionality" does not follow a similar time course, but appears to be a permanent effect of the lesion. A second and perhaps related component of the syndrome--increased shock-induced "fighting"--also has been shown to disappear within a week or two after surgery (1,96), and it is tempting to relate this to the temporary increase in shock sensitivity which has been described (78,79). Ahmad and Harvey (1) have argued that the time course of the two transient effects of septal lesions may be sufficiently different to rule out a causal relationship. We (97) have more recently observed a time course for overt reactions to foot shock (see Fig. 3) which indicated sharply increased reactivity to shock at a time (ten days after the lesion) when shock-induced fighting (96) had returned to normal (see Table 1). We (96) also found hat the effects of septal lesions on intra-species fighting in a food competition situation which does not involve painful shock, as well as interspecies attack behavior (muricide), follow a similar time course. When tested during the first 5-8 days after surgery, dominant male rats became submissive and refused to fight with another male (see Table 2) but killed all mice placed in their home cage (see Fig. 4). When tested 10-15 days after surgery, dominance and intraspecies fighting were normal and none of the rats killed mice. The time course for increased muricide is quite different from that which we (44) have seen after serotonin depleting lesions in the raphé nuclei of the brainstem. In this case, we find that the probability of killing increases during the first week after the lesion and remains elevated after 30 days. This suggests that muricide is increased only after serotonin has been

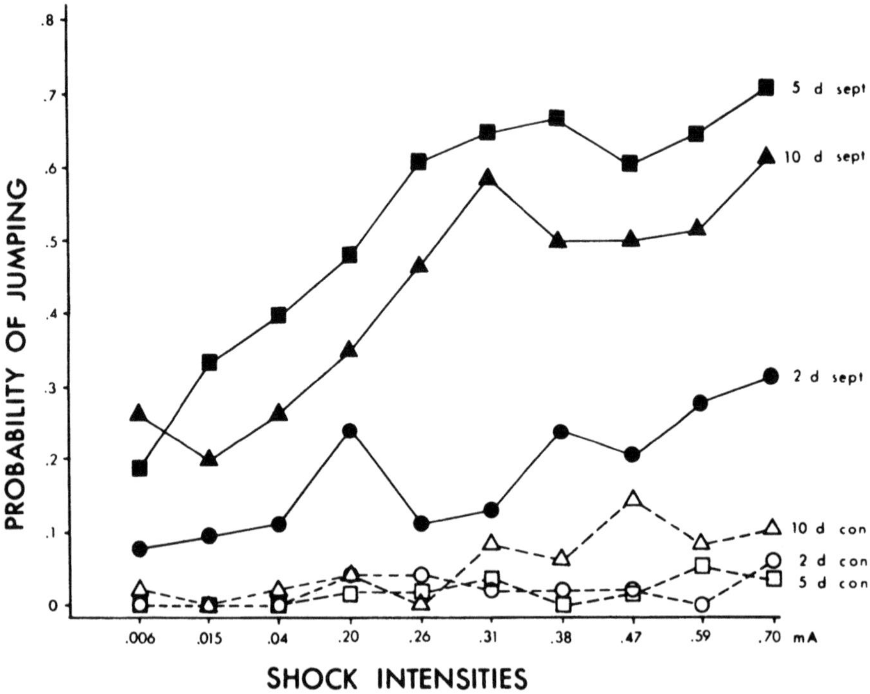

Figure 3. Mean probability of jumping in response to footshock of varying intensity in rats with septal lesions (solid symbols) and in controls (open symbols) tested 2, 5, or 10 days postoperatively. From Miczek, Kelsey, & Grossman (97). Reprinted by permission of the American Psychological Association, Washington, D. C.

TABLE 1: MEAN PROBABILITY OF SHOCK-INDUCED "FIGHTING" IN RATS WITH SEPTAL LESIONS, AVERAGED OVER 10 SHOCK INTENSITIES RANGING FROM 0.01 TO 0.40 mA

Experimental group	n	Days tested after operation	Fighting probability
Sham operated	8	5	.04
Septal lesion	8	5	.32*
Sham operated	8	10	.11
Septal lesion	8	10	.08
Septal lesion (retest)	8	10	.14

*p < .01

From Miczek and Grossman (96). Reprinted by permission of the American Psychological Association, Washington, D. C.

depleted from the brain and thus argues against a significant interaction with the effects of septal lesions on this behavior.

In an extension of this investigation, we (97) have observed the early postoperative time course of two other prominent components of the septal lesion syndrome. Somewhat to our surprise, we observed that the pronounced facilitation in CAR acquisition in a shuttle box which results from septal lesions was not seen two days after surgery but appeared full-blown five days after the lesion and persisted, unchanged, at ten days post-surgery (see Fig. 5). The impaired response to punishment (what has been called a "passive" avoidance deficit), on the other hand, was most pronounced on our test two days after the lesion and declined, although not to baseline levels, on tests given five or ten days after surgery (see Fig. 6). When we examined reactivity to footshock, a pattern emerged which might account for the time course of

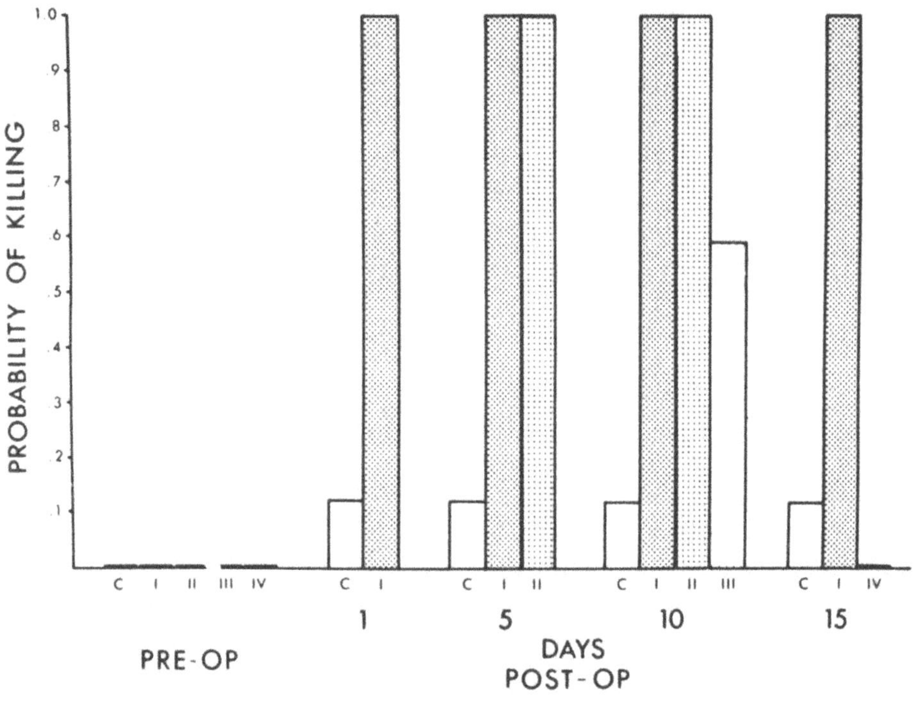

Figure 4. Mean probability of mouse killing by rats with septal lesions as a function of the lesion-test interval in Group 1 (tested at Days 1, 5, 10, and 15 postoperatively), 2 (tested at Days 5 and 10 postoperatively), 3 (tested at Day 10 postoperatively), 4 (tested at Day 15 postoperatively), and sham-operated control rats (Group C). From Miczek & Grossman (96). Reprinted by permission of the American Psychological Association, Washington, D. C.

TABLE 2: CHANGES IN DOMINANCE/SUBMISSIVENESS IN FOOD COMPETITION AFTER SEPTAL LESIONS WERE MADE IN THE DOMINANT RAT (GROUPS 1, 2, AND 3 DIFFERED ONLY WITH RESPECT TO THE SURGERY-TEST INTERVAL AS SHOWN)

		Postoperative		
Group	Preoperative	Day 5	Day 10	Day 15
1	Dominant	Submissive		Dominant
2	Dominant		Dominant	
3	Dominant			Dominant

From Miczek and Grossman (96). Reprinted by permission of the American Psychological Association, Washington, D. C.

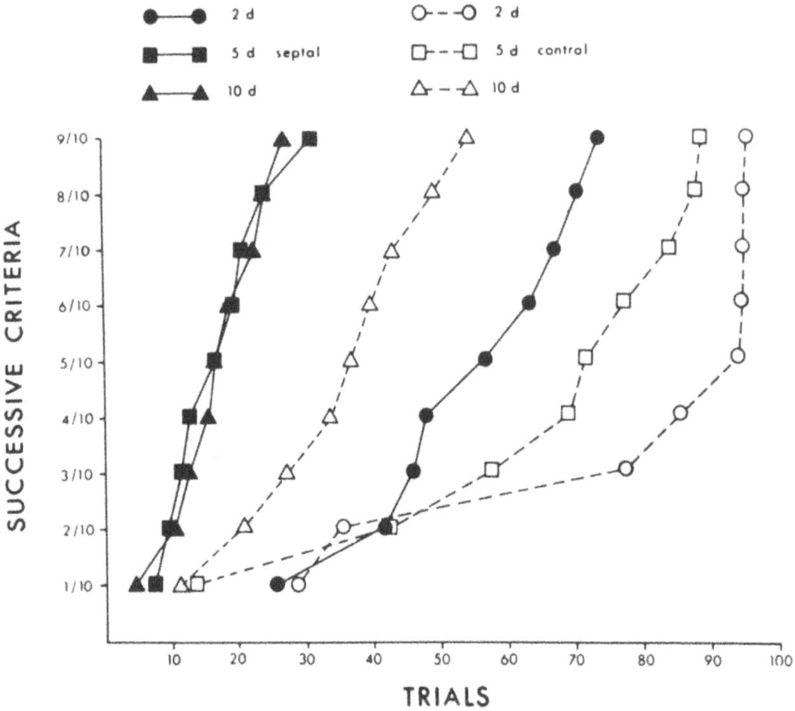

Figure 5. Mean number of trials to reach successive avoidance criteria (number of avoidances in blocks of 10 trials) during acquisition of a shuttle box CAR. (Solid symbols refer to the rats with septal lesions, tested 2, 5, or 10 days post-operatively; open symbols refer to the corresponding control groups.) From Miczek, Kelsey, & Grossman (97). Reprinted by permission of the American Psychological Association, Washington, D. C.

the CAR facilitation but not for the gradually decreasing response to punishment. When experimentally naive, unhandled animals were exposed to painful footshock after septal lesions, the probability of a gross overt reaction (a "jump") was clearly increased at five and ten days after surgery, but two days after there was only a slight increase (see Fig. 3). When we repeated this experiment with animals which had received handling comparable to that which the animals in the punishment experiment received, the response to footshock was comparably increased two and five days after the lesion, suggesting that animals which were not handled prior to surgery may have had a greater tendency to "freeze" in response to shock when they were tested very shortly after surgery.

Viewed together, the results of these experiments suggest that a number of components of the septal lesion syndrome which one might have viewed as manifestations of a common dysfunction appear to be independent. Few if any of our animals displayed overt signs of "rage," and it appears likely that none of the behaviors tested in our study is in any causal sense related to this aspect of the lesion syndrome. The CAR facilitation showed a time course which did not parallel the development of decreased reactivity to punishment, and neither of these effects showed any tendency to disappear, as all of our changes in intra- and inter-species aggressive behaviors did. It is unlikely that the lesions' temporary effects on shock sensitivity (78) accounts for the changes in CAR acquisition or reactivity to punishment since we (117) as well as others (76) have found marked effects in both situations long after the relatively brief period of postoperative hyperreactivity to footshock.

It may be interesting to contrast the transient changes in aggressive behavior after septal lesions with the overtly similar "rage" syndrome which is seen after ventromedial hypothalamic (VMH) lesions. We (50,124) as well as others (see MacLean (89) for review) consistently find that the exaggerated response to normal handling does not disappear nearly as soon or as readily as it does after septal damage (sometimes not at all), and I have found very marked and persistent changes in intra- and well as inter-species aggressive behaviors many weeks or months after ventromedial hypothalamic lesions (50) or after transection of some of the afferent or efferent connections of the VMH (48,51).

Attempts to Pharmacologically Manipulate Selected Aspects of the Septum

It is widely accepted that not all chemical synapses in the brain use the same neurohumoral transmitter, and there is considerable evidence that functionally related neurons within some geographically defined region may use a common transmitter, perhaps in order to assure concurrent activation and prevent "cross talk." This raises the possibility that functionally related neurons might be selectively activated or inhibited by microinjections of some compounds which do not affect the metabolism of other transmitters. Although the techniques available for the delivery of drugs to selected areas of the brain of unrestrained animals are still quite crude, this approach has been successful in a number of instances, including some investigations of septal functions (see Grossman (47,49,50) for more detailed review).

My first attempt to apply this approach to the septum (46) demonstrated that the application of carbachol, a powerful cholinomimetic stimulant, prevented the acquisition of shuttle box CAR

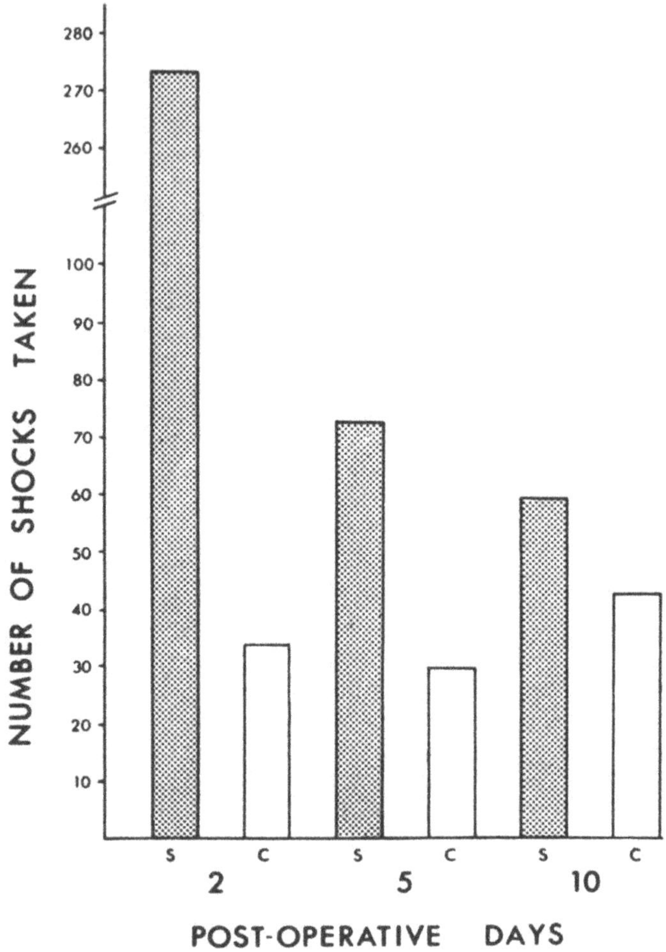

Figure 6. Mean number of punishing shocks taken by rats with septal lesions (S) and controls (C) as a function of the surgery-test interval. From Miczek, Kelsey, & Grossman (97). Reprinted by permission of the American Psychological Association, Washington, D. C.

(Fig. 7) and impaired the performance of a previously learned lever-press avoidance response (Table 3). Intraseptal injections of the anticholinergic compound atropine improved escape as well as avoidance behavior in the lever-press paradigm. This internally consistent pattern of effects fits well into the lesion literature and suggests that cholinergic components of the septum might be responsible for its influence on avoidance behavior.

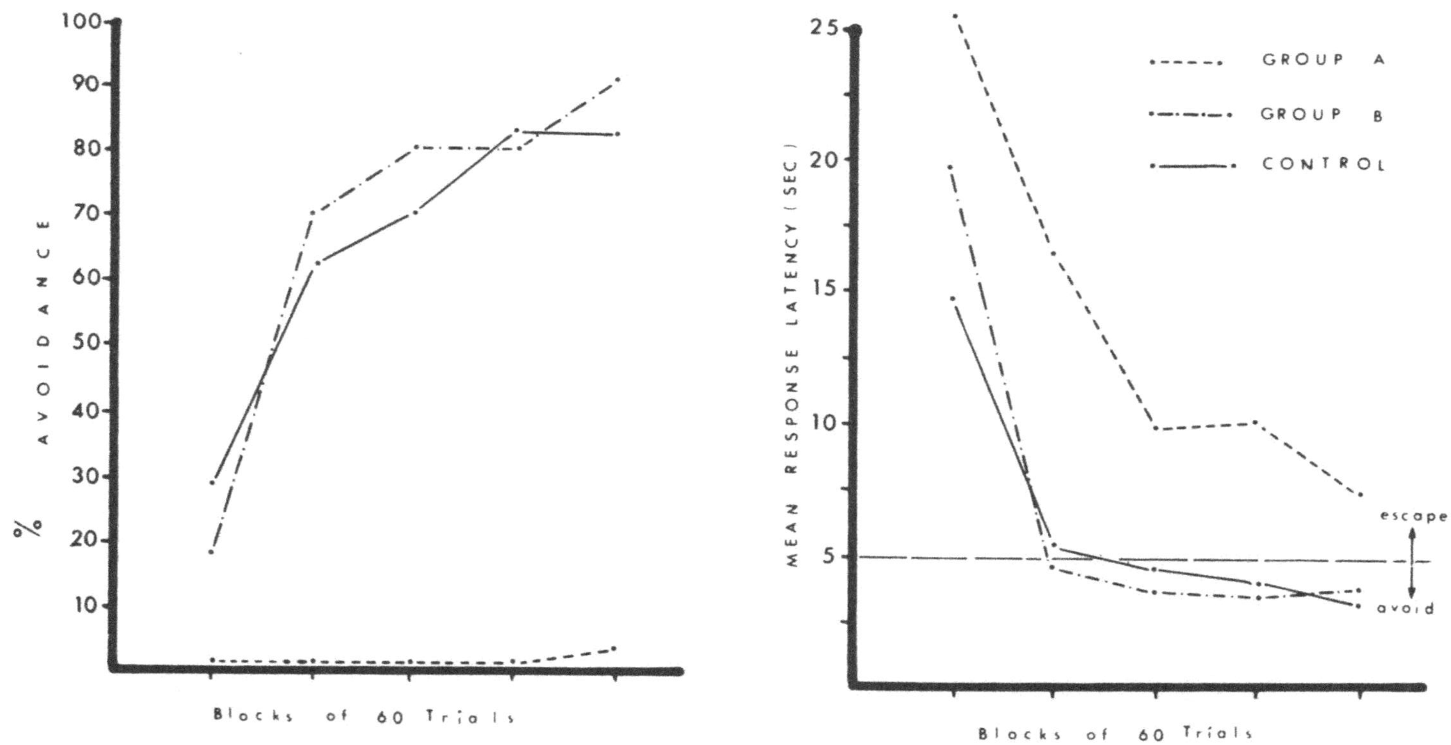

Figure 7. Acquisition of a shuttle box avoidance response by rats receiving cholinergic stimulation of the medial septal area (Group A); rats receiving cholinergic stimulation of an area immediately below the medial septal area (Group B); and an operated control group. From Grossman (46). Reprinted by permission of the American Psychological Association, Washington, D. C.

A number of subsequent investigations which were undertaken in my laboratory in collaboration with L. Hamilton and J. Kelsey have confirmed this interpretation and inciated that some other components of the septal lesion syndrome may also be related to cholinergic mechanisms. We found, for instance, that injections of the anticholinergic compound scopolamine into the septum of cats or rats almost precisely duplicated the effects of septal lesions in two quite different avoidance situations; the acquisition of a shuttle box CAR was facilitated (69) (see Fig. 8), but the learning of a shelf-jump (one-way) avoidance response was significantly impaired (54). We also found that both effects could be observed after systemic injections of scopolamine hydrobromide (which easily passes the blood-brain barrier) but not after injections of atropine methyl nitrate (which does not) (54), suggesting that the effects of systemically administered anticholinergics on avoidance behavior (77), may, at least in part, be due to a drug action on components of the septum. The specificity of the drug effec is emphasized by an observation by Hamilton, McCleary, and myself (56) that intraseptal atropine injections did not significantly modify the acquisition of an unusual one-way avoidance response which previous investigators had shown to be unaffected (85) or slightly facilitated (145) by septal lesions.

When avoidance responding was punished in a so-called "forced extinction" procedure, intraseptal injections of an anticholinergic compound mimicked the effects of septal lesions (56) but systemic injections of scopolamine did not (54). A similarly inconsistent picture emerged when we looked at other punishment effects. When feeding was punished in the classic "passive avoidance" paradigm popularized by McClearly (91), injections of atropine into the septal area of cats produced impairments similar to those seen after septal lesions (56). However, when we attempted to observe this effect in a situation where both active and "passive" avoidance responses were required, a clear drug effect on active avoidance responding was apparent, but the reaction to punishment seemed unaffected (54). In the rat, Kelsey (68) found that intraperitoneal injections of atropine did not increase punished licking. These inconsistencies suggest that the influence of septal mechanisms on reactions to punishment may itself have several components which may be differentially activated in different punishment paradigms.

We (56) also found that intraseptal atropine injections did not affect the reversal of a position habit (see Table 4), one of the classic tests for response "perseveration" in cats with septal lesions (146). This suggests that the lesion effects on "response inhibition" which many investigators (92) have considered to be the basic mechanism that accounts for most, if not all, of the behavioral consequences of septal lesions, may not be related to a disruption of cholinergic components of the septum.

TABLE 3: MEAN NUMBER OF ESCAPE AND AVOIDANCE RESPONSES OR RESPONSE FAILURES IN A SIGNALLED LEVER-PRESS AVOIDANCE SITUATION DURING A 30 MIN. TEST GIVEN 10 MIN. AFTER THE ADMINISTRATION OF VARIOUS COMPOUNDS TO THE SEPTAL AREA OF RATS. CONTROL DATA WERE OBTAINED DURING TESTS WHEN EMPTY CANNULAS WERE INSERTED

	Avoidance		Escape		Response Failure	
Drug	Control	Drug	Control	Drug	Control	Drug
Carbachol	20.0	15.7**	8.9	10.0	1.1	4.3*
Atropine	20.5	24.8**	8.4	4.2**	1.1	1.0
n-epinephrine	20.5	22.5*	7.4	5.8*	2.1	1.7
Dibenzyline	22.2	21.5	6.9	6.9	.9	1.6

*$p < .05$
**$p < .01$

After Grossman (46). Reprinted by permission of the American Psychological Association, Washington, D. C.

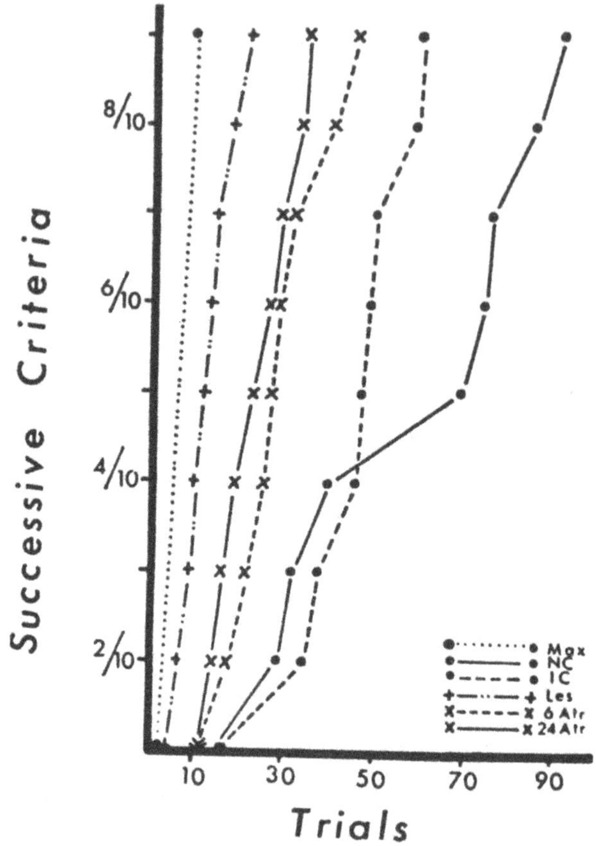

Figure 8. Acquisition of a shuttle box avoidance response by rats with septal lesions (Les) or cannula implants which permitted the application of low (6 Atr) or high (24 Atr) doses of atropine to the septum 7 min. before each training session. Both normal controls (NC) and rats with empty cannulas (IC) were used as controls. The line labeled Max identifies perfect performance. Modified after Kelsey & Grossman (69). Reprinted by permission of the American Psychological Association, Washington, D. C.

Kelsey and I (69) have found that intraseptal injections of atropine facilitate shuttle box CAR acquisition without producing the photophobia or hyperactivity which some investigators (25,122) believe play a possibly important role in the etiology of the effects of septal lesions on CAR acquisition. We also found that rats with septal lesions learned a reversal of the avoidance response faster than controls even though any tendency to perseverate in the previously correct behavior should have produced opposite effects on reversal learning. One might argue that in this case the effects of the lesion on avoidance behavior rather than any concurrent changes in inhibitory control determined behavior, but our drug tests do not support such an interpretation. Although intraseptal injections of atropine clearly mimicked the effects of the lesions on the original acquisition of the CAR, there were no reliable effects during the reversal phase of the experiment.

Kelsey and I (in preparation) have also looked at the effects of intraseptal and systemic injections of anticholinergics on behavior in the modified Sidman and DRI, paradigms where we had previously observed clear evidence of overresponding (70). We found that systemic injections of scopolamine reproduced the effects of septal lesions whereas eserine produced opposite effects on responding in both experimental paradigms. Destruction of the septum (which itself produced an increase in non-reinforced responding in both situations, as reported earlier) abolished the drug effects on responding in the DRL paradigm but did not prevent further significant increases in responding after systemic injections of scopolamine in the signalled Sidman avoidance situation. These observations are of significance because they indicate that the overresponding seen after septal lesions in the two situations is probably due to a disruption of quite different functions. We also found that intraseptal injections of cholinomimetic and anticholinergic compounds consistently and reliably modified behavior in the modified Sidman avoidance paradigm but not in the DRL situation. This suggests that the mechanà isms which control behavior in the Sidman avoidance situation may rely on a cholinergic synapse in the septum, whereas the mechanisms which determine responding in the DRL paradigm do not, although they appear to have a cholinergic link somewhere in the brain.

Before we leave our discussion of the cholinergic components of the septum, I would like to mention briefly their possible influence on water intake. I (46) as well as other (34) have found that intraseptal injections of cholinomimetics, such as carbachol, elicit drinking. Similar drinking responses can be obtained from many hypothalamic (45) and other extrahypothalamic sites (34), and one might dismiss the matter in the present context were it not for the fact that some septal lesions produce marked although sometimes transient hyperdipsia (8,58). The direction of the drug effect appears paradoxical (stimulation ought to produce effects opposite to those of a lesion). It is possible that this may be related to non-sepecific, disruptive effects exerted by the powerful cholinomimetic compound carbachol that is used in most of these studies, but intraseptal injections of atropine reduce water intake--a finding which is clearly consonant with the elicitation of drinking by carbachol but not with the lesion data. Matters are further complicated by Sorenson and Harvey's (130) (1971) observation that only those septal lesions which result in a significant decline in brain acetylcholine produce hyperdipsia. Clearly, cholinergic mechanisms seem to be involved in the phenomenon, but the nature of their influence remains obscure.

Partial Lesions

Rage. Harrison and Lyon (57) concluded from their investigation that the septal "rage" effect could be obtained from all portions of the septum and defied attempts at localization. Stark and Henderson (131) have more recently reported that lesions in the ventral quadrant of the septum were most effective in eliciting hyperreactivity to normal handling. Turner (136) has reported similar observations, and suggested that damage to the stria terminalis might be responsible, at least in part, for the septal rage.

We (98) have recently investigated the effects of lesions in the bed nucleus of the stria terminalis on intra-male aggressive behaviors and found that dominance and attack reactions in food competition situations are sharply reduced or abolished by such lesions (a pattern of effects which is similar to that seen in our laboratory after septal lesions), but shock-induced "fighting," which is sharply increased by septal lesions, was not affected by extensive damage to the bed nucleus of the stria terminalis.

Avoidance behavior. We (55) have consistently found that lesions anywhere in the septum result in facilitated CAR acquisition in a shuttle box (Fig. 12), and a number of other investigators (25,139) have come to similar conclusions. Hamilton (53) has reported the unusual observation that unilateral lesions in the septum of rats produced more pronounced effects on shuttle box CAR acquisition than larger bilateral lesions, but the implications of this observation are not yet clear. Facilitated shuttle box avoidance acquisition has also been seen after lesions in the nucleus accumbens septi (83,84) which is almost invariably involved in ventral septal lesions, and it is possible that an interruption of fibers of passage from either the medial forebrain bundle or the diagonal band of Broca may be responsible for this effect.

Overresponding in operant paradigms requiring inhibitory control. Carey (16) has reported that only lesions restricted to the anteroventral septum, including the dorsal fornix and diagonal band of Broca, produce a marked increase in DRL responding.

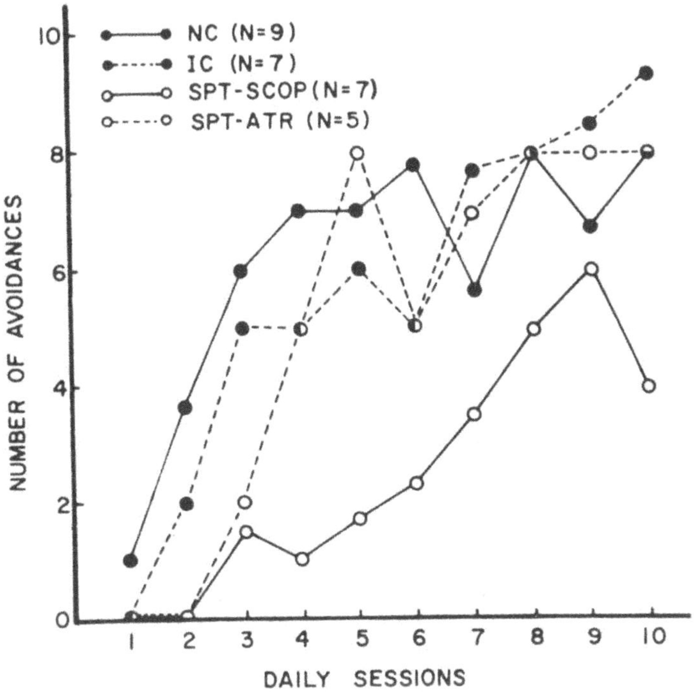

Figure 9. Effects of microinjections of cholinergic blocking agents into the septum of cats on the acquisition of shelf-jump (one-way) active avoidance behavior (NC = normal control; IC = implanted control; SPT-SCOP = scopolamine injections; SPT-ATR = atropine injections). From Hamilton & Grossman (54). Reprinted by permission of the American Psychological Association, Washington, D. C.

Lesions in the ventral and precommissural fornix produced more efficient DRL responding (i.e., fewer responses per reinforcement) and also decreased FR responding, suggesting that a general decrease in motivation-related functions might be responsible for the latter effects. Carey's conclusion that the "disinhibitory" effects of septal lesions might be due to an interruption of an afferent path which may originate in the midbrain and ascend via the medial septum and dorsal fornix to the hippocampus receives some support from the observation of MacDougall, Van Hoesen, & Mitchell (88) that large fornix lesions which included the fimbria produced marked overresponding in a DRL situation whereas lesions confined to the medial fornix did not.

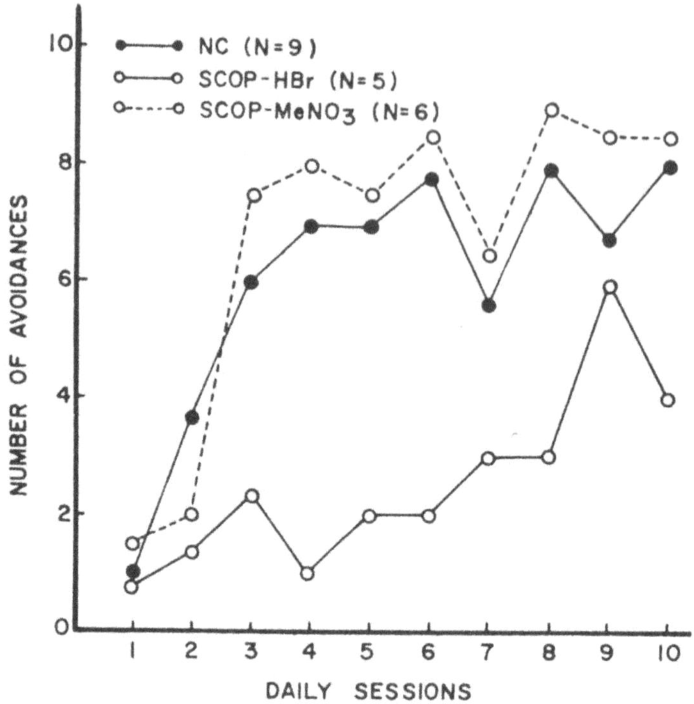

Figure 10. Effects of intraperitoneal injections of 0.05 mg/kg of scopolamine hydrobromide (SCOP-HBr) or 0.05 mg/kg of scopolamine methylnitrate (a quaternary form of scopolamine which does not readily cross the blood-brain barrier) on the acquisition of a shelf jump (one-way) avoidance response by cats. From Hamilton & Grossman (54). Reprinted by permission of the American Psychological Association, Washington, D. C.

Donovick (25) has reported that only lesions in the anterior septum which interrupt the fornix produced a deficit in position habit reversal. Fischman and McCleary (cited by McCleary (92) reported comparable results in the cat and noted that anterior lesions did not modify the animals' reactions to punishment. The latter observations have been replicated in the rat by Van Hoesen, MacDougall, and Mitchell (193). Further confirmation that the fornix may be specifically involved in the position habit reversal but not in the "passive avoidance" deficit has been obtained in my laboratory (55). We found that antero-medial septal lesions produced the most pronounced deficit in position habit reversal but had little effect on reactions to punishment. Indeed, dorsal septal lesions which did produce a sharply reduced response to punishment no longer did so when combined with lesions to the antero-medial fornix system. Lesions of the lateral nuclei impaired the animals' reactions to punishment and had little or no effect on position habit reversal (see Fig. 13 and Table 5).

Water intake and consumption of palatable and unpalatable fluids. Lubar and associates (86,87) have reported that lesions in the posterior septum produce the most marked hyperdipsia. Besch and Van Dyne (5) further localized the effect to the posteroventral septum and suggested that incidental damage to the periventricular nucleus of the hypothalamus (which contributes to the regulation of ADH release by the pituitary) may be a significant factor. The picture is, unfortunately, not as clear as these reports suggest. Carey (17) reported that lesions in the posterior septum reduced water-rewarded lever pressing on an FR schedule whereas lesions in the anterior septum had facilitatory effects. Rats with posterior septal damage also consumed significantly less quinine-adulterated water than controls or rats with anterior lesions. When water intake itself was compared, rats with anterior as well as posterior lesions drank more than controls and did not differ reliably from

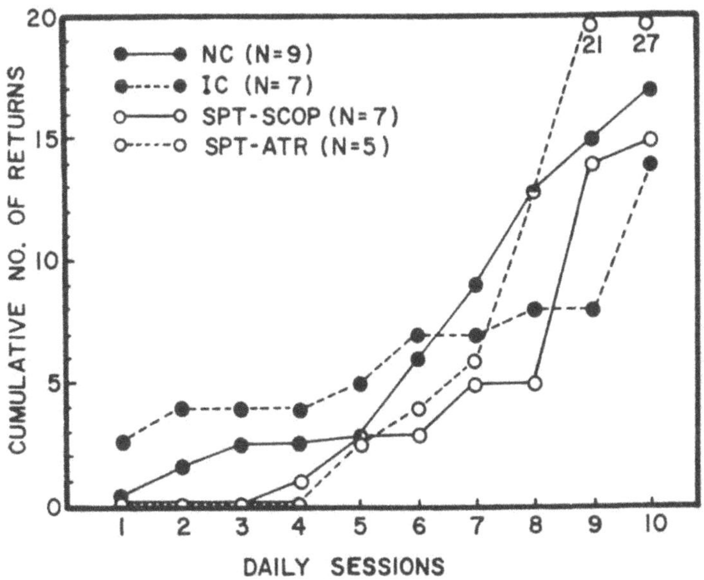

Figure 11. Effects of microinjections of atropine (SPT-ATR) or scopolamine (SPT-SCOP) on a punished feeding response (i.e., the number of returns to the food) observed in a paradigm where painful shock was signalled by a CS and avoidance responses were possible. (IC = implanted controls; NC = normal controls). After Hamilton & Grossman (54). Reprinted by permission of the American Psychological Association, Washington, D. C.

TABLE 4: MEAN NUMBER OF DAILY 20-TRIAL SESSIONS REQUIRED TO REACH A CRITERION OF 18 CORRECT RESPONSES ON 20 TRIALS IN A POSITION-HABIT PARADIGM USING CATS WITH CANNULA IMPLANTS IN THE SEPTAL AREA WHICH PERMITTED THE APPLICATION OF ATROPINE 7 MIN. BEFORE THE BEGINNING OF SOME "REVERSAL" TRAINING SESSIONS. THE GROUPS DIFFERED WITH RESPECT TO THE SCHEDULING OF ATROPINE TREATMENTS AS INDICATED

Group	Reversals			
	1	2	3	4
1	2.80	3.00	2.80	2.80
2	3.50	2.50[a]	3.75	2.50[a]
3	3.00[a]	2.50[a]	2.75[a]	2.75[a]

[a]Reversal training following atropine injections into the septum. After Hamilton, McCleary, and Grossman (56). Reprinted by permission of the American Psychological Association, Washington, D. C.

from each other. Chiaraviglio (19) has reported that postero-ventral lesions did not produce hyperdipsia but did increase the ingestion of a 1% sodium chloride solution. Sorenson and Harvey (130) have reported that they could not differentiate septal lesions which produced hyperdipsia from those which did not.

Comparisons of the Effects of Lesions in Structurally Related Regions of the Brain

Septal lesions typically involve components of several major fiber systems which originate, terminate, or pass through the region (113). The detailed anatomy of the septum and its connections has been reviewed elsewhere, and I will mention here only the major fiber systems which have received some experimental attention to date. The septum receives afferent inputs from the hippocampus via the fornix and fimbria, from the amygdala via the stria terminalis, and from the olfactory tubercle and lower brainstem via the medial forebrain bundle (MFB). All of these pathways also contain efferent projections, and there is an additional

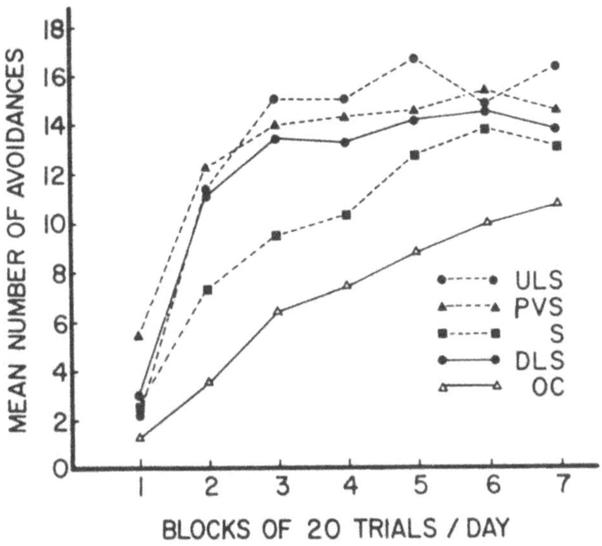

Figure 12. Mean number of avoidance responses made by rats with septal lesions during successive days of training in a shuttle box avoidance paradigm. Rats in groups ULS and DLS sustained lesions which were restricted to the lateral nuclei, rats in group PVS had lesions in the medial nuclei and columns of the fornix. Rats in group S sustained large lesions which included all aspects of the septum. After Hamilton, Kelsey, & Grossman (55).

significant septo-habenula projection via the stria medullaris. The septum is also traversed by numerous fibers which do not synapse in the area but interconnect the brainstem with the hippocampus, amygdala, and temporal lobe.

One approach to the important question of how these pathways might be related to the apparently diverse functions of the septum is to compare the effects of septal lesions with those produced by damage to the sites of origin and/or termination of some of these fiber systems. The literature on this issue is much too voluminous and complex to be reviewed in detail at this point, but a few directly relevant examples may place the present status of the problem into some perspective.

Hippocampal lesions produce effects on CAR acquisition and reactivity to punishment which are remarkably similar to those seen after septal lesions (63,74,106). This suggests that an interruption of septo-hippocampal connections may be responsible for these effects, but the specifically relevant literature does not fully support such an interpretation. Facilitated CAR acquisition has been observed after fornix lesions, but reactivity to punishment or the acquisition of one-way avoidance responses were not affected in these experiments (81,139).

It is possible that the septal influences on these behaviors may be carried by the stria medullaris since impaired reactions to punishment (139) have been seen after stria-habenula lesions. However, there is some indication that the reaction to punishment may be contingent on extensive unpunished experience with the positive reinforcer (99) and can be attenuated by increases in the appetitive drive state (60). It has consequently been suggested that motivational or "incentive-related" influences which are presumed to be mediated by MFB connections with the hypothalamus may be responsible for the effects of septal lesions on punished responding (58). This does not exhaust the possibilities since lesions in the stria terminalis of the cat (137) or in its origin in the amygdala of the rat (108) (Grossman, Grossman, & Walsh, in press) also produce impaired reactions to punishment.

Hippocampal lesions also produce "perseverative" effects on responding maintained on FI or DRL schedules of reinforcement and during extinction (4,20,73,109). There is some direct evidence that a disruption of septo-hippocampal connections may be responsible for the impairments in these behaviors after septal lesions. For instance, we (55) have found that lesions in the medial portion of the septum (which projects to the hippocampus via the fornix) but not lesions in the lateral septum (which has few if any projections to the hippocampus) produced disinhibitory effects

TABLE 5: NUMBER OF PUNISHED WATER-LICKING RESPONSES DURING A 17 MIN. TEST IMMEDIATELY FOLLOWING 3 MIN. OF UNPUNISHED ACCESS TO A DRINKING TUBE (TEST 1) AND LATENCY OF THE FIRST LICK DURING A TEST GIVEN 24 HOURS LATER IN RATS WITH SEPTAL LESIONS. RATS IN GROUPS DLS AND ULS SUSTAINED LESIONS RESTRICTED TO THE LATERAL NUCLEI; THOSE IN GROUP S HAD LARGE LESIONS WHICH INVOLVED MEDIAL AS WELL AS LATERAL ASPECTS OF THE SEPTUM.

Response Measures	Control:	Lateral Damage		Medial and Lateral Damage
	Group OC	Group ULS	Group DLS	Group S
First test day				
Average number of Shocks Received	15.6	20.4	44.8	16.9
Range	6-26	8-43	10-85	13-21
Second test day				
Average latency (sec.) to initiate drinking	6.8	2.3	1.3	2.8
Range	0.5-24.4	0.8-6.1	1.0-1.6	0.9-5.2

After Hamilton, Kelsey, and Grossman (55). Reprinted by permission of the American Psychological Association, Washington, D.C.

on position habit reversal (see Fig. 13). MacDougall, Van Hoesen, & Mitchell (88) as well as Carey (16) have reported that lesions in the anterior septum-fornix also resulted in DRL deficits.

Hippocampal lesions do not produce most of the changes in appetitive behavior which are seen after septal lesions. There is no evidence of hyperdipsia, increased operant responding for food or water rewards on CRF schedules, and no increase in the intake of palatable solutions (2,4,20). In conjunction with the many observations which have suggested that the MFB and its targets in the hypothalamus may be related to motivational (102) and/or reward-related processes (132), these results have suggested to some investigators that those effects of septal lesions which appear to be related to a change in motivational or incentive-related functions might be due to an interruption of components of the MFB.

Selective Transection of the Connections of the Septum

A few years ago, Ross and I decided to perform a systematic analysis of the functions of the principal connections of the septum using an encephalotome which Sclafani and I (123) had developed earlier to transect subcortical fibers of passage. The instrument consists of a retractable wire knife that is only 125 microns in diameter and thus permits the transection of fibers of passage with minimal direct damage to the cellular components of the region of the cut or to brain tissue along the approach of the knife guide. This work has resulted in several papers which are currently in press and has culminated in a fine Ph.D. dissertation for Dr. Ross (115). I would like to take this opportunity to summarize some of our principal findings at this time as a fitting end to my brief review of our research on the behavioral functions of the septum.

We (117) began our investigation by comparing the effects of transections of the fornix, just dorsal to the septum, and of the medial forebrain bundle, just ventral to the septum, with those of septal lesions on CAR acquisition in a shuttle box and in a one-way avoidance paradigm. After some experimentation, we were able to consistently produce cuts which transected the fornix in its entirety but produced little or no damage to the stria medullaris which runs parallel to it in the region of the cut (see Fig. 14A). This cut resulted in wide-spread degeneration in all areas of the brain which are known to be innervated by the fornix, including the septum, diagonal band, lateral preoptic area, lateral hypothalamus, mammillary bodies, anterior thalamic nuclei, zone incerta, and hippocampus. We also succeeded in transecting most of the ventral connections of the septum by a

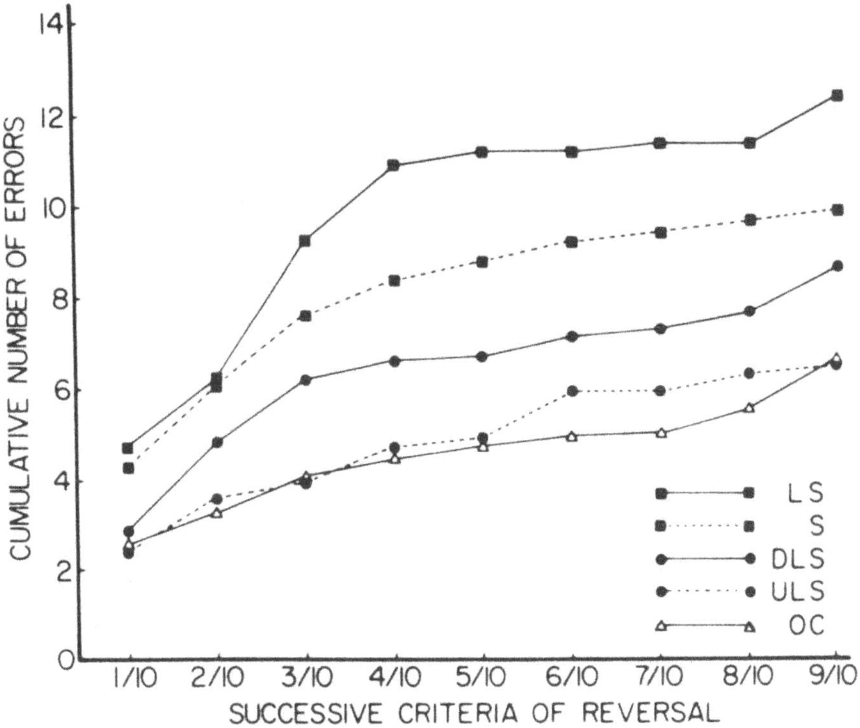

Figure 13. Cumulative number of errors committed by rats with septal lesions in reaching successive criteria of performance during the reversal of a T-maze position habit. Rats of groups ULS and DLS had damage restricted to the lateral nuclei, rats in groups S and LS had larger lesions which also involved the medial nucleus and columns of the fornix. After Hamilton, Kelsey, & Grossman (55). Reprinted by permission of the American Psychological Association, Washington, D. C.

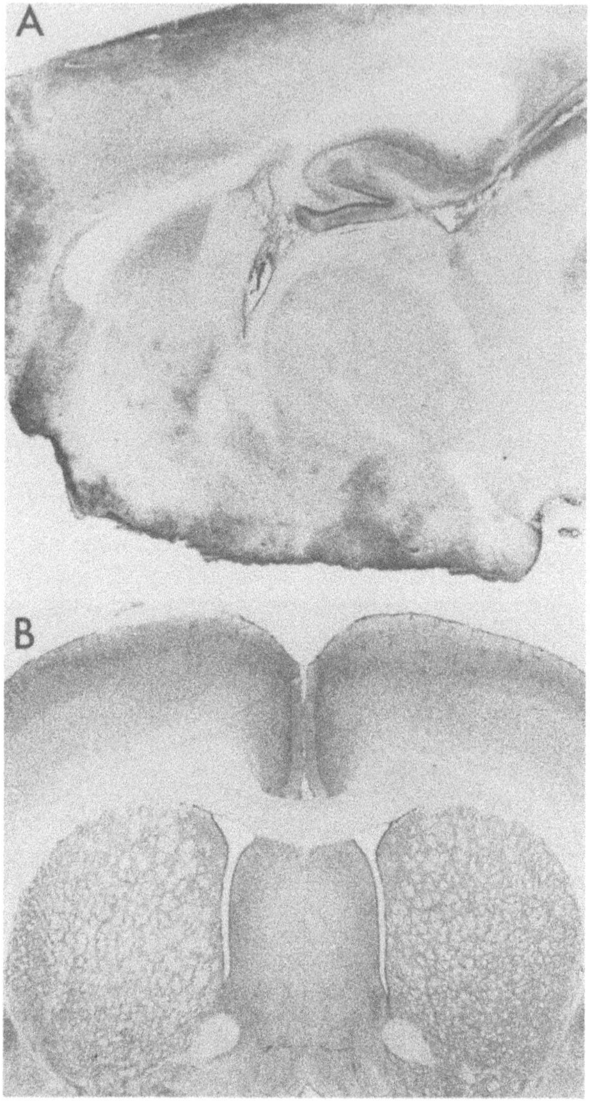

Figure 14. Photomicrographs of parasagittal sections through rats brains (stained with Cresyl violet) showing (A) the region of a fornix transection and (B) the region of a transection of the ventral connections of the septum as produced by a retractable wire encephalotome. After Ross, Grossman, & Grossman (117). Preprinted by permission of the American Psychological Association, Washington, D. C.

rotational cut between the rostral limbs of the anterior commissure (see Fig. 14B). This cut resulted in heavy degeneration in the diagonal band, lateral hypothalamus, medial septum, and parataenial nuclei of the thalamus, and relatively slight degeneration in the fimbria and hippocampus. Both of our cuts produced only very slight degeneration in the stria medullaris or stria terminalis.

In the shuttle box apparatus, all three experimental groups acquired the CAR reliably faster than the controls and there were no reliable differences between the effects of septal lesions, dorsal fornix cuts, or ventral MFB transections (Fig. 15). These results suggest that the effect may be related to an interruption of reticulo-septo-hippocampal interconnections which may, according to our pharmacological observations, have a cholinergic synapse in the septum. We observed hyperactivity in rats with fornicotomies or septal lesions but not in animals with ventral cuts, suggesting that this effect may not be related to the CAR deficit and that it may be due to an interruption of septo-hippocampal connections.

Somewhat to our surprise, we found that neither fornicotomy nor the transection of the ventral connections of the septum replicated the marked inhibitory effects of septal lesions on CAR acquisition in our one-way situation (Fig. 16). These observations are of special interest since they indicate that the facilitatory and inhibitory effects of septal lesions on CAR acquisition in different paradigms may be due to a disruption of entirely different functions.

Ross and I (116) have followed up these observations by investigating the effects of our dorsal fornix cut and the ventral MFB transection on CAR acquisition in a discriminated Sidman avoidance paradigm where the rat can avoid painful shock either by responding to a CS as in the traditional shuttle box istuation or by emitting the avoidance response spontaneously at least once every 28 sec. Kelsey and I (70) had previously shown that rats with septal lesions learn such a CAR faster than normals but perform it inefficiently by emitting many more spontaneous responses than necessary to consistently avoid punishment. Ross and I replicated this finding in the present investigation and demonstrated that, as in the classic shuttle box paradigm, rats with dorsal or ventral cuts also learned the discriminated Sidman CAR faster than controls. There were no significant differences in that respect between any of our three experimental groups. However, a closer look at the animals' response patterns revealed some interesting differences (Fig. 17). Rats with fornix cuts as well as rats with septal lesions avoided mainly by shuttling spontaneously from one compartment to the other, whereas the controls' as well as rats with ventral cuts made few such responses but learned, instead, to wait for the CS and

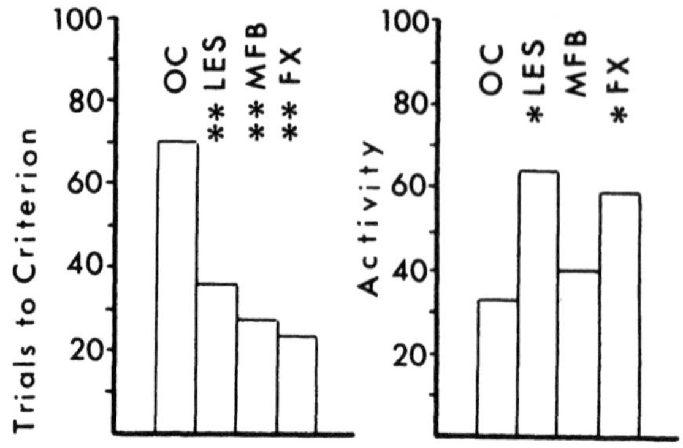

Figure 15. Number of trials required to reach a criterion of 10 avoidance responses on 10 trials in a shuttle box (left panel) and number of spontaneous shuttle responses during a 20 min. test prior to the introduction of the CS-UCS contingencies (right panel) of rats with septal lesions (LES); cuts of the ventral connections of the septum (MFB); or cuts of the dorsal connections of the septum (FX), compared to operated controls (OC). * indicates a difference from controls, p < .05; ** indicates a difference from controls, p < .01. Data from Ross, Grossman, & Grossman (117). Reprinted by permission of the American Psychological Association, Washington, D. C.

respond to it. These results may be of theoretical significance since they indicate that the enhanced CAR acquisition which is seen in shuttle box paradigms after septal lesions may not be secondary to a general inability to inhibit responding, as McCleary (92) and others have proposed. Our observations further suggest that the disinhibitory effects of septal lesions may be specifically related to a disruption of septo-himmocampal connections.

Before we pursue this point further, I would like to relate some further data on CAR acquisition which Ross (115) has collected in my laboratory. Since we had not been able to affect one-way CAR acquisition with knife cuts across the fornix or across the MFB, we decided to examine the effects of transections limited to the stria medullaris or stria terminalis. In the case of the former, data were used from six animals which sustained a complete transection of the stria medullaris bilaterally just beneath the lateral ventricle without incurring incidental damage to the fornix (see Fig. 18A). In the case of the stria terminalis,

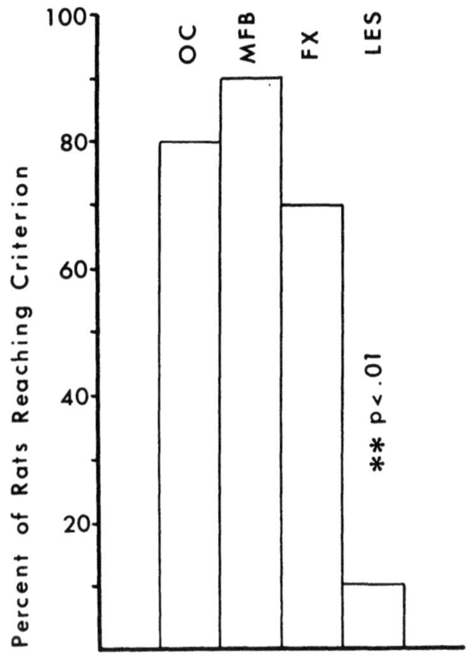

Figure 16. Percentage of rats with septal lesions (LES); cuts of the ventral connections of the septum (MFB); cuts of the dorsal connections of the septum (FX); or sham operations (OC) reaching a criterion of 10 avoidance responses on 10 trials in an automated one-way avoidance paradigm. (** indicates a difference from controls, $p < .01$). Data from Ross, Grossman, & Grossman (117). Reprinted by permission of the American Psychological Association, Washington, D. C.

data were used from six animals which sustained a clear transection of the stria terminalis just caudal to its termination in the bed nucleus of the stria terminalis (see Fig. 18B).

When the behavior of these animals was examined in our one- and two-way avoidance apparatus, it quickly became apparent that transection of the stria terminalis did not affect the acquisition of avoidance behavior in either of these paradigms, a finding which reinforces our belief that the effects of septal lesions on aggressive and avoidance behaviors are due to a disruption of entirely different functions. Transection of the stria medullaris also failed to affect shuttle box CAR acquisition (which we had previously found to be affected by both fornix and MFB transections) but reproduced the marked inhibitory effects of septal lesions in our one-way paradigm (Fig. 19). That stria medullaris

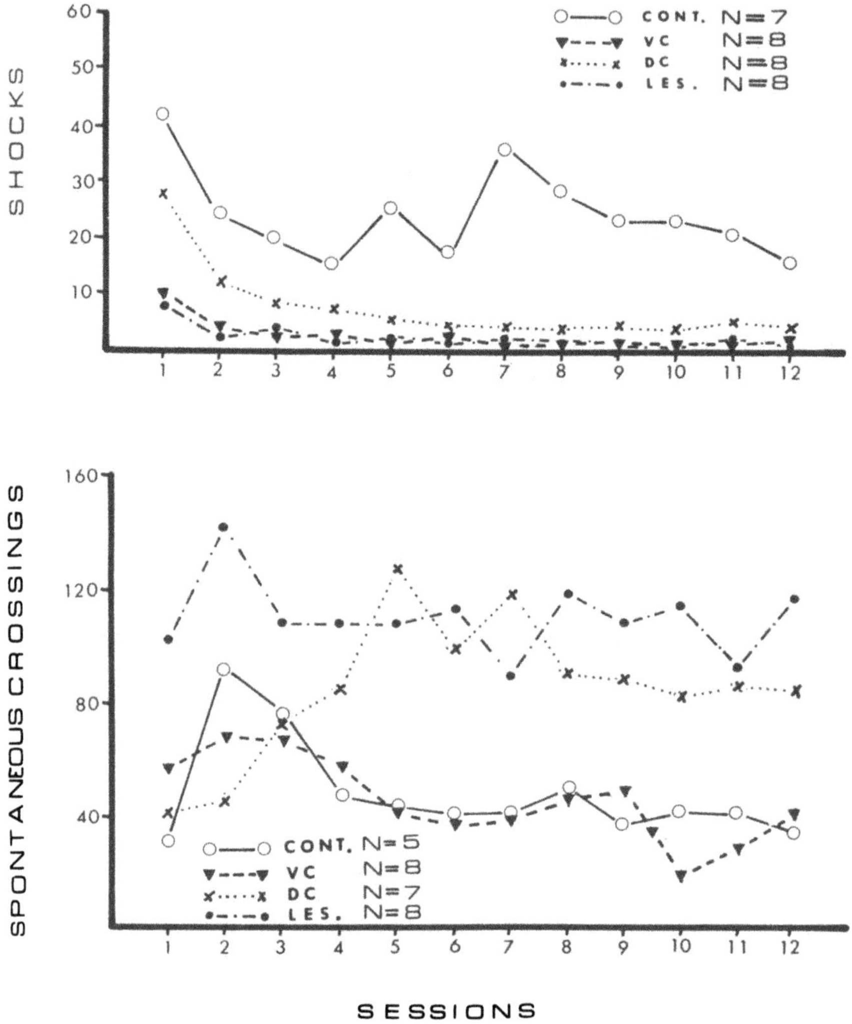

Figure 17. Number of shocks received (upper panel) and number of spontaneous avoidance responses (lower panel) made during the acquisition of a discriminated Sidman avoidance response in a shuttle box by rats with lesions in the septum (LES); cuts of the dorsal connections of the septum (DC); cuts of the ventral connections of the septum (VC) or sham operations (OC). From Ross & Grossman (116). Reprinted by permission of the American Psychological Association, Washington, D. C.

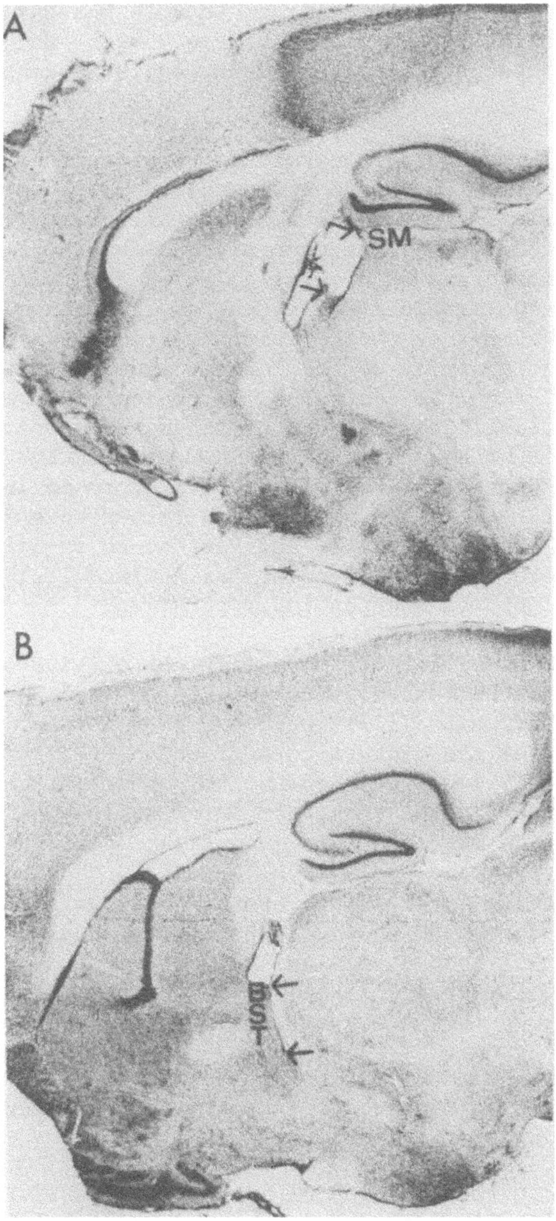

Figure 18. Photomicrographs of parasagittal sections through rat brains (stained with Cresyl violet) showing (A) the region of a typical stria medullaris transection and (B) the region of a typical stria terminalis transection as produced by a retractable wire encephalotome. After Ross (115).

transections did not also facilitate shuttle box CAR acquisition appears to conflict with a report by Van Hoesen, MacDougall, and Mitchell (138) that stria medullaris-habenula lesions produced the effect.

Ross (115) consequently examined the effects of stria medullaris lesions in the modified Sidman avoidance paradigm which has consistently shown facilitatory effects of septal lesions (70), fornix transections, and cuts of the ventral connections of the septum (117). Here, too, rats with verified transections of the stria medullaris were not reliably different on any of our measures (including the number of shocks tolerated, number of anticipatory responses, activity prior to the introduction of the shock contingency, etc.) from the controls. In conjunction with our similar data from the classic shuttle box paradigm, these data suggest that the results of habenula lesions as reported by Van Hoesen, MacDougall and Mitchell (193) may be related to an interruption of habenulo-interpeduncular connections rather than septo-habenular projections. Wilson, Mitchell, and Van Hoesen's (141) report of facilitatory effects of interpeduncular nucleus lesions on shuttle box avoidance supports such an interpretation.

Next, Ross and I tried to localize the origins of the effects of septal lesions on reactions to punishment (see Fig. 20). In our initial experiments (117), we found that neither fornicotomy nor transection of the ventral connections of the septum reproduced the markedly impaired response to punishment which we as well as many others have seen after large septal lesions. Ross's (115) dissertation extended this investigation and demonstrated that transection of the stria medullaris also does not affect punished responding in our experimental paradigm. Once again, this is not in harmony with the report by Van Hoesen, MacDougall, and Mitchell (193) that habenular lesions produced passive avoidance deficits. It appears likely that the effects of habenular damage on punished responding as well as on shuttle box CAR acquisition may be due to a disruption of habenulo-interpeduncular connections since Wilson, Mitchell, and Van Hoesen (141) also observed significant effects on reactivity to punishment after damage to the interpeduncular nucleus.

The marked effects of septal lesions on "passive avoidance" behavior were finally duplicated by transections of the stria terminalis (115)--an effect which supports Ursin's (137) observation that lesions which interrupt this pathway in the cat produced passive but not active avoidance deficits. This very clear dissociation of the origins of the septal lesion effects on one- and two-way active avoidance behavior and on the impaired response

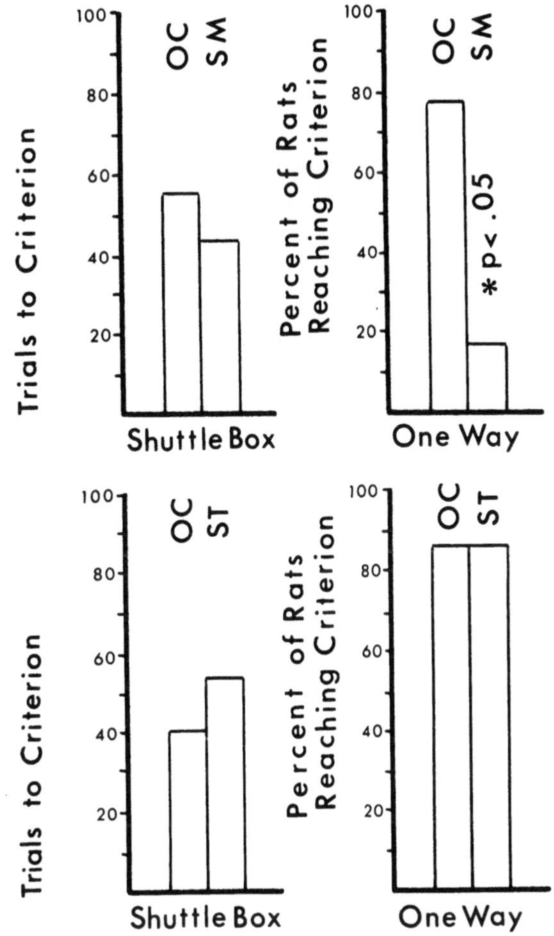

Figure 19. Number of trials required to reach a criterion of 10 avoidance responses in 10 trials in a shuttle box (left panel) and percentage of rats which reached a criterion of 10 avoidance responses in 10 trials in an automated one-way avoidance paradigm. OC = operated controls; ST = rats with transections of the stria terminalis; SM = rats with transections of the stria medullaris. Data from Ross (115).

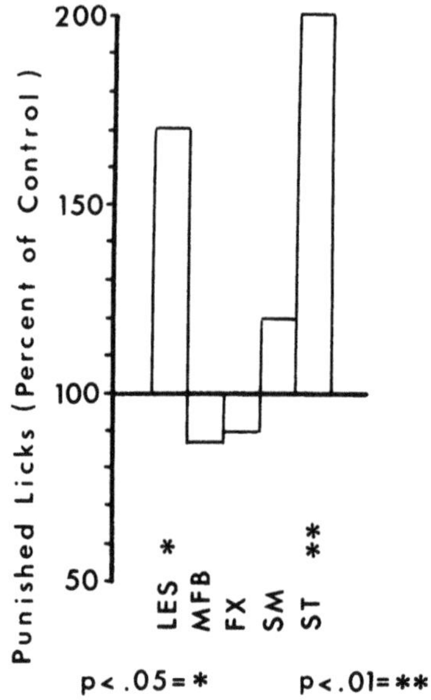

Figure 20. Number of punished licks during a 13 min. test immediately following 2 minutes of unpunished access to a drinking device containing sweetened milk. OC = operated controls; LES = septal lesions; MFB = cuts across the ventral connections of the septum; FX = cuts across the dorsal connections of the septum; SM = transections of the stria medullaris; ST = transections of the stria terminalis. The data are expressed as percentages of baselines established by operated control groups (one for the experiment involving LES, MFB, and FX animals; a second for the experiment involving SM animals; and a third for the experiment involving ST animals) because these experiments were not conducted at the same time and there were significant differences between the control baselines. The data from the LES, MFB, and FX animals are from Ross, Grossman, & Grossman, in press. Reprinted by permission of the American Psychological Association, Washington, D. C. The data from SM and ST animals are from Ross (115).

to punishment has important theoretical implications because it argues against a variety of interpretations which have sought to account for the effects of septal lesions on all kinds of avoidance behaviors in terms of a single functional disturbance, such as McCleary's (92) notion that the facilitated CAR acquisition in a shuttle box might be the result of a reduced reactivity to punishment (or the threat thereof).

The third major problem area to which we addressed ourselves was the apparently pervasive disinhibitory effects of septal lesions on operant responding in a variety of paradigms where high rates of responding are plainly inefficient or even punished. In the first series of experiments (116), we found that transection of the fornix replicated, at least to some extent, the disinhibitory effects of septal lesions in our modified DRL situation, whereas the ventral cuts across the MFB did not (see Fig. 21). Both rats with septal lesions and rats with fornicotomies received significantly fewer reinforcements than controls but showed no evidence of perseverative responding, an observation which replicates our earlier observations of the effects of septal lesions on DRL responding (70). That our ventral cuts did not reliably increase responding in the DRL paradigm argues against the suggestion (4) that the "disinhibitory" effects of septal lesions might, at least in part, reflect an increase in motivation- or incentive-related processes thought to be mediated by septo-diencephalic connections.

Since MacDougall, Van Hoesen, and Mitchell (88) have reported slight and transient disinhibitory effects of stria medullaris/habenula lesions in a DRL paradigm, Ross (115) extended our investigation by examining the effects of stria medullaris cuts on responding in this paradigm. The results confirm MacDougall, Van Hoesen, and Mitchell's conclusion that an interruption of the stria medullaris does not reproduce the massive and persistent deficits seen after septal lesions, and in our recent experiments, the transection of the fornix, but do have some minor and transient disruptive effects on DRL performance. In our situation, the rats with stria medullaris cuts made significantly <u>fewer</u> perseverative erros/reinforcement and slightly more anticipatory errors than controls but the number of reinforcements earned per session did not differ reliably from control data.

The pattern of effects seen in the DRL paradigm dovetails perfectly with the effects of our cuts on responding in the signalled Sidman avoidance situation (see above) which also reflects an animal's ability to perform efficiently by inhibiting potentially correct responses. Although in the Sidman paradigm there is no penalty for waiting until the CS is presented (in fact, it is the most parsimonious strategy), we find that rats with fornix transections as well as rats with septal lesions emit many more spontaneous avoidance responses than controls, thus expending a large amount of unnecessary effort. This behavior is similar to the overresponding seen in the DRL paradigm, and it would appear significant that none of our other surgical procedures duplicated these effects in either paradigm. I believe that we thus have strong evidence for the hypothesis (16,88) that the "disinhibitory"

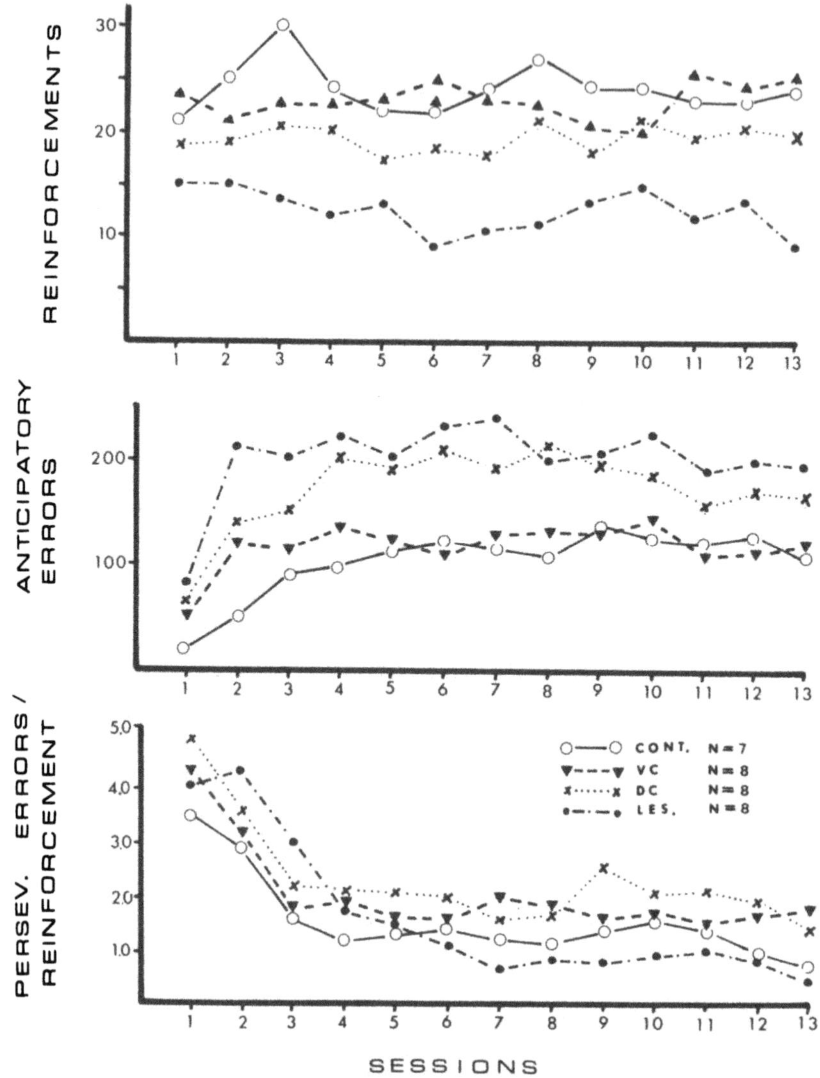

Figure 21. Number of reinforcements earned per session (upper panel); number of anticipatory errors (middle panel); and number of perseverative errors per reinforcement under modified DRL contingencies which required alternation between two levers in addition to the withholding of successive responses for a delay period of 30 sec. Cont. = controls; Les = rats with large septal lesions; VC = rats with cuts across the ventral (MFB) connections of the septum; and DC = rats with cuts across the dorsal (fornix) connections of the septum. From Ross & Grossman (116). Reprinted by permission of the American Psychological Association, Washington, D. C.

effects of septal lesions are due specifically to a disruption of septo-hippocampal connections.

This conclusion receives further support from the results of another experiment in which Ross and I (116) compared the effects of dorsal fornix cuts and ventral MFB transections with those of large septal lesions on lever pressing maintained by an FI contingency of reinforcement. We found that all three experimental groups showed a transient increase in responding during the normally quiet first half of the FI, but only the fornicotomy and the septal lesion produced persisting disinhibitory effects on responding (see Fig. 22). These persisting effects were confined to the second half of the FI and thus may reflect "anticipatory" rather than "perseverative" responses. This is in good agreement with our observation that septal lesions (70) as well as fornix cuts (117) increase anticipatory rather than perseverative erros in the DRL paradigm.

The last group of behaviors which Ross and I investigated in this series of experiments is ingestive behavior and simple continuously reinforced (CRF) operant responding for food rewards. In the first of these experiments, we (117) found that cuts which interrupted the ventral connections of the septum, but not fornicotomy, reproduced the sudden weight loss and hyperdipsia which has been seen after septal lesions (see Fig. 23). Ad libitum food intake was not reliably modified by the lesions or by either of the knife cuts, but the intake of sweetened milk during a brief test was reliably higher in all three experimental groups than in controls. Also, when the animals were permitted to obtain palatable pellet reinforcers by alternating responses to two levers placed at opposite ends of a runway, all three experimental groups emitted more responses during acquisition and reached a criterion of 40 or more reinforcements on each of two successive 10 min. tests reliably faster than controls (see Fig. 24). There were no significant differences between any of the groups during extinction of this response. Ross (115 repeated these measures in rats with stria medullaris transections and found their performance to be entirely normal in all of these tests. The intake of sweetened milk was also measured in rats with stria terminalis lesions and there was, once again, no reliable effect. Viewed in the context of our other tests, these results indicate that the regulatory deficits in body weight and water intake which are often seen after septal lesions may be specifically due to an interruption of septo-diencephalic connections. The marked effects of septal lesions on the intake of palatable solutions or on the animal's willingness to work for palatable rewards, on the other hand, do not seem to be peculiar to an interruption of these pathways, as Beatty and Schwartzbaum (3,4) have suggested, since we observed significant effects of fornicotomy in all situations.

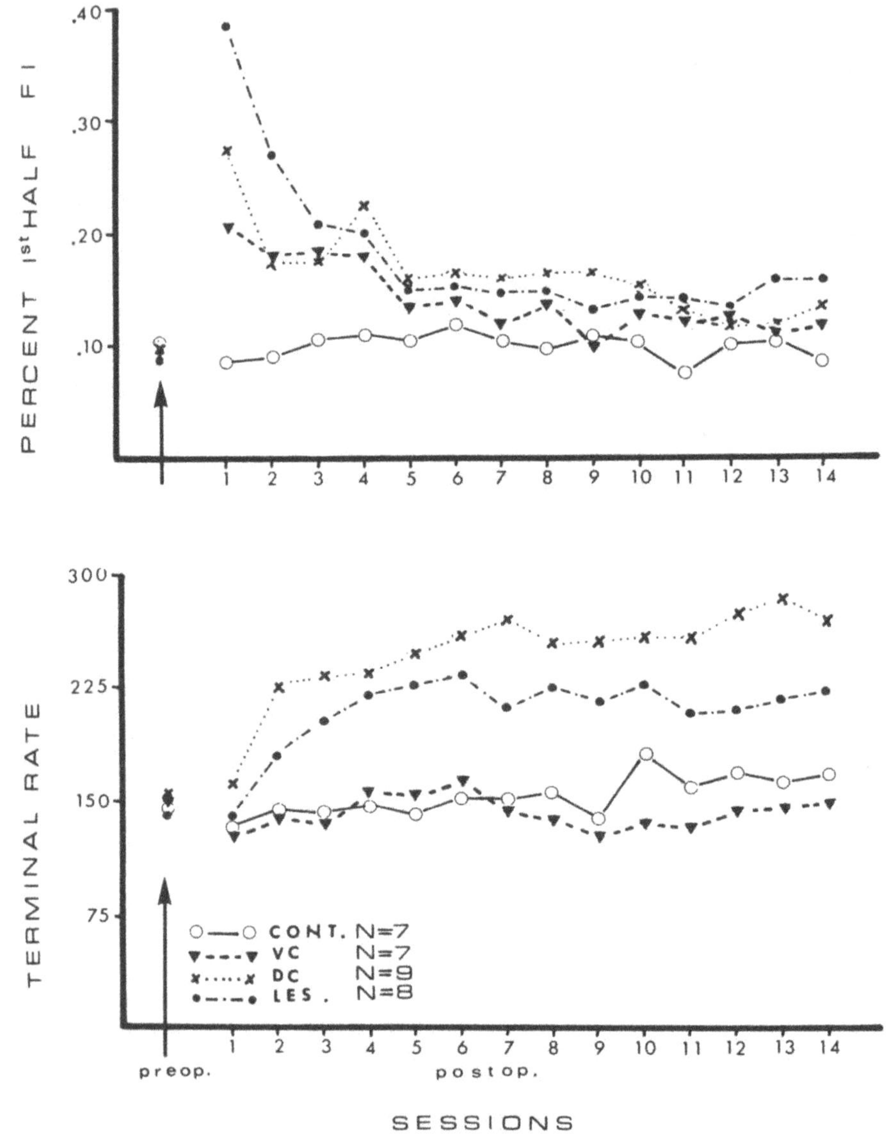

Figure 22. The percentage of lever presses emitted in the first half of a 60 sec. FI period (upper panel) and the number of responses emitted in the last 15 seconds (lower panel) of that FI, before and after large septal lesions (LES); cuts across the ventral connections of the septum (VC); and cuts across the dorsal (fornix) connections of the septum (DC); or control operations (OC) were made in rats. From Ross & Grossman (116). Reprinted by permission of the American Psychological Association, Washington, D. C.

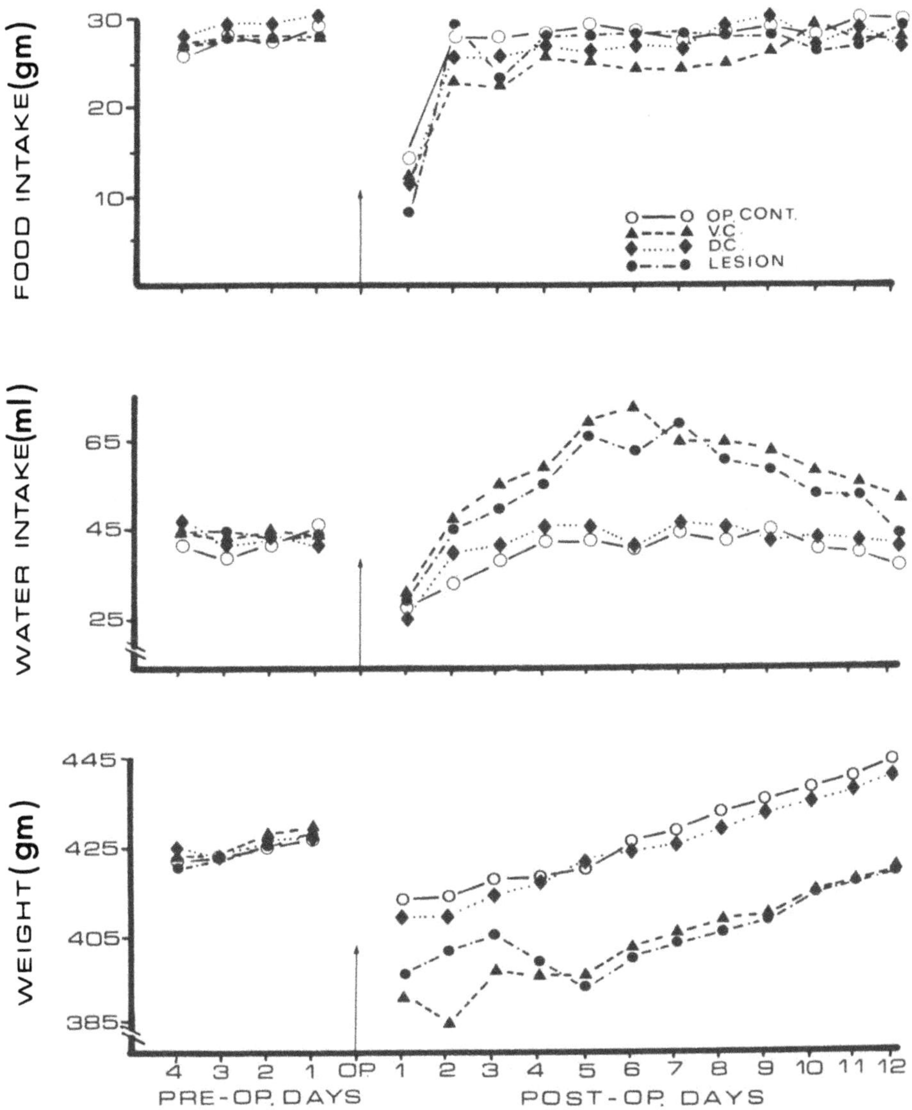

Figure 23. Food intake, water intake, and body weight before and after large septal lesions or transections of the ventral (VC) or dorsal (DC) connections of the septum of rats. From Ross, Grossman, & Grossman (117). Reprinted by permission of the American Psychological Association, Washington, D. C.

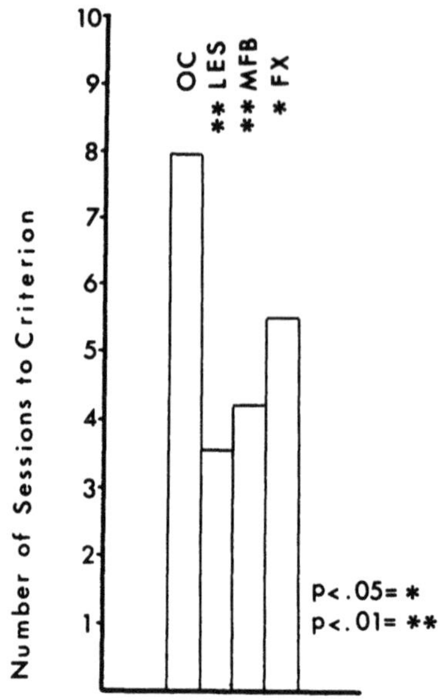

Figure 24. Number of 10 min. training sessions required to average 40 reinforcements in two consecutive daily sessions in a situation where food reinforcements were contingent on successive reponses to levers at alternate ends of a straight runway. OC = operated controls; LES = large septal lesions; MFB = transections of the ventral connections of the septum; FX = transections of the dorsal (fornix) connections of the septum. Data from Ross, Grossman, & Grossman (117). Reprinted by permission of the American Psychological Association, Washington, D. C.

CONCLUSIONS

I would like to end this review on the cheerful note that now that we can be reasonably sure that septal lesions produce a variety of seemingly unrelated behavioral dysfunctions, we can proceed unhindered by our proclivity to seek the most parsimonious interpretation in our search for the multifaceted answer to the question which has been in our minds for the past 10 or 15 years--what in the world is it that the septum does to influence behavior in so many situations.

We have not ruled out the possibility that all of the different influences which appear to be carried by the various pathways which project to or from the septal area may, after all, be subject to a fundamentally important process of integration or "digestion" in the septum so that its influence on behavior is, indeed, more than the sum of the various influences that seem to play on it. Yet, every major behavioral dysfunction which has been seen after septal lesions has been observed, in our investigations, after disruption of specific connections of the septal area, a pattern of effects which makes it seem at least possible that this region may act mainly as a relay station between subcortical and cortical mechanisms which are more specifically concerned with the different behaviors under study. What keeps one from embracing this appealingly simple answer is not only the obvious lack of parsimony but also the appaling lack of evidence that other regions of the brain do, indeed, subserve specific aspects of the functions which appear to be disrupted by septal lesions. There is a good deal of corroborating evidence from studies of the hippocampal formation, and it would seem likely that septo-hippocampal connections may indeed carry information relevant to many of the behaviors that we have just considered. A closer look at the many discrepancies between the effects of septal and hippocampal lesions nonetheless suggests that one must be careful even in this very tempting territory not to force generalizations which may not, in fact, exist.

REFERENCES

1. Ahmad, S. S., and Harvey, J. A. Long-term effects of septal lesions and social experience on shock-elicited fighting in rats. J. Comp. Physiol. Psychol., 66 (1968) 596-602.

2. Beatty, W. W., and Schwartzbaum, J. S. Enhanced reactivity to quinine and saccharine solutions following septal lesions in the rat. Science, 8 (1967) 483-484.

3. Beatty, W. W., and Schwartzbaum, J. S. Consummatory behavior for sucrose following septal lesions in the rat. J. Comp. Physiol. Psychol., 65 (1968) 93-102.

4. Beatty, W. W., and Schwartzbaum, J. S. Commonality and specificity of behavioral dysfunctions following septal and hippocampal lesions in rats. J. Comp. Physiol. Psychol, 66 (1968) 60-68.

5. Besch, N. F., and Van Dyne, G. C. Effects of locus and size of lesion on consummatory behavior in the rat. Physiol. and Behav., 10 (1973) 379-384.

6. Black, S. L., and Morgenson, G. J. The regulation of serum sodium in septal-lesioned rats: A test of two hypotheses. Physiol. and Behav., 10 (1973), 379-384.

7. Blanchard, R. J., and Blanchard, D. C. Limbic lesions and reflexive fighting. J. Comp. Physiol. Psychol, 66 (1968) 603-605.

8. Blass, E. M., and Hanson, D. G. Primary hyperdipsia in the rat following septal lesions. J. Comp. Physiol. Psychol, 70 (1970) 87-93.

9. Brady, J. V., and Nauta, W. J. H. Subcortical mechanisms in emotional behavior: Affective changes following septal forebrain lesions in the albino rat. J. Comp. Physiol. Psychol., 46 (1953) 339-346.

10. Brady, J. V., and Nauta, W. J. H. Subcortical mechanisms in emotional behavior: The duration of affective changes following septal and habenular lesions in the albino rat. J. Comp. Physiol. Psychol., 48 (1955) 412-420.

11. Bunnell, B. N., and Smith, M. H. Septal lesions and aggressiveness in the cotton rat, Sigmodon hispidus. Psychon. Sci., 6 (1966) 443-444.

12. Bunnell, B. N., Bemporad, J. R., and Flesher, C. K. Septal forebrain lesions and social dominance behavior in the hooded rat. Psychon. Sci., 6 (1966) 207-208.

13. Caplan, M., and Stamm, J. DRL acquisition in rats with septal lesions. Psychon. Sci., 8 (1967) 5-6.

14. Carey, R. J. Independence of effects of septal ablation on water intake and response inhibition. Psychon. Sci., 8 (1967) 3-4.

15. Carey, R. J. Contrasting effects of increased thirst and septal ablations on DRL responding in rats. Physiol. and Behav., 2 (1967) 287-290.

16. Carey, R. J. A further localization of inhibitory deficits resulting from septal ablation. Physiol. & Behav., 3 (1968) 645-649.

17. Carey, R. J. Contrasting effects of anterior and posterior septal injury on thirst motivated behavior. Physiol. & Behav., 4 (1969) 759-764.

18. Carlson, N. R. Two-way avoidance behavior of mice with limbic lesions, J. Comp. Physiol. Psychol, 70 (1970) 73-78.

19. Chiaraviglio, E. Effect of lesions in the septal area and olfactory bulbs on sodium chloride intake. Physiol. & Behav., 4 (1969) 693-697.

20. Clark, C. V. H., and Isaacson, R. L. Effect of bilateral hippocampal ablation on DRL performance. J. Comp. Physiol. Psychol., 59 (1965) 137-140.

21. Clody, D. E., and Carlton, P. L. Behavioral effects of lesions of the medial septum in rats. J. Comp. Physiol. Psychol, 67 (1969) 344-351.

22. Corman, C. D., Meyer, P. M., and Meyer, D. R. Open-field activity and exploration in rats with septal and amygdaloid lesions. Brain Res., 5 (1967) 469-476.

23. Dalby, D. A. Effect of septal lesions on the acquisition of two types of active-avoidance behavior in rats. J. Comp. Physiol. Psychol, 73 (1970) 278-283.

24. Dallard, T. Response and stimulus perseveration in rats with septal and dorsal hippocampal lesions. J. Comp. Physiol. Psychol., 71 (1970) 114-118.

25. Donovick, P. J. Effects of localized septal lesions on hippocampal EEG activity and behavior in rats. J. Comp. Physiol. Psychol., 66 (1968) 569-578.

26. Donovick, P. J., and Wakeman, K. A. Open-field luminance and septal hyper-emotionality. Anim. Behav., 17 (1969) 186-190.

27. Donovick, P. J., Burright, R. G., and Gittelson, P. L. The effects of septal lesions on saccharine choice as a function of water deprivation. Physiol. & Behav., 3 (1968) 677-681.

28. Donovick, P. J., Burright, R. G. and Gittelson, P. L. Body-weight and food and water consumption in septal lesioned and operated control rats. Psychol. Rep., 25 (1969) 303-310.

29. Donovick, P. J., Burright, R. G., and Zuromski, E. Localization of quinine aversion within the septum, habenula and interpeduncular nucleus of the rat. J. Comp. Physiol. Psychol, 71 (1970) 376-383.

30. Douglas, R. J., and Raphelson, A. C. Septal lesions and activity. J. Comp. Physiol. Psychol., 62 (1966) 465-467.

31. Eichelman, Burr S., Jr. Effect of subcortical lesions on shock-induced aggression in the rat. J. Comp. Physiol. Psychol., 74 (1971) 331-339.

32. Ellen, P., and Powell, E. W. Temporal discrimination in rats with rhinencephalic lesions. Exp. Neurol., 6 (1962) 538-547.

33. Ellen, P., Wilson, A., and Powell, E. Septal inhibition and timing behavior in the rat. Exp. Neurol., 10 (1964) 120-132.

34. Fisher, A. E., and Coury, J. Cholinergic tracing of a central neural circuit underlying the thirst drive. Science, 138 (1962) 691-693.

35. Flaherty, C. F., and Hamilton, L. W. Responsitivity to decreasing sucrose concentrations following septal lesions in the rat. Physiol. & Behav., 6 (1971) 431-437.

36. Fox, C. A. Certain basal telencephalic centers in the cat. J. Comp. Neurol., 72 (1940) 1-62.

37. Fox, S. S., Kimble, D. P., and Lickey, M. E. Comparison of caudate nucleus and septal-area lesions on two types of avoidance behavior. J. Comp. Physiol. Psychol, 58 (1964) 380-386.

38. Fried, P. A. Effect of pre- and post-operative treatment in septal lesioned rats. Canad. Psychol., 10 (1969) 203.

39. Fried, P. A. Effects of septal lesions on conflict resolution in rats. J. Comp. Physiol. Psychol., 69 (1969) 375-380.

40. Fried, P. A. Pre- and post-operative approach training and conflict resolution by septal and hippocampal lesioned rats. Physiol. & Behav., 5 (1970) 975.

41. Gittelson, P. L., and Donovick, P. J. The effects of septal lesions on the learning and reversal of a kinesthetic discrimination. Psychon. Sci., 13 (1968) 137-138.

42. Gotsick, J. Factors affecting spontaneous activity in rats with limbic system lesions. Physiol. & Behav., 4 (1969) 587-593.

43. Grace, J. E. Central nervous system lesions and saline intake in the rat. Physiol. & Behav., 3 (1968) 387-393.

44. Grant, L. D., Coscina, D. V., Grossman, S. P., and Freedman, D. X. Muricide after serotonin depleting lesions of midbrain raphé nuclei. Pharmac. Biochem. & Behav., 1 (1973) 77-80.

45. Grossman, S. P. Eating or drinking elicited by direct adrenergic or chlinergic stimulation of hypothalamus. Science, 132 (1960) 301-302.

46. Grossman, S. P. Effects of chemical stimulation of the septal area on motivation. J. Comp. Physiol. Psychol, 58 (1964) 194-200.

47. Grossman, S. P. The physiological basis of specific and non-specific motivational processes. In The Nebraska Symposium on Motivation. W. J. Arnold, Ed. Lincoln, Neb.: University of Nebraska Press, 1969, pp. 1-46.

48. Grossman, S. P. Avoidance behavior and aggression in rats with transections of the lateral connections of the medial or lateral hypothalamus. Physiol. & Behav., 5 (1970) 1103-1108.

49. Grossman, S. P. Modification of emotional behavior by intracranial administration of chemicals. In Physiological Correlates of Emotion. P. Black, Ed., New York: Academic Press, 1970, pp. 73-93.

50. Grossman, S. P. Cholinergic synapses in the limbic system and behavioral inhibition. Neurotransmitters. Res. Publ., Assn. Res. Nerv. Ment. Dis., 50 (1972) 315-326.

51. Grossman, S. P., and Grossman, L. Surgical interruption of the anterior or posterior connections of the hypothalamus: Effects on aggressive and avoidance behavior. Physiol. & Behav., 5 (1970) 1313-1317.

52. Hamilton, L. W. Active avoidance impairment following septal lesions in cats. J. Comp. Physiol. Psychol., 69 (1969) 420-431.

53. Hamilton, L. W. Behavioral effects of unilateral and bilateral septal lesions in rats. Physiol. & Behav., 5 (1970) 855-859.

54. Hamilton, L. W., and Grossman, S. P. Behavioral changes following disruption of central cholinergic pathways. J. Comp. Physiol. Psychol, 69 (1969) 76-82.

55. Hamilton, L. W., Kelsey, J. E., and Grossman, S. P. Variations in behavioral inhibition following different septal lesions in rats. J. Comp. Physiol. Psychol, 70 (1970) 79-86.

56. Hamilton, L. W., McCleary, R. A., and Grossman, S. P. Behavioral effects of cholinergic septal blockade in the cat. J. Comp. Physiol. Psychol., 66 (1968) 563-568.

57. Harrison, J. M., and Lyon, M. The role of septal nuclei and components of the fornix in the behavior of the rat. J. Comp. Neurol., 108 (1957) 121-137.

58. Harvey, J. A., and Hunt, H. F. Effect of septal lesions on thirst in the rat as indicated by water consumtpion and operant responding for water reward. J. Comp. Physiol. Psychol., 59 (1965) 49-56.

59. Harvey, J. A., and Lints, C. E. Lesions in the medial forebrain bundle: Relationship between pain sensitivity and telencephalic content of serotonin. J. Comp. Physiol. Psychol., 74 (1971) 28-36.

60. Harvey, J. A., Lints, C. E., Jacobson, L. W., and Hunt, H. F. Effects of lesions in the septal area on conditioned fear and discriminated instrumental punishment in the albino rat. J. Comp. Physiol. Psychol., 59 (1965) 37-48.

61. Heller, A., and Moore, R. Y. Control of brain serotonin and norepinephrine by specific neural systems. Adv. Pharmac., 64 (1968) 191-206.

62. Hothersall, D., Johnson, D. A., and Collen, A. Fixed-ratio responding following septal lesions in the rat. J. Comp. Physiol. Psychol, 73 (1970) 470-476.

63. Isaacson, R. L., Douglas, R. J., and Moore, R. Y. The effect of radical hippocampal ablation on acquisition of avoidance responses. J. Comp. Physiol. Psychol, 54 (1961) 625-628.

64. Kaada, B. R. Somatomotor, automatic and electrocorticographic responses to electrical stimulation of "rhinencephalic" and other structures in primates, cat and dog: A study of responses from the limbic, subcallosal, orbital insula, pyriform and temporal cortex, hippocampus, fornix and amygdala. Acta Physiol. Scand., 24 (1951) (Supplement 83), 1-285.

65. Kaada, B. R., Rasmussen, E. W., and Kveim, O. Impaired acquisition of passive avoidance behavioral by subcallosal, septal, hypothalamic and insular lesions in the rat. J. Comp. Physiol. Psychol., 55 (1962) 661-670.

66. Karli, P. Septum, hypothalamus postérieur et agressivité interspécifique rat-souris. J. Physiol., 52 (1960) 135-136.

67. Karli, P., Vergnes, M., and Didergeorges, F. Rat-mouse interspecific aggressive behaviour and its manipulation by brain ablation and by brain stimulation. In Aggressive Behaviour. S. Garattini and E. B. Siggs, Eds., New York: John Wiley, 1969.

68. Kelsey, J. The rleations between septal lesions, scopolamine, the pituitary-adrenocortical system, and behavior. Unpublished Ph.D. Dissertation, University of Chicago, 1971.

69. Kelsey, J. E., and Grossman, S. P. Cholinergic blockade and lesions in the ventromedial septum of the rat. Physiol. & Behav. 4 (1969) 837-845.

70. Kelsey, J. E., and Grossman, S. P. Nonperseverative disruption of behavioral inhibition following septal lesions in rats. J. Comp. Physiol. Psychol., 75 (1971) 302-311.

71. Kemble, E. D., Levine, M. S., Gregoire, K., Koepp, K., and Thomas, T. T. Reactivity to saccharin and quinine solutions following amygdaloid or septal lesions in rats. Behav. Biol., 7 (1972) 503-512.

72. Kenyon, J., and Krieckhaus, E. E. Decrements in one-way avoidance learning following septal lesions in rats. Psychon. Sci., 3 (1965) 113-114.

73. Kimble, D. P., and Kimble, P. J. Hippocampectomy and response preseveration in the rat. J. Comp. Physiol. Psychol, 3 (1965) 474-476.

74. Kimura, D. Effects of selective hippocampal damage on avoidance behavior in the rat. Can. J. Psychol., 12 (1958) 213-218.

75. King, F. A. Effects of septal and amygdaloid lesions on emotional behavior and conditioned avoidance responses in the rat. J. Nerv. Ment. Dis., 126 (1958) 57-63.

76. Krieckhaus, K., Simmons, H., Thomas, G., and Kenyon, J. Septal lesions enhance shock avoidance behavior in rats. Exp. Neurol., 9 (1964) 107-113.

77. Leaf, R. C., and Muller, S. A. Central cholinergic response inhibition during massed free-operant and discrete-trial avoidance acquisition. In Proceedings of the Fifth Internationale Congress of the Collegium Internationale Nuero-Psycho-Pharmacologicum. H. Brill, J. O. Cole, P. Deniker, H. Hippicus, and P. B. Bradley, Eds., Amsterdam: Exerpta Medica, 1966.

78. Lints, C. E. Changes in sensitivity to electric shock following central nervous system lesions in the rat. Unpublished Ph.D. dissertation, University of Chicago, 1965.

79. Lints, C. E., and Harvey, J. A. Altered sensitivity to footshcok and decreased brain content of serotonin following brain lesions in the rat. J. Comp. Physiol. Psychol, 67 (1969) 23-31.

80. Liss, P. The role of the hippocampus and septum in response inhibition. Unpublished Ph.D. dissertation, McGill University, 1965.

81. Liss, P. Avoidance and freezing behavior following damage to the hippocampus or fornix. J. Comp. Physiol. Psychol., 66 (1968) 193-197.

82. Lorens, S. A. Activity and consummatory and avoidance behavior following lesions in the nucleus accumbens septi. Proc. 78th Annual APA Convention, Washington, D.C., 1970, pp. 209-210.

83. Lorens, S. A., and Kondo, C. Y. Effects of septal lesions on food and water intake and operant responding for food. Physiol. & Behav., 4 (1969) 729-732.

84. Lorens, S. A., Sorenson, J. P., and Harvey, J. A. Lesions in the nuclei accumbens septi of the rat: Behavioral and neurochemical effects. J. Comp. Physiol. Psychol., 73 (1970) 284-290.

85. Lubar, J. F. Effect of medial cortical lesions on the avoidance behavior of the cat. J. Comp. Physiol. Psychol., 58 (1964) 38-46.

86. Lubar, J. F., Boyce, B. A., and Schaefer, C. F. Etiology of polydipsia and polyuria in rats with septal lesions. Physiol. & Behav., 3 (1968) 289-292.

87. Lubar, J. F., Schaefer, C. F., and Wells, D. G. The role of the septal area in the regulation of water intake and associated motivational behavior. Ann. N. Y. Acad. Sci., 157 (1969) 875-893.

88. MacDougall, J. M., Van Hoesen, G. W., and Mitchell, J. C. Anatomical organization of septal projections in maintenance of DRL behavior in rats. J. comp. physiol. Psychol., 68 (1969) 568-575.

89. MacLean, P. D. The hypothalamus and emotional behavior. In The Hypothalamus. W. Haymaker, E. Anderson, and W. J. H. Nauta, Eds., Springfield, Ill.: Charles C. Thomas, 1969, pp. 659-678.

90. Matalka, E. S., and Bunnell, B. N. Septal ablation and CAR acquisition in the golden hamster. Psychon. Sci., 12 (1968) 27-28.

91. McCleary, R. A. Response specificity in the behavioral effects of limbic system lesions in the cat. J. Comp. Physiol. Psychol., 54 (1961) 605-613.

92. McCleary, R. A. Response-modulating functions of the limbic system: Initiation and suppression. In Progress in Physiological Psychology, Vol. I. E. Stellar and J. M. Sprague, Eds., New York: Academic Press, 1966, pp. 209-272.

93. McGowan, B. K., Garcia, J., Ervin, F. R., and Schwartz, J. Effects of septal lesions on bait-shyness in the rat. Physiol. & Behav., 4 (1969) 907-911.

94. McMullen, M., and Slotnick, B. M. Fighting behavior in mice following septal forebrain lesions. Amer. Zool., 7 (1967) 793 (Abstract).

95. McNew, J. J., and Thompson, R. Role of the limbic system in active and passive avoidance conditioning in the rat. J. Comp. Physiol. Psychol., 61 (1966) 173-180.

96. Miczek, K. A., and Grossman, S. P. Effects of septal lesions on inter- and intra-species aggression in rats. J. Comp. Physiol. Psychol., 79 (1972) 37-45.

97. Miczek, K. A., Kelsey, J. E., and Grossman, S. P. Time course of effects of septal lesions on avoidance, response suppression, and reactivity to shock. J. Comp. Physiol. Psychol., 79 (1972) 318-327.

98. Miczek, K. A., Brykczynski, T., and Grossman, S. P. Differential effects of lesions in the amygdaloid cortex or stria terminalis on aggressive behaviors in rats. J. Comp. Physiol. Psychol., in press.

99. Middaugh, L. D., and Lubar, J. F. Interaction of septal lesions and experience on the suppression of punished responses. Physiol. & Behav., 5 (1970) 233-238.

100. Moore, R. Y. Effects of some rhinencephalic lesions on retention of conditioned avoidance behavior in cats. J. Comp. Physiol. Psychol., 57 (1964) 65-71.

101. Morgan, J. M., and Mitchell, J. C. Septal lesions enhance delay of responding on a free operant avoidance schedule. Psychon. Sci., 16 (1969) 10-11.

102. Morgane, P. J. Medial forebrain bundle and "feeding centers" of the hypothalamus. J. Comp. Neurol., 117 (1961) 1-25.

103. Negro-Vilar, A., Gentil, C. G., and Covian, M. R. Alterations in sodium chloride and water intake after septal lesions in the rat. Physiol. & Behav., 2 (1967) 167-170.

104. Neill, D. B., Ross, J. F., and Grossman, S. P. Comparison of the effects of frontal, striatal, and septal lesions in paradigms thought to measure incentive motivation or behavioral inhibition. Physiol. & Behav., 13 (1974) 297-305.

105. Nielson, H. C., McIver, A. H., and Boswell, R. S. Effect of septal lesions on learning, emotionality, activity and exploratory behavior in rats. Exp. Neurol., 11 (1965) 147-157.

106. Olton, D. S., and Isaacson, R. L. Hippocampal lesions and active avoidance. Physiol. & Behav., 3 (1968) 719-724.

107. Papez, J. W. A proposed mechanism of emotion. Arch. neurol. Psychiat. (Chicago), 38 (1937) 725-743.

108. Pellegrino, L. Amygdaloid lesions and behavioral inhibition in the rat. J. Comp. Physiol. Psychol, 65 (1968) 483-491.

109. Peretz, E. Extinction of a food-reinforced response in hippocampectomized cats. J. Comp. Physiol. Psychol., 60 (1965) 182-185.

110. Porter, R. H., and Garreffa, L. F. Septal lesions affect agonistic behavior in bobwhite quail (Colinus virginianus). Comm. Behav. Biol., 5 (1970) 19-22.

111. Porter, R. W., Conrad, D. B., and Brady, J. V. Some neural and behavioral correlates of electrical self-stimulation of the limbic system. J. Exp. Anal. Behav., 2 (1959) 43-55.

112. Pubols, L. M. Changes in food motivated behavior of rats as a function of septal and amygdaloid lesions. Exp. Neurol., 15 (1966) 240-254.

113. Raisman, G. The connexions of the septum. Brain, 89 (1966) 317-348.

114. Rich, I., and Thompson, R. Role of the hippocampo-septal system, thalamus, and hipothalamus in avoidance conditioning. J. Comp. Physiol. Psychol., 59 (1965) 66-72.

115. Ross, J. F. A neuroanatomical analysis of the septal lesion syndrome. Unpublished Ph.D. dissertation, University of Chicago, 1974.

116. Ross, J. F., and Grossman, S. P. Septal influences on operant responding in the rat. J. Comp. Physiol. Psychol., in press.

117. Ross, J. F., Grossman, L., and Grossman, S. P. Some behavioral effects of transection of ventral or dorsal fiber connections of the septum. J. Comp. Physiol. Psychol., in press.

118. Schwartzbaum, J. S., and Donovick, P. J. Discrimination reversal and spatial alternation associated with septal and caudate dysfunction in rats. J. Comp. Physiol. Psychol., 65 (1968) 83-92.

119. Schwartzbaum, J. S., and Gay, P. E. Interacting behavioral effects of septal and amygdaloid lesions in the rat. J. Comp. Physiol. Psychol., 61 (1966) 59-65.

120. Schwartzbaum, J. S., and Spieth, T. M. Analysis of the response inhibition concept of septal functions in "passive-avoidance" behavior. Psychon. Sci., 1 (1964) 145-146.

121. Schwartzbaum, J. S., Kellicutt, M. H., Spieth, T. M., and Thompson, J. B. Effects of septal lesions in rats on response inhibition associated with food-reinforced behavior. J. Comp. Physiol. Psychol., 58 (1964) 217-224.

122. Schwartzbaum, J. S., Green, R. H., Beatty, W. W., and Thompson, J. B. Acquisition of avoidance behavior following septal lesions in the rat. J. Comp. Physiol. Psychol., 63 (1967) 95-104.

123. Sclafani, A., and Grossman, S. P. Hyperphagia produced by knife cuts between the medial and lateral hypothalamus in the rat. Physiol. and Behav., 4 (1969) 533-538.

124. Sclafani, A., and Grossman, S. P. Reactivity of hyperphagic and normal rats to quinine and electric shock. J. Comp. Physiol. Psychol., 74 (1971) 157-166.

125. Seggie, J. Effect of somatosensory stimulation on affective behavior of septal rats. J. Comp. Physiol. Psychol., 66 (1968) 820-822.

126. Sheard, M. H. The effect of p-chlorophenylalanine on behavior in rats: Relation to brain serotonin and 5-hydroxyindolacetic acid. Brain Res., 15 (1969) 524-528.

127. Singh, D., and Meyer, D. R. Eating and drinking by rats with lesions of the septum and the ventromedial hypothalamus. J. Comp. Physiol. Psychol., 65 (1968) 163-166.

128. Sodetz, F. J. Septal ablation and free-operant avoidance behavior in the rat. Physiol. & Behav., 5 (1970) 773-778.

129. Sodetz, F. J., Matalka, E. S., & Bunnell, B. N. Septal ablation and affective behavior in the golden hamster. Psychon. Sci., 7 (1967) 189-190.

130. Sorenson, J. P., and Harvey, J. A. Decreased brain acetylcholine after septal lesions in rats: Correlation with thirst. Physiol. & Behav., 6 (1971) 723-725.

131. Stark, P., and Henderson, J. K. Increased reactivity in rats caused by septal lesions. Int. J. Neuropharm., 5 (1966) 379-384.

132. Stein, L. Chemistry of purposive behavior. In Reinforcement and Behavior. J. T. Tapp, Ed., New York: Academic Press, 1969.

133. Thomas, G. J., Moore, R. Y., Harvey, J. A., and Hunt, H. F. Relations between the behavioral syndrome produced by lesions in the septal region of the forebrain and maze learning of the rat. J. Comp. Physiol. Psychol., 52 (1959) 527-532.

134. Thompson, R., and Langer, S. Deficits in position reversal learning following lesions of the limbic system. J. Comp. Physiol. Psychol., 56 (1963) 987-995.

135. Trafton, C. L. Effects of lesions in the septal area and cingulate cortical areas on conditioned suppression of activity and avoidance behavior in rats. J. Comp. Physiol. Psychol., 63 (1967) 191-197.

136. Turner, B. H. Neural structures involved in the rage syndrome of the rat. J. Comp. Physiol. Psychol, 71 (1970) 103-113.

137. Ursin, H. Effect of amygdaloid lesions on avoidance behavior and visual discrimination in cats. Exp. Neurol., 11 (1965) 298-317.

138. Vanderwolf, C. H. Effect of combined medial thalamic and septal lesions on active avoidance behavior. J. Comp. Physiol. Psychol., 58 (1964) 31-37.

139. Van Hoesen, G. W., MacDougall, J. M., and Mitchell, J. C. Anatomical specificity of septal projections in active and passive avoidance behavior in rats. J. Comp. Physiol. Psychol., 68 (1969) 80-89.

140. Votaw, C. L. Study of septal stimulation and ablation in the macaque monkey. Neurology, 10 (1960) 202-209.

141. Wilson, J. R., Mitchell, J. C., and Van Hoesen, G. W. Epithalamic and ventral tegmental contributions to avoidance behavior in rats. J. Comp. Physiol. Psychol, 78 (1972) 442-449.

142. Winocur, G., and Mills, J. A. Effects of caudate lesions on avoidance behavior in rats. J. Comp. Physiol. Psychol., 68 (1969) 552-557.

143. Wishart, T. B., and Morgenson, G. J. Effects of food deprivation on water intake in rats with septal lesions. Physiol. & Behav., 5 (1970) 1481-1486.

144. Yutzey, D. A., Meyer, P. M., and Meyer, D. A. Emotionality changes following septal and neocortical ablations in rats. J. Comp. Physiol. Psychol., 58 (1964) 463-467.

145. Zucker, I. Effect of lesions of the septal-limbic area on the behavior of cats. J. Comp. Physiol. Psychol., 60 (1965) 344-352.

146. Zucker, I., and McCleary, R. A. Perseveration in septal cats. Psychon. Sci., 1 (1964) 387-388.

BEHAVIORAL CORRELATES AND FIRING REPERTOIRES OF NEURONS IN SEPTAL NUCLEI IN UNRESTRAINED RATS[1]

James B. Ranck, Jr.

Department of Physiology, The University of Michigan
Ann Arbor, Michigan 48104

INTRODUCTION

The behavioral correlates and firing repertoires of neurons in hippocampal formation have already been studied and published (32). In earlier work the basic strategy of the experimentation has been described, including many of its pitfalls. Briefly, the idea is to find the behavioral correlates of neurons in a fairly wide range of behavior. If one knows the behavioral correlates of all the neurons which are inputs and outputs of a region of brain, then one can determine the transformation of behavioral correlates that occurs in the region. This can be interpreted in terms of the information processing of the region. Some of the cellular mechanisms by which this processing may occur often follow as a simple interpretation from known anatomy and physiology.

This approach is applied in this paper to the septal nuclei. We do not have data on all the inputs, so the approach cannot be applied in as strong a form as it could in hippocampus. Nevertheless, certain major patterns are apparent. It will be shown that there is some general agreement with the interpretation of this approach and that from rats with lesions of septal nuclei.

One of the major conclusions of this study is the agreement between the anatomical regions of the septal nuclei as determined

---

[1]Ann Maxwell did the surgery, made the electrodes, trained the rats, perfused the rats, did the histology, made some of the figures and did many other important parts of the study. This work was supported by a grant from NSF GB26184 and a grant from NIH NS 10970.

by behavioral correlates and firing repertoires and those determined by cytoarchitecture, anatomical connections, and electrophysiology. The anatomical terminology which will be used in this paper is based on that used by Swanson and Cowen (34) largely derived from Fox (12). The major divisions are shown in Fig. 1.

Some of the results of this work on septal nuclei have been reported in preliminary form elsewhere (32,33).

METHODS

All of the rats were male Sprague-Dawley rats weighing 350 to 450 grams. The methods of recording from single neurons in unrestrained rats have been described previously (31,32). Briefly, recording was from a tungsten wire with an etched tip insulated to within 5 uM of the tip attached to a 1-72 screw. Turning the screw lowered the electrode. Recordings from the microelectrode were

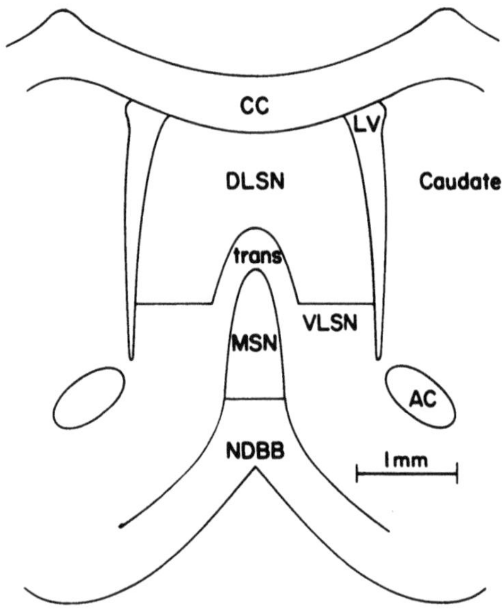

Figure 1. The septal nuclei described in this paper, DLSN = dorsal part of lateral septal nucleus; VLSN = ventral part of lateral septal nucleus; MSN = medial septal nucleus; NDBB = nucleus of the diagonal band of Broca; trans = transition zone between DLSN and MSN; LV = lateral ventricle; AC = anterior commissure; CC = corpus callosum. The division between MSN and NDBB is at the level of the anterior commissure. The division between DLSN and VLSN is 1.0-1.5 mm dorsal to the anterior commissure. Other septal nuclei are listed in Table I.

made with a narrow band filter for action potentials (flat 1000 to 5000 Hz, falling off 8.6 Db per octave below Hz). Slow waves from the microelectrode were recorded with a broad band filter (flat 1 to 20,000 Hz). The indifferent electrode was in neocortex above septum. Neocortical EEG, slow waves from hippocampus, AC coupled electro-oculograms (for eye movements) and EMG from dorsal neck muscles were also recorded. Eight FET's in source follower configurations were mounted on the head of the rat and all recordings were through these FET's. This eliminated almost all movement artifact. Polygraph recordings were taken of neocortical EEG, slow waves from hippocampus, slow waves from septum, dorsal neck muscle EMG, and EOG. The pulse output of a window discriminator was also written on the polygraph. A continuous record from a leaky integrator of the discriminator output was written out on the polygraph for some cells.

The protocol of behavior testing was the same as that used for a study of hippocampal formation (32). Briefly, rats were trained to eat and drink immediately upon presentation of food or water. All rats were run after overnight food, water, and sleep deprivation. All observations were made in a 60 x 30 x 30 cm. box. Observations of neuronal firing were made while the rats were in the following behavioral states: sleeping, smelling, responding to crude visual inputs, responding to auditory inputs, responding to touch, grooming, eating, drinking, responding to novel stimuli, and exploring. The rats were also picked up in the experimenter's hand. The experimenter blew in the rats faces and the rats tails were pinched. The rats were startled by a loud noise. The most important observations were those of spontaneous behavior in the cage.

There were two stages to the study. In the first the various behavioral types of the neurons were described, often retrospectively. In the second stage all neurons were classified by behavioral type at the time of study, and informal checks on the ability of the behavioral type to predict the firing of the cell were made. Further details of these two stages and the significance is given elsewhere (32).

T.V. tapes, showing the behavior of the rats and the neuronal firing on an oscilloscope face on a split T.V. screen, and with the output of the unit cahnnel on the audio of the T.V. were made of seven cells.

The rats were perfuse fixed the day after the last recording from the microelectrode. All tracts were verified histologically. There was only one track per rat, except in two cases where two tracts were made. A small lesion was made at the end of each tract. Recording sites were determined by measuring back from the lesion. The microelectrode went through the corpus callosium where there are no cell bodies and where no action potentials could be recorded.

This gave another landmark from which to measure. The two methods were always in good agreement.

It was much harder to record from septal nuclei than from hippocampus. The action potentials of neurons were smaller in amplitude, harder to isolate, and harder to hold. Most data were from action potentials with negative spike of 150 to 300 uV, held for only 30 to 60 minutes. In order to get satisfactory recording the electrode had to be insulated to within less than 5 uM of the tip. As the microelectrode was lowered through neocortex on the way to septum, action potentials of at least 700 uV had to be recorded from some neurons in neocortex to record any action-potentials in septum as large as 200 uV.

In the first 12 rats the microelectrode was angled laterally 10° or 20°. In six of the rats the electrode track crossed the midline in the medial septal nucleus. In four of these six there were pale infarcts of septal nuclei caudal, superior, and lateral to the point where the electrode crossed the midline. Because of the vasculature of the septal nuclei (8), it was assumed that the electrode had damaged septal arteries where they are dense in the midline. Therefore in the next 12 rats the microelectrode was angled 30° laterally. In ten of these rats the microelectrode track crossed the midline in the medial septal nucleus. Two of these rats had pale infarcts affecting regions caudal, superior, and lateral to the site of crossing the midline. The arterial supply to the septal nuclei is apparently very easily damaged in the midline in the medial septal nucleus. (There may be far more vascular damage than many researchers are aware of. A pale infarct is not noticeable histologically in perfused material within a few hours of the infarct. Even a few days later the lesion is easily missed using the standard Nissl stains).

RESULTS

Table I lists the number of neurons studied in each of the septal nuclei.

<u>Approach-orient cells</u> (Fig. 2 and 3). The major group of neurons in the dorsal part of the lateral septal nucleus were "approach-orient" neurons. (These were called "neck movement" cells in an earlier publication (32). There were two ways in which the firing of approach-orient nuerons was related to the behavior of the rats. Sometimes the rate of firing of the neurons increased for 100 to 500 msec. during some approach or orienting movements. At other times the rate of firing of these cells averaged over a second or more did not change much. Rather, any firing which did occur within a second or so of some approach or orienting movements

	First Stage		Second Stage		Total	
	neurons	rats	neurons	rats	neurons	rats
DLSN	62	11	90	10	152	21
VLSN	0	0	15	3	15	3
MSN	13	2	24	7	37	12
SF	16	2	0	0	16	2
Trans.	8	5	31	9	39	14
NDBB	19	5	11	3	30	8
BNST	16	3	0	0	16	3
NA	6	1	0	0	6	1
NT	0	0	0	0	0	0
BNAC	0	0	0	0	0	0
AH	0	0	0	0	0	0
	140	12	171	11	311	23

TABLE 1. Abbreviations as in Figure 1. Also SF = septofimbrial nucleus; BNST = bed nucleus of the stria terminals; NA = nucleus accumbens; NT = nucleus triangularis; BNAC = bed nucleus of the anterior commissure; AH = anterior continuation of the hippocampus.

occurred during the movement synchronized with the movement. While the rats were inactive or engaged in a consummatory behavior the rate of firing of the neuron was similar to the rate during approach or orienting movements, but there was no relation between the firing of the neuron and any movement of the rats or any events external to the rats. The duration of the correspondence between firing and approach-orient movements was usually 100 to 500 msec. With sustained approach, or sustained exploration there was no sustained relation to firing. There was, however, often a relation between firing of the neuron and details of movements for 100 to 500 msec. during sustained approach or exploring behavior. Many of the movements took longer than 100 to 500 msec., but the firing was only associated with part of the movement. There was often a pause in firing for almost one half second after movement related firing. The relation of firing to behavior was especially likely to occur during presumed changes in attention, or during orienting. (Attention and orienting are not proper behavioristic concepts, but a behavioristic approximation is clear most of the time.) Cells which fired less than once a second most of the time when the rats were awake usually were related to behavior by an increase in rate of firing. Cells which fired more rapidly than once a second usually were related to behavior by synchronizing with the behavior. However, both modes were sometimes seen in the same cell, and there is no sharp line of distinction between these modes.

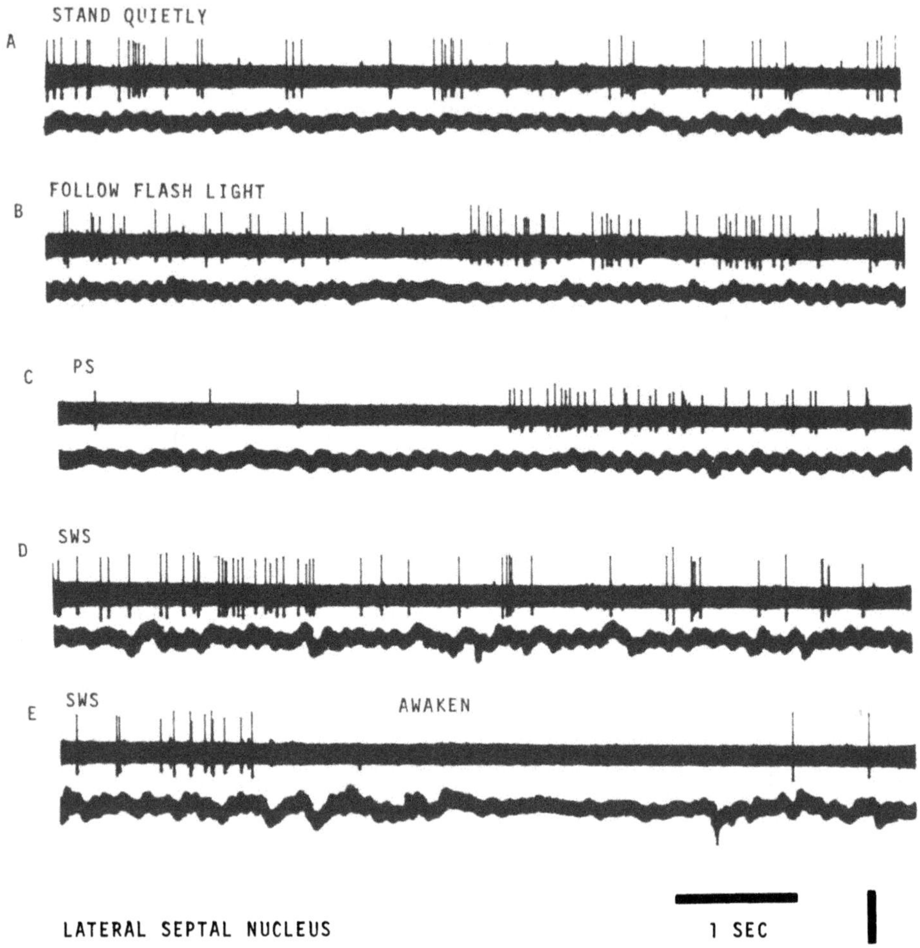

Figure 2. An approach orient cell. In this figure and in Figure 7, 8, 9, 10 and 11 each pair of traces are both from the microelectrode. The top trace is narrow band, the lower trace broad band (see methods). None of these figures has been retouched in any way. Negative is up. Part E was picked to show the longest period of no firing seen. Note the relatively constant rate of firing averaged over a few seconds. Same cell as in Figure 3 and 5.

The firing of approach-orient cells was associated with only about half of approach or orienting movements. A great deal of attention was given to the problem of finding any further specificity to these cells, i.e., was there any difference between these movements which were associated with firing and those which were

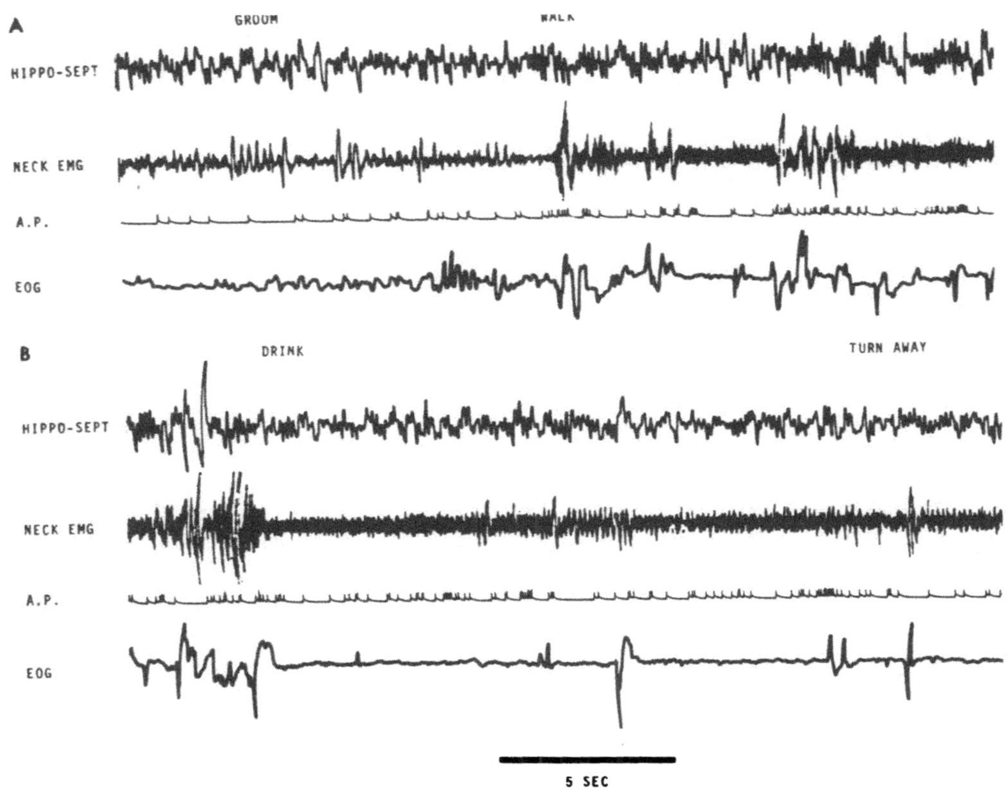

Figure 3. An approach-orient cell, the same cell as in Figure 2 and 5. The trace labeled A.P. is the output of a discriminator triggered by the action potential. Note that there is some relation of firing to both neck EMG and EOG but that this relation is not striking. Note the relatively constant rate of firing averaged over a few seconds.

not. The neurons fired during movements for which there was no identifiable external stimulus and did not fire if there was an identified external stimulus. Notice that this is not a behavioristic distinction. It could be stated more honestly that if I thought I knew what the rat was approaching or orienting to it was unlikely that the cell would show firing related to behavior. If I did not think I knew what the rat was approaching or orienting to, then the cell would probably increase rate of firing synchronously with behavior. In particular, these cells usually did not fire when the rat approached or oriented to objects introduced by the experimenter (for instance, food, water, pencils, the experimenter's hand, flashlights). Some cells decreased their rate of firing when orienting to something which the experimenter could identify. These cells did not increase rate of firing synchronoulsy with escape behavior.

Rats were picked up in the experimenter's hand and the rats' necks were moved from side to side and up and down. At first the rats resisted such movements, then they relaxed. Thus we can test both motor and proprioceptive aspects of pure neck movement. These were never associated with the firing.

These neurons fired irregularly, almost always fired slower than ten per second, and most fired slower than three per second most of the time. This meant that one had to record for many seconds to count more than thirty action potentials in order to determine a rate of firing. Only certain behaviors were sustained for more than five seconds. Rates of firing were determined in slow wave sleep, paradoxical sleep, quiet arousal, drinking, and grooming in 55 cells in lateral septal nucleus. These are plotted in Fig. 4. Some behaviors are plotted more than once to facilitate comparisons. In most of these cells rates could only be determined in two or three of these behaviors. It is clear from Fig. 4 that almost all rates of firing are between 0.2 and 10 per second. Furthermore, for most cells the rate of firing does not change much from one behavior to another. Even for behaviors that were not sustained for more than a few seconds, and hence could not have rates of firing properly counted, the rates of firing were in the same range. Maximum rates of firing were plotted only if a rate higher than any others were sustained for three seconds or longer; only nine cells had maximum rates defined in this way, and these rates were all less than ten per second.

Some cells changed rates of firing with some specific consummatory behaviors. Ten cells decreased rates of firing to once every five seconds or less during drinking (Fig. 4). Other cells increased the rates of firing during drinking. Furthermore, the change was specific to drinking - drinking is a salient behavioral category in the firing of these cells. The firing of two cells

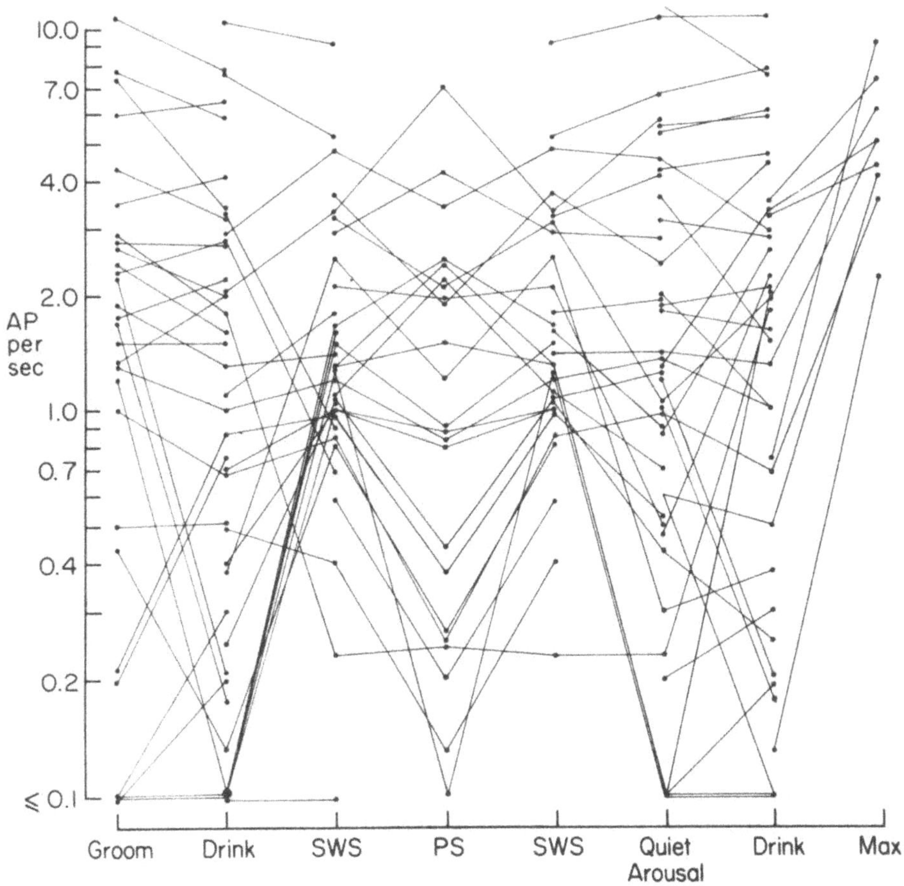

Figure 4. Firing rates of neurons in the dorsal part of the lateral septal nucleus. Data from some behavioral states are plotted more than once to allow for more comparisons. All sleep data are from at least 30 action potentials or 20 sec. Data from other states are for at least 10 sec. or 20 action potentials. Note the log scale for the firing rate.

was recorded continuously while the rats drank to satiety. (Each rat had had about half his daily intake of water when the recordings began.) There was no change in the rate of firing when drinking to satiety. All rats were distracted while they were drinking by touching them, pulling on their tails, moving the water bottle or water jar, or moving some object in front of them. As long as drinking continued, there was no change in the firing of the cell.

Four approach-orient cells stopped firing during freezing behavior of the rats. The firing of other cells was unchanged during freezing. There was no change in the firing of these cells associated with being startled. A few cells increased firing rates during grooming. Bridge (4) has found neurons in lateral septal nucleus which changed their firing rates during eating. Firing during eating usually could not be observed in this study because of chewing artifact.

These approach-orient cells of dorsal part of the lateral septal nucleus differ from appetitive cells in fascia dentata and Ammon's horn in that approach-orient cells do not change overall rate of firing much, while appetitive cells do; approach-orient cells fire with about half of all approach or orient movements, while appetitive cells fire with less than 10%; approach-orient cells show no specificity, while appetitive cells do; approach-orient cells are synchronized with behavior for periods of about 100 to 500 msec., while appetitive cells can be related for many seconds.

There are differences between different approach-orient cells - differences in firing in drinking, grooming, and in different states of consciousness (see below), differences in relation to approach-orient behavior (rate vs. synchrony), and differences in overall rate. No pattern to these various differences was found.

Sleep. Six of the neurons in the dorsal part of the lateral septal nucleus on which there was sleep data were of undetermined behavioral type because to get sleep data the cell usually had to be recorded from for a half-hour or more. The rats were not awakened until all the sleep data had been obtained. In some cases isolation of the neuron was lost before the cell could be observed during the active behaviors, which were used to classify the cell. Since most of the neurons in the dorsal part of the lateral septal nucleus are approach-orient cells, most of these unidentified cells were probably approach-orient cells.

Slow wave sleep is not a proper behavioral correlate for the firing of cells in dorsal part of lateral septal nucleus. The relation of the firing of these cells to slow wave sleep is the same as in Ammon's horn and fascia dentata (32). Along the spectrum of motionless behaviors between motionless arousal to slow wave sleep with the largest slowest slow waves, but excluding freezing and paradoxical sleep, there is some threshold point at which the firing of a neuron will change. This point is constant for a given neuron and is the same whether the state of consciousness of the rat is becoming deeper or lighter. At all states of consciousness deeper than this threshold point the firing of the neuron is the same and the firing is homogenous averaged over more than 5 seconds. In some cells this threshold was at the point where the largest slow waves appear in both neocortex and hippocampus. In other cells the threshold was when the rats simply lie down with a low voltage fast neocortical EEG. All intermediate stages were seen. All of the data labeled slow wave sleep in Fig. 4 was taken while the rats were in slow wave sleep and also below the threshold of change in firing. The rate was not necessarily characteristic of all stages of slow wave sleep for that neuron, nor did the behavioral state associated with this rate of firing necessarily include only slow wave sleep.

Quiet arousal is not an adequate category for describing behavioral correlates of neuronal firing in lateral septal nucleus for the same reason as slow wave sleep. It was also hard to get rats to lie or stand quietly for more than five seconds without slow waves appearing in the neocortex. Quiet arousal was not a homogenous behavior in all cells. In some neurons there was a difference in firing between standing and sitting in quiet arousal. Two episodes of sitting in quiet arousal were not associated with the same rate of firing in some cells. The data of Fig. 4 give averaged firing rates for all episodes which were observed.

The firing of many neurons in the dorsal part of the lateral septal nucleus was not homogenous in paradoxical sleep and the changes if firing were unrelated to the behavior of the animals or environmental changes. Relations to phasic episodes of paradoxical sleep were specifically looked for in two ways. The characteristic movement of vibrissae which occurs during phasic episodes was observed while listening to the firing of the cell. No relations of firing to phasic episodes were seen with either method. The data reported are rates of firing averaged over the entire duration of paradoxical sleep observed. This had to be at least 20 seconds and in most cases was greater than 60 seconds.

The major point of these qualifications on the use of slow wave sleep, paradoxical sleep, and quiet arousal as categories during which to measure rate of firing of a neuron is that one cannot properly impose behavioral categories on neuronal firing a priori. One must study the neurons first to see what categories are salient and homogenous for that neuron. A behavioral category may be very well defined as behavior of the rat, but it may not be salient for a neuron. The names for the conventional categories of different states of consciousness will be used with the qualifications on their use described above.

The firing during paradoxical sleep was slower than slow wave sleep in 16 neurons, and faster in 7 neurons. The more slowly firing neurons seem to decrease firing in paradoxical sleep and more rapidly firing neurons seem to increase. Because of the inhomogeneity of firing in paradoxical sleep a simple statistical test would not be proper. There are no dramatic patterns in changes between paradoxical sleep and slow wave sleep. Fig. 5 shows the firing of a cell in slow wave sleep and paradoxical sleep in which there is little difference in rate. In this and some other neurons perhaps there are differences in pattern, but quantitative analyses were made.

Twelves neurons fired more rapidly in slow wave sleep than in quiet arousal, eight neurons fired more slowly, and two fired at the same rate. Those which fired more rapidly in slow waves sleep had a greater difference in rate of firing compared to quiet arousal (2.9 times), than those which were slower in slow wave sleep (1.2 times).

<u>Motion punctuate cells</u>. Motion punctuate cells fired one to three action potentials at the <u>end</u> of an approach or orienting movement. Some of these cells fired less than once a second most of the time and effectively increased their rate of firing associated with the end of these movements. Other motion punctuate cells did not change their rates of firing much in different behaviors, but at the end of an approach or orienting motion the firing was synchronized with the behavior.

A motion punctuate cell fired with the end of about half of these movements. (Motion punctuate cells in Ammon's horn and fascia dentata fire at the end of about 10% of orienting movements). These cells were thus like approach-orient cells except that the firing occurred at the end of the movement, not during it. Some cells fired like both approach-orient and motion punctuate cells. However, a given cell showed predominately one type of firing or the other and would clearly fall into one type or the other. Like approach-orient cells, these motion punctuate cells showed no relation to non-voluntary movements during consummatory behaviors. They also fired at the end of movements for which no external stimulus could be identified just like approach-orient cells.

Just as for approach-orient cells, there were some changes in firing associated with grooming, drinking, and different motionless behaviors. It was common for some motion punctuate cells to be off during most movements and fire at one to five per second during all motionless behaviors. They were motion-punctuate-motionless cells.

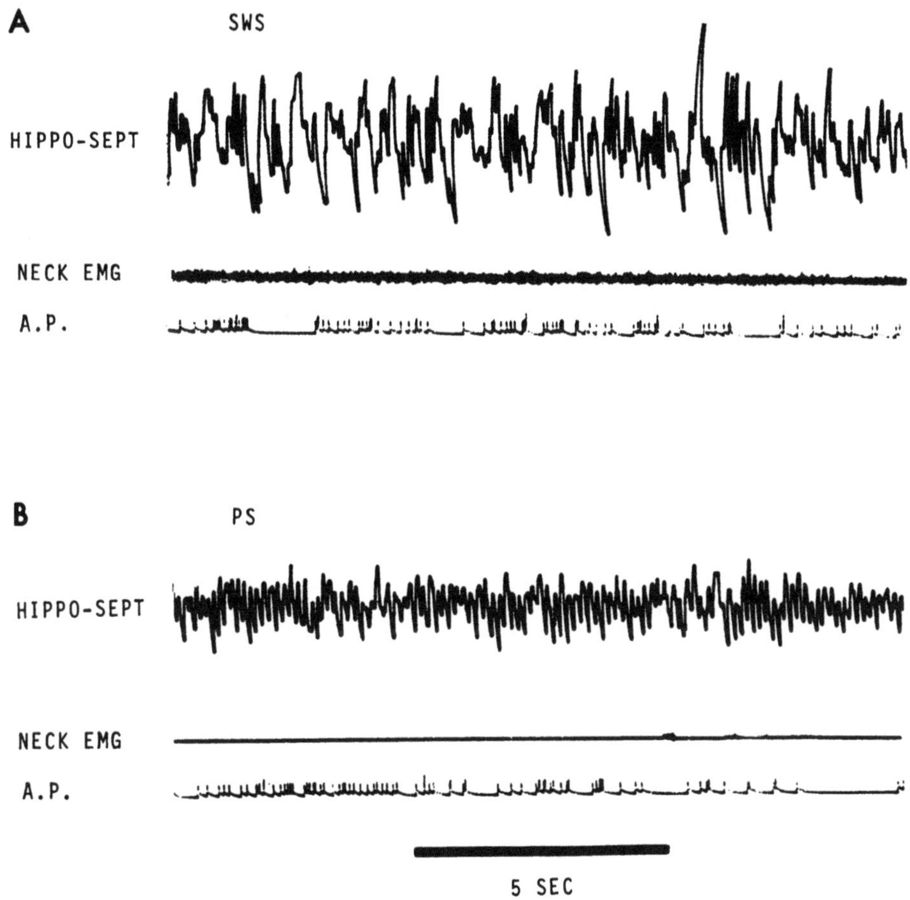

Figure 5. Firing during sleep of an approach-orient cell. The same cell as in Figure 2 and 3. Traces as in Figure 3.

Constant firing cells. Throughout all the septal nuclei constant firing cells fall into two groups. Those which fire faster than 20 per second fire very regularly and change very little (see below). Those which fire more slowly than 20 a second are not as regular and show more variability in rate.

"Sleep cells" were neurons which only fired faster than once every five seconds in slow wave sleep or paradoxical sleep.

Ten cells with "other" behavioral correlates were seen in the dorsal part of the lateral septal nucleus. These were an escape cell; a cell which ceased firing during any approach and consummatory behavior; a cell which fired during any consummatory behavior;

two cells which fired during sniffing; one cell which only fired during slow movements; a cell which only fired when the rat was motionless; and a cell which only fired when the rat was motionless or sniffing.

Nine cells were seen in which the firing of the neuron changed, but no relation between the firing of the neuron and the behavior of the rat could be determined. This is a much larger proportion than in Ammon's horn and fascia dentata where behavioral correlates could not be found in only two cells out of 310.

Distribution of neurons. One hundred fifty-one of the neurons in lateral septal nucleus were in the dorsal part and only 13 in the ventral part (Table 1). However, the ventral part of the nucleus was clearly different from the dorsal part. Fig. 6 gives the distribution of the various types seen in these two parts. The dividing line between dorsal and ventral parts was 1 to 1.5 mm. above the anterior commissure. Table II gives the number of cells of each type seen in each track. As can be seen a given track was likely to have cells predominately of the same behavioral type. Furthermore, those cells of the same behavioral type were often similar to each other, in usual firing rates, changes associated with drinking, grooming, and states of consciousness. There were only two tracks in which no approach-orient cells were found. Both of these tracks were in the ventral part of the lateral septal nucleus, the nucleus which Andy and Stephan (2) call the ventral group of nuclei. All the other tracks went through regions which Andy and Stephan call dorsal septal group or the nucleus septalis medialis of their medial septal group. In spite of the differences in behavioral correlates in different tracts, no other systematic anatomical groups were found. Specifically the other subdivisions which Andy and Stephan describe had no counterparts in this study.

The following correlations were looked for and not seen in any class of cells. In spite of the relation to movements, there were no particular movements related to the firing except in one cell. Recordings were made of dorsal neck EMG and action potentials of the neuron. There were some relations seen, but nothing striking or which could not be seen more clearly by simple observation of the rats. There were no sensory features related to firing. (Vision, hearing, and touch were tested). Only two cells showed spatial relations. There was no relation to some imagined emotional state of the rats, in particular fear. There was no relation to escape or avoidance behavior except in one cell. No theta cells were seen in lateral septal nucleus nor any cells which fired in phase with the hippocampal slow wave theta rhythm by simple visual inspection. However, no computer analyses were made to look for less obvious relations.

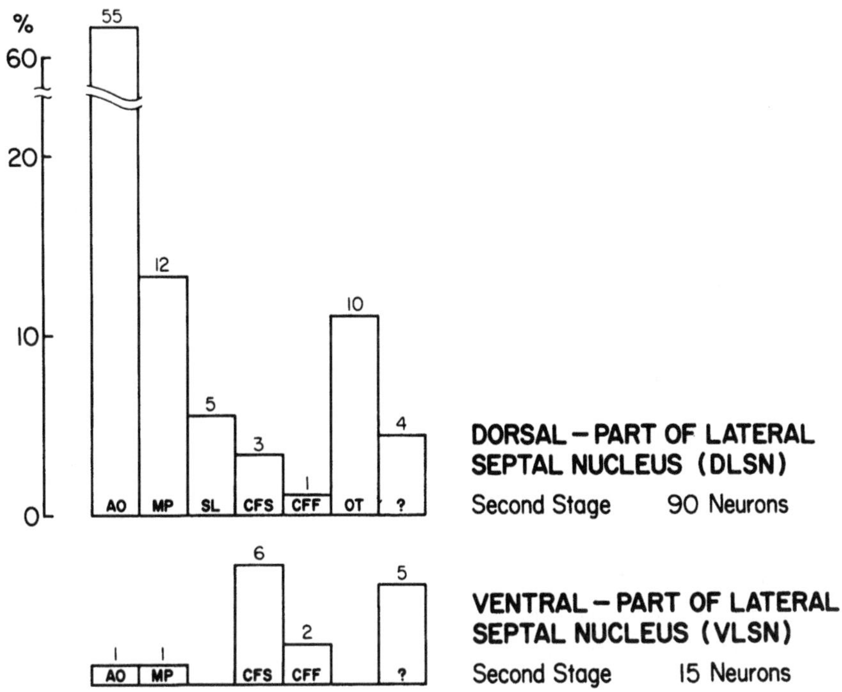

Figure 6. Numbers of different behavioral types of neurons seen in the second stage of the study in the dorsal and ventral parts of the lateral septal nucleus. AO = approach-orient cell; MP = motion punctuate cell; SL = sleep cell; CFS = constant firing slow cell; CFF = constant firing fast cell; OT = other types; ? = cells in which no relation to behavior could be determined.

Four cells were seen which sometimes fired complex spikes (as defined for hippocampal neurons (32). Three of these were approach-orient cells, and in one no relation to behavior could be determined. The duration of the extracellular negativity was measured in 16 cells using the narrow band filter. In one case it was 200 usec. (an approach-orient cell) and in 15 cases it was betwee- 250 and 600 usec. (The equivalent of a negative-positive spike between 0.6 to 1.0 msec. on a broad band amplifier.)

Medial Septal Nucleus

It was especially difficult to record from the medial septal nucleus. The spike heights were small, usually less than 150 μV, isolation was difficult, and cells could usually not be held for longer than 15 minutes. In any study one tends to find the cells one is looking for. This was especially true in medial septal

Rats	AO	MP	SL	CFS	CFF	Other	?	Total
148 D	6	0	1	2	0	3	2	14
150 D	1	0	0	0	0	0	1	2
158 D	3	0	0	0	1	2	0	6
159 D	5	1	0	0	0	1	1	8
159 V	0	0	0	0	0	0	4	4
164 D	1	0	0	6	0	0	1	8
165 D	6	6	0	0	0	0	0	12
168 D	8	0	3	1	0	1	0	13
168 V	0	1	0	0	2	0	0	3
173 D	5	3	0	0	0	0	0	8
184 D	1	0	0	0	0	1	0	2
185 D	9	0	1	0	0	2	0	12
204 D	11	2	0	0	0	0	0	13
TOTAL	56	13	5	9	3	10	9	105

TABLE II. Numbers of neurons of different behavioral types seen in each track in the dorsal and ventral parts of the lateral septal nucleus in the second stage of the study. There was one track in each rat. A "D" or a "V" after the number of the rat indicates if the track or part of the track was in the dorsal or ventral part. Abbreviations as in Fig. 6.

nucleus because of the striking patterns of firing of two types of cells and because of the difficulties in recording. There were five major groups of neurons seen. Two of these, theta cells and tight group cells, had striking firing repertoires which made their identification easy. Three other groups, motion punctuate cells, approach-orient cells, and constant firing cells had been discovered in lateral septal nucleus. In any event, the relative proportion of the various types is less reliable in medial septal nucleus than in the other areas that have been reported from this laboratory.

Theta cells. There were four types of theta cells. A "proper theta cell" was one which was just like a theta cell in Ammon's honr or fascia dentata. It increased its rate of firing if and only if there was a slow wave theta rhythm in hippocampus. Seven proper theta cells were seen in the second stage of the observations.

A "continuous theta cell" (Fig. 7) fired in the theta pattern almost always, whether or not there was a slow wave theta rhythm in hippocampus. The frequency of groups of spikes varied from 5 to 10 per second, and was in phase with the slow wave theta rhythm in hippocampus when it occured. Two of these cells were seen in the second stage of the observations. One of these cells was seen in the first stage.

A "selective theta cell" (Fig. 8) only increased its rate of firing when there was a hippocampal slow wave theta rhythm, but it did not increase its rate of firing every time there was a hippocampal slow wave theta rhythm, only with selective theta mode behaviors. I was not able to find any specificity which distinguished

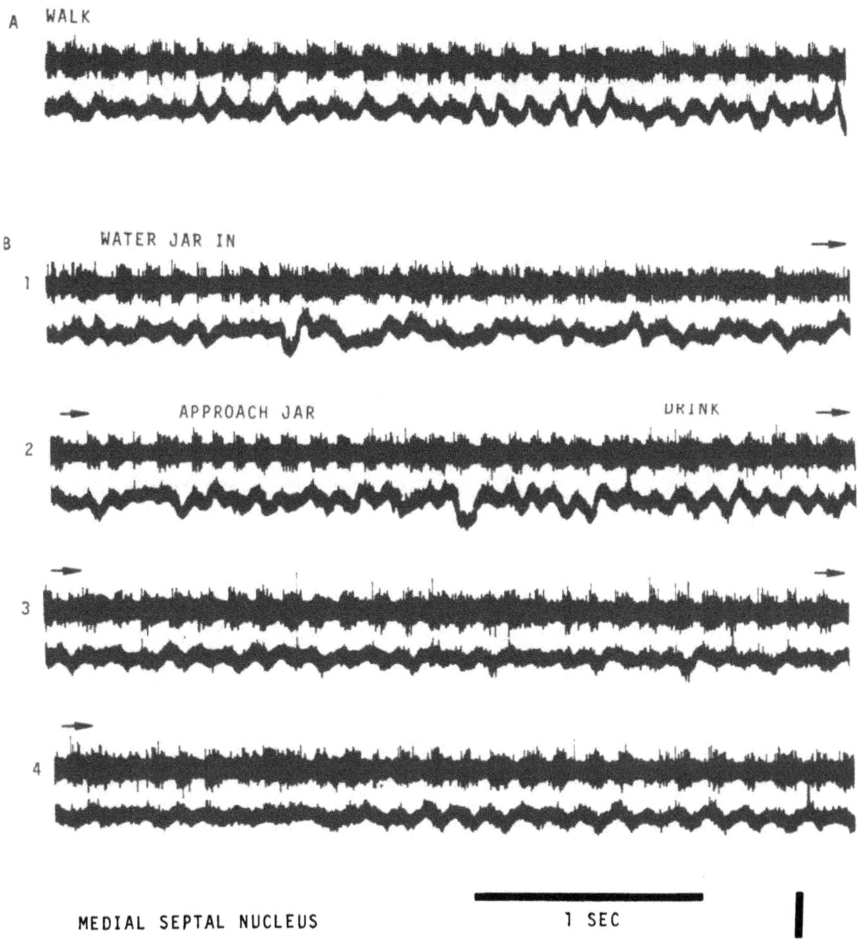

Figure 7. A continuous theta cell. Note the theta mode firing and the local slow wave theta rhythm during drinking. Drinking is never associated with a hippocampal slow wave theta. An arrow at the end of a pair of traces and another arrow at the beginning of the next pair of traces indicates a continuous recording in this figure and in Figure 8.

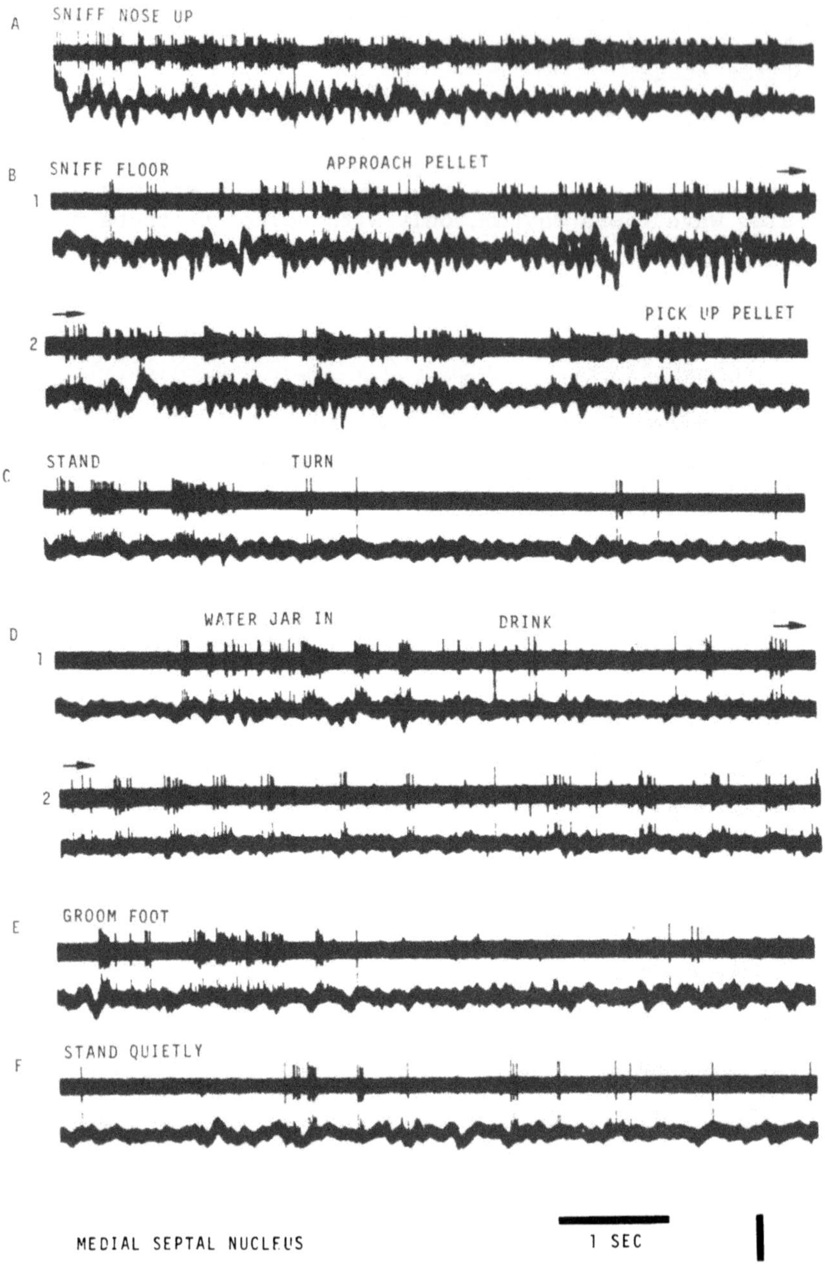

Figure 8. A selective theta cell. (There is a tight group cell present here also). This theta cell fires in the theta mode in phase with the slow waves during sniffing (A), approach to the water jar (D1), but only during the first part (B1) of the approach to a pellet, not the later part (B2), and not during turning (C). During non-theta mode behaviors drinking (D), grooming (E), and standing quietly (F) it fires in a non-theta mode.

those theta mode behaviors which were associated with increased rate of firing from those which were not. However, only a few of these cells were seen and I did not look for a specificity very carefully. Two of these cells were seen in the second stage and two in the first stage of the observations.

"Mixed approach-orient-theta cells" ("mixed theta cells") showed features of both the approach-orient cells of lateral septal nucleus and theta cells. During orienting or changes in attention the cell often fired like an approach-orient cell, and with sustained voluntary motor behavior, the cell fired like a theta cell. Three of these cells were seen in the second stage of the observations.

Perhaps selective theta cells and mixed theta cells should be grouped together as cells which often fire in phase with the slow wave theta rhythm, but not always, and which may fire rapidly under some other circumstances. Perhaps these two categories should be subdivided even further. The data is too limited to decide at present.

Six theta cells were seen in the second stage of the observations which were not studied carefully enough to place them in subclass.

The duration of the negative spike was observed in 13 theta cells in the second stage. In 12 of these the duration was between 0.2 to 0.3 msec. with the narrow band filter (equivalent to 0.4 to 0.6 msec. negativeity on a broad band). These included cells from each of the subtypes. One theta cell had an extracellular negativity of 0.4 msec. (Almost all hippocampal theta cells have an extracellular negativity of 0.2 to 0.25 msec. on the narrow band filter). Only one non-theta cell was seen in any of the septal nuclei which had an action potential with such a short duration.

No sleep data was taken from theta cells. These have been previously studied by Morales, Roig, Monti, Macadar, and Budelli (22). The cells they studied were almost all theta cells as defined here. These cells behaved like theta cells in Ammon's horn and fascia dentata in sleep (31).

"Tight group cells" (Figs. 9 and 10)were intentionally not given a name which described a behavioral oorrelate. The tight group pattern of firing was striking and the behavioral correlates are still only partially described. These cells fired a group of up to 35 action potentials with interspike intervals of from 1.0 to 5 msec. Two cells were seen with 1.0 msec. interspike intervals. They also fired single action potentials. This "tight group" pattern differed from the pattern of a complex spike seen in hippocampal formation in that in tight group cells the amplitude of the action potential may not decrease during the group, and when it did decrease, did so over a larger number of spikes. A tight group often had ten or more spikes, and it was very rare to see more than seven spikes in a complex spike (one complex spike with 14 spikes has been seen in CA1 but no others with more than ten have been seen). When one of these cells was firing tight groups, it usually fired a group from once every three seconds to three per second. Single action potentials occurred at about this range of frequencies also. Two tight group cells were seen which usually fired in doublets.

A tight group cell fired predictably during a specific consummatory behavior. Grooming, drinking, and automatic sniffing cells have been seen. They also fired less commonly during some appetitive behaviors, but this behavioral correlate has not been worked out.

These cells were hard to record from. This is presumably why they have never been described by others who have recorded from medial septal nucleus. The striking tight group pattern of firing was easily heard on a loudspeaker and seen on an oscilloscope even if the amplitude was 70 to 100 μV. Most of the observations were from cells with spikes of this small amplitude. The spikes in Figs. 8 and 9 are exceptionally large. Because of the small amplitude of these spikes it was difficult to know if a single spike had occurred and most of the observations of behavioral correlates of these cells were made for the tight group pattern, not for single spikes. Most of these cells could not be recorded from for more than ten minutes. A rapid burst of firing from a cell which can be held for less than ten minutes suggests that the firing is due to injury. However, three of these cells were held for over an hour during which no change in firing repertoire or behavioral correlate was observed. All these cells had similar behavioral correlates. Therefore, it seems that these observations are not just an artifact.

Only two tight group cells were studied in slow wave sleep and paradoxical sleep. In one case the rate of firing was 1.3 per second over a 60 second period in paradoxical sleep and 2.5 per second over a 40 second period in slow wave sleep. In the other case the firing was four per second over a 30 second period in slow wave sleep, one per second for 73 seconds in paradoxical sleep and then nine per second for 11 seconcs, unrelated to a phasic episode. During slow firing in paradoxical sleep the tight group cells did not fire in the tight group mode, but did during slow wave sleep and the fast firing in paradoxical sleep.

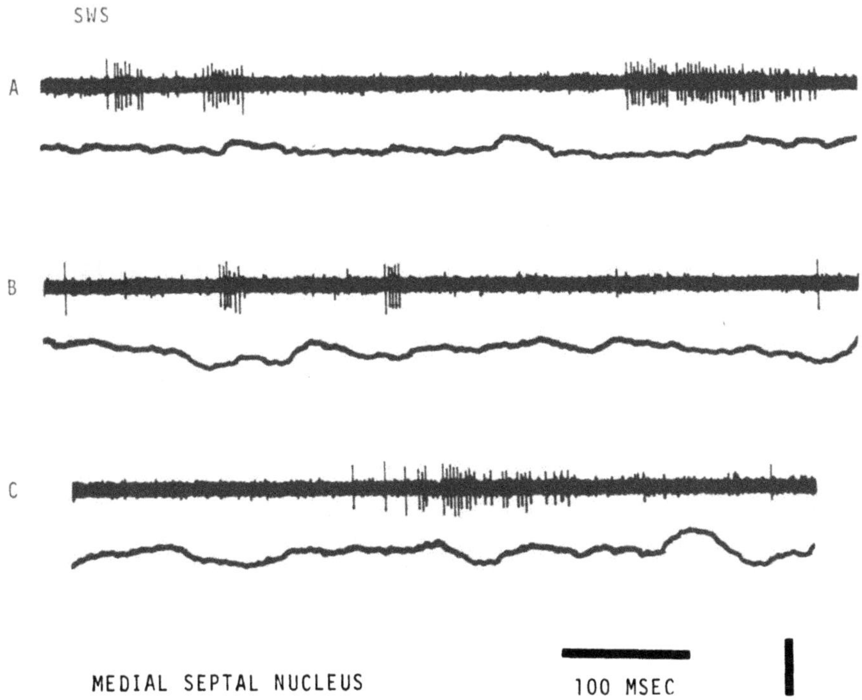

Figure 9. A tight group cell fast sweep. A and B are the same firing shown in Fig. 10C at a slower sweep. Note the variability in number of spikes per group and the amplitude of the spikes. These are all the instances of the firing of the cell in ten second period. The spikes of this tight group cell were the largest seen.

Distribution of neuronal types in medial septal nucleus (Fig. 11). Neurons in a transition zone between medial septal nucleus and the dorsal part of the lateral septal nucleus included behavioral types characteristic of both of these nuclei. This zone was 350 μM wide. The lateral and dorsal border of this transitional zone was considered to be the first theta cell or tight group cell encountered. In most cases these first cells were much too small to be studied and are not included in the data. The pattern of firing of theta and tight group cells is striking enough that they can be clearly identified even if only 40 or 50 μV amplitude.

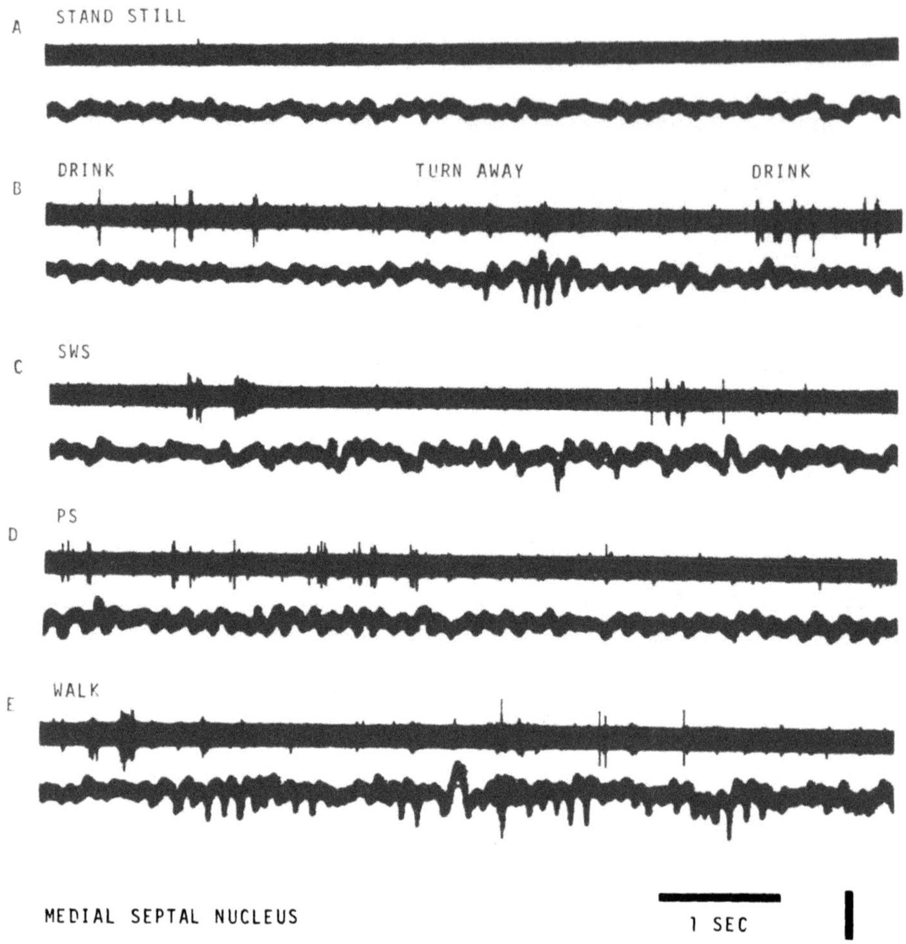

Figure 10. A tight group cell, the same cell as in Figure 9. This cell fired specifically during drinking. Note the pause in firing in B when the rat turns away from the bottle. This cell also fired at some other times (E) for which no common pattern has been discovered. Tight group cells also fire during sleep (C and D).

Twenty out of 24 neurons in the center of the medial septal nucleus seen in the second stage of the study were either theta cells or tight group cells. As discussed above, these types of neurons may be over represented in the sample.

Six cells were seen in the second stage which did not fit into the categories above. One cell fired during sniffing, and two cells seemed to fire during any movement. In three cells no behavioral correlate could be determined.

Theta cells and tight group cells are clearly the most characteristic cells of medial septal nucleus. In the second stage of the study theta cells were seen in all nine tracts through medial septal nucleus. In the second stage tight group cell were seen in eight out of nine tracts through medial septal nucleus. Theta cells tended to appear in groups. In one tract only theta cells were seen for over 700 μM. Tight group cells also were frequently bunched together.

Nucleus of the Diagonal Band of Broca

Figs. 11 and 13 give the distribution of various behavioral types of neurons in the nucleus of the diagonal band of Broca in the first and second stage of the study. The neurons in the first stage were more ventral than those in the second stage. Although the numbers are small this suggests that theta cells are predominately in the dorsal part of the nucleus, and constant firing fast cells predominately in the ventral part. The pattern of firing of both theta cells and constant firing fast cells is so striking that first stage data are reliable. Notice that no tight group cells were seen in this nucleus. The boundary between the medial septal nucleus and the nucleus of the diagonal band of Broca was the anterior commissure.

A constant firing fast cell fired from 20 to 50 per second, fairly regularly, with little change in rate or pattern of firing no matter what the behavior. Fig. 12 shows that some variation in firing was seen but no correlation to behavior was observed. Similar cells have also been observed in hippocampus. A rapid constant rate of firing may be due to damage. However, no change in any of these cells was noted over the time of observation, which in some cases lasted over an hour.

Other Septal Nuclei

<u>There were two tracts through the septofimbrial nucleus</u>. Even though both tracts were in the first stage of the study, they were both at the end of the first stage and the results are clear

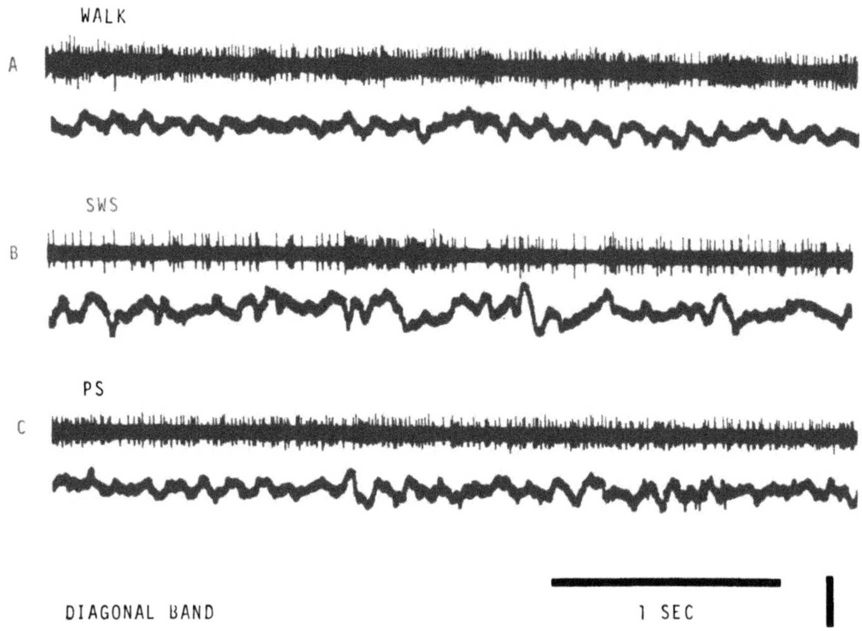

Figure 11. A constant firing cell in the nucleus of the vertical limb of the diagonal band of Broca. The difference between walking (A) and slow wave sleep (B) was the greatest difference seen in this cell. Note there is some variability during walking, but no correlations with behavior were noted. This cell was not generally faster during theta mode as opposed to non-theta mode behaviors.

(Fig. 13). Sixteen cells were studied. Three were theta cells (one of which was a selective theta cell); eight were tight group cells; one was a constant firing slow cell of 10 to 20 per second; two were approach-orient cells; and two were "other" types. Thus neurons in the septofimbrial nucleus had the same behavioral correlates as the neurons in medial septal nucleus. This result was unexpected. The septofimbrial nucleus has very different connections from the medial septal nucleus. Fox (12) considers the septofimbrial nucleus an extension of the lateral septal nucleus. Care was taken to be sure that these tracts were not in nucleus triangularis, which Fox considered to be an extension of the medial septal nucleus.

There were four tracts through the bed nucleus of stria terminalis, all in the first stage of the study. Sixteen cells were seen. Three were theta cells, four were constant firing cells (two fast, two slow), six seemed to be related to specific consummatory behaviors, two were related to appetitive behavior, and one only

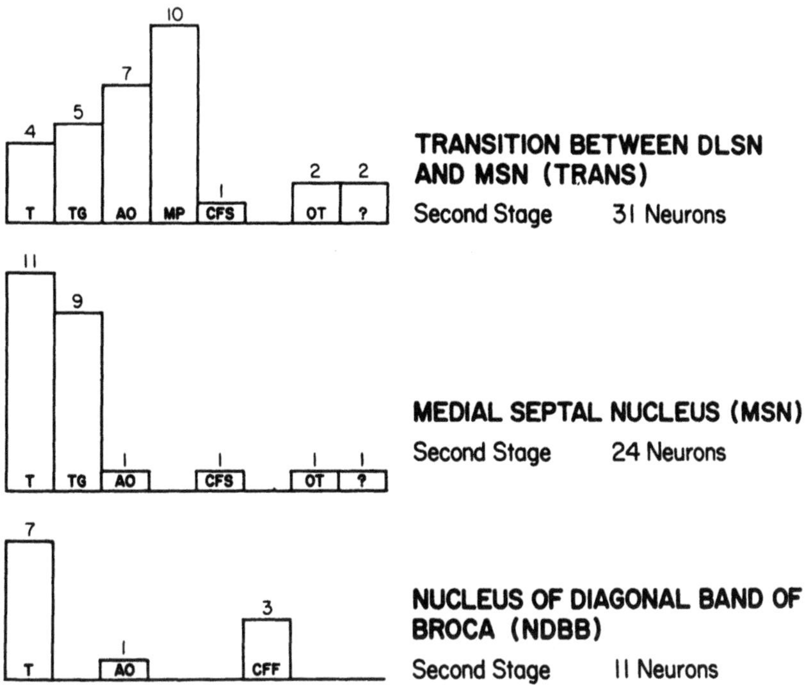

Figure 12. Numbers of different behavioral types of neurons seen in the second stage of the study in the transition zone between dorsal part of lateral septal nucleus and medial septal nucleus, in the medial septal nucleus, and nucleus of the diagonal band of Broca. T = theta cell; TG = tight group cell; AO = approach-orient cell; MP = motion punctuate cell; CFS = constant firing slow cell; CFF = constant firing fast cell; OT = other types; ? = cells for which no behavioral correlates could be found.

fired in slow wave sleep. No unusual firing repertoires were seen. The salient behavioral categories of neurons related to consummatory and appetitive behaviors have not been worked out. We can only conclude that this nucleus is not like any of the others studied.

Data from nucleus accumbens is too fragmentary to report. We can only conclude it is not like any of the other septal nuclei studied.

There is no data from nucleus triangularis, from the bed nucleus of the anterior commissure, or from the anterior extension of the hippocampus.

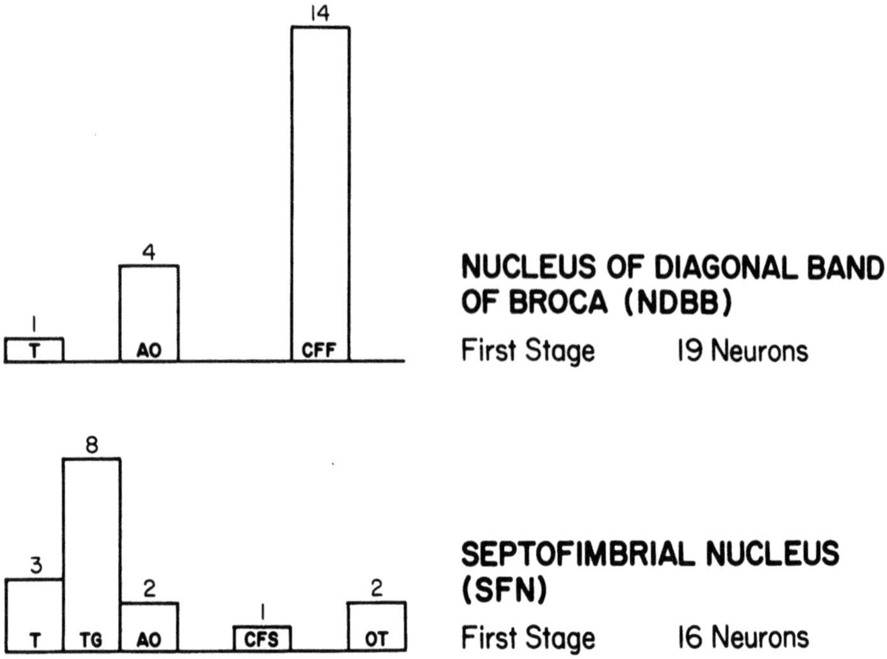

Figure 13. Numbers of different behavioral types seen in first stage of the study in the nucleus of the diagonal band of Broca and septofimbrial nucleus. Abbreviations as in Figure 12.

DISCUSSION

General Comments

As pointed out previously (32), there is no reason to assume that a neuron will have a behavioral correlate of its firing. The most important finding of this study is that there are behavioral correlates of the firing of neurons of septal nuclei, and that the behavioral correlates and firing repertoire of these neurons are stable over periods of up to three hours.

The behavioral correlate of some approach-orient and motion punctuate cells involves synchronized firing with certain movements, and no relation to other movements with little overall change in rate of firing. This is quite a different kind of relation to behavior from that seen in other neurons where patterns or rates of firing are associated with certain behaviors. This is a different way of coding information.

Other Studies of Neurons in Septal Nuclei

Bridge and Hatton (5) found neurons of anesthetized rats in both medial and lateral septal nuclei which fired more rapidly during dehydration (water deprivation or injections of hypertonic salin) than during full hydration. There is no comparable data in this study. Bridge (4) found neurons in septal nuclei in unrestrained rats which increased firing rates and others which decreased firing rates during drinking. Some approach orient cells decreased firing during drinking, and others increased. Some tight group cells increased firing during drinking. Bridge also found some neurons which increased their firing during grooming, just as was found in this study. He found some small changes with eating, Bridge had less chewing artifact than was found in this study, where observations on eating could not be made. Bridge found a relation of firing to swallowing in some cells, which was not seen in this study, but which was not looked for carefully. He did not find any relation to visual, auditory, olfactory, or tactile stimuli in most cells. He did find changes in firing during gustatory stimuli. Curiously Bridge and Hatton (5) saw firing related to external stimuli in urethane anesthetized rats. In general the results of Bridge's study and this are in good agreement.

Hayat and Feldman (18) have recorded from neurons in medical nucleus and nucleus of the diagonal band "in response to" various sensory stimuli in anesthetized or paralyzed cats. They found neurons which increased and others which decreased rates of firing during presentation of sensory stimuli. None of the cells in the present study changed rates of firing in response to sensory stimuli. Many neurons did change rates of firing associated with some behavioral responses to stimuli and associated with change in "behavioral state." These data from unrestrained animals demonstrate that a change in firing after a presentation of a sensory stimulus does not have a simple interpretation. I do not know how to interpret the Hayat and Feldman data. Mays and Best (23) have analyzed a similar case of "response to stimuli" in hippocampus.

Petsche, Stumpf, and Gogolak (30) discovered cells in the medial septal nucleus which fired in phase with eserine induced slow wave hippocampal theta rhythm. They called these B cells. These are clearly the same as the theta cells reported here. Others have also recorded from cells in medial septal nucleus which are clearly theta cells (3,15,16,22,26,29,35,36). I know of no previous reports of cells which seem likely to be the same as the right group cells of this study. I have not been able to find obvious correspondence between the non-theta cells of these previous studies and the non-theta cells of this report.

Some Suggested Synaptic Interactions

Lateral septal nucleus: Some of the major inputs to lateral septal nucleus are from CA1 and CA3. (CA1 is considered to project to lateral septal nucleus because of the data of Swanson and Cowan (this volume) and in spite of the data of Andersen (1). Fox and Ranck (13) have shown that in Ammon's horn the class of pyramidal cells and the class of non-theta cells (most of which are complex spike cells) have a very large overlap. The most common behavioral type of pyramidal cells in dorsal CA1 are approach-consummate-mismatch, approach-consummate, and appetitive cells. The most common behavioral types of pyramidal cells in dorsal CA3 are approach-consummate and appetitive cells. All these behavioral types in Ammon's horn fire during the kind of behavior where there is increased rate of firing or synchrony between movement and firing seen in approach-orient cells of dorsal lateral septal nucleus. DeFrance, Kitai and Shimono (8,9) and McLennon and Miller (25) have shown that axons from Ammon's horn are excitatory to lateral septal nucleus neurons. The relation to neck movement and the short duration of increased rate or synchrony of firing was not seen in cells in dorsal Ammon's horn. The behavioral correlates of cells in dorsal Ammon's horn had a lot of specificity which was not seen in approach-orient cells. These findings suggest that there is a lot of convergence from cells of Ammon's horn onto approach-orient cells, which washes out the specificity. The short duration of behaviorally correlated firing suggests that approach-orient cells show adaptation. The fact that the overall rate of firing of approach-orient cells does not change much is also consistent with the idea of convergence of many different neurons from Ammon's horn, each with a somewhat different specificity. The effect of averaging overall these inputs produces a fairly constant rate of firing.

There are no pyramidal cells in Ammon's horn which decrease their rate of firing during drinking. The decrease in rate of firing during drinking seen in some approach-orient cells is presumably due to inhibition. This and other changes in rate during consummatory behaviors could be from Ammon's horn through an interneuron, or from tight group cells, with or without an interneuron, or from other unknown inputs.

There are motion punctuate cells in Ammon's horn. These fire at the end of many fewer movements than do the motion punctuate cells of lateral septal nucleus. This suggests that motion punctuate cells of Ammon's horn converge onto septal motion punctuate cells. These septal motion punctuate cells are off during many approach or orienting movements and often fire at the end of these movements. This suggests that septal motion punctuage cells are inhibited by approach-orient cells, and fire as rebound excitations.

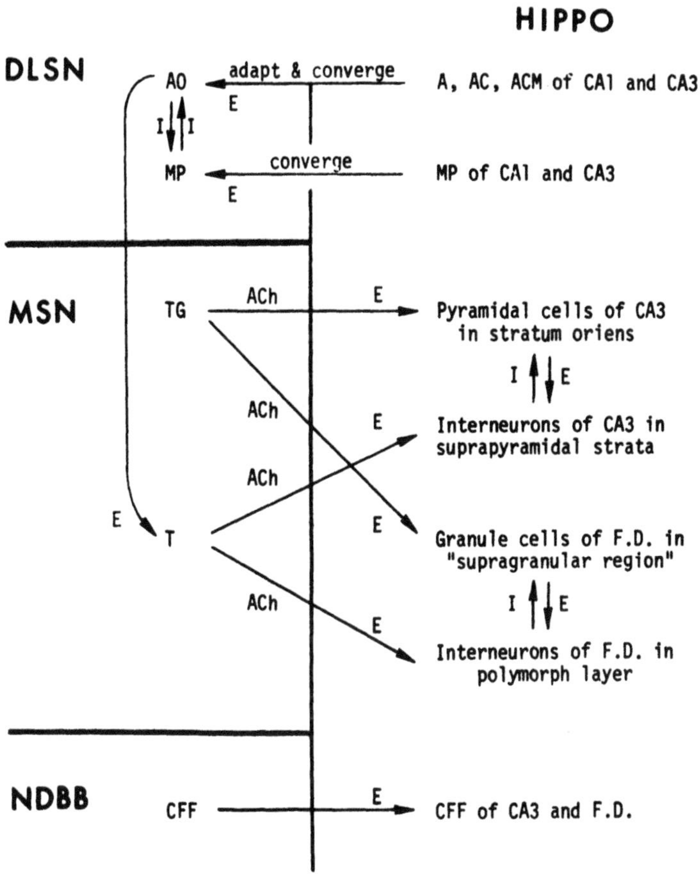

Figure 14. Summary of some suggested synaptic interactions. DLSN = dorsal part of lateral septal nucleus; MSN = medial septal nucleus; NDBB = nucleus of the vertical limb of the diagonal band of Broca; Hippo = hippocampal formation; AO = approach-orient neuron; A = appetitive neuron; AC = approach-consummate neuron; ACM = approach-consummate-mismatch neuron; MP = motion punctuate neuron; TG = tight group neuron; T = theta neuron; CFF = constant firing fast neuron; f.d. = fascia dentata; E next to an arrow indicates an excitatory synapse and I indicates an inhibitory one. ACh indicates a cholinergic neuron.

Approach-orient cells are off when motion punctuate cells fire. This suggests that motion punctuate cells inhibit approach-orient cells. DeFrance, Kitai, and Shimono (8,9) and McLennon and Miller (24,25) have shown that there are inhibitory interactions between neurons in lateral septum, as would be required for these suggested interactions. Inhibitory rebound and adaptation have not been seen, but do not seem to have been carefully looked for. The changes in rate associated with consummatory behaviors has the same possible sources as approach-orient cells.

The divisions of the lateral septal nucleus into dorsal and ventral nucleus is in agreement with the divisions of DeFrance, et. al. (10), for the cat. They also show some agreement with the cytoarchitectural division of Andy and Stephan (2) and Swanson and Cowan (34).

Medial septal nucleus. With the use of three dimensional slow wave analysis Winson (38) has shown that there are two different theta rhythm mechanisms in hippocampus. Vanderwolf (37) has also shown two different mechanisms, using behavioral and pharmacological analysis. In the awake rat, Vanderwolf's type I theta (voluntary movement and paradoxical sleep) comprises by far the largest amount of theta mode behavior. Perhaps the proper theta cells are related to Vanderwolf's type I theta, and the selective theta cells are related to Vanderwolf's type II behavior. (This is roughly related to motionless attention).

It is not clear if the medial septum is the pacemaker for the hippocampal theta rhythm, or if the theta rhythm is due to the interaction of medial septum and hippocampus. Petsch, Stumpf, and Gogolak (29) have seen theta cells which continue to fire in a theta mode during a hippocampal seizure, and use the existance of these cells to argue for a septal pacemaker. The findings of continuous theta cells in medial septum is even stronger evidence that indeed the medial septum is a pacemaker. Continuous theta cells would be gated in when a theta rhythm is present. However, I do not think these data are strong enough to establish the pacemaker hypothesis. One would like to see continuous theta cells in a rat with the fornix transected.

The presence of mixed approach-orient cells suggest that approach-orient cells project to some theta cells. They may help maintain an ongoing theta rhythm.

Tight group neurons are presumably the source of consummatory behavioral correlates seen in Ammon's horn and fascia dentata. The tight group mode of firing may be the source of the complex spikes seen in CA3 and granule cells of hippocampal formation. The interspike interval of the two groups is almost the same. The greater number of spikes seen in tight group cells may be due to the fact that all of the action potentials are not conducted down the axons of tight group cells. Perhaps pyramidal and granule cells cannot keep up with the rapid firing of tight group cells. If we count the occurrance of a tight group or a complex spike as a single event, then the frequency of these events is similar in tight group cells and in complex spike cells of hippocampal formation. Since there do not seem to be any septal projections to CA1 in rats complex spikes in CA1 presumably have another basis. One imagines they may be driven by complex spike cells in CA3. Fox and Ranck (13) give evidence that the complex spike mode of firing is more common in dorsal CA3 than dorsal CA1.

Proper theta cells in septum have firing repertoires and behavioral correlates which I can not distinguish from theta cells in hippocampal formation. The other types of septal theta cells have many similarities to hippocampal theta cells. Tight group cells have firing repertoires and behavioral correlates with many similarities to those of complex spike cells of hippocampal formation.

In another paper Fox and Ranck (13) have shown that in Ammon's horn and fascia dentata the class of theta cells and the class of interneurons have a very large overlap, and that the class of non-theta cells and projection cells (pyramidal and granule cells) have a very large overlap. These data on neurons in media septal nuclei suggest that septal theta cells project to interneurons and tight group cells project to projection cells in fascia dentata and CA3. The septal projections to fascia dentata go to two regions, "the supragranular region" (the outermost part of stratum granulosum and innermost part of stratum moleculare) and the polymorph layer (27). There are very few interneurons in the stratum moleculare and not many in the stratum granulosum (13), suggesting supragranular septal afferents go to granule cells. There are many interneurons in polymorph layer, and no dendrites of granule cells (but there are a few CA4 pyramidal cell dendrites (20) suggesting that these septal afferents go to interneurons. In CA3, septal afferents go to stratum oriens (where there are very few interneurons (12), but many basal dendrites of pyramidal cells) and to the suprapyramidal strata, but not to stratum pyramidale. There are many interneurons in the suprapyramidal strata of CA3 (13). This suggests that septal afferents in stratum oriens are to pyramidal cells, and those in the suprapyramidal strata go to interneurons. If so, then septal theta cells project to polymorph layer of fascia dentata and suprapyramidal strata of CA3. Tight groups cells project to the supragranular region of fascia dentata, and stratum oriens of CA3. All four of these sites of septal terminations in hippocampal formation show high levels of cholinesterase which disappears with fornix section (26). This suggests that both septal theta cells and tight group cells are cholinergic.

Theta cells may have much larger extracellularily recorded action potentials than tight group cells. There are large and medium sized neurons in medial septal nucleus. This suggests that

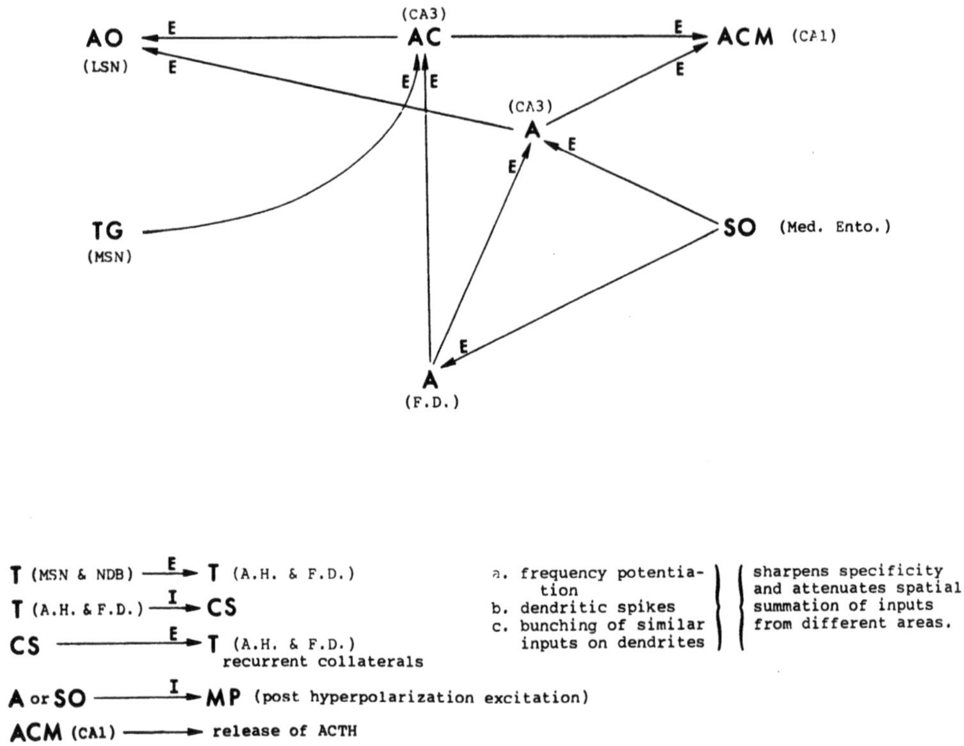

Figure 15. Summary of some suggested synaptic interactions. A = appetitive cell; AC = approach-consummate cell; ACM = approach-consummate-mismatched cell; A.H. = Ammon's horn; AO = approach-orient cell; CF = constant firing cell; CS = complex spike cell; F.D. - fascia dentata; LSN = lateral septal nucleus; Med. ento. = medial entohinal cortex; MP = motion punctuate cell; MSN = medial septal nucleus; NDB = nucleus of the vertical limb of the diagonal band of Broca; SO = specific orient cell; T = theta cell; TG = tight group cell; E next to an arrow indicates an excitatory synapse and I indicates an inhibitory one.

theta cells are the large neurons and tight group cells the medium sized ones. The short duration of extracellularily recorded actionpotentials seen in both septal and hippocampsl theta cells suggests a basic electrophysiological similarity.

Nucleus of the diagonal band of Broca. There are constant firing fast cells in Ammon's horn and fascia dentata (14 out of 310) which are very similar in rate and pattern of firing to the constant firing cells of diagonal band (32). This suggests that these diagonal band cells project to their counterpart in Ammon's horn and fascia dentata. There are also constant firing neurons in CA1. It is not clear how these are driven except perhaps from their counterparts in CA3 through the Schaffer collaterals.

These suggestions are summarized in Fig. 14. Fig. 15 gives the overall relation of neurons of various behavioral types in septal nuclei and hippocampal formation.

Global Function

As pointed out previously (32) we can not properly discuss the function of a region of brain using only data on the behavioral correlate of neurons with cell bodies in that region. Rather, we must know the behavioral correlates of all the inputs to the region and see how these are transformed in the region. The function of the region is reflected in these transformations. We know the behavioral correlates of hippocampal inputs to septum, but we do not know the behavioral correlates of other inputs. Therefore, we can not translate these data on behavioral correlates into the function of various septal nuclei as was possible in hippocampal formation. A few suggestions will nevertheless be made.

There are many possible explanations for the finding that approach-orient and motion punctuate neurons are more likely to fire if the movement had no apparent external stimuli. The firing may be associated with movements using olfactory cues. The firing may be associated with movements driven by internal cues as opposed to external cues. The firing may be associated with "self generated" behavior as opposed to responses to external stimuli. The behavioral changes associated with lesions of the septal nuclei have been interpreted to show "hyperactivity to exteroceptive stimulus events" (6), or show "an inability to effectively utilize response-produced proprioceptive feedback" (21). (However, they relate this to lesions of medial septal nucleus and abolition of the hippocampal slow wave theta rhythm). Fried (14) suggests a lesion "impairs the ability to alter a response on the basis of proprioceptive cues." (He relates this to damage to medial septal nucleus). Thus some interpretations of lesion data are consistent with these data on behavioral correlates. These views are all similar to that of Hirsh (19) and Nadel and O'Keefe (28) for hippocampus, who point out that hippocampal lesions produce an animal who is well described by classical stimulus-response psychological theory, and suggest a cognitive function for hippocampus. (This category of "self generated" behavior without external stimuli is different from Vanderwolf's type I voluntary motor behavior. Vanderwolf's voluntary behavior often occurs with extensive use of external stimuli, for instance walking across the cage to the water bottle).

No behavioral correlates have been seen in neurons of the septal nuclei which are related to emotion. Perhaps the "emotional" changes seen after lesions in septal nuclei are related to hyperactivity and loss of self generated behavior. Since all behavior is competing behavior, a loss of self generated behavior also makes it more difficult for an animal not to respond to external stimuli. The "emotional" changes after septal lesions in rats are short lived, making interpretation even more difficult. A loss of behavioral inhibition which has sometimes been suggested may have a similar basis in loss of "self generated" behavior.

There are three notable ways which correlates of neurons in the dorsal part of lateral septum differ from those in Ammon's horn: the loss of specificity, the short duration of correlation with behavior, and the relation to behavior with no apparent external cues.

Because of the lack of specificity many neurons in lateral septal nuclei of a given type seem to be the same. It was disappointing to not find stronger characteristics which could distinguish one approach-orient cell from another, or one motion punctuate cell from another. Perhaps the observer missed some significant category of behavioral correlates. Perhaps there are differences which are not reflected in behavioral correlates. On the other hand, perhaps the function of lateral septal nucleus is to transform information about specific appetitive behavior into non-specific information. As discussed above, this could be done with simple convergence of inputs.

The short duration of correlation with behavior may reflect differentiation in time, perhaps a signaling of change in attention. The preferential relation to perhaps self generated behavior suggests that a function of lateral septal nuclei neurons is to sort out from specific appetitive behaviors signaled by Ammon's horn the self generated from those generated by external cues (or whatever the appropriate distinction may be). The other inputs to lateral septum which were not studied surely would affect this sort of transformation.

The combination of these three suggestions is that the neurons in the dorsal part of lateral septal nuclei make a transformation from inputs which signal specific appetitive behavior and other inputs which have not been studied to an output which signals non-specific, self generated changes in attention.

I feel uneasy with this conclusion; this whole transformation could be done with ten or twenty neurons in lateral septal nucleus. A major dimension which distinguishes one neuron from another may have been missed which, if known, would alter the whole interpretation. Another uneasiness stems from the fact that the observations on behavioral correlates of neurons in hippocampal formation were done before the observations on neurons in septal nuclei and hippocampal neurons have not been re-examined. These neurons should now be re-examined, specifically looking for firing during behavior characteristic to neurons in septal nuclei.

The global significance of the theta rhythm system is best discussed in terms of the global function of Ammon's horn. This has been done in earlier papers (10,11,32) and will not be repeated here. The hippocampal slow wave theta rhythm does not change or even lose a beat when there is a change in attention from one non-automatic behavior to another. Therefore, while firing of neurons in lateral septal nucleus may help maintain the theta thythm, it does not affect the rhythm. Many of the effects of lesions of the septal nuclei are similar to effects of lesions of hippocampus. A discussion of the relation of hippocampal lesions to the behavioral correlates seen has been given previously (32).

The very fact that constant firing neurons do not change their firing means we have no data from which to guess their function. However, this property of constant firing in the face of widely varying behaviors is just the kind of characteristic that would be valuable for signaling some slowly changing internal state (e.g., osmolarity, glucose concentration, temperature, time spent awake or asleep). None of these neurons were observed long enough or with a known change in internal state to give any evidence for or against this suggestion.

There are changes in drinking behavior associated with lesions to some of the septal nuclei (17). Some tight group and approach-orient cells have changes in firing associated with drinking. Other changes in consummatory behavior area less obvious but also occur after lesions in septal nuclei. Behavioral correlates related to other consummatory behavior are seen in tight group, approach-orient, and motion punctuate cells.

There are thus many areas of agreement between data from lesions of septal nuclei and data from behavioral correlates of the neurons. However, a warning should be repeated. One cannot properly discuss global function of an area using data on behavioral correlates unless the behavioral correlates of all the inputs are known. We do not have adequate data for a proper discussion at present, although some of the cellular mechanisms by which these functions are performed are strongly suggested by the data of this study.

SUMMARY

1. The behavioral correlates and firing repertoires of 311 neurons in septal nuclei of 23 unrestrained rats were determined. The firing of neurons was examined during eating, drinking, grooming, during presentation of sensory stimuli, while the rats were picked up in the experimenter's hand, when the rats' tails were pinched, during different states of consciousness, while the rats explored the cage, during presentation of novel objects, and during distraction from ongoing consummatory behavior.

2. In the dorsla part of the lateral septal nucleus most of the neurons were approach-orient cells. These cells usually fired between one and ten per second. They either increased their rate of firing or synchronized firing during approach or orienting movements for 100 to 500 msec. The association of firing and movement occurred in about half of these movements, in general those approach or orienting movements for which no external stimulus was apparent. Some of these neurons decreased and others increased their rates of firing during drinking or grooming. There were only minor changes in firing during different states of consciousness. Motion punctuate cells were also present in the dorsal part of the lateral septal nucleus. These cells fired at the end approach or orienting movements for which no external stimulus was apparent. They also changed rates of firing associated with some consummatory behavior.

3. The neurons in the ventral part of the lateral septal nucleus had different behavioral correlates from the dorsal part. The behavioral correlates were not well characterized.

4. In medial septal nucleus there were two major types of cells, theta cells and tight group cells. There were four types of theta cells. Proper theta cells were just like theta cells in hippocampal formation (they increase their rates of firing if and only if there is a slow wave theta rhythm in hippocampus). Continuous theta cells fired at a theta rhythm almost all the time. Selective theta cells only increased their rates of firing some of the time during hippocampal slow wave theta rhythms but not at other times. Mixed approach orient-theta cells had components of both approach-orient and theta cells. Tight group cells fired up to thirty-five spikes with 1.0 to 5 msec. interspike intervals. This firing occurred during specific consummatory behaviors and at other times not yet determined.

5. In the 350 μM transition zone between medial septal nucleus and dorsal part of lateral septal nucleus neurons characteristic of both these nuclei were seen.

6. In the nucleus of the diagonal band of Broca there were theta cells and constant firing fast cells. Constant firing fast cells fired from 20-50 per sec. with little change no matter what the behavior. Theta cells were most common in the dorsal part of the nucleus, constant firing fast cells in the ventral part.

7. Neurons in the septofimbrial nucleus were like those in medial septal nucleus. Neurons in the bed nucleus of the stria terminalis were different from others seen but were not well characterized.

8. The behavioral correlates of approach-orient cells are consistent with the known converging excitatory inputs from pyramidal cells of dorsal CA1 and CA3. Motion punctuate cells may be driven by motion punctuate cells of dorsal Ammon's horn. Approach-orient and motion punctuate cells may have an inhibitory relation. Septal theta cells presumably drive theta cells in dorsal fascia dentata and CA3 (most of which are interneurons). Tight group cells presumably drive complex spike ce-1s of dorsal fascia dentata and CA3.

9. One cannot properly analyze the global function of any of the septal nuclei from data on behavioral correlates, because data on some important inputs does not exist. Nevertheless, these data on behavioral correlates are consistent with many of the suggestions of functions derived from lesions of septal nuclei.

REFERENCES

1. Andersen, P., Bland, B. H. and Dudar, J. D. Organization of the hippocampal output, Exp. Brain Res., 17 (1973) 152-168.

2. Andy, O. J. and Stephan, H. The Septum of the Cat, Charles C. Thomas, Springfield, Ill., 1964.

3. Apostol, B. and Creutzfeldt, O. D. Crosscorrelation between the activity of septal units and hippocampal EEG during arousal, Brain Research, 67 (1974) 65-75.

4. Bridge, J. G. Unit activity in the septal nuclei during water deprivation, drinking, and rehydration, Ph.D. dissertation at Michigan State University, 1974.

5. Bridge, J. G. and Hatton, G. I. Septal unit activity in response to alterations in bleed volume and osmotic pressure, Physiol. Behav., 10 (1973) 769-774.

6. Caplan, M. An analysis of the effects of septal lesions on negatively reinforced behavior, Behavioral Biology, 9 (1973) 129-167.

7. Coyle, P. Arterial and venous vessel patterns in rat septal structures, (this volume).

8. DeFrance, J. F., Kitai, S. T. and Shimono, T. Electrophysiological analysis of the hippocampal-septal projections: I. Response and topographical characteristics, Exp. Brain Res., 17 (1973) 447-462.

9. DeFrance, J. F., Kitai, S. T. and Shimono, T. Electrophysiological analysis of the hippocampal-septal projections: II. Functional characteristics, Exp. Brain Res., 17 (1973) 463-476.

10. DeFrance, J. F., Kitai, S. T., Hatada, K. and Christensen, C. Contrasting effects of fimbria stimulation in the septum of the cat. Brain Res., 58 (1973) 240-244.

11. Federa, R. and Ranck, J. B., Jr. Studies on single neurons in dorsal hippocampal formation and septum in unrestrained rats, II. Hippocampal slow waves and theta cell firing during bar pressing and other behaviors, Exptl. Neurol., 41 (1973) 532-555.

12. Fox, C. A., Certain basal telencephalic centers in the cat, J. Comp. Neurol., 72 (1940) 1-62.

13. Fox, S. and Ranck, J. B., Jr. Localization and identification of theta and complex spike ce-ls in dorsal hippocampal formation of rats, Exptl. Neurol., in press, (1965).

14. Fried, P. A. Septum and Behavior: A Review, Psychological Bulletin, 78 (1972) 292-310.

15. Gogolak, G., Stumpf, C., Petsche, H., and Sterc, J. Septum cell activity in the rabbit under reticular stimulation, Brain Research, 5 (1967) 508-510.

16. Gogolak, G., Stumpf, C., Petsche, H., and Sterc, J. The firing pattern of septal neurons and the form of the hippocampal theta wave. Brain Research, 7 (1968) 201-207..

17. Grossman, S. P. (A review of behavioral effects of septal lesions, this volume), 1976.

18. Hayat, A. and Feldman, S. Effects of sensory stimuli on single cell activity in the septum of the cat, Exptl. Neurol., 43 (1974) 298-313.

19. Hirsh, R. The hippocampus and contextual retrieval of information from memory: A theory, Behavioral Biology, 12 (1974) 421-444.

20. Lorente de no, R. Studies of the structure of the cerebral cortex, II. Continuation of the study of the Ammonic system, J. Psychol., 46 (1934) 113-117.

21. Lubar, J. F. and Numan, R. Behavioral and physiological studies of septal function and related medial cortical structures, J. of Behavioral Biology, 8 (1973) 1-25.

22. Macadar, O., Roig, J. A., Monti, J. M. and Budelli, R. The functional relationship between septal and hippocampal unit activity and hippocampal theta rhythm, Physiol. Behav., 5 (1970) 607-624.

23. Mays, L. E. and Best, P. J. Hippocampal unit responses to tonal stimuli during arousal from sleep in awake rats, Exptl. Neurol., 47 (1965) 268-279.

24. McLennan, H. and Miller, J. J. The hippocampal control of neuronal discharges in the septum of the rat, J. Physiol., 237 (1974) 607-624.

25. McLennan, H. and Miller, J. J. α-Aminobutyric acid and inhibition in the septal nuclei of the rat, J. Physiol., 237 (1974) 625-633.

26. Morales, R. R., Roig, J. A., Monti, J. M., Macadar, O. and Budelli, R. Septal unit activity and hippocampal EEG during the sleep-wakefulness cycle of the rat, Physiol. Behav., 6 (1971) 563-567.

27. Mosco, S., Lynch, G. and Cotman, C. W. The distribution of septal projections to hippocampus of the rat, J. Comp. Neurol., 152 (1973) 163-174.

28. Nadel, L. and O'Keefe, J. The hippocampus in pieces and patches: an essay on modes of explanation in physiological psychology. In R. Bellairs and E. G. Gray (Eds.), Essays on the Nervous System, 1974, pp. 367-390.

29. Petsche, H., Gogolak, G. and van Zwieten, R. A. Rhythmicity of septal cell discharges at various levels of reticular excitation, Electroencephalography and Clinical Neurophysiology, 19 (1965) 25-33.

30. Petsch, H., Stumpf, Ch. and Gogolak, G. The significance of the rabbit's septum as a relay station between the midbrain and the hippocampus. I. The control of hippocampus arousal activity by the septum cell, Electroencephalography and Clinical Neurophysiology, 14 (1962) 202-211.

31. Ranck, J. B., Jr. A movable microelectrode for recording from single neurons in unrestrained rats. In M. I. Phillips (Ed), Brain Unit Activity During Behavior, Univ. of Iowa Press, Iowa City, 1973a, pp. 76-79.

32. Ranck, J. B., Jr. Studies on single neurons in dorsal hippocampal formation and septum in unrestrained rats, Part I. Behavioral correlates and firing repertoires, Exp. Neurol., 41 (1973b) 461-555.

33. Ranck, J. B., Jr. Behavioral correlates and firing repertoires of neurons in dorsal hippocampal formation and septum of unrestrained rats. In R. L. Isaacson and K. H. Pribram (Eds.), The Hippocampus: A Comprehensive Treastise, Plenum Publ. Co., New York, in press, 1975.

34. Swanson, L. W. and Cowan, W. M. (Connections of the septal nuclei. This volume).

35. Stumpf, Ch., Petsche, H. and Gogolak, G. The significance of the rabbit's septum as a relay station between the midbrain and the hippocampus. II. The differential influence of drugs upon both the septal cell firing pattern and the hippocampus theta activity, Electroenceph. Clin. Neurophysiol., 14 (1962) 212-219.

36. Tombol, T. and Petsche, H. The histological organization of the pacemaker for the hippocampal theta rhythm in the rabbit, Brain Research, 12 (1969) 414-426.

37. Vanderwolf, C. H. Neocortical and hippocampal activation in relation to behavior: Effects of atropine, eserine, phenothiazines, and amphetamine, J. Comp. Physiol. Psychol., 88 (1975) 300-323.

38. Winson, J. Patterns of hippocampal theta rhythm in the freely moving rat, Electroinceph. Clin. Neurophysiol., 36 (1974) 291-301.

TOPOGRAPHIC PATTERNS OF HIPPOCAMPAL THETA RHYTHM IN FREELY MOVING RAT AND RABBIT

Jonathan Winson

The Rockefeller University, New York, New York

INTRODUCTION

In 1954 Green and Arduini (13) reported the presence of approximately sinusoidal, high-amplitude, slow waves of 3-7 cps in the hippocampus of both curarized and freely moving rabbits. In the curarized rabbit, the slow waves appeared in response to natural sensory stimulation of electrical stimulation of certain subcortical structures. The waves appeared in the freely moving animal when it was alert and interested in its surroundings. The hippocampal slow waves were termed theta rhythm in keeping with the designation that had previously been established for human EEG components in the 4-7 cps frequency range, and the theta rhythm was characterized as an arousal reaction of the hippocampus.

After this initial investigation, Green and his colleagues (14) turned to the determination of the neurophysiological source and nature of theta activity. Using the curarized rabbit, they placed a fixed electrode in the hippocampus on one side of the animal and recorded theta rhythm on the opposite side with a movable microelectrode. The microelectrode was lowered in a dorsoventral path, the theta rhythm being recorded at successive depths. Signals from the fixed electrode served as a reference to which the microelectrode signals could be compared. In this study, a sharp phase reversal of theta rhythm was found in the stratum radiatum of CA_1, the point of phase reversal coinciding with a null of theta activity. Further dorsoventral penetration showed a maximum point of theta amplitude in the distal region of the apical dendrites of the CA_1 pyramidal cells. This type of amplitude and phase profile of theta rhythm, characterized by a phase reversal with a coincident null in stratum radiatum and a more ventral peak of theta rhythm amplitude, is termed a Type I profile (Fig. 1). On the basis of their findings, Green et al. (14) suggested that theta rhythm was the reflection of postsynaptic potentials generated in CA_1 pyramidal cells. A later study, carried out in the curarized rabbit by Artemenko (4), essentially confirmed Green's results.

A number of investigators have considered the question of the neural circuitry involved in theta rhythm generation. Their results have suggested that the medial septal nucleus is an essential element in that circuitry and that periodic bursting of medial septal cells probably serves to pace the rhythm (11, 12, 28). It has also been proposed that the generating mechanism involves phasic lateral septal influences on the medial septum (19). The medial septal projection to the hippocampus is believed to be cholinergic (9, 18, 20, 21); and therefore there may be a cholinergic link in the theta-generating network.

In addition to neurophysiological studies, numerous behavioral studies have been carried out to determine the behavioral conditions in which theta rhythm occurs. The situation in this regard has proven to be more complicated than appeared initially. Theta rhythm has been found to occur in both sleeping and waking states. Further, the waking behavioral correlates of theta rhythm appear to be species-specific (34). In sleep, theta rhythm occurs throughout the rapid eye movement (REM) state but not at all in slow-wave sleep. This has been seen in all mammalian species in which theta rhythm has been detected and which exhibit REM sleep. Such species include the opossum (33), mole (1), mouse, rat, hamster, squirrel, and chinchilla (32), rabbit (15), cat (8), dog (26), and tree shrew (5). In the waking state, the correlation of theta rhythm with arousal seen in the rabbit (13) has not been found to be true across species. In the rat, for example, theta rhythm appears during "voluntary" movement (31) of the type that occurs during exploration. There is no theta rhythm when the animal is aroused but motionless, even following such intense stimulation as a foot shock. Studies in three species, the rat, rabbit, and cat (34), indicate that there is a distinct set of waking behavioral correlates of theta rhythm in each of these species.

Previous studies of the spatial patterns of theta rhythm had been carried out in acute preparations. It was decided to study these patterns under more natural conditions, both to eliminate any possible effect of using an acute preparation and to determine whether the spatial patterns of theta rhythm were dependent on behavior. An experiment was undertaken in the freely moving rat (36), in which depth profiles of theta rhythm were determined during voluntary motion and REM sleep, the two behaviors in which theta rhythm occurs naturally in this species. The objective was to study the sources of generation of theta rhythm in the freely moving rat and to ascertain whether they were the same or different during the two theta-correlated behaviors.

METHODOLOGY

The methodology employed in the experiment was similar to that used by Green et al. (14) in that recordings were taken at successive depths from a microelectrode which was lowered through the neocortex and dorsal hippocampus of the animal. There were two differences, however. In the present experiment, two recordings were taken at each microelectrode depth, one during an episode of voluntary movement and another during an episode of REM sleep. Secondly, two fixed electrodes, rather than one, were implanted in the contralateral hippocampus to serve as references to which signals recorded from the microelectrode could be compared. It had been found previously that theta rhythm exhibited a phase reversal in the stratum radiatum of CA_1 (Fig. 1). In this experiment, one fixed electrode was implanted either in or dorsal to the pyramidal layer of CA_1 and another ventral to the hippocampal fissure. The inset of Fig. 3 shows a typical placement of a pair of fixed electrodes, as well as a representative microelectrode path. It was expected that the signals from these two fixed electrodes would be phase-reversed with respect to one another, as they proved to be. The upper and lower records of each panel of Fig. 2 show such phase-reversed signals from a pair of fixed reference electrodes. The purpose of using pairs of fixed electrodes straddling the region of phase reversal was to provide a reference signal that was in phase with the signal from the microelectrode whether the microelectrode was recording from a position above or below the phase reversal point. The availability of in-phase reference signals allowed more precise determination of phase and amplitude profiles of theta rhythm than would be possible otherwise. Other aspects of the methodology are described elsewhere (35,36).

RESULTS

The major results of this experiment were the following. The depth profiles of phase and amplitude found during the two behavioral states of voluntary movement and REM sleep were essentially the same, implying that the neural mechanisms responsible for generating theta activity were basically the same in the two behaviors. However, the profiles were not the Type I that were reported in the curarized rabbit. Dorsoventral microelectrode penetration in the freely moving rat did not reveal either a sudden phase reversal or a null of theta activity in the stratum radiatum. Rather, a slowly developing phase shift was detected by the descending microelectrode, the shift starting in the stratum radiatum

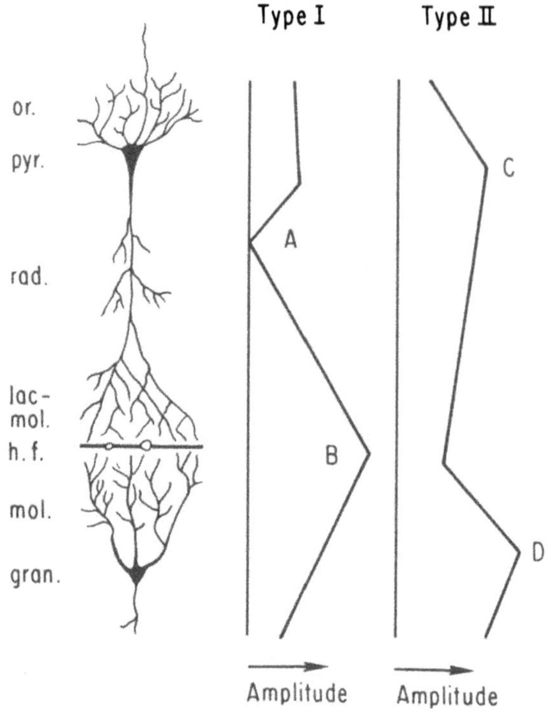

Figure 1. Depth profiles of theta rhythm. Left: strata oriens, pyramidale, radiatum, and lacunosum-moleculare of CA_1, hippocampal fissure, and strata moleculare and granulosum of dorsal blade of dentate gyrus. Center: Type I profile seen in curarized rabbit. Null and sudden phase reversal at A, amplitude peak B ventral to null region. Peak, shown in this Figure as occurring at hippocampal fissure, was reported within CA_1 in curarized rabbit (4,14). Right: Type II profile seen in freely moving rat (36). Peak C in or near str. pyramidale, slowly developing phase reversal without null from C to hippocampal fissure. Higher peak D in dorsal blade of the dentate gyrus.

and reacing phase reversal at the hippocampal fissure some 175-350 μ ventral to the depth at which the phase shift was first detected. A clear theta rhythm was present throughout the phase shift region. Fig. 2 shows the slowly developing phase reversal. In this Figure, the upper records are from a fixed electrode placed dorsal to the stratum radiatum, the middle records are from the microelectrode and the lower records are from a ventrally placed fixed electrode in the dentate gyrus. In all four panels, the signals from the two fixed electrodes are phase-reversed with

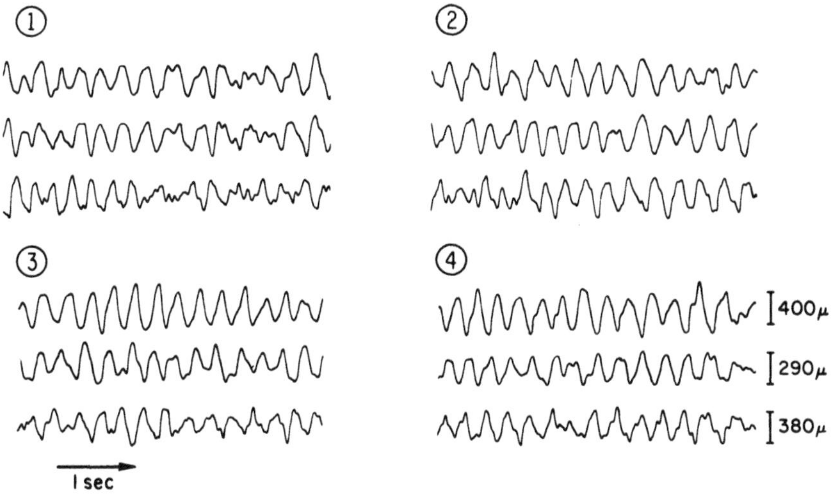

Figure 2. Records showing development of slow phase reversal in Type II profile. Upper record: dorsal fixed electrode. Lower record: ventral fixed electrode. Middle record: microelectrode. Fixed electrode signals are phase-reversed. (1)-(4): Four successively more ventral positions of the microelectrode, 88 μ apart. (1) Microelectrode signal in phase with signal from dorsal macroelectrode. Wave shapes are similar. (2) Microelectrode signal leads by a mean angle of 38°. There is still some resemblance between the signals. (3) Mean lead angle now 85°. (4) Growing similarity in shape between signals from microelectrode and ventral fixed electrode. Mead lead angle is 139°.

respect to one another. In panel 1, the microelectrode signal is in phase with the dorsal fixed electrode and both signals have the same wave shapes. As the microelectrode is advanced, the microelectrode signal begins to lead the dorsal electrode signal and the wave shapes become dissimilar. In panel 4, the microelectrode signal is approaching an in-phase relationship with the signal from the ventral electrode and the wave shapes of these two signals are now growing similar. The phase reversal was complete at the hippocampal fissure and no further change of phase occurred in the hippocampus.

The amplitude of theta rhythm in the freely moving rat showed two distinct peak values, one located approximately in the stratum pyramidale of CA_1, the second in the dorsal blade of the dentate gyrus (Fig. 1). This is in contrast to the single amplitude peak in the ventral part of CA_1 characteristic of the Type I profile reported in the curarized rabbit. The theta rhythms recorded at the two positions in the brain at which the two peaks occurred in the rat were phase-reversed with respect to one another, the slow phase occurring between them over the length of the stratum radiatum of CA_1 as described. The phase and amplitude profile found in the freely moving rat is called Type II (Fig. 1).

The Type II profile suggests that there are two generators of theta rhythm in the dorsal hippocampus of the freely moving rat, one associated with the CA_1 pyramidal cells and the other with the granule cells of the dorsal blade of the dentate gyrus. The generators appear to be closely coupled and are phase-reversed with respect to one another.

Profiles in the Curarized Rat

The finding of a Type II profile in the freely moving rat, in contrast to the Type I profile reported in the curarized rabbit, suggested that the difference might be due to curare and so a study was made of the curarized rat. Profiles were determined in several curarized rats in a preliminary experiment (36) in which theta rhythm was produced by systemic injection of eserine (28). It was necessary to produce theta rhythm in this way because theta rhythm is difficult to evoke in the curarized rat by natural means. In these curarized and eserinized rats, microelectrode penetrations showed a Type I profile. The point of sudden phase reversal and null was in the proximal stratum radiatum as in the curarized rabbit. As is the case in the Type I profile, there was an amplitude peak ventral to the point of phase reversal. The position of the peak in the brain was not determined.

A more complete investigation (37) was then undertaken, with the objective of accurately determining the theta rhythm profile in the curarized rat, using natural rather than pharmacological means to elicit theta rhythm. As was mentioned, it was found that, in contrast to the curarized rabbit, theta rhythm was not readily evoked by natural means in the curarized rat. Visual and auditory stimuli were completely ineffective. Brief episodes of theta rhythm could be evoked by tactile stimuli but they were insufficiently sustained for accurate data analyses. A natural means was found to produce clear and long-lasting theta rhythm. The animal was placed on a turntable 24 inches in diameter, air for respiration being brought in through a bearing in the base of the device. Rotation of the table at a speed of 15-30 rpm produced an immediate train of theta waves. The effect habituated with continued turning but could be reinstituted by reversing the direction.

A study was carried out eliciting theta rhythm in the manner just described, and the results verified the presence of a Type I theta profile in the curarized rat. Fig. 3 shows records from a typical penetration. The inset illustrates the positions of the fixed reference electrodes and the path of the microelectrode in this experiment. In panels (a) - (c), the upper record is from the dorsal electrode (electrode 2), the middle record is from the microelectrode, and the lower record is from the ventral electrode (electrode 1). In all recordings, the signals from the fixed reference electrodes were approximately phase-reversed. Dorsal to the stratum radiatum, the signal from the microelectrode was approximately in phase with the signal from the dorsal reference electrode and the wave forms of the two signals were similar. This is shown in the upper two records of panel (a). As the microelectrode penetrated the proximal stratum radiatum, the theta rhythm abruptly disappeared and was replaced by a null signal (Fig. 3b). The null signal was present for a dorsoventral distance of about 90 μ, below which the theta rhythm reappeared, now approximately in phase with, and similar in form to, the signal from the ventral fixed electrode (lower two records of Fig. 3c). As is characteristic of the Type I profile, a peak of theta amplitude was found ventral to the point of phase reversal. The location of this peak was determined by terminating a number of penetrations immediately ventral to the peak, and noting the position of the tip of the electrode histologically in each case. The peak of the amplitude profile was found to be close to the hippocampal fissure within a span running from 70 μ above the fissure to 50 μ below it.

The finding of a Type I profile in the curarized rat, taken in conjunction with the presence of a Type II profile in the freely moving animal, offered the opportunity to follow the time course of the effect of curare on the theta profile. A microelectrode penetration was made in the curarized rat to the null point of the Type I profile (point A in Fig. 1). The microelectrode was then left in place and the animal allowed to recover. At this location in the brain one would expect to record a substantial theta rhythm when the animal had recovered and the profile had reverted to Type II (see Fig. 1). Fig. 4 illustrates an experiment of this kind.

The upper records are from a dorsal reference electrode and the lower records from the microelectrode left in position at the null point in a curarized rat. Panel (a) is a record taken approximately 2 hours after injection of d-tubocurarine chloride (2 mg/kg, i.p.). At this time the animal was still being artificially respired but was showing the first signs of recovering voluntary muscular activity. The null signal at the microelectrode was the same as that recorded upon initial penetration to the null point some 1 3/4 hours earlier. Panel (b) is a record taken 45 min later. The animal was sufficiently active to free itself from the face mask and was breathing unaided. The null signal was growing

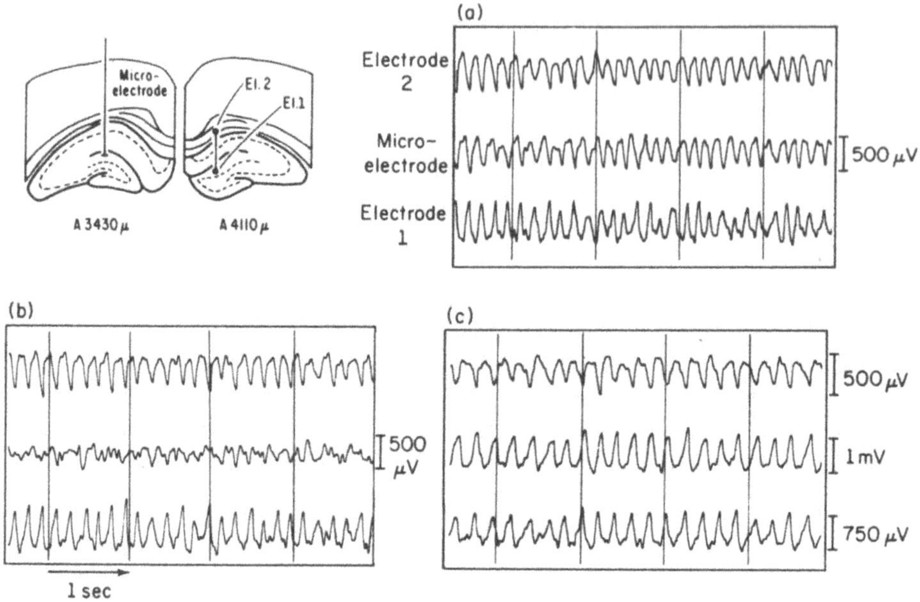

Figure 3. Records from penetration in the curarized rat. Electrode positions shown in inset. Location designations refer to König and Klippel's atlas (17). Upper record is from dorsal reference electrode, middle record is from microelectrode, and lower record is from ventral reference electrode. Amplifier gain from microelectrode signal is reduced by a factor of 2 in (c). Signals from electrodes 2 and 1 are phase-reversed. Microelectrode is dorsal to str. radiatum in (a); signal is approximately in phase with electrode 2. Signals have similar wave shapes. Microelectrode is in proximal str. radiatum in (b) and records a null signal. In (c) microelectrode is in ventral str. radiatum, signal is in phase with electrode 1. Wave shapes are similar.

in amplitude. Eight minutes later (panel [c]) the rat was moving in its cage with no effects of paralysis evident. The signal at the microelectrode continued to grow in amplitude and regularity, stabilizing as a steady, high amplitude, theta rhythm some 30 min later (panel [d]). The effect of curare on theta rhythm outlasted the visible effect on muscular paralysis by about 30 min.

Generators in the Freely Moving Rabbit

The presence of a Type II profile in the freely moving rat but a Type I profile in both the curarized rat and the curarized rabbit raised several possibilities. It was possible that the

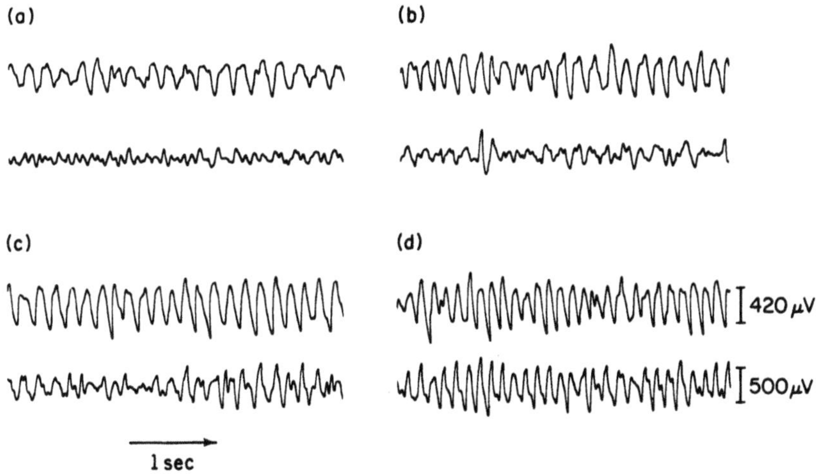

Figure 4. Records from curarized rat during recovery. Upper records are from fixed reference electrode in str. oriens. Lower record is from microelectrode locked in place at null position in the curarized animal. Voltage scales in (d) apply to all records. In (a) animal had been immobilized for 2 hr and was showing first signs of voluntary movement. Null signal was as observed in initial penetration. In (b), 45 min later, animal was breathing unaided. In (c), 8 min after (b), animal was moving with no behavioral effects of curare evident. Although fixed electrode shows continuous theta rhythm, microelectrode rhythm is sporadic. In (d), recorded 30 min after (c), the microelectrode signal has reverted to clear theta thythm.

normal profile in the freely moving rabbit was Type II, as in the freely moving rat, but that curare changed the rabbit profile to Type I as it did in the rat when curare was injected. Alternatively, the type of profile found in both the rat and the rabbit might be dependent upon the behavioral condition in which theta rhythm was generated. In the rabbit, theta rhythm occurs normally during REM sleep, voluntary action, and during a condition of arousal in which the animal while motionless is presented with a novel sensory stimulus. The studies in the curarized rabbit in which the Type I profile was reported used only one natural condition to elicit theta rhythm, namely, the application of sensory stimuli; in the freely moving rat, on the other hand, the Type II profile was found during REM sleep and voluntary movement. It was possible that the Type I profile was generated during sensory stimulation and the Type II during REM sleep and voluntary movement.

As another alternative, it was possible that there might be a species difference in theta generation between the rat and the rabbit, in which the normal profile was Type II in the rat and Type I in the rabbit in all behavioral conditions in which the theta rhythm appears. An investigation (38) was carried out in the freely moving rabbit to resolve this question.

The methodology used in this study of the freely moving rabbit was the same as that used to investigate the freely moving rat with the exception that profiles were determined in three behavioral conditions: REM sleep, voluntary movement, and the application of sensory stimulation rather than the two behavioral conditions studied in the rat. In the microelectrode penetrations, recordings were taken during all three behavioral conditions at each depth of the microelectrode in the brain. The profiles found were essentially the same for all behavioral conditions. The common profile was Type I with a phase reversal and null in the proximal stratum radiatum of CA_1. This is normal for the Type I profile. However, the peak of the amplitude profile was not in CA_1 as reported in the curarized rabbit but was located in the lower two-thirds of the molecular layer of the dorsal blade of the dentate gyrus.

Fig. 5 shows a representative profile. The inset illustrates the positions of the two reference electrodes straddling the stratum radiatum and the path of the microelectrode. On the abscissa, left to right represents dorsoventral microelectrode penetration in the brain. The depth of penetration is indicated in units of 352 μ (one turn of the microelectrode-advancing screw). The zero position on the abscissa is arbitrary. The ordinate is amplitude of theta rhythm. Points are plotted for all behavioral conditions. The profile shown in typically Type I. Several penetrations were terminated in the vicinity of the null of the profile to determine its position unequivocally in the brain. The null was located in the proximal stratum radiatum, 50-150 μ below the pyramidal layer. It is seen in Fig. 5 that dorsal to the stratum radiatum the profile is flat. Recordings in this region show microelectrode signals that are in phase with and have similar wave shapes to the signals from electrode 2. The term isomorphic is used in Fig. 5 to describe signals that are approximately in phase with one another and have similar wave shapes. The sudden null in the proximal stratum radiatum occurs at the same microelectrode depth in all behavioral conditions. Ventral to the null, the microelectrode signals are isomorphic to the signals from electrode 1. There is a rapid growth of amplitude below the null and a peak in the dentate gyrus.

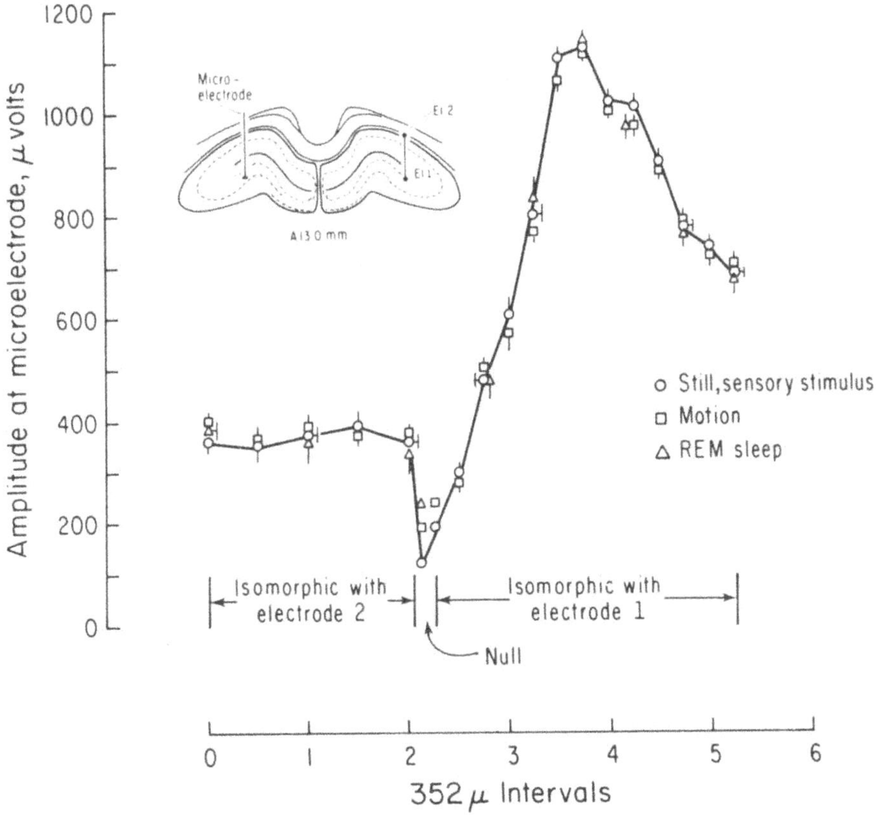

Figure 5. Amplitude profile: Normalized amplitude of theta rhythm versus depth in brain for three behavioral conditions. Inset shows electrode positions. Location designation refers to Urban and Richard's atlas (30). Left to right: Dorsoventral penetration of microelectrode in units of 352 μ (one turn of microelectrode-advancing screw). Zero chosen for abscissa is arbitrary. Ordinates were normalized by dividing the signal at the microelectrode by the simultaneously recorded signal at an in-phase reference electrode. For sake of comparison, final curves were relabeled so that the value in microvolts at the peak was the same for all three behavioral conditions and was the actual value recorded during REM sleep at this point in the brain (38). Vertical bars in this Figure and in Fig. 6 indicate standard errors of the mean.

Fig. 6 is representative of several penetrations carried out to locate the position of the amplitude peak in the brain. In this penetration, the electrode was advanced in steps of 44 μ in the vicinity of the peak. The penetration was terminated 88 μ after the peak was reached. The amplitude profile is shown in the lower right-hand panel. Left to right indicates dorsoventral penetration of the brain. The upper panel shows the area of penetration and the lower left panel shows the point of termination of the traverse as marked by a Prussian Blue reaction. Such penetrations located the peak in the lower two-thirds of the molecular layer of the dentate gyrus granule cells.

The finding of a Type I profile with its peak in the dentate gyrus suggests the presence in the freely moving rabbit of two generators, closely coupled and phase-reversed with respect to one another, one associated with the cells of dentate gyrus and the other with the overlying CA_1 pyramids. The generators appear to be the same during REM sleep, voluntary motion, and the application of sensory stimuli.

DISCUSSION

The investigation of theta rhythm in the freely moving rat (36) summarized in this chapter found a Type II theta profile during REM sleep and voluntary movement and suggested the existence of two generators of theta rhythm in the dorsal hippocampus, one in CA_1 and the other in the dentate gyrus. Recently, Bland et al. (7) investigated theta rhythm in the lightly anesthetized rabbit. In that study, theta rhythm was elicited either by electrical stimulation of the hypothalamus or by application of sensory stimulation. The investigators found a Type I profile with the peak amplitude of the profile in the dentate gyrus, thus suggesting the presence of CA_1 and dentate gyrus generators in the anesthetized rabbit as well as in the freely moving rat. Further, these workers provided additional evidence of the existence of separate dentate gyrus and CA_1 generators by selectively eliminating the CA_1 generator, using cooling or surgical intervention in the CA_1 region. The study of the freely moving rabbit (38) just described found a Type I profile with the amplitude peak in the dentate gyrus during REM sleep, voluntary movement, and in the condition of arousal induced by sensory stimulation. It would appear then, that in both the anesthetized and the freely moving rabbit and also in the freely moving rat there are two generators of theta rhythm in the dorsal hippocampus. The generators are closely coupled and phase-reversed with respect to one another. One generator seems to be associated with the granule cells of the dentate gyrus and the second with the overlying CA_1 layer. The mechanism of theta rhythm generation is unknown as is the possible role of this mechanism in hippocampal function. The potentials of theta rhythm

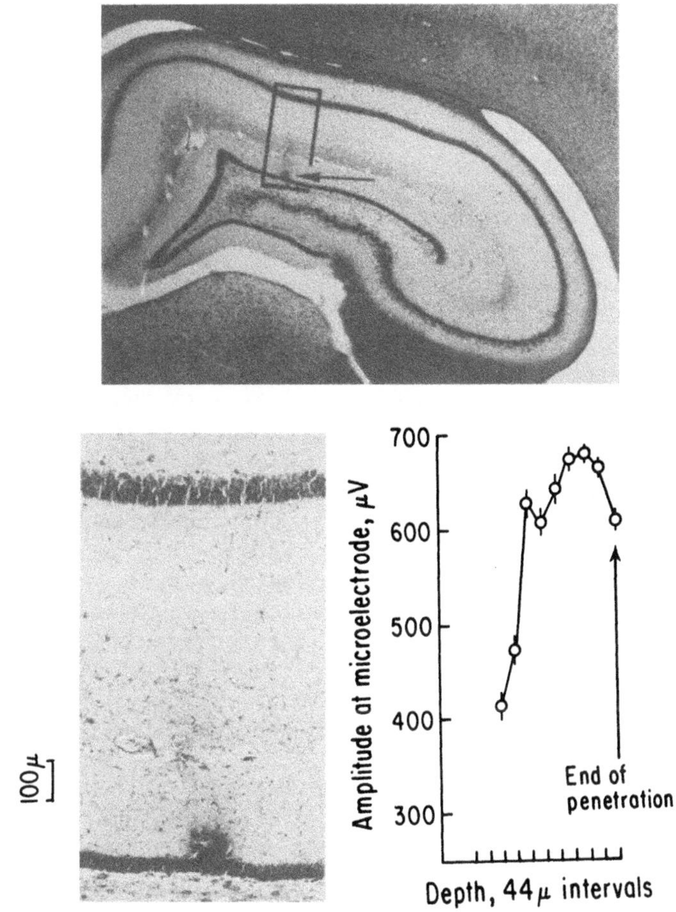

Figure 6. Penetration to determine location of peak. Lower right: Plot of normalized amplitude of theta rhythm versus depth in brain. Recordings were made during movement. Dorsoventral penetration is represented from left to right on the abscissa. Electrode was advanced in 44 μ steps and the penetration was terminated 88 μ ventral to the peak. Upper photograph shows location of penetration. Enlargement at lower left shows position of microelectrode tip at termination, directly dorsal to the granule cell layer.

are probably a reflection of synchronous postsynaptic potentials occurring in populations of hippocampal neurons (4,10). In particular, Andersen and Schwartzkroin (3) have recently reported synaptic potentials in dentate gyrus cells that were synchronous with the local theta activity.

As has been noted, the theta rhythm occurs during species-specific waking behaviors in both rat and rabbit as well as during REM sleep. The foregoing results indicate that the neural mechanism responsible for theta rhythm generation is probably the same in each of these species during REM sleep and the theta-correlated waking behaviors in the species. Further research may elucidate this relationship between the hippocampus, REM sleep, and behavior.

The finding of a Type II profile in freely moving rat and a Type I profile in the freely moving rabbit indicates a species difference, primarily it would seem in the CA_1 generator. There may be a neuroanatomical basis for this difference. One such neuroanatomical difference among species may exist in the medial septal-CA_1 projection. Degeneration studies in the rat have variously reported a minor projection from the medial septal nucleus to CA_1 (20,21) or have found no projection at all (24). The minimal nature of this pathway has been confirmed by an autoradiographic investigation reported at this conference (29). On the other hand, a substantial medial septum - CA_1 projection has been noted in the cat (16,25), and a medial septal - CA_1 pathway has been demonstrated electrophysiologically in the rabbit (2). Differences of acetylcholinesterase distribution in the rat and rabbit hippocampus have also been reported (27).

It has been shown (37) that systemic injection of curare in the rat causes a change in type of profile from a Type II profile to a Type I profile. It is not known whether the curare is acting centrally. However, the fact that the effect on profile outlasts the paralytic effect suggests that the profile change is not a result of paralysis. Central effects of systemic injection of curare have been reported (23). As has been indicated the medial septal - hippocampal projection is believed to be cholinergic. An iontophoretic study has shown the presence in the hippocampus of cells selectively blocked by curare (6). With the detailed circuitry involved in theta rhythm generation unknown at this time, the mechanics of the curare effect cannot be delineated. However, since the action of curare changes the normal Type II profile in the rat to a Type I profile such as is seen in the rabbit, it is possible that a species difference in the neural circuitry underlying theta rhythm generation is responsible for the difference of theta profiles between the freely moving rat and the freely moving rabbit, and that the difference in the neural circuitry involves nicotinic cells, so that the blockade of these cells by curare in the rat reduces the different systems in the two species to systems that are functionally the same. Clarification requires further investigation of the theta-generating mechanism.

REFERENCES

1. Allison, T., and Van Twyver, H., Sleep in moles, Scalopus aquaticus and Candylura crestata, Exp. Neurol., 27 (1970) 564-578.

2. Andersen, P., Bruland, H., and Kaada, B. R., Activation of the field CA_1 of the hippocampus by septal stimulation, Acta physiol. scand., 51 (1961) 29-40.

3. Andersen, P., and Schwartzkroin, P. A., Mechanism of dentate theta generation in rabbits. In preparation (1975).

4. Artemenko, D. P., Role of hippocampal neurons in theta-wave generation, Nierofiziologiya, 4 (1972) 531-539.

5. Berger, R. J., and Walker, J. M., A polygraphic study of sleep in the tree shrew (Tupaia glis), Brain Behav. Evol., 5 (1972) 54-69.

6. Biscoe, T. J., and Straughan, D. W., Micro-electrophoretic studies of neurones in the cat hippocampus. J. Physiol. (Lond.), 183 (1966) 341-359.

7. Bland, B. H., Andersen, P., and Ganes, T., Two generators of hippocampal theta activity in rabbits. In preparation (1975).

8. Brown, B. B., Frequency and phase of hippocampal theta activity in the spontaneously behaving cat, Electroenceph. clin. Neurophysiol., 24 (1968) 53-62.

9. Dudar, J. D., The effect of septal nuclei stimulation on the release of acetylcholine from the rabbit hippocampus, Brain Research, 83 (1975) 123-133.

10. Fujita, Y., and Sato, T., Intracellular records from hippocampal pyramidal cells in rabbit during theta rhythm activity, J. Neurophysiol., 27 (1964) 1011-1025.

11. Gogolák, G., Petsche, H., Sterc, J., and Stumpf, C., Septum cell activity in the rabbit under reticular stimulation, Brain Research, 5 (1967) 508-510.

12. Gogolák, G., Stumpf, C., Petsche, H., and Sterc, J., The firing pattern of septal neurons and the form of the hippocampal theta wave, Brain Research, 7 (1968) 201-207.

13. Green, J. D., and Arduini, A. A., Hippocampal electrical activity in arousal, J. Neurophysiol., 17 (1954) 533-557.

14. Green, J. D., Maxwell, D. S., Schindler, W. J., and Stumpf, C., Rabbit EEG "theta" rhythm: Its anatomical source and relation to activity in single neurons. J. Neurophysiol., 23 (1960) 403-420.

15. Harper, R. M., Frequency changes in hippocampal electrical activity during movement and tonic immobility, Physiol. Behav., 7 (1970) 55-58.

16. Ibata, Y., Deisraju, T., and Pappas, G. D., Light and electron microscopic study of the projection of the medial septal nucleus to the hippocampus of the cat, Exp. Neurol., 33 (1971) 103-122.

17. König, J. F. R., and Klippel, R. A., The Rat Brain, Kreiger, Huntington, N.Y., 1970, 162 pp.

18. Lewis, P. R., and Shute, C. C. D., The cholinergic limbic system: Projection to hippocampal formation, medial cortex, nuclei of the ascending cholinergic reticular system and the subfornical organ and supra-optic crest, Brain, 90 (1967) 521-537.

19. McLennan, H., and Miller, J. J., The hippocampal control of neuronal discharges in the septum of the rat, J. Physiol. (Lond.), 237 (1974) 607-624.

20. Mellgren, S. I., and Srebro, B., Changes in acetylcholinesterase and the distribution of degenerating gibres in the hippocampal region after septal lesions in the rat, Brain Research, 52 (1973) 19-36.

21. Mosko, S., Lynch, G., and Cotman, C. W., The distribution of septal projections to the hippocampus of the rat, J. comp. Neurol., 152 (1973) 163-174.

22. Nafstad, P. H. J., An electron microscope study on the termination of the perforant path fibres in the hippocampus and the fascia dentata, Z. f. Zellforsch., 76 (1967) 532-542.

23. Purpura, D. P., and Grundfest, J., Nature of dendritic potentials and synaptic mechanisms in cerebral cortex of cat, J. Neurophysiol., 19 (1956) 573-595.

24. Raisman, G., Cowan, N. M., and Powell, T. P. S., The extrinsic afferent, commissural and association fibres of the hippocampus, Brain, 88 (1965) 963-996.

25. Siegel, A., and Tassini, J. P., Differential efferent projections of the lateral and medial septal nuclei to the hippocampus in the cat, Brain Behav. Evol., 4 (1971) 201-219.

26. Shimazoni, Y., Horie, T., Tanagisawa, Y., Hori, N., Chikazawa, S., and Shozuka, K., The correlation of the rhythmic waves of the hippocampus with the behavior of dogs, Neurol. Med. Chir., 2 (1960) 82-88.

27. Storm-Mathisen, J., Quantitative histochemistry of acetylcholinesterase in rat hippocampal region correlated to histochemical staining, J. Neurochem., 17 (1970) 739-750.

28. Stumpf, C., Drug action on the elctrical activity of the hippocampus, Internat. Rev. Neurobiol., 8 (1965) 77-138.

29. Swanson, L. W., The organization of the afferent and efferent connections of the rat septum. Paper presented at International Symposium on the Septal Nuclei, Detroit, Michigan, October, 1974.

30. Urban, I., and Richard, P., A Stereotaxic Atlas of the New Zealand Rabbit's Brain, Charles C. Thomas, Springfield, Ill., 1972, 86 pp.

31. Vanderwolf, C. H., Hippocampal electrical activity and voluntary movement in the rat, Electroenceph. clin. Neurophysiol., 26 (1969) 407-418.

32. Van Twyver, H., Sleep patterns in five rodent species, Physiol. Behav., 4 (1969) 901-905.

33. Van Twyver, H., and Allison, T., Sleep in the opossum Didelphis marsupalis, Electroenceph. clin. Neurophysiol., 29 (1970) 181-189.

34. Winson, J., Interspecies differences in the occurrence of theta, Behav. Biol., 7 (1972) 479-487.

35. Winson, J., A compact movable microelectrode assembly for recording from the freely-moving rat, Electroenceph. clin. Neurophysiol., 35 (1973) 215-217.

36. Winson, J., Patterns of hippocampal theta rhythm in the freely moving rat, Electroenceph. clin. Neurophysiol., 36 (1974) 291-301.

37. Winson, J., Hippocampal theta rhythm. I. Topographical patterns in the curarized rat. Submitted manuscript (1975).

38. Winson, J., Hippocampal theta rhythm. II. Topographic patterns in the freely moving rabbit. Submitted manuscript (1975).

Thanks are given to Elsevier Publishing Company for permission to reproduce the following figures :

Fig. 3 from Electroencephal. Clin. Neurophysiol. (1974), Vol. 36, pp. 291-301.

Figs. 3 and 7 from Brain Research (1976) Vol. 103, pp. 57-70.

Figs. 3 and 4 from Brain Research (1976) Vol. 103, pp. 71-79.

IMMEDIATE AND LONG-TERM EFFECTS OF SEPTAL AND FRONTAL ABLATIONS ON THE SPECIES-TYPICAL BEHAVIOR OF THE RAT[1]

Thomas F. Herrmann and Joel F. Lubar

Department of Psychology, U. of Guelph, Guelph, Ontario
Department of Psychology, U. of Tennessee, Knoxville, Tennessee

INTRODUCTION

It has been accepted that the role of the septum is that of being part of the neural substrate for mediation and expression of behavior's emotional or motivational component (24). For the most part, explanations of this role have been couched in descriptors connotating emotional behavior which should be recognized in gross behavioral displays. Examples of such terms are hyper-emotionality (18,30,48), rage reactions (14,15), increased or hyper-reactivity (36), hyper-activity (25,38), hyper-sensitivity (33,52), increased boldness (30,58,62), disinhibition (27,42,50), decreased fear (50,53), and altered species-specific defense reactions (2,54,48,56,59).

Such states of behavior have often been found as dissipating over time (55,63), depending on the post-operative exposure of the subject to the stimulus events used in the testing situation (47) or the introduction of intervening variables, such as handling (36).

Other investigators have suggested that behavior displays result from changed stimulus mediation properties (33), such as a deficit in the ability to utilize proprioceptive stimuli or exaggerations (63) of reinforcing effectiveness of stimuli (24), again indicating gross behavioral deficits (46).

---

[1]This research was supported by National Institute of Mental Health Training Grant # MH 10513-08, to Joel F. Lubar.

While evidence has accumulated for gross deviancies in septal-operated subjects, so has evidence accumulated for specific deficits such as water intake (10,26,28,43,51). From all of the evidence, one imagines that a septal animal must present itself as a highly unusual, hyperemotional, reactive, bold, fearless, disinhibited, altered species-specific defense reactor who is most interested in exteroceptive stimuli and drinks a lot.

The question remains: Is this indeed the case? All of the descriptors mentioned have been the result of behavior-restrictive laboratory tasks. Each task is highly selective in the behavioral displays. Differences between operations, apparatus, and procedures make any general description tenuous. Questions about which task-specific deficits are related to a general syndrome and which are stimuli-specific are difficult to answer from the available evidence (44,65).

While septal ablated animals should be interesting in themselves, the question of identifying a subject who has lesion damage which affects its behavior is intrinsically interesting. At the present time there is little known about the effect of frontal damage on species-typical behavior in the rat (16,19,32, 35,37,49,60,64). McCleary (66) has suggested an antagonistic relationship of cortex to septum. This antagonism should be displayed in different patterns than those exhibited by septally-lesioned subjects.

The studies reported here examined a large number of behaviors in a semi-natural environment. This was a fluid situation, allowing an examination of the subjects' adaptations to the milieu and the adjustment of others in the milieu to the individual subject. The use of the semi-naturalistic environment for studying behavioral changes could facilitate the interpretation of results of experimentally-induced brain damage. Hypotheses derived from such situations may be useful in explaining the types of deficits that are commonly reported in the more restrictive laboratory test situation.

METHOD

To investigate these phenomena regarding their importance or relevance as indicators of gross behavior, a series of experiments has been conducted using what we are calling a "free response choice social" situation. In this situation, behavior was measured by observational counts of species-typical behaviors. The situation was composed of five inter-connected chambers housed in a temperature-regulated room.

Apparatus

The apparatus (Fig. 1) was constructed of plywood, with the exception of the front wall of each chamber: this was made of clear plexiglass and served as an observation window. Access to the external chambers was obtained via the central chamber by using wire mesh ramps. Each chamber contained food and water sources. A 24-hour reversed diurnal cycle was maintained by use of two 25-watt incandescent bulbs in each chamber. One bulb was red and remained on at all times. The other bulb was white and was controlled by a commercial timer. The white bulb was turned on to simulate the 12-hour daylight portion and off to simulate night. Each outer chamber contained a nesting box filled with paper strips. A 2-inch layer of sanitary material covered the floor of each box.

Observations

Observations of behaviors were made during the red light only (i.e., night cycles) during 3-hour counterbalanced blocks such that both peak and lesser activity periods received equal treatment.

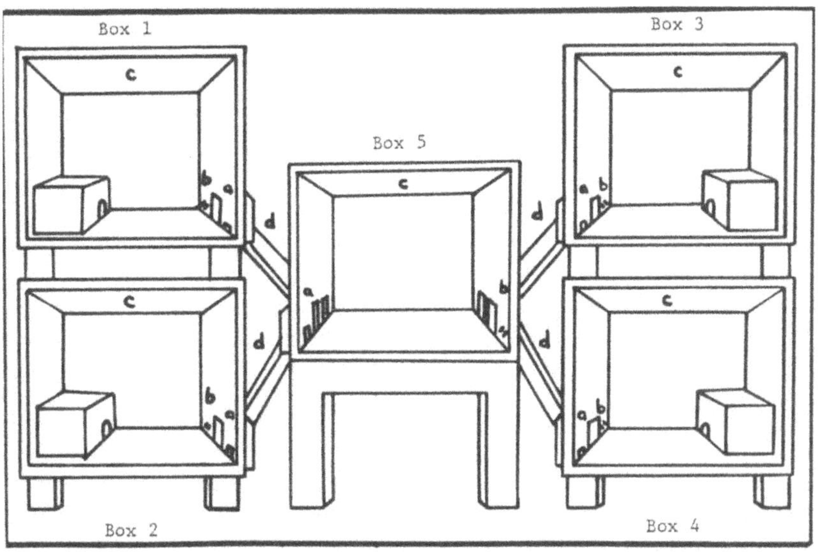

Figure 1. Diagram of Free-Choice Social Situation. The milieu consisted of five chambers (122 x 92 x 76 cm), each containing a food source (a), a water source (b), and an internal lighting system (c). The four exterior boxes (designated 1 - 4) each contained a nesting box and was connected to the central chamber (box 5) by means of wire ramps (d).

Behavioral counts consisted of postures, position, and locomotions obtained from an extensive review of the ethological literature, typified by Barnett (4,5,6,7,8), Calhoun (20,21,22, 23), Dewsberry (31), Grant (39,40,41), Ewer (34), and MacIntosh (40). Over the years, these sources have provided a basic catalog of rat behavior. Descriptions of the behaviors were then modified by pilot research to the point where high inter-observer reliability was achieved. The present observational set consists of over 160 recognizable postures and locomotions which, for ease of interpretation, have been reduced in number and grouped loosely into eight categories which refer to identifiable bodies of the literature (Table 1).

Counts of behavior (i.e., frequency and order of occurrence) were made daily on each animal through the use of a 5-minute each focal-animal technique (9,29,44,45,57,65). During the daily period, each action and interaction of each animal was noted by use of a code whispered into a tape recorder. Five different observers have participated in these experiments and have achieved inter-rater reliabilities of greater than .9 throughout. In addition to behavior, a location suffex was added to detail the "where" of each behavior (Fig. 2).

Procedures

The experimental design chosen was a repeated measures multi-group type. Each experiment consisted of two phases, one pre- and the other post-surgery. Pre-operative phases consisted of ear marking, randomly assigning a group designation, and placing 30 Long Evans rats (15 males, 15 females, divided equally into treatment groups) into the central chamber and, for 4 days, allowing them to habituate to the apparatus, to each other, and to the presence and behaviors of the observers. This habituation period was followed immediately by the observation period, which lasted for either 28 or 14 days. At the end of the observational period, subjects were removed from the apparatus and placed in individual cages. Surgery under sodium pentobarbital anesthesia was performed on the following day. Septal ablations consisted of 1.5 ma anodal current applied at the bilateral co-ordinates of 1 mm anterior to bregma, 5 mm lateral to midline, @ 25° from midline lowered 5.5 mm below dura for 20 seconds. Frontal ablations (32,49) consisted of bilateral aspirations from 2 mm anterior to bregma forward to the tip of the frontal poles. Laterally, the lesions extended approximately 2 mm to a depth of 4 - 5 mm below dura. Sham surgery consisted of complete operative procedure except that no electrode placement was made. Upon recovery from anesthesia, subjects were returned to individual cages. In experiments which studied the immediate effects of lesions, the subjects were returned to the milieu on the second day post-surgery

TABLE 1

LISTING OF BEHAVIORS AND THEIR DEFINITIONS BY CATEGORIES

Category	Behavior
Eating	"Source" - self-explanatory
	"Scattered" - eating of food anywhere other than source
	"Debris" - eating of material other than food
	"Carrying" - self-explanatory
	"Manipulation" - maneuvering of food with forepaws
	"Stealing" - taking food away from the possession of another rat
	"Cannibal" - eating of pups or of deceased rats
	"Hoarding" - self-explanatory
Drinking	"Drinking" - self-explanatory
Explanatory	"Jump" - self-explanatory
	"Upright" - standing or sitting on hind legs
	"Investigative Sniff" - standing at halt or moving slowly with nose close to substrate
	"Frozen Attention" - remaining motionless while oriented to some stimulus
	"Crawl" - a low-profile slink
	"Walk" - self-explanatory
	"Burst" - quick run
Nesting	"Fraying" - drawing material through mouth by means of forepaws, usually shredding material
	"Collecting" - carrying nesting material in mouth
	"Arranging" - self-explanatory
	"Pulling-In" - pulling material into nest site with forepaws
	"Digging" - self-explanatory
	"Tunnelling" - self-explanatory

TABLE 1 (continued)

Category	Behavior
Sleeping	"Watchful" - lying on side or abdomen, eyes open or partially open
	"Lactating" - sleeping on side with breast area exposed
	"Semicircle" - sleeping on side and curled into a ball
	"Extended" - sleeping on abdomen outstretched
	"Huddle" - group of rats sleeping in a pile
Self-Grooming	"Shake" - self-explanatory
	"Scratch" - self-explanatory
	"Claw Clean" - licking and cleaning of claws with tongue
	"Head Wash" - use of both paws to clean head
	"Genital Grooming" - licking and cleaning of genital area
	"Grooming" - wiping of fur with forepaws and licking of fur with tongue
Aggressive	"Aggressive Approach" - circling another rat with flank presented and back arched
	"Escape" - quick burst away from another rat who has made an aggressive move
	"Submissive Upright" - sitting upright, head pointed upward, forepaws held rigidly
	"Full Submissive" - laying rigidly upon back with throat exposed
	"Full Aggressive" - rat above another rat with head pointed at exposed throat
	"Boxing" - rearing back on hind legs and pawing opponent
	"Fight" - self-explanatory
	"Move Away" - both opponents break combat and move away
	"Chase" - self-explanatory
	"Chased" - self-explanatory
Social	"Social Approach" - head to head approach with mutual sniffing
	"Avoid" - moving away from another rat with quick gait or with quick head movement
	"Rooting" - lifting another rat with head and shoulders
	"Social Grooming" - grooming another rat
	"Resting" - being groomed socially

TABLE 1 (continued)

Category	Behavior
Social (continued)	"Squiring" - walking in contact with other rats
	"Grouping" - collection of rats in close contact
	"Crawl Under" - self-explanatory
	"Crawl Over" - self-explanatory
Mating (FEMALE)	"Mounted" - self-explanatory
	"Mating Approach" - female approaches male, presents rear end, and moves away with short, quick gait
	"Rejection" - tossing a mounting animal off
	"Lordosis" - upon being mounted, female raises rear haunches and moves tail out of way
(MALE)	"Mount" - self-explanatory
	"Intromission" - thrusting of hips following mount
	"Rejected" - self-explanatory
	"Dismount" - self-explanatory
	"Attempted Mount" - self-explanatory
	"Incorrect Mount" - self-explanatory
(COMMON)	"Homomount" - male (female) mounting a male (female)
	"Homomounted" - self-explanatory
	"Sex Cleaning" - cleaning of genitals following a mount

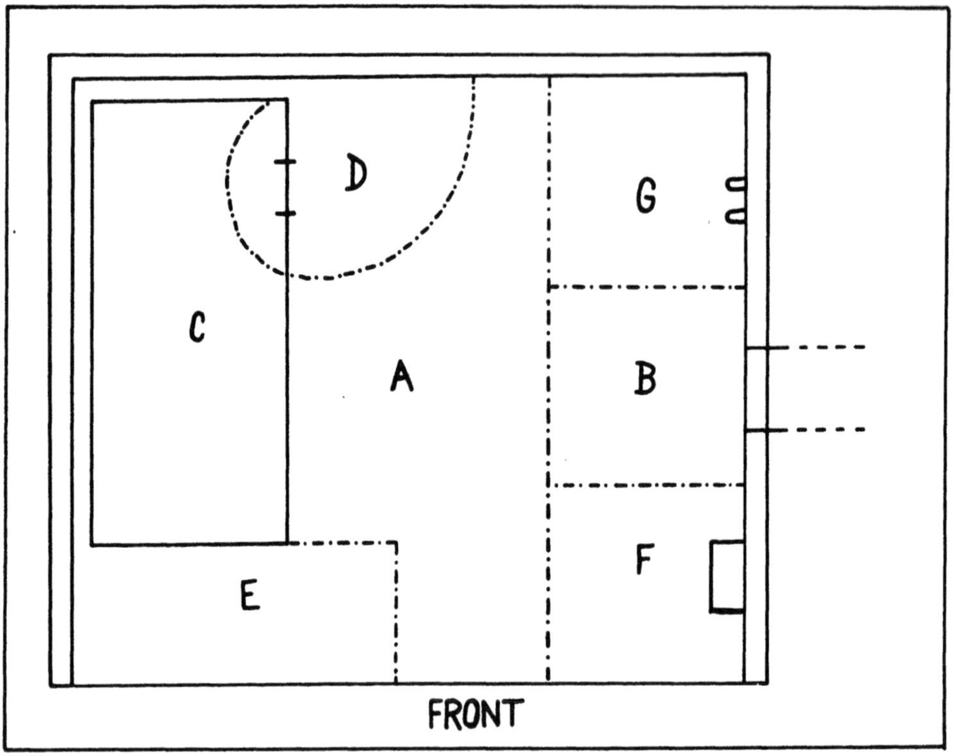

Figure 2. Behavioral Locations. View of chamber from above. Behavioral locations are: Middle of Box (A); Entrance to Box, including ramp (B); Nesting Box (C); Entrance to Nesting Box (D); Between Glass and Nesting Box (E); Near Food Source (F); Near Water Source (G).

and allowed to rehabituate for 4 days. Observations were then made for the succeeding 14 days. In experiments which studied the long-term effects, subjects were allowed 9 days recovery in individual cages; 4 days rehabituation, and 28 days of observation.

Histology

Septal lesions in each experiment were highly consistent. The septal groups sustained almost complete destruction of the medial and lateral nuclei, with little variation between subjects in the extent of damage. Involvement of adjacent structures, such as the caudate nucleus of the pre-optic nucleus, was minor.

Frontal lesions displayed the same high degree of consistency, with only minor damage to adjacent olfactory and motor areas.

RESULTS

Data (i.e., frequency of occurrence) were analyzed by using an unequal-n, repeated measures, within-subject, multi-factor analysis of variance. All meaningful combinations of variables (group, phase, category, behavior, and sex) were analyzed. To expedite interpretations, Thigpen's modification of Duncan's multiple range test was used to identify differences that could account for statistical significance. Attempts have been made to analyze the data sets sequentially by using a variety of techniques (cross-lag correlations, cannonical correlations, lambda analysis, transitional analysis, trend analysis, and tetrachloric analysis) to further clarify the results.

The major findings of these experiments have been reported elsewhere (54) and were not the identification of differences in patterns of behavior based on lesion, but rather were differences based on the sex of the subjects and on the fact that operations had been performed. Differences in behavior attributable to specific lesions have been difficult to sort out; only after repeated examinations of the data have patterns emerged. It is important to realize that the pervading behavior displayed was an adjustment by all of the subjects to a changed situation. The overall tenor of behavior reflects this adjustment (Figs. 3 and 4). When compared to the pre-operative baseline phase, the post-operative period demonstrates a radically altered behavioral milieu. Such changes have been discussed elsewhere and are not the basic subject matter of this report 54. Rather, this report concerns the subtle changes following septal or frontal ablations.

Immediate Lesion Effects

Eating behaviors increased generally (4.82 F, 2/248 df, $p < .01$), septal males preferring to eat material which was scattered (4.88 F, 2/10 df, $p < .03$) rather than food that was available at specified sources.

Male septal subjects did drink more than other groups (12.93 F, 2/11 df, $p < .002$), the drinking occurring at the expected times, such as following sleeping periods and eating. In addition, drinking behavior was more probable ($r = .68$ lambda) when male septal subjects were in the vicinity of the water source. Female subjects and control subjects maintained approximately the same level of drinking behavior as demonstrated pre-operatively.

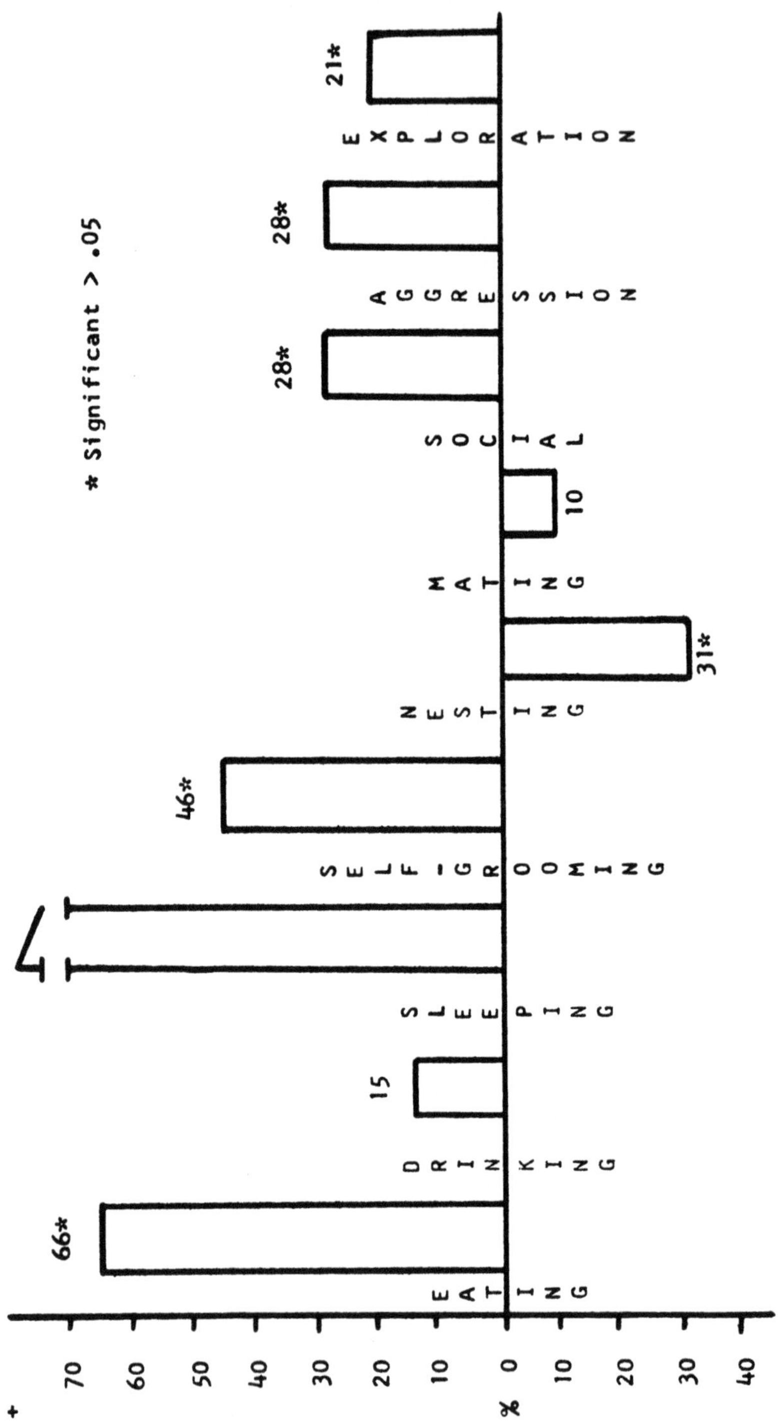

Figure 3. Comparison by Categories of Post-Operative Behaviors to Pre-Operative (Baseline) Levels (Immediate Lesion Effects).

Frontal males engaged in less drinking behavior than they had pre-operatively (8.02 F, 1/9 df, $p < .02$).

Overall, there was a general increase in exploratory behaviors (8.07 F, 2/248 df, $p < .001$). Male septal subjects had the greatest increase in situation exploratory behavior, i.e., increased investigative sniffs (25.42 F, 5/12 df, $p < .0001$) and walks (44.60 F, 5/12 df, $p < .0001$), while engaging in the fewest bursts (24.63 F, 5/12 df, $p < .0001$) and crawls (4.75 F, 5/12 df, $p < .02$). Female septals differed from other females in having the greatest number of investigative sniffs (19.35 F, 5/12 df, $p < .0007$). Frontal males demonstrated the general pattern of increase with little variation. Frontal females, alternatively, did vary in that they crawled less than other female groups post-operatively (19.57 F, 5/12 df, $p < .0007$). Control males increased in each exploratory behavior, but the largest increases were reactive, i.e., alert behaviors of frozen attention (6.21 F, 5/12 df, $p < .01$), crawling (5.27 F, 5/12 df, $p < .01$), and bursts (4.64 F, 5/12 df, $p < .03$). Control females differed from other females in these same behaviors, i.e., most uprights (16.46 F, 5/12 df, $p < .003$), crawls (4.96 F, 5/12 df, $p < .03$), and bursts (12.47 F, 5/12 df, $p < .008$). An additional detail which makes the data more interesting is that lesioned animals and non-dominant control males post-operatively reduced the amount of area in which most of their activity took place. During the pre-operative phase, 80% of the entire apparatus was used by the majority of the subjects. A dominant male controlled access to a single external chamber. During the post-operative phase, the generally used area was reduced to less than 30% of the apparatus. The dominant male controlled the remaining portions. This pattern has not been observed in control experiments in which sham operations have been used instead of lesions.

Sleeping behaviors increased generally (6.64 F, 1/48 df, $p < .02$), frontal males demonstrating the least amount of increase (13.14 F, 2/9 df, $p < .007$).

Sleeping postures resulted in an interesting localization set (Fig. 5). Male and female subjects located in a crossing pattern. Males were more likely to choose sleeping locations near the water source (16.97 F, 1/48 df, $p < .0003$), in the middle of the box (9.61 F, 1/48 df, $p < .004$), and between the nesting box and walls (39.86 F, 1/48 df, $p < .0001$). Females chose sleeping locations near the food source (35.28 F, 1/48 df, $p < .001$), the entrance to the nesting box (15.95 F, 1/48 df, $p < .0004$), or in the nesting box (4.03 F, 1/48 df, $p < .05$). This pattern was further divided by the tendency of control subjects to choose

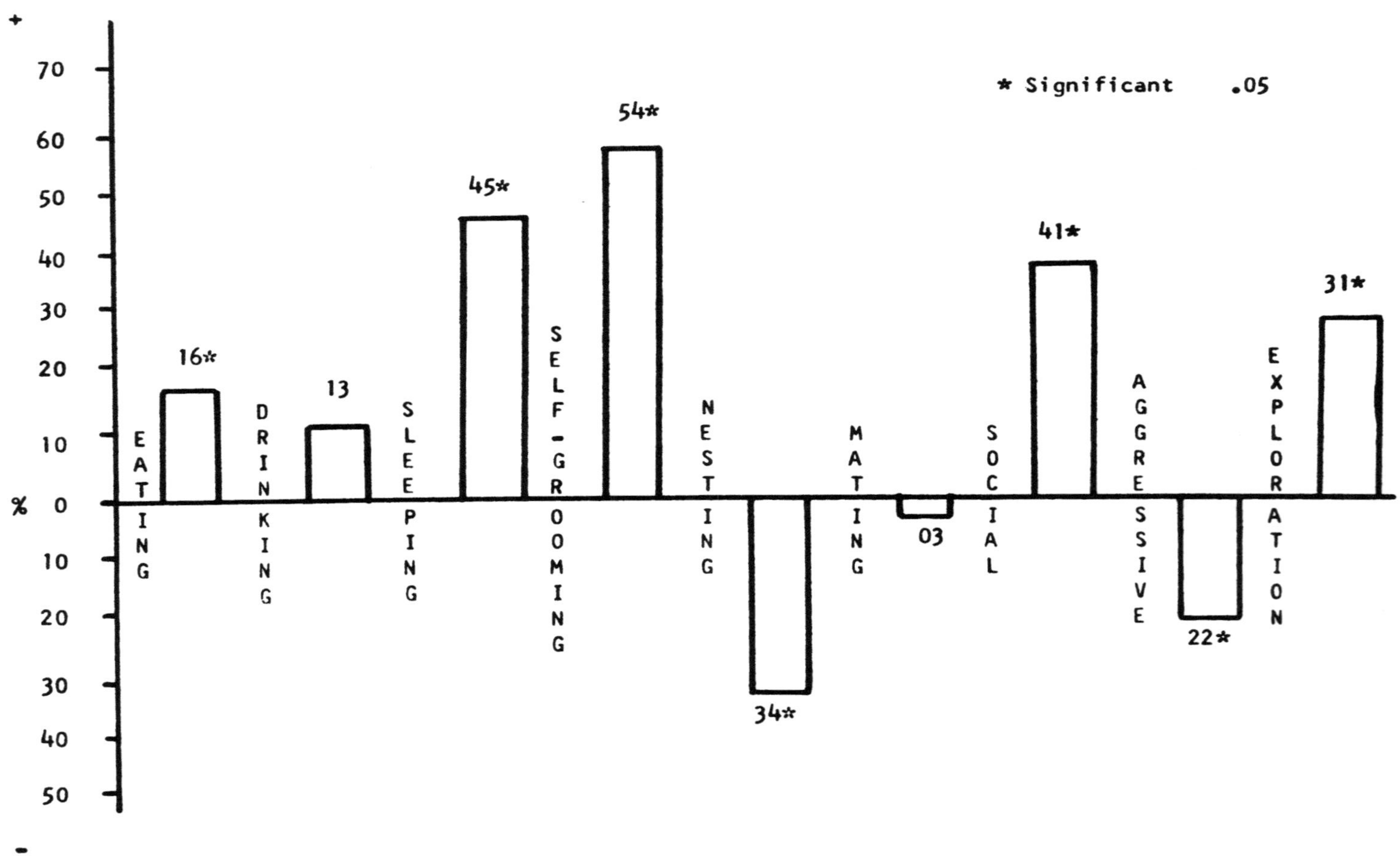

Figure 4. Comparison by Categories of Post-Operative Behaviors to Pre-Operative (Baseline) Levels (Long-Term Lesion Effects).

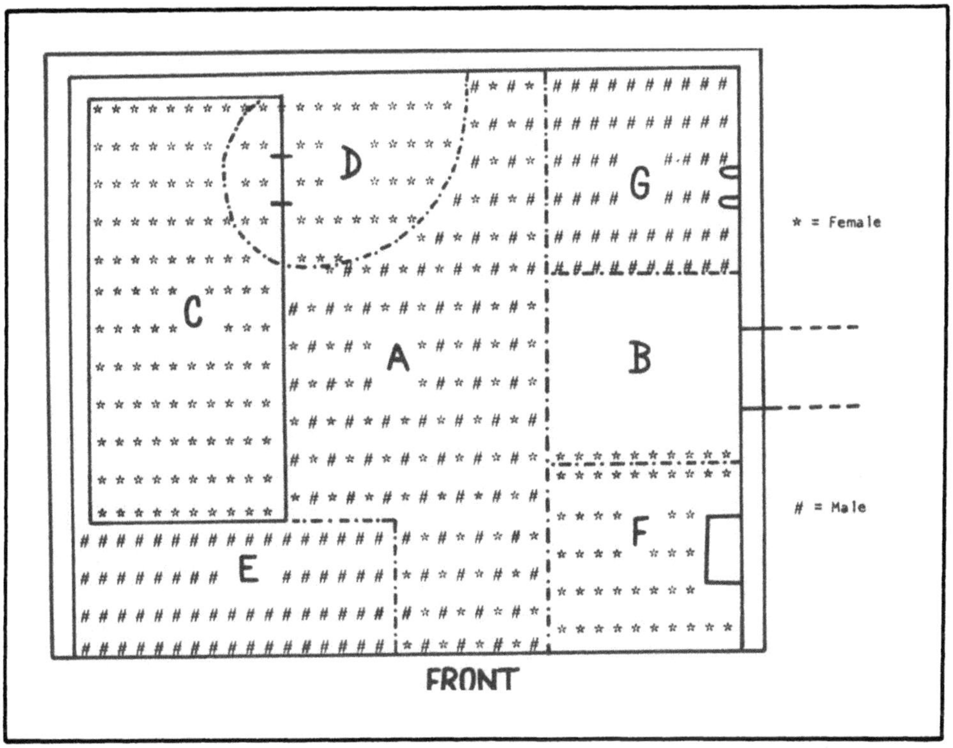

Figure 5. Post-Operative Sleeping Locations

the most secure area, such as in the nesting box (19.18 F, 2/48 df, $p < .0001$) or between the nesting box and walls (24.57 F, 2/48 df, $p < .0001$).

Social behavior increased across subjects (4.89 F, 1/48 df, $p < .03$). There was a tendency for female septal subjects to be the most gregarious, i.e., most social approaches (11.78 F, 5/15 df, $p < .0005$), flank sniffs (16.92 F, 5/15 df, $p < .0001$), avoids (16.60 F, 5/15 df, $p < .0001$), groupings (12.10 F, 5/15 df, $p < .0004$), and squirings (10.26 F, 5/15 df, $p < .0007$). Frontal females, conversely, decreased several social behaviors, namely, social approaches (10.62 F, 5/15 df, $p < .0007$), avoids (11.16 F, 5/15 df, $p < .0005$), and squirings (7.94 F, 5/15 df, $p < .005$). Control females took particular interest in grooming frontal males (4.51 F, 1/26 df, $p < .05$). Normal males were the only males to deviate from the general increase trend by decreasing both in the number of social approaches (6.92 F, 5/15 df,

$p < .008$) and flank sniffs (9.41 F, 5/15 df, $p < .007$). Septal males were the most social of all the males (5.31 F, 1/26 df, $p < .03$).

Mating behaviors generally declined for control and for septal males (7.64 F, 2/17 df, $p < .01$), with frontal males maintaining approximately the same level of mating behavior as they had pre-operatively. Septal females were the greatest recipients of mating behaviors displayed by males, even though they were not presenting esterous-typical behavioral displays (4.48 F, 2/17 df, $p < .04$).

During the pre-operative phase, the females in the colony had constructed a series of nests and burrows, using the provided paper strips, within the territory of the dominant male. Post-operatively, control females continued this pattern of construction and the number of behaviors displayed. Septal females demonstrated a slight increase in the number of nesting behaviors displayed (4.86 F, 2/7 df, $p < .05$), but the locations of these nests were no longer in the relatively safe areas defended by the dominant male. Instead, septal females built nests in the most densely populated area. These locations resulted in repeated rebuilding efforts. Frontal females drastically reduced their nesting behaviors, the result being that few nests were built and those that were attempted did not establish a pttern of localization.

Aggression in the post-surgical period was different from that observed in the pre-operative phase. Male frontal subjects were the most aggressive (14.51 F, 2/12 df, $p < .01$). Female subjects maintained lower levels of aggression except septal females. The increase by septal females, however, was mainly due to the mating encounters coming after rejectings ($r = .43$ lag) and not to portions of specific aggressive encounters. Male septals were the least ritualistic in their encounters, showing a pattern of not taking up submissive postures unless knocked over on their backs into a full submissive. Upright submissive and boxing postures were unlikely ($r = -.52$ lambda), fights ending with bites being viven to septal males (7.57 F, 2/12 df, $p < .01$).

After surgery, there was an expected general increase in self-grooming by every subject. Among septals, however, there was a difference in the pattern of grooming. Grooming among controls and frontals often followed a caudal to rostral pattern; the head, face, body, and genital areas were cleaned in a somewhat sequential manner. Among septals, rudiments of this pattern were observable but, in general, attention was paid to a particular area during any one grooming session ($r = -.27$ lag).

Long-term Lesion Effects

Eating behaviors did not increase across groups, with the exception that male septal subjects did increase their consumatory responses (9.62 F, 2/12 df, $p < .005$). This was especially true for those behaviors away from the food source, such as eating of scattered (7.11 F, 2/10 df, $p < .01$) and eating of debris (5.96 F, 2/10 df, $p < .02$).

Male septal subjects drank more than other subjects (9.44 F, 2/11 df, $p < .005$). This behavior was more probable for septal males than for other groups when they were in the vicinity of the water source ($r = .57$ lambda). Frontal males drank less post-operatively than they had pre-operatively (13.02 F, 1/3 df, $p < .05$). Female and control subjects demonstrated individual rather than group trends post-operatively.

The finding that behavior in general, and exploratory behavior in particular occurred in a restricted area post-operatively was again in evidence over the long term as it had been immediately after surgery. All of the male subjects, with the exception of the dominant male, were reduced in area of activity to less than 25% of the entire apparatus. Operated female subjects expended most of their behaviors in this same reduced space, while control females free-ranged throughout the entire apparatus.

Control male subjects decreased their exploratory behavior (4.31 F, 5/12 df, $p < .03$). Control females maintained the same level of behavior post-operatively as pre-operatively, with the exceptions of decreased frozen attentions (4.03 F, 5/12 df, $p < .03$) and walks (9.61 F, 5/12 df, $p < .001$). Subjects having undergone brain damage increased all of their exploratory behaviors. Female septal subjects had the greatest increase in investigative sniffs (52.00 F, 5/12 df, $p < .0001$) and walks (9.61 F, 5/12 df, $p < .001$), while frontal females increased the number of bursts (28.87 F, 5/12 df, $p < .001$) and crawls (24.85 F, 5/12 df, $p < .0001$) to a greater degree than other groups.

The number of sleeping postures taken post-operatively increased across subjects, with no significant differences between lesion or sham subjects (9.72 F, 1/48 df, $p < .01$). Location of sleeping postures resulted in the same crossing pattern as found in the immediate effects study. Males were more likely to choose sleeping positions near the water source (34.82 F, 1/48 df, $p < .0001$), in the middle of the box (12.77 F, 1/48 df, $p < .005$), and between the nesting box and walls. Females chose sleeping locations near the food source (5.84 F, 1/48 df, $p < .04$), the entrance to the nesting box (9.04 F, 1/48 df, $p < .01$), and in the nesting box. Control subjects again preferred the most secure locations--in the nesting box (4.45 F, 2/48 df, $p < .03$) or between the nesting box and walls (11.73 F, 2/48 df, $p < .005$).

Social behaviors increased across subjects (8.38 F, 1/248 df, $p < .01$). Males subjects had the greatest overall increase (8.09 F, 1/15 df, $p < .04$). The only lesion-related differences were that septal males increased the number of squirings (6.25 F, 5/15 df, $p < .006$) and groupings (5.27 F, 5/15 df, $p < .01$) more than any other group.

Mating behaviors continued at the same level as they had pre-operatively, excepting that female septal subjects did receive significantly more attention from males than they had pre-operatively (4.53 F, 5/17 df, $p < .01$). Control subjects engaged least often in any behavior which could be termed homosexual (4.25 F, 2/17 df, $p < .03$).

Control females post-operatively continued their pre-operative pattern of building nests and tunnels within the area defended by the dominant male. Over the long term, the number of such behaviors were less than during the pre-operative period (8.53 F, 2/7 df, $p < .04$). Septal females maintained the same level of nesting behaviors as they had pre-operatively, except that the location of nesting sites were within the most heavily populated areas, resulting in repeated rebuilding efforts and few finished nests. Frontal females engaged in the least amount of post-operative nesting behavirs (4.83 F, 2/7 df, $p < .05$), the locations of the same not being localized within a particular area. No completed nest built by a frontal female was observed over the long term period.

Aggressive behaviors observed post-operatively did differentiate between male subjects according to surgical treatment. Frontal subjects engaged in the most aggressive approaches (5.73 F, 2/12 df, $p < .03$), boxings (5.42 F, 2/12 df, $p < .04$), fightings 4.71 F, 2/12 df, $p < .05$), submissive uprights (8.01 F, 2/12 df, $p < .008$), full submissives (5.31 F, 5/12 df, $p < .03$), full aggressives (7.02 F, 5/12 df, $p < .01$), and bites given (6.92 F, 5/12 df, $p < .01$). Control males evidenced the same levels of aggressive behaviors as they had pre-operatively. Septal males decreased aggressive approaches (31.30 F, 5/12 df, $p < .0001$), boxings (5.76 F, 5/12 df, $p < .03$), submissive uprights (6.92 F, 5/12 df, $p < .01$), and move-aways. At the same time they increased the number of fightings (5.12 F, 5/12 df, $p < .04$) and reception of bites (5.72 F, 5/12 df, $p < .03$). Female subjects demonstrated the same levels of aggressive behavior as they had pre-operatively, with the exception that female septals were more likely to fight ($r = .32$ lag) after being mounted than the other females.

The self-grooming of all subjects increased after surgery. Control subjects increased to a lesser degree than lesioned subjects (4.82 F, 5/12 df, $p < .01$). Septal subjects again did not follow the caudil to rostril pattern during grooming sessions that was demonstrated by the other subjects ($r = -.26$ lag), repeatedly grooming instead a particular body location.

DISCUSSION

The clearest findings from this series of experiments were that the sex of the subject and the fact that surgery had been performed were more important variables than lesion effects alone. When lesion effects were found, they usually were differentiated by sex differences. This overall lack of grossly deviant behavioral changes identifying particular lesion treatments is analagous to the case of the "silent tumor" in which the pathology may go unnoticed unless sensory or motor structures such as the caudate nucleus was extremely limited. The observations were performed in a blind manner so that the observers did not know which treatments individual animals had received. "Guesstimates" by the observers as to which subject belonged to which group proved to be highly inaccurate. Agreement as to group membership proved accurate only once during pilot experimentation; this occurred when several of the subjects had undergone severe caudate damage and so displayed noticeable deviant motor behavior.

This is not to say that this method did not find indications that specific subjects had undergone similar surgical treatments. What did happen is that many common laboratory task-specific evidences of gross behavioral deviancies, such as hyper-emotionality or hyper-sensitivity, were not shown in this situation. In the laboratory situation, apparent behavioral deficits are obtained as a result of controlled contingencies in which the desired behavioral repetoire is extremely restricted and directed. For example, an unnatural form of punishment, shock, is used in avoidance conditioning. In the case of appetitive tasks, subjects first are partially deprived of food and water and then are required to develop or maintain a response in order to obtain reinforcement. At all times in the laboratory situation, there are severe constraints placed upon the possible behaviors which can be displayed. Proportionally, these constraints did not exist in the free choice social situation. This condition allowed the full repertoire of behaviors to occur. Lack of constraint also allowed for a compensatory adjustment for both operated and control subjects, resulting in a stable social organization. Such a milieu demonstrated surgically induced changes in altered patterns or strategies of behavioral respondability. Indicators of differences between lesioned subjects were available (post hoc). The increased drinking by septal males finds support in the studies evidencing increased drinking by septals. The high level of activity of septals over time (while not hyper) suggests that destruction of the septal area produces a dependence on external stimulation rather than interoceptive control of behavior.

Septal subjects demonstrated this strategy of adjustment by eating material which was in their general vicinity rather than going to the food source, by drinking whenever they passed a water source, by displaying information-seeking behaviors (such as sniffing and walking) even though the area being investigated has been greatly reduced, by engaging in close contact social behaviors, by fighting rather than engaging in stereotyped encounters, and by disengaging from fighting only when forced onto their backs or when bitten. The lack of interoceptive control is shown by the lack of reactive behaviors (such as burst or crawls), improper location of nesting sites, decreased areas of activity, lack of stereotyped aggressive displays, and repeated grooming of particular body locations rather than complete grooming sessions.

Frontal-operated subjects indicated a different strategy of coping with the changed milieu. They displayed little adjustment to the overall situation, except for those behaviors which are under some degree of interoceptive control. This is demonstrated by the high degree of stereotyped aggressive behavior, reactive behavior (such as bursts and crawls), and the deletion of nest building by the frontal females.

Control subjects, over time, indicated a strategy of adjustment which minimized their contact with other subjects by decreasing their ambient movement behaviors and by choosing the best-protected sleeping locations.

All of the subjects adopted a strategy of coping behavior of general increases and decreases both in categories and in particular behaviors as an overall adjustment to a changed milieu.

Bolles (12,13) has made the point that an orgamism's survival strategies often depend upon the display of species-specific reactions arranged in a hierarchial manner whenever there is any stimulus change in the environment. In these experiments, the normal rat's behavioral responses did occur in a hierarchial fashion prior to surgery, but not following surgery. At that subsequent time, many expected or previously-evidenced stereotyped patterns of behavior were disrupted. This disruption, such as a decrease in freezing by the septals, was not replaced by fleeing or fighting as would be expected from hierarchical arguments, but rather by a non-reactivity to the situation. Initiators of aggressive encounters decreased their approaches to septals. This indicates that new patterns did not seem to be related to a hierarchy, but reflected instead a change in adopted strategy toward the situation. The change minimized inter-subject interference and maximized those particular cues necessary to maintain a homeostatic environment.

Over the long term, a milieu developed such that approximately the same number of pups survived as in control studies in which no lesion conditions existed and in which the size, weight and general condition of the subjects was nearly the same. What was different was the tenor of behavior. The presence of lesion subjects resulted in a re-ordering of behavior probabilities across subjects and in a minimization of between-subject differences, while identifying the composition of the population. Whether the particular adopted strategies of particular groups of subjects are of general applicability or are specific to the composition of this milieu is now under investigation. The introduction of different combinations of lesions and varying stressors, such as food or water scarcities, should identify the applicability of the strategies observed in this instance.

The value of this research may have been to reindicate the importance of sex as a variable in discussions of lesion deficit behavior and to reinforce the concept of the lack of generality of task-specific findings.

SUMMARY

Studies of ablations of the septal or frontal areas of the rat's brain, designed to examine subsequent alteration in species-typical social behaviors, were conducted by using a free-choice social situation. Few particular gross behavioral differences were evidenced by particular lesion treatment groups. Instead, extremely subtle patterns of adjustment were found for brain-damaged and non-brain-damaged subjects, varying according to sex. Septal subjects adopted a strategy of adjustment based on maximizing the availability of exteroceptive information. Frontal subjects demonstrated a pattern of utilizing proprioceptive information. Control subjects' patterns minimized behavioral interaction with either group. These patterns combined over time into a general pattern of behavior evidenced by all of the subjects which minimized the opportunity for identification of particular subject damage and yet which provided an altered behavioral milieu when compared to control experiments.

It is argued that this adjustment may be an important indicator of the presence of brain-damaged individuals in a social situation rather than being an indicator of behavioral deficit.

REFERENCES

1. Ahmad, S. and Harvey, J. A., Long-term effects of septal lesions and social experience on shock-elicited fighting in rats, J. Comp. Physiol. Psych., 66 (1968) 596.

2. Allikemets, L. K. H. and Dietrich, M. D., Effects of lesions of the limbic system on emotional reactions and conditioned reflexes in the rats, Fed. Proc., 24 (1965) T1003 - T1007.

3. Baenninger, L. P., Social dominance orders in the rats: spontaneous food and water consumption, J. Comp. Physiol. Psych., 71 (197) 202-209.

4. Barnett, S. A. and Spencer, M. M., Feeding, social and inter-specific competition in wild rats, Behaviour, 3 (1951) 229-242.

5. Barnett, S. A., Behaviour components of the feeding of wild and laboratory rats, Behaviour, 9 (1957) 24-43.

6. Barnett, S. A., An analysis of social behavior in wild rats, Proc. Zool. Soc. London, 130 (1958) 107-152.

7. Barnett, S. A. The Rat: A Study in Behavior, Aldine, Chicago, 1963.

8. Barnett, S. A., Grouping and dispersive behaviour among wild rats. In S. Garattini and E. B. Sigg (Eds.), Aggressive Behavior, Wiley, New York, 1969.

9. Beach, F. A., Hormones and Behavior, Hoeber, New York, 1948.

10. Beatty, W. W. and Schwartzbaum, J. S., Enhanced reactivity to quinine and saccharine solutions following septal lesions in the rat, Psychon. Sci., 8 (1967) 483-484.

11. Blanchard, R. J., Blanchard, D. Limbic lesions and reflexive fighting, J. Comp. Physiol. Psychol, 66 (1968) 603-606.

12. Bolles, R. C., Species-specific defense reactions and avoidance behavior, Psychol. Rev., 77 (1971) 32-48.

13. Bolles, R. C., and Woods, P. J., The ontogeny of behavior in the albino rat, Anim. Behav., 7 (1965) 427-441.

14. Brady, J. V. and Nauta, W. J. H. Subcortical mechanisms in emotional behaviors: affective changes following septal forebrain lesions in the albino rat, J. Comp. Physiol. Psychol, 46 (1953) 339-346.

15. Brady, J. V. and Nauta, W. J. H. Subcortical mechanisms in emotional behavior: the duration of the affective changes following septal and habenular lesions in the albino rat, J. Comp. Physiol. Psychol, 48 (1955) 412-420.

16. Brutkowski, S. Functions of the pre-frontal cortex in animals, Physiol. Rev., 45 (1965) 721-746.

17. Bunnell, B. N., Bemporad, J. R., and Flesher, C. K. Septal forebrain lesions and social dominance behavior in the hooded rat, Psychon. Sci., 6 (1966) 207-208.

18. Bunnell, B. N. and Smith M. H. Septal lesion and aggressiveness in the cotton rat, "Sigmodon hispidus", Psychon. Sci., 6 (1966) 443-444.

19. Butler, S. R. and Engers, J. T. The role of frontal cortex in the performance of a conditional reaction, Physiol. and Beh., 4 (1969) 847-852.

20. Calhoun, J. B., The social aspects of population dynamics, J. of Mammal., 33 (1952) 139-159.

21. Calhoun, J. B. Social wlefare as a variable in population dynamics, Cold Spring Harbor Symposia on Quantitative Biology, 22 (1957) 339-366.

22. Calhoun, J. B. Population density and social pathology, Sci. Amer., 206 (1961) 139-148.

23. Calhoun, J. B., The ecology and sociology of the Norway rat, P H S publication 1008, 1968.

24. Caplan, M. An analysis of the effects of septal lesions on negatively reinforced behavior, Behav. Biol., 9 (1973) 129-167.

25. Caplan, M. and Stamm, J., DRL acquisition in rats with septal lesions, Psychon. Sci., 8 (1967) 5-6.

26. Carey, R. J., Contrasting effects of increased thirst and septal ablations in DRL-responding rats, Physiol. and Beh., 2 (1967) 287-290.

27. Carey, R. J. A further localization of inhibitory deficits resulting from septal ablation, Physiol. and Beh. 4 (1968) 759-764.

28. Carrey, R. J. Contrasting effects of anterior and posterior septal injury on thirst-motivated behavior, Physiol. and Beh., 4 (1969) 749-764.

29. Cloudsley-Thompson, J. L. Animal Conflict and Adaptation, Dufour, Chester-Springs, Pa., 1965.

30. Corman, C. D., Meyer, P. M., and Meyer, D. R. Open-field activity and exploration in rats with septal and amygdaloid lesions, Brain Research, 5 (1967) 469-476.

31. Dewsberry, D. A. A quantitative description of ghe behavior of rats during copulation, Behaviour, 29 (1967) 154-177.

32. Divac, I. Frontal lobe system and spatial reversal in the rat, Neuropsychologia, 9 (1971) 175-183.

33. Ellen, P. and Butler, J. External cue control of DRL performance in rats with septal lesions, Physiol. and Beh., 4 (1969) 1-6.

34. Ewer, R. F. The biology and behavior of a free-living population of black rats (rattus rattus), Anim. Beh. Mono., 4 (1971) 127-173.

35. French, G. M. Cortical Functioning in Behavior, Scott-Foresman, Glenview, Ill., 1973.

36. Fried, P. A. Hyper-reactivity and the spetum, Brit. J. of Psychol., 86 (1973) 418-473.

37. Girgis, M. The orbital surface of the frontal lobe of the brain and mental disorders, Acta Psychiatrica Scandinavia, 1971.

38. Gotsick, J. E. Factors affecting spontaneous activity in rats with limbic system lesions, Physiol. and Beh., 4 (1969) 587-593.

39. Grant, E. C. An analysis of the social behavior of the male laboratory rat, Behaviour, 21 (1963) 260-281.

40. Grant, E. C. and Mackintosh, J. H. A comparison of the social postures of some common laboratory rodents, Behaviour, 21 (1963) 246-258.

41. Grant, E. C. and Chance, M. R. A. Rank order in caged rats, Animal Behavior, 6(3) (1959) 183-194.

42. Hamilton, L. W., Kelsey, J. E., and Grossman, S. P. Variation in behavioral inhibition following different septal lesions in rats, J. Comp. Physiol. Psychol., 70 (1970) 79-86.

43. Harvey, J. A. and Hunt, H. F., Effects of septal lesions on thirst in rats as indicated by water consumption and operant responding for water reward, J. Comp. Physiol. Psychol, 59 (1965) 49-56.

44. Hutt, S. J. and Hutt, C., Direct Observation and Measurement of Behavior, Charles Thomas, Springfield, Ill., 1970.

45. Johnson, R. N., DeSisto, Jr., and Koenig, A. B. Social and developmental experience and interspecific aggression in rats, J. Comp. Physiol. Psychol., 70(2) (1972) 237-242.

46. Jonason, K. R. and Enloe, L. J. Alterations in social behavior following septal and amygdaloid lesions in the rat, J. Comp. Physiol. Psychol., 75 (1971) 286-301.

47. Kaada, B. R., Rasmussen, E. W., and Kveim, O. Impaired acquisition of passive avoidance behavior by sub-callosal, septal, hypothalamic and insular lesions in rats, J. Comp. Physiol. Psychol., 55 (1962) 661-670.

48. King, F. A. Effect of septal and amygdaloid lesions on emotional behavior and conditioned avoidance response in rats, Journal of Nervous and Mental Disorders, 126 (1956) 57-73.

49. Leonard, C. M. The pre-frontal cortex of the rat: Cortical projections of the medial-dorsal nucleus efferent connections, Brain Research, 12 (1969) 321-343.

50. Lubar, J. F. The effect of medial cortical lesions on the avoidance behavior of the cat. Unpublished doctoral dissertation, University of Chicago, 1963.

51. Lubar, J. F., Schaefer, C. F., and Wells, D. G. The role of the septal area in the regulation of water intake and associated motivational behavior, Annals of the New York Academy of Sciences, 157 (1969) 875-893.

52. Lubar, J. F., Brener, J. M., Deagle, J. H., Numan, R., and Clemens, W. The effect of septal lesions on detection threshold and unconditioned response to shock, Physiol. and Beh., 5(4) (1970) 459-463.

53. Lubar, J. F. and Numan, R. L. Behavioral and physiological studies of septal function and related medial cortical structures, Behav. Biol., 8(1) (1973) 1-25.

54. Lubar, J. F., Herrmann, T. F., Shouse, M., and Moore, D. Effect of septal and frontal ablations on the species typical behavior of the rat, J. Comp. Physiol. Psychol., 83 (1973) 260-270.

55. Lynch, G., Ballantine II, P. C., Differential rates of recovery following frontal cortical lesions in rats, Physiol. and Beh., 7 (1971) 737-741.

56. MacLean, P. D. The limbic system with respect to self-preservation and the preservation of species, Journal of Nervous and Mental Disorders, 35 (1958) 416-432.

57. Mathews, S. R. and Finger, F. W. Direct observation of the rat's activities during food deprivation, Physiol. and Beh., 1 (1966) 85-88.

58. McCleary, R. A. Response specificity in the behavioral effects of limbic system lesions in the cat, J. Comp. Physiol. Psych., 54 (1961) 605-613.

59. McIntire, R. W. Extinction of the freezing component of the CER after frontal ablation, Psychological Reports, 19 (1966) 187-193.

60. Meyer, P. M. The behavior of rats following cortical and subcortical lesions, Ammon's Horn, Spring, 1967, 1-7.

61. Morrison, B. J. and Thatcher, K. Overpopulation effects on social reduction of emotionality in the albino rat, J. Comp. Physiol. Psychol. 69(5) (1969) 658-662.

62. Seigel, D. and Leaf, R. C. Effects of septal and amygdaloid brain lesions in rats on mouse killings. Paper presented at the Eastern Psychological Association, Philadelphia, 1969.

63. Sodetz, F. J., Matalka, E. S., and Bunnell, B. N. Septal ablation and affective behavior in the golden hamster, Psychon. Sci., 7 (1967) 189-190.

64. Speigel, E. A., Miller, H. R., and Oppenheimer, M. J. Forebrain and rage reactions, J. Neurophysiol., 3 (1940) 538-548.

65. Webb, E. J., Campbell, D. T., Schwartz, R. D., and Sechrest, L. *Unobtrusive Measures: Non-Reactive Research in the Social Sciences*, Rand-McNally and Company, Chicago, 1966.

66. Zucker, I. and McCleary, R. A. Perseveration in septal cats, *Psychon. Sci.*, 1 (1964) 387-388.

BEHAVIORAL, ANATOMICAL, AND BIOCHEMICAL ASPECTS OF SEPTAL HYPERREACTIVITY[1]

David S. Olton and Fred H. Gage III

Department of Psychology, The Johns Hopkins University
Baltimore, Maryland 21218

INTRODUCTION

Following septal lesions, rats develop a very characteristic hyperreactivity which appears as an exaggerated reaction to normally nonnoxious stimuli. We will report here the results of some recent experiments concerning the behavioral, anatomical, and biochemical aspects of this hyperreactivity.

METHODOLOGY AND RESULTS

Behavior

In testing for hyperreactivity, we present four different stimuli to the rat and rate the animal's reactions on a scale from 0 to 3. A complete description of our rating procedure is presented in Gage and Olton (3). On the first day after a septal lesion, rats are more reactive to these stimuli than they were preoperatively, a finding that has been well documented before (5).

We would like to make a distinction between two different types of hyperreactivity, a distinction which to our knowledge has not been made before. One of the types of hyperreactivity is an exaggerated attack response. When a glove or a stick is presented

---

[1]This project was supported by research grant #1-R01-MH24213-01 from the National Institute of Mental Health. We express appreciation to Hoffman-LaRoche for donation of the decarboxylase inhibitor.

to the rat, the rat will orient towards the stimulus and then attack the stimulus in a characteristic way. The attack may be a single bite, usually made with the eyes closed and the ears flattened against the skull, followed by a leap away from the stimulus, or the attack may be a series of chewing and pulling movements that persist for several seconds. On the first day after the septal lesion, these attack behaviors often appear as soon as the stimulus is placed in the cage with the rat. As the hyperreactivity gradually decreases over days, the attack behaviors are elicited only when the stimulus actually touches the rat. The second type of hyperreactivity that we have observed is an exaggerated escape response. When the stimulus object is presented to the rat, the rat moves away from the stimulus and constantly attempts to avoid contact. The rat may back into a corner away from the stimulus or leap up to the top of the testing apparatus. On the first day after the septal lesion, these responses may appear as soon as the stimulus enters the cage, even prior to contact with the rat. On subsequent days, as the hyperreactivity decreases, actual contact with the rat is usually necessary to elicit the hyperreactive escape response.

We are currently unable to specify the variables that determine whether the hyperreactivity following a septal lesion will be exhibited as an exaggerated escape response or an exaggerated attack response. Each rat, however, has a very consistent pattern of responding and the difference between the escape and the attack behaviors is striking. Other experiments have demonstrated that septal lesions can produce either attack or escape responses depending on the testing situation. Our data indicate that even in the same testing situation there may be substantial individual differences among animals and that either type of hyperreactivity may be exhibited.

Neuroanatomy

Many studies have attempted to localize the areas of the septum which must be destroyed in order to elicit hyperreactivity. A review of the literature has been provided in Olton and Gage (5). In general, these studies have focused on septal nuclei and utilized only a single lesion. Our approach has been to focus on septal fiber tract connections and utilize two lesions. The general paradigm is as follows: Rats are first tested for reactivity as normals. Then one of the fiber tracts connecting the septum with other brain structures is severed and the rat tested to determine if there is any change in reactivity. Next, a lesion is placed in the septum and the rat tested again for reactivity. The question of interest is: does the preceeding fiber tract lesion influence the magnitude or duration of the hyperreactivity observed following the subsequent septal lesion?

A previous study (5) demonstrated that complete transection of the body of the fornix as it courses between the septum and the hippocampus 8 days prior to a septal lesion completely blocked the appearance of hyperreactivity expected following the septal lesion. Rats with a control operation instead of a fornix lesion exhibited hyperreactivity following the septal lesion. These data demonstrate that the fibers of the fornix play an integral role in the elaboration of hyperreactivity following septal lesions.

Our current research is designed to determine which components of the fornix are responsible for the blocking of septal hyperreactivity. The general testing paradigm is the same as described above, with the first lesion placed in one of the structures related to the fornix (followed by behavioral testing) and the second lesion placed in the septum (again followed by behavioral testing). The data we will report here concerns the effects of lesions in the precommissural fornix, postcommissural fornix, anterior hippocampus, and posterior hippocampusentorhinal area on hyperreactivity following subsequent septal lesions. There was a 16 day interval between the first and the second lesions; rats were tested on the first, second, and last day of this interval. Following the second (septal) lesion, rats were tested for at least four days and until their reactivity scores were 3 or less on two consecutive days. Some rats were not tested for two days during the experiment; in such cases no data are presented for these days. The maximum reactivity score that could be obtained on any reactivity test was 14. Normal animals usually received a score of 3 or less. Complete details of the testing procedure are presented in Gage and Olton (3).

Fig. 1 presents the individual histological and behavioral data from rats receiving a control operation followed by a septal lesion. In this figure, as in all subsequent ones, each row presents the data from a single animal. Across each row are presented in order: the animal's number, the histology from the first operation, the behavioral data from the 16 days following the first operation, the histology from the second operaton, and the behavioral data following the second operation.

As can be seen from Fig. 1, the control operation had little effect on reactivity. On the first postoperative day, reactivity scores ranged from 0 to 1. The septal lesions were large and bilateral, destroying major segments of all septal nuclei. Following the septal lesions, reactivity increased markedly. On the first postoperative day, reactivity scores ranged from 7 to the maximum of 14. The reactivity gradually decreased over days, and after about ten days reactivity scores were approximately the same as observed in normal rats. Since the minimum reactivity score on the first postoperative day following the septal lesion was 7, any

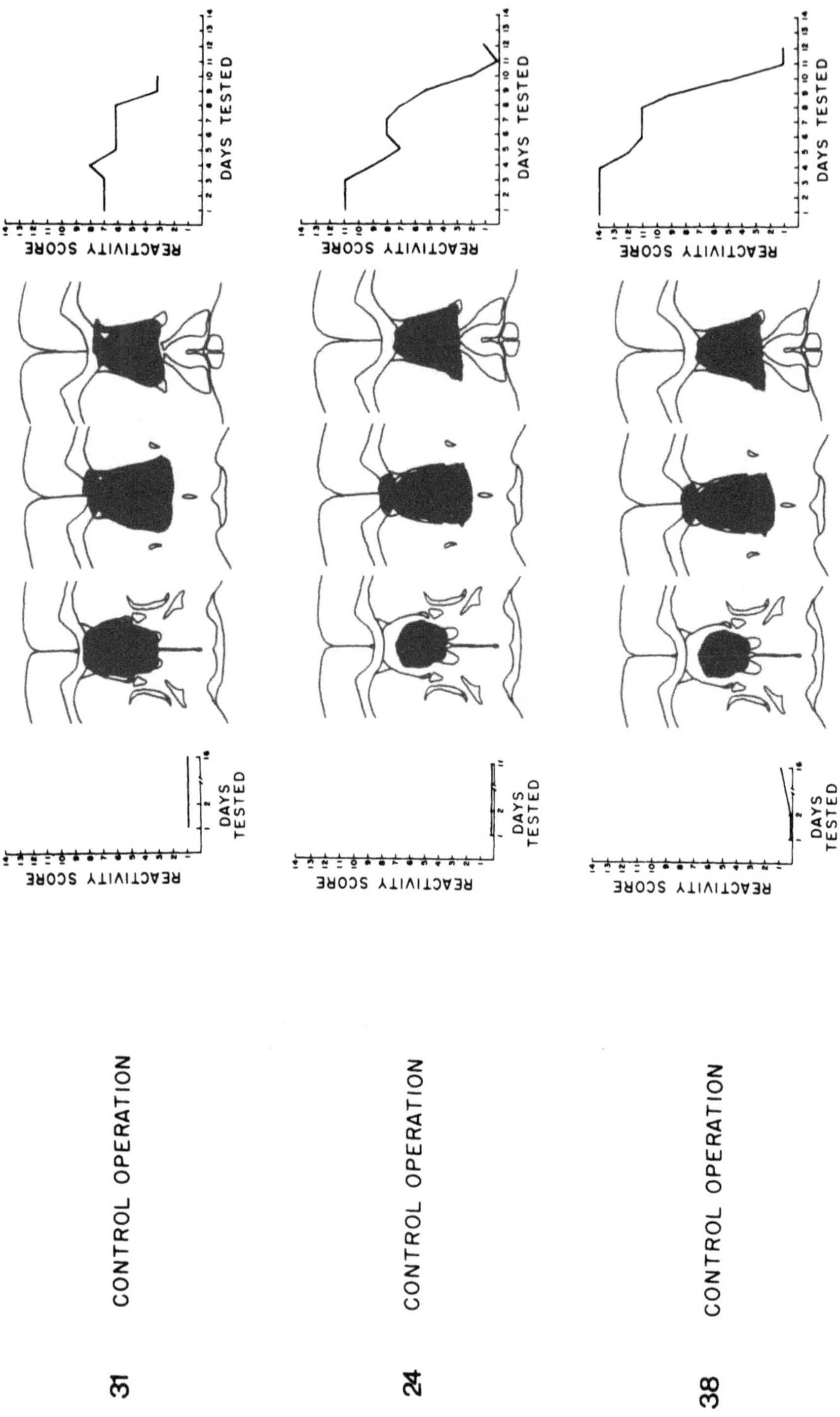
31
CONTROL OPERATION
24
CONTROL OPERATION
38
CONTROL OPERATION
REACTIVITY SCORE
DAYS TESTED

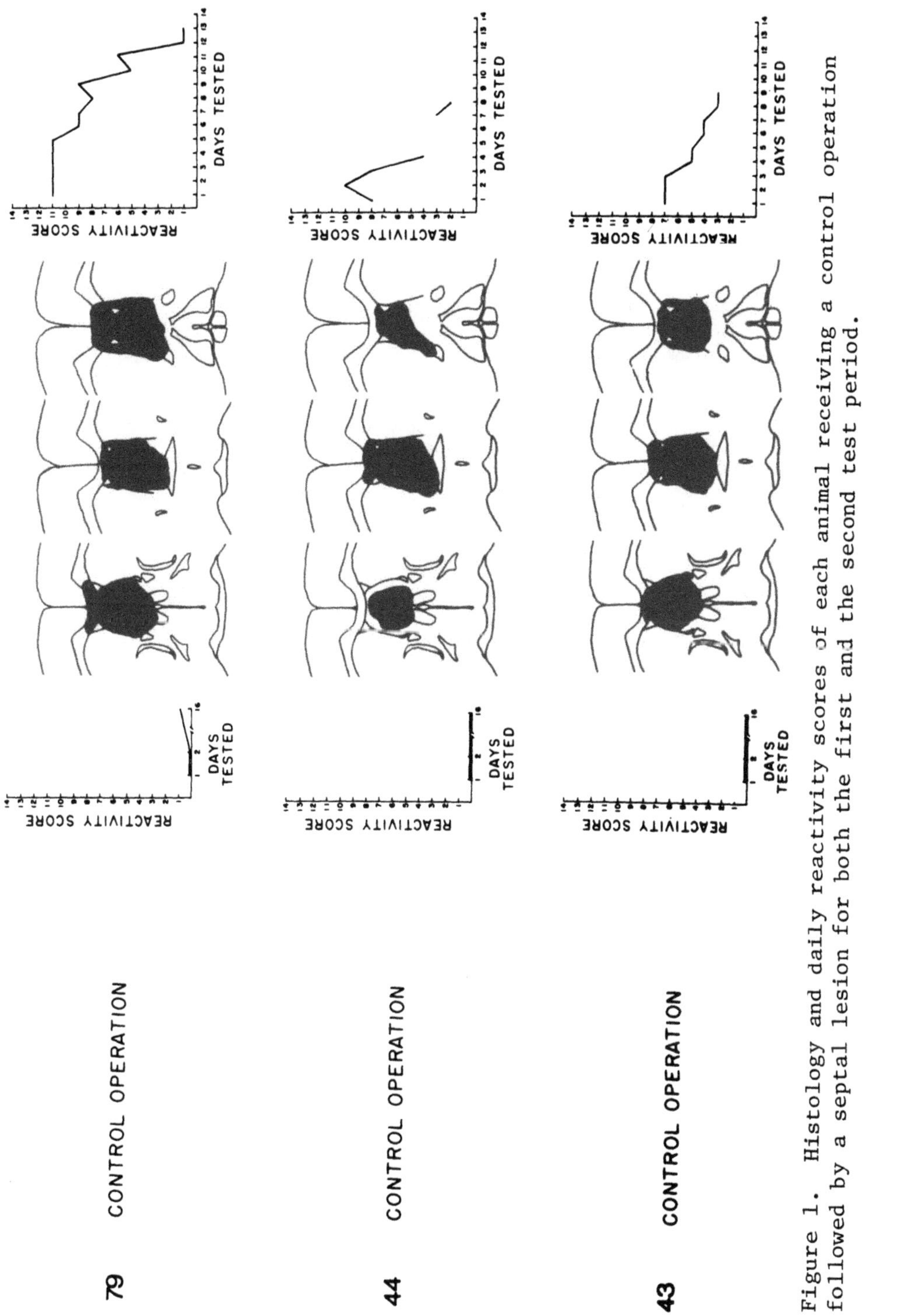

Figure 1. Histology and daily reactivity scores of each animal receiving a control operation followed by a septal lesion for both the first and the second test period.

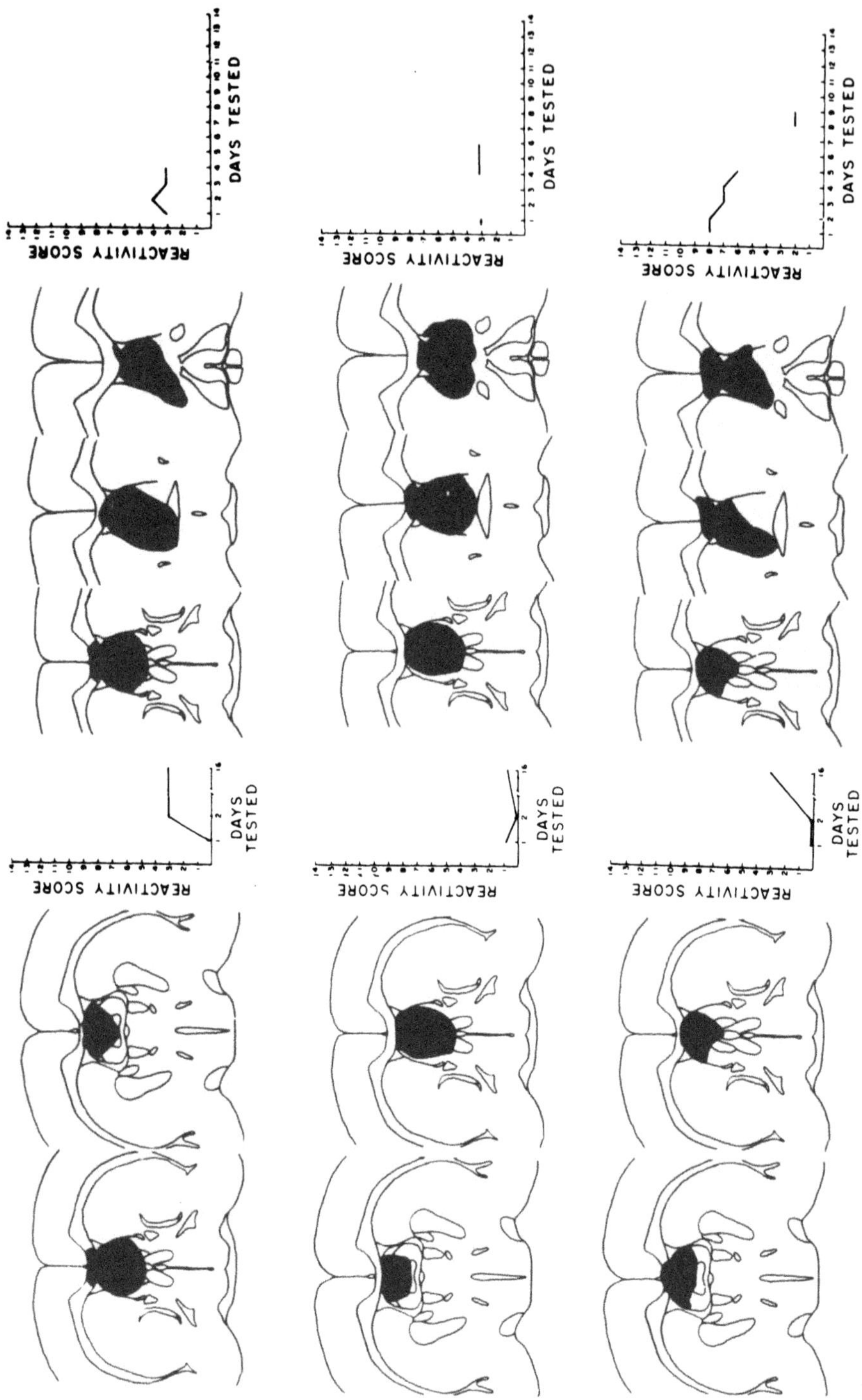
REACTIVITY SCORE
DAYS TESTED
REACTIVITY SCORE
DAYS TESTED
REACTIVITY SCORE
DAYS TESTED
REACTIVITY SCORE
DAYS TESTED
REACTIVITY SCORE
DAYS TESTED
REACTIVITY SCORE
DAYS TESTED

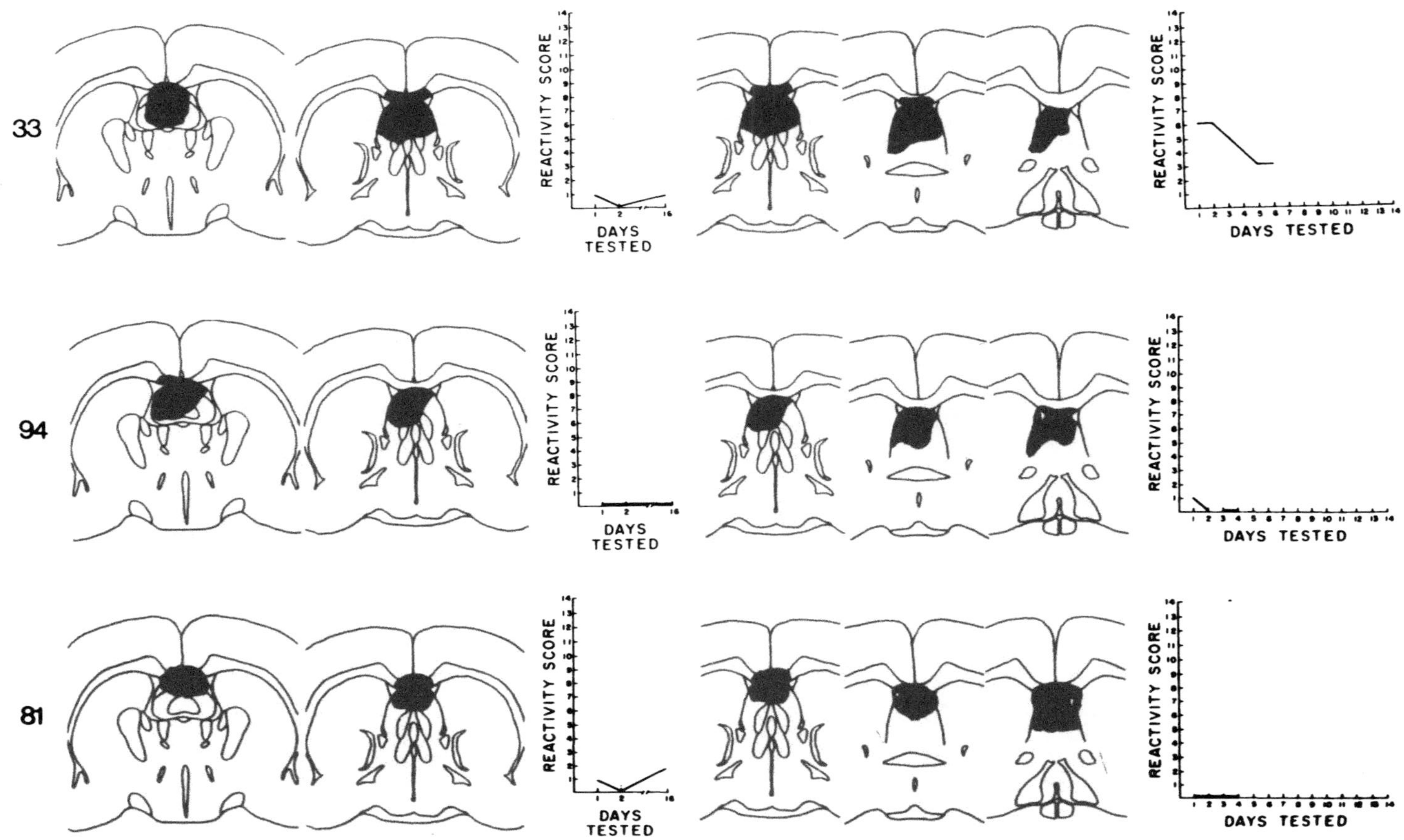

Figure 2. Histology and daily reactivity scores of each animal receiving a precommissural fornix lesion followed by a septal lesion for both the first and second test period.

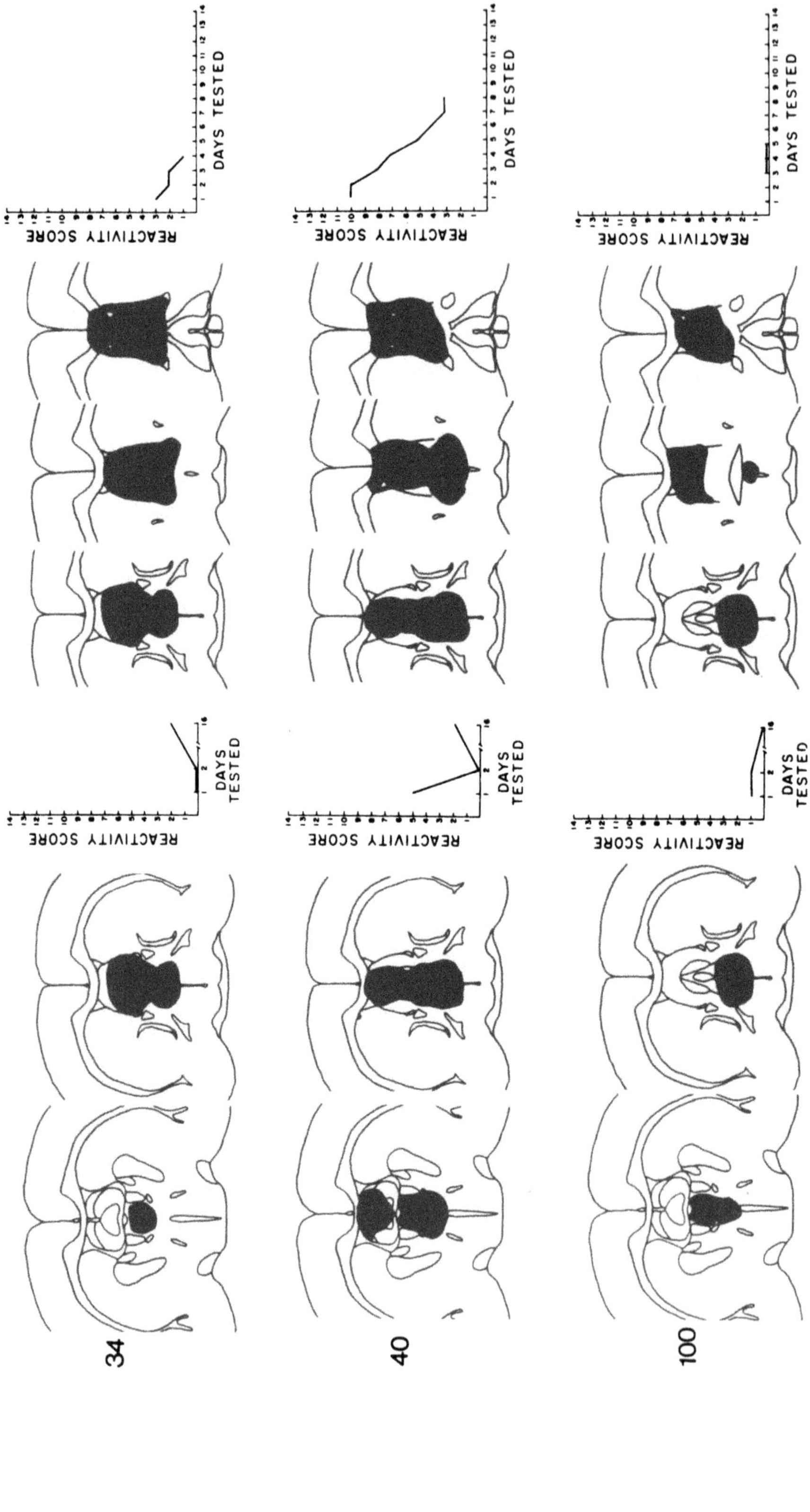
34
40
100
REACTIVITY SCORE
DAYS TESTED

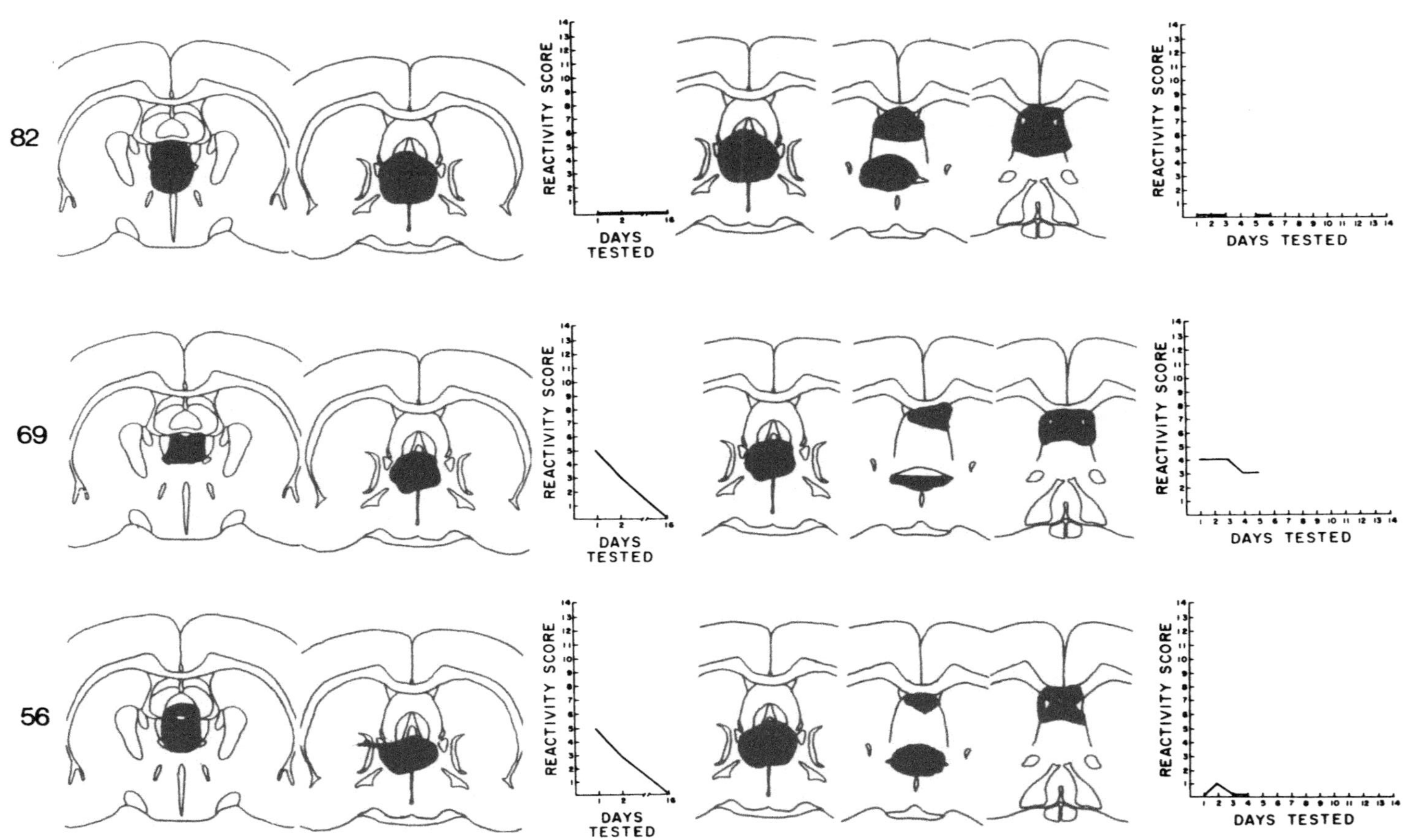

Figure 3. Histology and daily reactivity scores of each animal receiving a postcommissural fornix lesion followed by a septal lesion for both the first and second test period.

animal receiving a score of 7 or greater will be considered hyperreactive in the following discussion.

Fig. 2 presents the individual histological and behavioral data from animals receiving a precommissural fornix lesion prior to a septal lesion. The precommissural fornix lesion destroyed major portions of the precommissural fornix and caused slight damage to the adjacent septal nuclei. Following the precommissural fornix lesion there was no systematic alteration in reactivity. The septal lesions were similar to those described previously. Following the septal lesions, however, only one animal (#68) was hyperreactive. On the first postoperative day, four rats (#'s 39,7,94,81) were essentially unreactive to the stimuli; one rat (#33) had a score of 6, which although higher than the scores usually observed in normal rats is still outside the range of the rats with control operations followed by septal lesions. These data demonstrate that destruction of the precommissural fornix 16 days prior to a septal lesion can suppress the hyperreactivity usually observed after the septal lesion.

Fig. 3 presents the individual histological and behavioral data from animals receiving a postcommissural fornix lesion prior to a septal lesion. The postcommissural fornix lesion was large and completely transected the columns of the postcommissural fornix as well as damaging adjacent thalamic structures. Following the postcommissural fornix lesion, there was little change in reactivity. The septal lesions were similar to those described previously. Following the septal lesion, only one rat (#40) was hyperreactive. On the first postoperative day, reactivity scores of the remaining animals ranged from 0 to 3. These data demonstrate that destruction ofthe postcommissural fornix 16 days prior to a septal lesion can effectively suppress the hyperreactivity usually observed after a septal lesion.

Fig. 4 presents the individual histological and behavioral data from animals receiving an anterior hippocampal lesion prior to a septal lesion. The anterior hippocampal lesion effectively destroyed almost all of the anterodorsal hippocampus while in most cases sparing the fimbria. Following the anterior hippocampal lesion, there was little change in reactivity. The septal lesions were similar to those described previously. Following the septal lesion, only two animals (#'s 12,60) received a score above 7 on the first postoperative day, and in both these cases the reactivity rapidly disappeared. The other three animals exhibited no hyperreactivity. These data demonstrate that an anterior hippocampal lesion 16 days prior to a septal lesion can suppress the hyperreactivity usually observed following a septal lesion, but not as effectively as a precommissural or postcommissural fornix lesion.

Fig. 5 presents the individual histological and behavioral data from animals receiving a posterior hippocampal - entorhinal cortex lesion prior to a septal lesion. The posterior hippocampal - entorhinal lesions destroyed the most posteroventral portions of the hippocampus and the ventral aspects of the entorhinal area adjacent to the hippocampus. Following the hippocampal - entorhinal lesions, there was little change in reactivity. The septal lesions were similar to those described previously. Following the septal lesion, five of the six animals were markedly hyperreactive. On the first postoperative day, reactivity scores of these animals ranged from 9 to 11. One rat (#31) did not exhibit any signs of hyperreactivity. These data suggest that posterior hippocampal - entorhinal lesions have little or no effect on the hyperreactivity usually observed after septal lesions.

Comparison of the histology presented in Figs. 1, 2 and 3 indicates that destruction of a particular brain area might result from either a septal lesion on the one hand or a precommissural or postcommissural fornix lesion on the other. In animals where the posterior extent of the septal lesion encroached upon the anterior extent of the fornix lesion, assignment of destruction to one lesion or the other was not possible on the basis of histological examination. Thus some sections were presented for both the septal lesion and the preceeding precommissural or postcommissural fornix lesion. Although precise delineation of the boundaries of the above lesions was not possible in some cases, this fact does not compromise the argument made here. In some animals, lesions were well localized so that there were areas of undamaged brain tissue between them. In these animals, precise lesion localization was possible, and their behavior did not differ from those animals where the borders were not so clear. Thus although the precise anterior-posterior boundaries of some of the lesions remain in question, the data still clearly demonstrate that either a precommissural fornix lesion or a postcommissural fornix lesion is sufficient to block the appearance of hyperreactivity following a subsequent septal lesion.

There still remains the question of the time parameters of the fornix and the septal lesions. In our previous study, simultaneous destruction of the septal area and the body of the fornix did not suppress hyperreactivity to the extent that sequential destruction did, indicating that the time interval between the two lesions had a substantial effect on the ability of the fornix lesion to block the hyperreactivity expected from the septal lesion. Similar results were found here in that some of the animals with control lesions preceeding septal lesions had septal destruction which extended posteriorly to include major destruction of the precommissural or postcommissural fornix. Yet these animals were still hyperreactive, even though animals with similar fornix destruction produced 16 days prior to the septal destruction were not. Thus the time interval between the fiber tract destruction

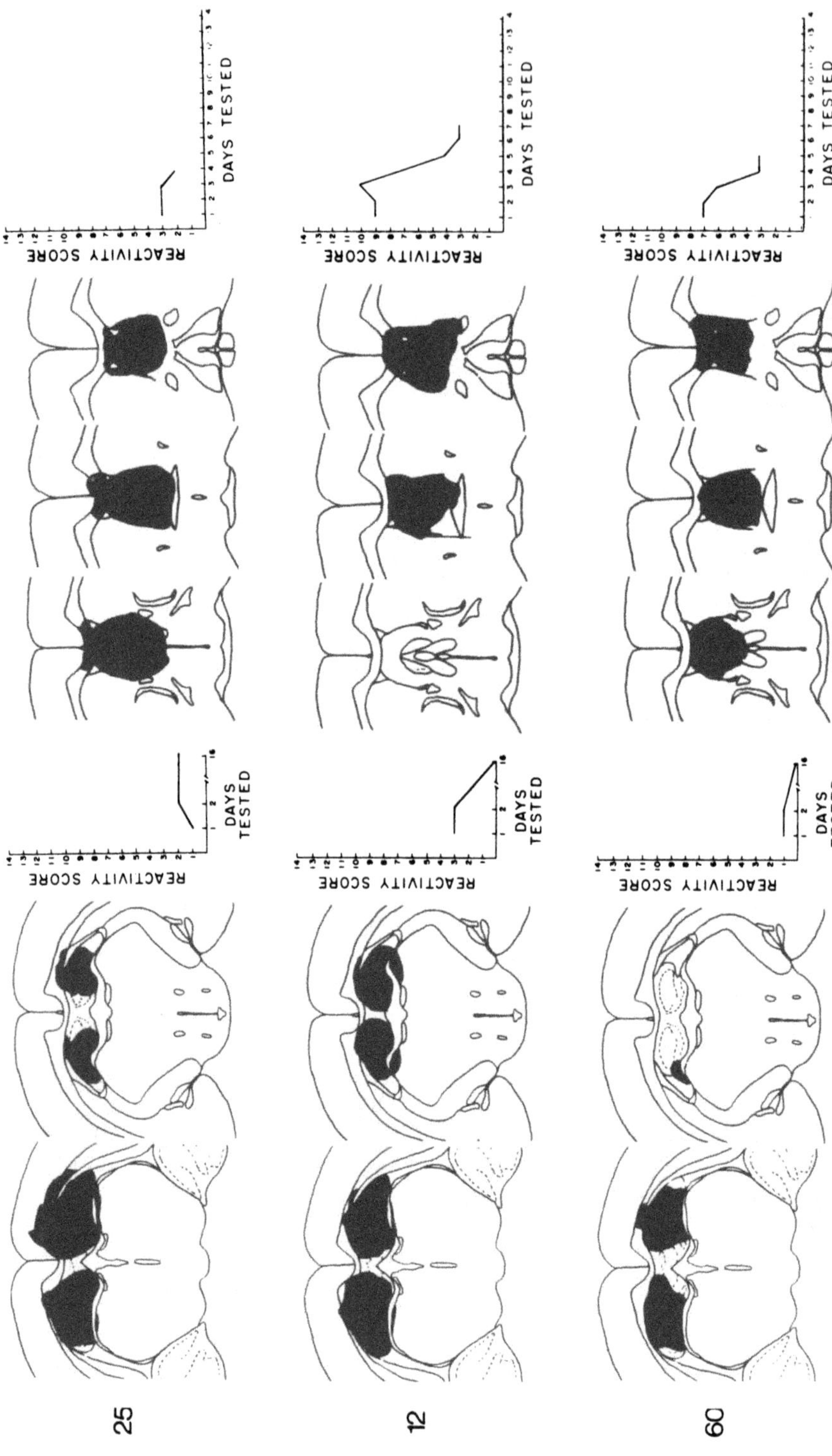
25
12
60
REACTIVITY SCORE
DAYS TESTED

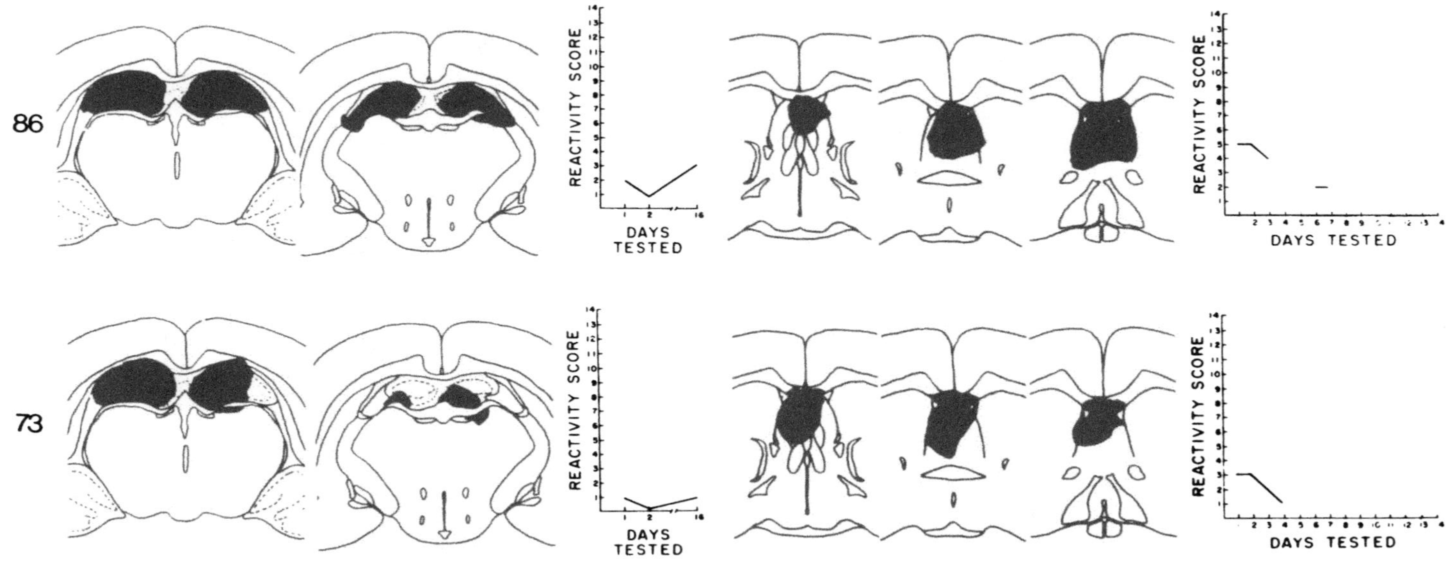

Figure 4. Histology and daily reactivity scores of each animal receiving an anterior hippocampal lesion followed by a septal lesion for both the first and second test period.

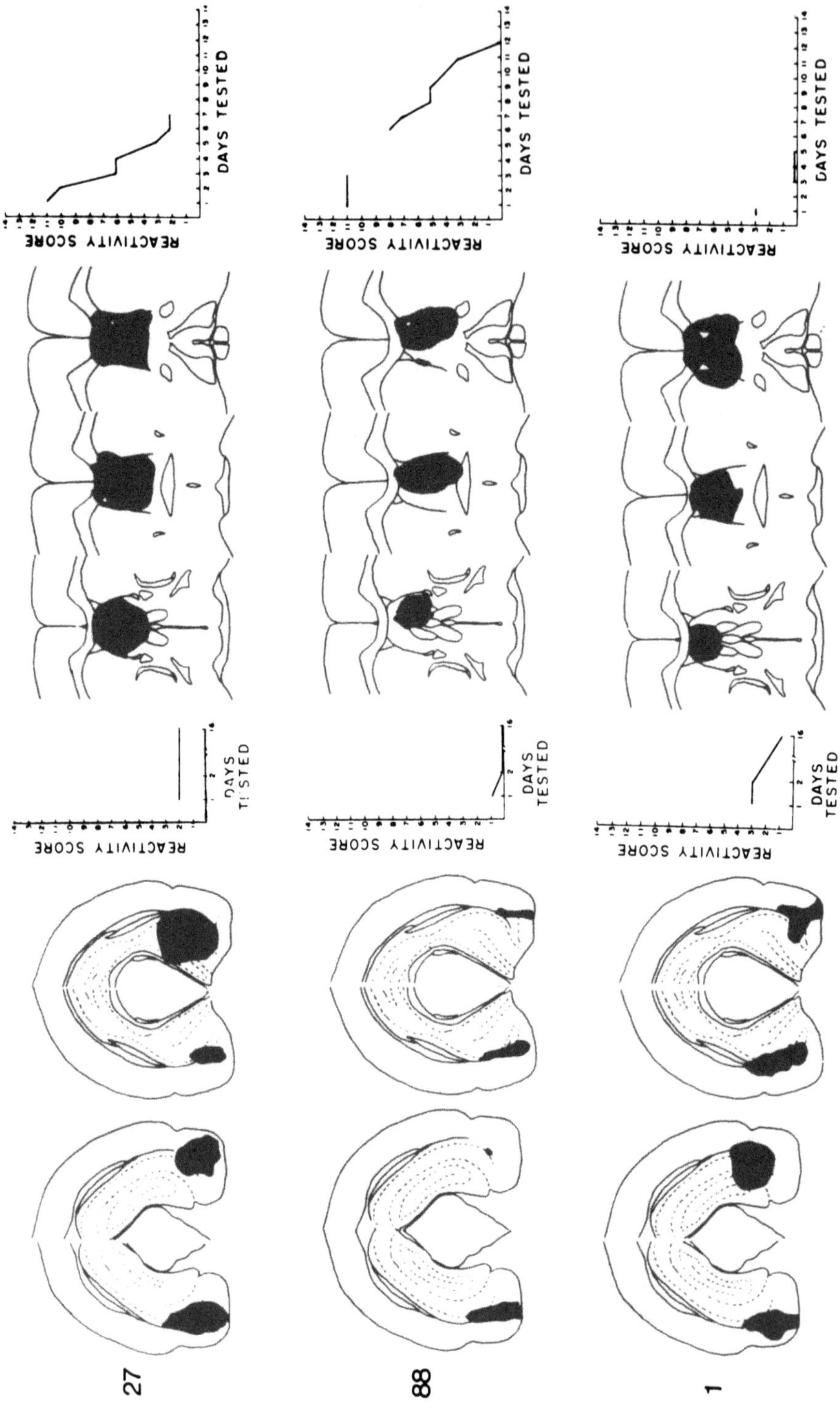
27
88
1
REACTIVITY SCORE
DAYS TESTED

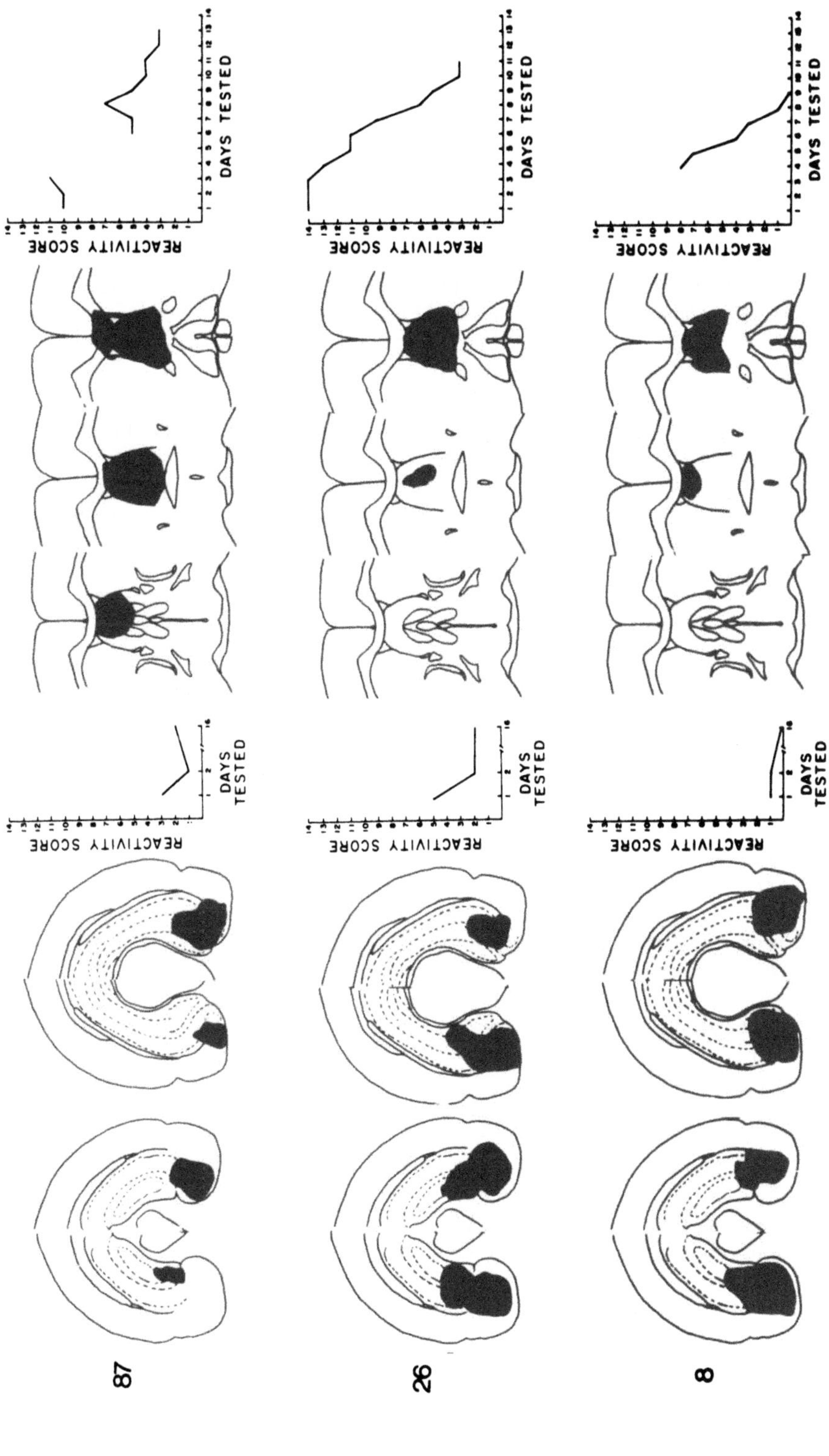

Figure 5. Histology and daily scores of each animal receiving a posterior hippocampal-entorhinal cortex lesion followed by a septal lesion for both the first and the second test period.

and the septal lesion is critical in determining the extent of hyperreactivity suppression that will be observed following the septal lesion.

The evidence from animals with hippocampal lesions suggests that terminations of the fornix in the anterior hippocampus are crucial for the appearance of hyperreactivity while terminations in the posterior hippocampus are not. Since posterior hippocampal lesions did not block hyperreactivity following the septal lesion, a fornix-hippocampalentorhinal connection appears to be unimportant in the mediation of hyperreactivity. On the other hand, since anterior hippocampal lesions did block the hyperreactivity, a connection of the fornix with the anterior hippocampus must be involved.

As a result of this study, further specification of the neuronal circuit necessary for the appearance of hyperreactivity following septal lesions is possible. Since damage to the precommissural fornix, postcommissural fornix, or anterior hippocampus completely suppressed the hyperreactivity as effectively as total transection of the body of the fornix, connections of the fornix with the septum, anterior hippocampus, and hypothalamus are involved. Since damage to the posterior hippocampus and entorhinal cortex did not effectively block the hyperreactivity, connections of the fornix through the posterior hippocampus to the entorhinal cortex are probably not involved. When the evidence is considered together, the following septal-hippocampal-hypothalamic circuits emerge as being important for the occurence of hyperreactivity following septal lesions:

1) From the septum to the anterior hippocampus via the precommissural fornix.

2) From the anterior hippocampus to the hypothalamus via the postcommissural fornix.

Our suggestion is that complete disruption of any of these circuits prior to a septal lesion will block the development of hyperreactivity.

Biochemistry

Previous research has closely linked hyperreactivity with catecholamines. Intraventricular injections of 6-hydroxydopamine (a drug that selectively destroys adrenergic nerve terminals) produced a hyperreactivity similar to that observed following septal lesions; the degree of hyperreactivity was inversely related to the levels of brain norepinephrine (4). Intraventricular injections of 6-hydroxydopa (a drug that selectively destroys norepinephrine nerve terminals but does not affect dopamine nerve terminals) produced hyperreactivity (6); the return of normal behavior following the 6-hydroxydopa injections was correlated with the return of normal brain levels of norepinephrine. Yutzey (personal communication, this symposium) has demonstrated a decrease in brain norepinephrine following septal lesions.

The above results suggest that the hyperreactivity observed after septallesions may be caused in part by a depletion of brain norepinephrine and that the gradual disappearance of hyperreactivity may be caused by the recovery of normal brain norepinephrine levels. If this suggestion is correct, then a more rapid repletion of brain norepinephrine ought to produce a more rapid disappearance of hyperreactivity. In the present study, we tested this hypothesis by injecting rats with L-Dopa (a drug that is the precursor to dopamine and norepinephrine) one day after a septal lesion and testing the animals for hyperreactivity on subsequent days. The rating scale used to measure the hyperreactivity was the same as that described previously (3). The maximum hyperreactivity score for an animal on a given test day was 14. Normal animals and animals following control operations usually obtain scores between 0 to 4. Animals following septal lesions usually obtain scores between 7 and 14 on the first postoperative day.

All rats were tested preoperatively and obtained reactivity scores of 3 or less. Each animal was then give a septal lesion and tested for hyperreactivity on the next day. Animals receiving scores of 6 or below were discarded from the experiment. All other animals were divided into drug dose groups so that animals with high and low reactivity scores were distributed evenly throughout the groups. Each animal was lightly anesthetized with ether because the hyperreactivity made handling very difficult. A peripheral decarboxylase inhibitor (Hoffman-LaRoche, RO-4602, seryl-trihydrocybenzylhydrazine) was injected IP, 50 mg/Kg, in order to inhibit peripheral decarboxylation of the L-Dopa (1). Each animal received either a vehicle injection (saline) or one of the following doses of L-Dopa methyl ester hydrochloride, IP: 100 mg/Kg, 225 mg/Kg, 225 mg/Kg, 300 mg/Kg, 400 mg/Kg, 475 mg/Kg. Following the injection, all cages were rearranged and numbered differently from the animals so that the experimenter did not know the surgical or drug history of the animal. Reactivity testing occurred on the day following the drug injection and continued daily for at least 6 postoperative (5 postdrug) days.

Following the vehicle and decarboxylase inhibitor injections alone, there was essentially no change in reactivity; all animals continued to be hyperreactive on the day after the injection and the maximum decrease in reactivity score from the previous day was 1. On subsequent days, the hyperreactivity gradually declined, following a time course similar to that observed in other animals that were not anesthetized or injected with the decarboxylase inhibitor. Thus neither the anesthetization process nor the decarboxylase inhibitor appeared to affect the magnitude or the time course of reactivity following septal lesions.

In contrast to control animals which exhibited little change in reactivity on the second day of postoperative testing, animals receiving L-Dopa tended to show a substantial decrease in reactivity scores on the first day after the drug. This decrease in reactivity scores on the first day after the drug. This decrease in reactivity scores from the first postoperative (predrug) day to the second postoperative (first postdrug) day was significantly greater for the animals in all L-Dopa groups than for the animals receiving control injections (Mann-Whitney U test, U - 188.5, P < .01). The effectiveness of the L-Dopa in lowering reactivity scores was dependent on the dose, and the animals receiving the three higher doses of L-Dopa showed a significantly greater decrease in reactivity scores from the first postoperative day to the second postoperative day than did animals receiving the two lower doses of L-Dopa (Mann-Whitney U test, U = 93, P < .01). The drug injections also affected the rapidity with which animals returned to normal reactivity. By the seventh day of postoperative testing, 19 of 20 animals in the L-Dopa groups had attained a reactivity score of 3 or less for two consecutive days (the criterion of normal reactivity) as compared to only 2 of 8 animals in the control group (Chi Square test, X = 11.9, P < .01). These data are summarized in Table I.

The data from the present study indicate that injections of L-Dopa are capable of rapidly and permanently abolishing the hyperreactivity observed after septal lesions, adding further support to the idea that the appearance of hyperreactivity following septal lesions is due to a depletion of brain catecholamines and that the return to normal reactivity is due to the repletion of brain norepinephrine. Other experiments have suggested that hyperreactivity following septal lesions can be reduced by depletion of serotonin with parachlorophenylalanine (2) or by repletion of acetylcholine with physostigmine (7). The serotonin findings are difficult to reconcile with evidence suggesting that septal lesions themselves produce a depletion of serotonin; why a further depletion of serotonin should eliminate the hyperreactivity following septal lesions is not obvious. The aceylcholine findings do demonstrate a reduction in hyperreactivity following treatment with physostigmine, but the reduction is very small and no information is provided on the duration of the decrease. In contrast, the L-Dopa treatment in the present experiment produced a major reduction in hyperreactivity that was permanent. In any case, the present data demonstrate a major role of catecholamine depletion in septal

TABLE 1

Drug	Rat #	Postoperative Days											
		1 ↓	2	3	4	5	6	7	8	9	10	11	12
Saline	5	7	6	6	6	5	3	3					
	12	10	10	10	10	7	6	5	4	4	3	3	
	18	11	11	10	9	8	7	6	5	3	3		
	24	7	6	5	4	1	2	2					
	32	9	9	9	9	9	9	7	6	5	5	3	3
	35	7	7	7	7	7	4	6	6	4	2	3	
	62	9	9	8	8	7	7	5	6	3	4	3	3
	23	11	10	10	8	6	6	4	4	3	3		
100 mg/Kg	3	10	10	7	3	2	1						
	17	7	2	5	1	4	4	1	1				
	65	8	5	5	4	6	7	7	7	4	4	3	3
	85	13	8	4	3	3	3						
225 mg/Kg	8	13	13	9	8	6	5	2	3				
	11	8	4	3	3	1	0						
	16	9	8	3	1	0	0						
	2	13	7	4	4	1	3						
300 mg/Kg	9	10	3	2	1	1	1						
	15	11	8	5	5	3	2						
	66	9	4	3	3	3	2						
	67	10	4	3	4	2	2						
400 mg/Kg	4	7	2	2	0	1	1						
	22	9	2	1	1	1	1						
	31	9	4	3	4	2	2						
	48	7	1	2	2	1	1						
475 mg/Kg	50	14	5	1	2	3	2						
	60	11	9	6	4	2	3						
	64	9	3	3	4	2	2						
	84	10	6	6	6	4	4						

Postoperative reactivity scores of all animals. On day 1 the animal was tested after the septal lesion but before any drug injection. The drug indicated in the left hand side of the table was injected at the arrow between Day 1 and Day 2. Testing was continued daily for at least 6 days and to a criterion of two successive days with reactivity scores of 3 or less.

hyperreactivity; further experimentation may indicate that other neurotransmitters are also involved.

DISCUSSION

Many questions regarding septal hyperreactivity still remain, and the most important ones probably concern the interrelationship of the behavioral, neuroanatomical, and biochemical data presented here. We are currently working on these questions, and hope to have answers to them soon. For the present, we would like to make three suggestions concerning the hyperreactivity following septal lesions: (1) Behaviorally, the syndrome may be divided into a hyperreactive escape and a hyperreactive attack component. (2) Anatomically, the syndrome appears to be dependent on the integrity of a precommissural fornix - anterior hippocampus - postcommissural fornix circuit. (3) Biochemically, the syndrome seems to result from a depletion of brain catecholamine and diminish over time as catecholamine levels return to normal.

REFERENCES

1. Butcher, L. L., Engel, J., Fuxe, K. Behavioral, biochemical, and histochemical analyses of the central effects of monoamine precursors after peripheral decarboxylase inhibition. Brain Research, 41 (1972) 387-411.

2. Dominguez, M. and Longo, V. G. Effects of p-chlorophenylalanine, α-methylaparatyrosine and of other indol- and catechol-amine depletors on the hyperirratability syndrome of septal rats. Physiology and Behavior, 5 (1970) 607-610.

3. Gage, Fred H. III, and Olton, David S. Hippocampal influence on septal hyperreactivity. Unpublished manuscript.

4. Nakamura, K. and Thoenen H. Increased irritability: A permanent behavior change induced in the rat by intraventricular administration of 6-hydroxydopamine. Psychopharmacologia (Berl.), 24 (1972) 359-372.

5. Olton, David S., and Gage, Fred H. III. Role of the fornix in the septal syndrome. Physiology and Behavior, 13 (1974) 269-279.

6. Richardson, J. S. and Jacobowitz, D. M. Depletion of brain norepenephrine by intraventricular injection of 6-hydroxydopa: A biochemical, histochemical and behavioral study in rats. Brain Research, 58 (1973) 117-133.

7. Stark, P., and Henderson, J. K. Central cholinergic suppression of hyper-reactivity and aggression in septal-lesioned rats. Neuropharmacology, 11 (1972) 839-847.

PROGRAM

Thursday, October 3rd

OPENING REMARKS
Clement A. Fox, Chairman of Anatomy
Wayne State University

SESSION ON ANATOMY
Moderator: Clement A. Fox

THE MORPHOLOGY AND DEVELOPMENT OF THE RAT SEPTUM

W. Maxwell Cowan
Washington University

SEPTUM DEVELOPMENT IN PRIMATE

Orlando J. Andy
University of Mississippi

ACETYLOCHOLINESTERASE HISTOCHEMISTRY IN THE STUDY OF THE SEPTAL REGION IN THE RAT

Bolek Srebro
University of Bergen

THE ORGANIZATION OF THE AFFERENT AND EFFERENT CONNECTIONS OF THE RAT SEPTUM

Larry W. Swanson
Washington University

THE SEPTO-HIPPOCAMPAL SYSTEM: THE SIGNIFICANCE OF THE SUBICULUM

Robert Chronister
University of South Alabama

ORGANIZATION OF THE HIPPOCAMPAL-SEPTAL AXIS

Allan Siegel
New Jersey Medical School

VASCULAR PATTERNS OF THE RAT SEPTAL REGION

Peter Coyle
The University of Michigan

SOME SIMILARITIES AND SOME DIFFERENCES BETWEEN DEAFFERENTATION INDUCED SYNAPTOGENESIS IN THE SEPTAL NUCLEI AND IN THE SUPERIOR CERVICAL SYMPATHETIC GANGLION

Geoffrey Raisman
University of Oxford

SESSION ON ENDOCRINOLOGY
Moderator: Glenn I. Hatton
Michigan State University

NEUROENDOCRINOLOGICAL APPROACH TO THE INVESTIGATION OF THE SEPTAL FUNCTION

Nobuyoshi Hagino
Southwest Foundation for
Research and Education

THE SEPTUM, A TARGET AREA FOR STEROID HORMONES

Walter E. Strumpf
University of North Carolina

THE EFFECTS OF ABLATION OF THE SEPTAL NUCLEI IN THE RAT ON CIRCADIAN VARIATION AND STRESS RESPONSE PATTERN OF CORTICOSTERONE, GROWTH HORMONE AND PROLACTIN

Joe Seggie
Clarke Institute of Psychiatry

OLFACTORY INFLUENCE ON RAT'S BLOOD PRESSURE AFTER SEPTAL LESIONS

Robert J. Rifkin
Methodist Hospital of Brooklyn

Friday, October 4th

SESSION ON PHYSIOLOGY AND PATHOLOGY
Moderator: Stephen T. Kitai
Wayne State University

SEPTAL-HYPOTHALAMIC RELATIONSHIPS

Gordon J. Mogenson
University of Western Ontario

ELECTROPHYSIOLOGICAL STUDIES OF THE CAT SEPTUM

Jon F. DeFrance
Wayne State University

FUNCTIONAL ASPECTS OF THE HIPPOCAMPAL-SEPTAL AXIS

Henry Edinger
New Jersey Medical School

SEPTAL UNIT RESPONSES TO HIPPOCAMPAL AND HYPOTHALAMIC STIMULATION IN THE CHLORALOSE-ANAESTHETIZED OR CERVEAU ISOLE' CAT

John S. McKenzie
University of Melbourne

UNIT ACTIVITY IN THE SEPTAL NUCLEI DURING WATER DEPRIVATION, DRINKING, AND REHYDRATION

John G. Bridge
Vassar College

SOME PRELIMINARY OBSERVATIONS ON SEPTAL AND RELATED BRAIN ENZYME ACTIVITIES IN DECEASED MENTALLY NORMAL, CHRONIC SCHIZOPHRENIC AND ORGANIC BRAIN SYNDROME PATIENTS

Edward F. Domino
The University of Michigan
Lafayette Clinic

Luncheon

SESSION ON BEHAVIOR
Moderator: Foster K. Redding
Wayne State University

BEHAVIORAL FUNCTIONS OF THE SEPTUM: A RE-ANALYSIS

Sebastian P. Grossman
University of Chicago

CORTICOSTERONE LEVELS DURING ACTIVE AVOIDANCE LEARNING IN RATS WITH SEPTAL LESIONS

Holger Ursin
University of Bergen

BEHAVIORAL CORRELATES AND FIRING REPERTOIRES OR NEURONS IN MEDIAL AND LATERAL SEPTAL NUCLEI OF RAT

James B. Ranck, Jr.
University of Michigan

TOPOGRAPHIC PATTERNS OF HIPPOCAMPAL THETA RHYTHM IN FREELY MOVING RAT AND RABBIT

Jon Winson
Rockefeller University

SOME EFFECTS OF SEPTAL LESIONS ON MATERNAL AND AGGRESSIVE BEHAVIOR

Burton Slotnick
American University

EFFECT OF SEPTAL LESIONS ON SPECIES TYPICAL BEHAVIOR IN THE RAT

Thomas F. Herrmann
University of Guelph

Saturday, October 5th

Workshops

Anatomy and Physiology
Moderator: Larry Swanson

Behavior
Moderator: James Ranck

ROSTER OF PARTICIPANTS

Dr. Orlando J. Andy
Department of Neurosurgery
University of Mississippi
Jackson, Mississippi 39216

Dr. James Boston
American Cyanamid Company
Pearl River, New York 10965

Mr. Ziad Boulos
Graduate Student
Northeastern University
Boston, Massachusetts 02115

Dr. John G. Bridge
Assistant Professor
Department of Psychology
Vassar College
Poughkeepsie, New York 12601

Dr. W. McIntyre Burnham
Department of Neurophysiology
University of Toronto
Toronto, Ontario Canada

Dr. Robert B. Chronister
Assistant Professor
Department of Psychology
Division of Neuroscience
University of South Alabama
Mobile, Alabama 36688

Dr. Jerzy Cytawa
Department of Psychology
University of Illinois
Champaign, Illinois 61820

Dr. W. Maxwell Cowan
Department of Anatomy
Medicine School
St. Louis, Missouri 63110

Dr. Peter Coyle
Department of Anatomy
The University of Michigan
Ann Arbor, Michigan 48104

Dr. Jon F. DeFrance
Assistant Professor
Department of Anatomy
School of Medicine
Wayne State University
Detroit, Michigan 48201

Dr. Edward F. Domino
Department of Pharmacology
The University of Michigan
and Lafayette Clinic
Ann Arbor, Michigan 48104

Dr. William G. Drew
Department of Psychiatry
University of Kentucky
Medical Center
Lexington, Kentucky 40506

Dr. Henry Edinger
Assistant Professor
Department of Physiology
New Jersey Medical School
Newark, New Jersey 07103

Dr. Clement A. Fox, Professor
Department of Anatomy
School of Medicine
Wayne State University
Detroit, Michigan 48201

Dr. James E. Gotsick
Department of Psychology
Morehead State University
Morehead, Kentucky 40351

Dr. Harry J. Gould, III
University of Cincinnati
Medical Center
Cincinnati, Ohio 45219

Dr. Sebastan P. Grossman
Department of Psychology
University of Chicago
Chicago, Illinois

Dr. Nobuyoshi Hagino
Department of Anatomy
University Medical School
Pecs, Hungary

Ms. Lucy B. Halboth
Research Fellow in Training
Department of Neurology
Albert Einstein College of Medicine
Bronx, New York 10461

Dr. Glenn Hatton
Michigan State University
East Lansing, Michigan

Dr. James C. Hazlett
Assistant Professor
Department of Anatomy
School of Medicine
Wayne State University
Detroit, Michigan

Dr. Thomas F. Herrmann
Assistant Professor
Department of Psychology
University of Guelph
Guelph, Ontario Canada

Dr. J. Eric Holmes
Associate Professor
Department of Neurology
School of Medicine
University of Southern California
Los Angeles, California 90033

Dr. John E. Kelsey
Assistant Professor
Department of Psychology
Indiana University
Bloomington, Indiana 47401

Dr. Yong Kie Kim
Department of Neurosurgery
Broughton Hospital
Morganton, North Carolina 28655

Dr. Stephen Kitai
Professor
Department of Anatomy
School of Medicine
Wayne State University
Detroit, Michigan 48201

Dr. Barry R. Komisarik
Professor of Zoology
Institute of Animal Behavior
Rutgers University
Newark, New Jersey 07102

Dr. Duncan Kennedy
Assistant Professor
Department of Anatomy
School of Medicine
Wayne State University
Detroit, Michigan 48201

Dr. Joel F. Lubar
Professor
Department of Psychology
University of Tennessee
Knoxville, Tennessee 37916

Dr. Maryann Marrazzi
Assistant Professor
Department of Pharmacology
School of Medicine
Wayne State University
Detroit, Michigan 48201

Dr. W. J. McBride
Assistant Professor
Institute of Psychiatric
Research
Indiana University Medical
Center
Indianapolis, Indiana 46260

Dr. Duncan A. McCarthy
Director, Pharmacological
Research
Parke, Davis & Company
Ann Arbor, Michigan 48106

Dr. John S. McKenzie
Senior Lecturer
Department of Physiology
University of Melbourne
Parkville, Victoria Australia

Dr. John R. McLean
Section Director
Neuropharmacology
Parke, Davis & Company
Ann Arbor, Michigan 48106

Dr. Marcel Mesulam
Neurological Unit
Boston City Hospital
Boston, Massachusetts

Dr. J. J. Miller
Assistant Professor
Department of Physiology
University of British Columbia
Vancouver, British Columbia Canada

Dr. Jerry A. Mitchell
Assistant Professor
Department of Anatomy
School of Medicine
Wayne State University
Detroit, Michigan 48201

Dr. James C. Mitchell
Professor
Department of Psychology
Kansas State University
Manhattan, Kansas 66506

Dr. Gordon J. Mogenson
Department of Physiology
University of Western Ontario
London, Ontario Canada

Mr. Auri Naggar
Graduate Student
Institute of Animal Behavior
Rutgers University
Newark, New Jersey

Dr. David S. Olton
Assistant Professor
Johns Hopkins University
Baltimore, Maryland 21218

Dr. Geoffrey Raisman
Department of Human Anatomy
University of Oxford
Oxford, England

Dr. James B. Ranck, Jr.
Department of Physiology
The University of Michigan
Ann Arbor, Michigan 48104

Dr. Foster K. Redding
Department of Anatomy
School of Medicine
Wayne State University
Detroit, Michigan 48201

Dr. Robert J. Rifkin
Chief, Endocrine Laboratory
Methodist Hospital
Brooklyn, New York 11215

Dr. Douglas K. Rogers
Graduate Student
Department of Physiology
University of Melbourne
Parkville, Victoria Australia

Dr. Robert Schwyn
Department of Neurology
Harper Hospital
School of Medicine
Wayne State University
Detroit, Michigan 48201

Dr. Jo Seggie
Research Scientist
Clarke Institute of Psychiatry
Toronto, Ontario, Canada

Ms. Kazue Senba
Graduate Student
Institute of Animal Behavior
Rutgers University
Newark, New Jersey

Dr. Allan Siegel
Associate Professor
Department of Anatomy
and Neuroscience
New Jersey Medical School
Newark, New Jersey 07103

Dr. Burton M. Slotnick
Department of Psychology
American University
Washington, D.C. 20016

Dr. Bolek Srebro
Institute of Physiology
University of Bergen
Bergen, Norway

Dr. Walter E. Stumpf
Professor
Department of Anatomy
and Pharmacology
University of North Carolina
Chapel Hill, North Carolina 27514

Dr. Larry W. Swanson
Department of Biology
School of Medicine
Washington University
St. Louis, Missouri 63110

Dr. Philip S. Ulinski
Assistant Professor
Department of Anatomy
Loyola University
School of Dentistry
Maywood, Illinois 60153

Dr. Holger Ursin
Institute of Physiology
University of Bergen
Bergen, Norway

Ms. Linda Walsh
Graduate Student
University of Chicago
Chicago, Illinois 60637

Ms. Mary P. White
Graduate Student
Department of Neuroscience
The University of Michigan
Ann Arbor, Michigan 48104

Dr. Jon Winson
Departments of Physiology
and Psychology
Rockefeller University
New York, New York 10021

Dr. Henry Yoshihara
Department of Anatomy
School of Medicine
Wayne State University
Detroit, Michigan 48201

Dr. David A. Yutzey
Associate Professor
Department of Psychology
University of Connecticut
Storrs, Connecticut 06268

Mr. Josef Zaide
Graduate Student
Department of Psychology
McMaster University
Hamilton, Ontario Canada

INDEX

www.ingramcontent.com/pod-product-compliance
Ingram Content Group UK Ltd.
Pitfield, Milton Keynes, MK11 3LW, UK
UKHW051449290926
15471UKWH00008B/1092
* 9 7 8 1 4 6 8 4 3 0 8 5 1 *